JN416613

SIXTH EDITION

Operational Amplifiers and Linear Integrated Circuits

Robert F. Coughlin
Wentworth Institute of Technology

Frederick F. Driscoll
Wentworth Institute of Technology

Prentice Hall

Upper Saddle River, New Jersey

Columbus, Ohio

Original edition, entitled OPERATIONAL AMPLIFIERS AND LINEAR INTEGRATED CIRCUITS, 6th Edition, 0130149918 by COUGHLIN, ROBERT F.; DRISCOLL, FREDERICK F., published by Pearson Education, Inc, publishing as Prentice Hall, Copyright © 2001.

All rights reserved. No part of this book may be reproduced or transmitted in any form or by any means, electronic or mechanical, including photocopying, recording or by any information storage retrieval system, without permission from Pearson Education, Inc.

KOREA edition published by PEARSON EDUCATION KOREA LTD, Copyright © 2016.

This edition is manufactured in KOREA, and is authorized for sale only in KOREA.

ISBN 978-89-450-0130-6
ISBN 89-450-0130-1

To Our Partners in Ballroom Dancing
and
Our Lifetime Partners,
Barbara and *Jean*

As
We Grow Older
We Grow Closer

Contents

Preface

The authors' intention in all previous editions of *Operational Amplifiers and Linear Integrated Circuits* has been to show that operational amplifiers and other linear integrated circuits are easy to use and fun to work with. This sixth edition has kept that basic philosophy. For the fundamental circuits, we have continued to use devices that are readily available, easy to use, and forgiving if a wiring error is made. Newer devices are introduced where the application requires it. We have preserved our original objective of simplifying the process of learning about applications involving signal conditioning, signal generation, filters, instrumentation, timing, and control circuits. This edition continues to reflect the evolution of analog circuits into applications requiring transducer signals that must be conditioned for a microcontroller's analog-to-digital input.[1] We have kept circuit simulation using OrCAD® PSpice®. A laboratory manual is now available to accompany

[1] A detailed procedure on how to design circuits that interface between the physical world and microcontrollers is presented in *Data Acquisition and Process Control with the M68HC11 Microcontroller, 2nd Edition,* by F. Driscoll, R. Coughlin, and R. Villanucci, published by Prentice Hall (2000).

this sixth edition.[2] It includes both detailed hardware and simulation exercises. Some exercises are step-by-step; others are design projects. The exercises follow the text material.

Chapters 1 through 6 provide the reader with a logical progression from op amp fundamentals to a variety of practical applications without having to worry about op amp limitations. Chapter 7 shows how op amps combined with diodes can be used to design ideal rectifier circuits as well as clamping and clipping circuits. PSpice models and simulations are included in these chapters.

Chapter 8 shows applications that require measuring a physical variable such as temperature, force, pressure, or weight and then having the signal conditioned by an instrumentation amplifier before being input into a microcontroller's A/D converter. Instrumentation amplifiers are required when a designer has to measure a differential signal, especially in the presence of a larger noise signal.

As previously mentioned, in order not to obscure the inherent simplicity and overwhelming advantages of using op amps, their limitations have been left for Chapters 9 and 10. Dc limitations are studied in Chapter 9 and ac limitations are covered in Chapter 10. An expanded discussion on common-mode rejection ratio has been included in this edition. Many limitations have been made negligible by the latest generations of op amps, as pointed out in these chapters.

Active filters, low-pass, high-pass, band-pass, and band-reject, are covered in Chapter 11. Butterworth-type filters were selected because they are easy to design and produce a maximally flat response in the pass band. Chapter 11 shows the reader how to design a variety of filters easily and quickly.

Chapter 12 introduces a linear integrated circuit known as the multiplier. The device makes analysis and design of AM communication circuits simpler than using discrete components. Modulators, demodulators, frequency shifters, a universal AM radio receiver, and analog divider circuits all use a multiplier IC as the system's basic building block. This chapter has been retained because instructors have written to say that the principles of single-side band suppressed carrier and standard amplitude-modulation transmission and detection are clearly explained and quite useful for their courses.

The inexpensive 555 IC timer is covered in Chapter 13. This chapter shows the basic operation of the device as well as many practical applications. The chapter also includes a timer/counter unit.

In previous editions, analog-to-digital and digital-to-analog converters have been covered in a single chapter. This edition separates these topics into two chapters so that more device specifications can be included as well as practical applications. Chapter 14 deals only with analog-to-digital converters, while the new Chapter 15 covers digital-to-analog converters. A serial ADC connected to a Motorola microprocessor is shown (with assembly language code) in Chapter 14.

Chapter 16 shows how to design a regulated linear power supply. This chapter begins with the fundamentals of unregulated supplies and proceeds to regulated supplies. It shows how IC regulators are used for building low-cost 5 V and ± 15 V bench supplies.

[2] *Laboratory Manual to Accompany Operational Amplifiers and Linear Integrated Circuits, 6th Edition,* by R. Coughlin, F. Driscoll, and R. Villanucci published by Prentice Hall (2001).

This edition has more than enough material for a single-semester course. After the first three chapters, instructors often take chapters out of sequence depending on the class interest, need to complement another course (such as a design course), or availability of lab equipment or class time. Therefore, Chapters 4 through 16 have been written as stand-alone chapters for this very reason. The circuits have been tested in the laboratory by the authors and the material is presented in a form useful to students or as a reference to practicing engineers and technologists. Each chapter includes learning objectives and problems, and most chapters have PSpice simulations. The reader should refer to the accompanying laboratory manual for lab exercises and additional simulation exercises.

ACKNOWLEDGMENTS

We acknowledge with gratitude the advice of Professor Robert Villanucci, who is also a co-author of the laboratory manual, and two highly respected engineers, Dan Sheingold of Analog Devices and Bob Pease of National Semiconductor. A special thanks goes to Libby Driscoll for assisting in the preparation of the manuscript. We thank the following reviewers of the manuscript: Warren Hioki, Community College of Southern Nevada; Gregory M. Rasmussen, St. Paul Technical School; Michael W. Rudisill, Northern Michigan University; Rod Schein, Edmonds Community College, ATTC; and Andrew C. Woodson.

Finally, we thank our students for their insistence on relevant instruction that is immediately useful and our readers for their enthusiastic reception of previous editions and their perceptive suggestions for this edition.

CHAPTER 1

Introduction to Op Amps

LEARNING OBJECTIVES

Upon completing this introductory chapter on op amps, you will be able to:

- Understand why analog circuitry using op amps is still required in computer-based systems.
- Draw the circuit symbol for a general-purpose op amp such as the 741 and show the pin numbers for each terminal.
- Name and identify at least three types of package styles that house a general-purpose op amp.
- Identify the manufacturer, op amp, and package style from the PIN.
- Correctly place an order for an op amp.
- Identify the pins of an op amp from the top or bottom view.
- Identify the power supply common on a circuit schematic, and state why you must do so.
- Breadboard an op amp circuit.

1-0 INTRODUCTION

One of the most versatile and widely used electronic devices in linear applications is the operational amplifier, most often referred to as the op amp. Op amps are popular because they are low in cost, easy to use, and fun to work with. They allow you to build useful circuits without needing to know about their complex internal circuitry. Op amps are usually very forgiving of wiring errors because of their self-protecting internal circuitry.

The word *operational* in operational amplifiers originally stood for mathematical operations. Early op amps were used in circuits that could add, subtract, multiply, and even solve differential equations. These operations have given way to digital computers because of their speed, accuracy, and versatility. However, digital computers were not the demise of the op amp.

1-1 IS THERE STILL A NEED FOR ANALOG CIRCUITRY?

1-1.1 Analog and Digital Systems

You often hear an expression similar to "It is a digital world." This usually is followed by a statement such as "Is there a reason for studying analog circuitry, including op amps and other linear integrated circuits, when so many applications use a computer?" It is true that more and more functions are being done and problems are being solved by microcomputers, microcontrollers, or digital signal processing chips and systems today than ever before. This trend of going digital will continue at an even faster pace because software packages are better and easier to use, computers are faster and more accurate, and data can be stored and transferred over networks. However, as more digital systems are created for data acquisition and process control, more interface circuits using op amps and other linear integrated circuits are also required. These integrated systems now require designers to understand the principles of both the analog and the digital world in order to obtain the best performance of a system at a reasonable cost.

In the past, op amps were studied as separate entities and entire analog systems were developed using only analog circuitry. In some specialized real-time applications, this is still true but most systems that find their way to the marketplace are a combination of analog and digital. A typical data acquisition system block diagram is shown in

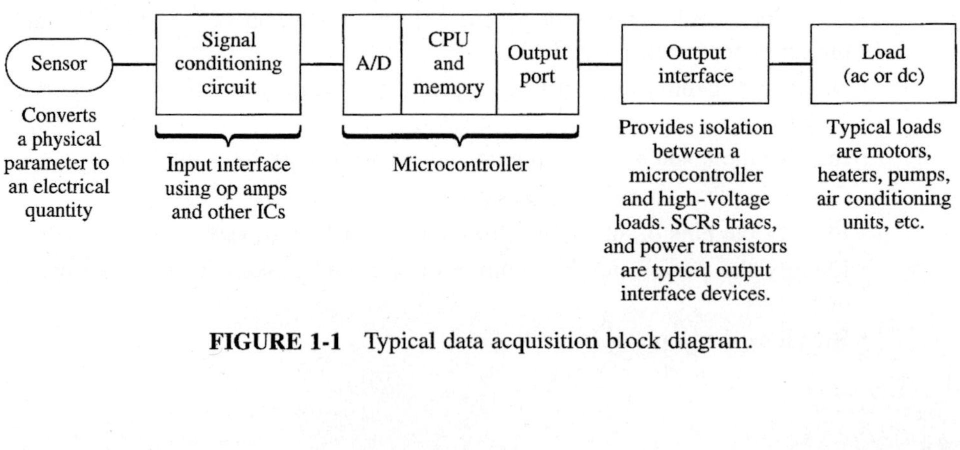

FIGURE 1-1 Typical data acquisition block diagram.

Fig. 1-1. It uses a sensor to convert a physical parameter (such as temperature, pressure, or flow) into an electrical parameter (such as voltage, current, or resistance). Unfortunately, sensors rarely produce an output whose electrical parameter or value is suitable for direct input into the computer through an analog-to-digital (A/D) converter. Thus an input interface circuit using op amps or other linear ICs is needed to condition the signal for the computer's A/D. Similarly, at the computer's output another analog circuit is needed to interface and isolate the computer's low voltage from a high-voltage ac or dc load. This text is designed to show applications of op amps and other linear integrated circuits in these combined analog and digital systems.

1-1.2 Op Amp Development

Op amps are designed using a wide variety of fabrication techniques. Originally they contained only bipolar transistors, but now there are a host of devices that use field-effect transistors within the op amp. Junction field-effect transistors at the input draw very small currents and allow the input voltages to be varied between the power supply limits. MOS transistors in the output circuitry allow the output terminal to go within millivolts of the power supply limits.

Op amps designed with bipolar inputs and complementary MOS outputs, appropriately named BiMOS, are faster and have a higher frequency response than the general-purpose op amps. Manufacturers have also designed dual (2) and quad (4) op amp packages. Hence, the package that once housed a single op amp can now contain two or four op amps. In the quad package, all four op amps share the same power supply and ground pins.

1-1.3 Op Amps Become Specialized

Inevitably, general-purpose op amps were redesigned to optimize or add certain features. Special function ICs that contain more than a single op amp were then developed to perform complex functions.

You need only to look at linear data books to appreciate their variety. Only a few examples are

1. High current and/or high voltage capability
2. Sonar send/receive modules
3. Multiplexed amplifiers
4. Programmable gain amplifiers
5. Automotive instrumentation and control
6. Communication ICs
7. Radio/audio/video ICs
8. Electrometer ICs for very high input impedance circuits
9. ICs that operate from a single supply
10. ICs that operate from rail to rail

General-purpose op amps will be around for a long time. However, more complex integrated circuits on a single chip are being developed. These devices combine analog with digital circuitry. In fact, with improved very large scale integrated (VLSI) technology, entire systems are being fabricated on a single large chip.

A single-chip computer is today's reality. A single-chip TV set will happen eventually. Before learning how to use op amps, it is wise to learn what they look like and how to buy them. As previously stated, the op amp's greatest use will be as a part in a system that interfaces the real world of analog voltage with the digital world of the computer, as will be shown throughout this text. If you want to understand the system, you must understand the workings of one of its most important components.

1-2 741 GENERAL-PURPOSE OP AMP

1-2.1 Circuit Symbol and Terminals

The 741 op amp has been "around" for a number of years. However, it still is a great device to begin with because it is inexpensive, rugged, and easy to obtain. The op-amp symbol in Fig. 1-2 is a triangle that points in the direction of signal flow. This component has

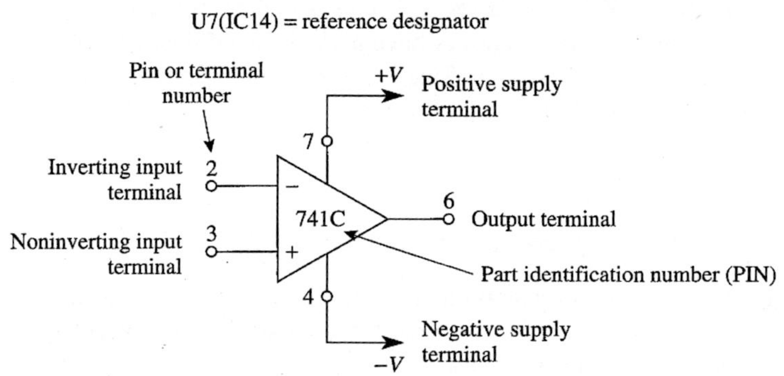

FIGURE 1-2 Circuit symbol for the general-purpose op amp. Pin numbering is for an 8-pin mini-DIP package.

a *part identification number* (PIN) placed within the triangular symbol. The PIN refers to a particular op amp with specific characteristics. The 741C op amp illustrated here is a general-purpose op amp that is used throughout the book for illustrative purposes.

The op amp may also be coded on a circuit schematic with a *reference designator* such as U7, IC14, and so on. Its PIN is then placed beside the reference designator in the parts list of the circuit schematic. All op amps have at least five terminals: (1) The positive power supply terminal V_{CC} or $+V$ at pin 7, (2) the negative power supply terminal V_{EE} or $-V$ at pin 4, (3) output pin 6, (4) the inverting ($-$) input terminal at pin 2, and (5) the noninverting ($+$) input terminal at pin 3. Some general-purpose op amps have additional specialized terminals. (The pins above refer to the 8-pin mini-DIP case discussed in the following section.)

1-2.2 *Simplified Internal Circuitry of a General-Purpose Op Amp*

General-purpose op amps are multistage systems. As shown in Fig. 1-3(a), the basic op amp consists of an input stage with two input terminals, an output stage with one output terminal, and an intermediate stage that connects the output signal of the input stage to the input terminal of the output stage.

Dc power is applied from a bipolar supply to the op amp's external power supply terminals and thus to each internal stage of the op amp. Depending on the application, input signals, $V_{(+)}$ and $V_{(-)}$ can be positive, negative, or zero. The resulting output voltage is measured across the load resistor R_L, which is connected between the op amp's output terminal and common. The output voltage, V_o, depends on the input signals and characteristics of the op amp.

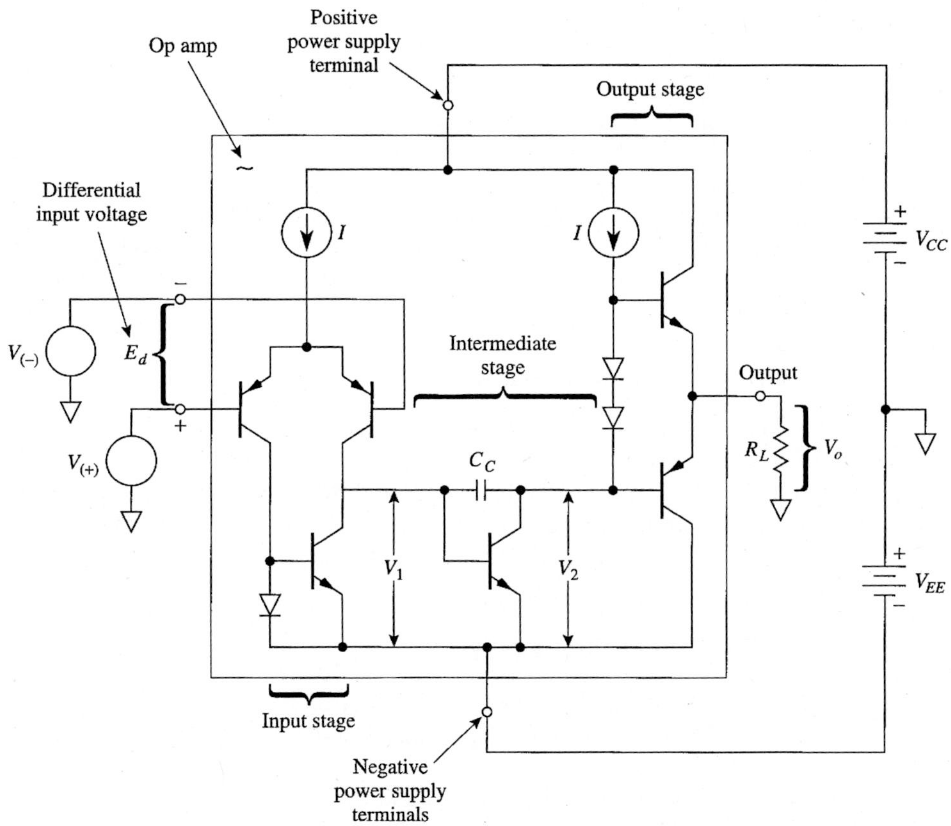

(a)

FIGURE 1-3 (a) Simplified block diagram of a general-purpose operational amplifier with external connections; (b) external connections using the op amps circuit symbol.

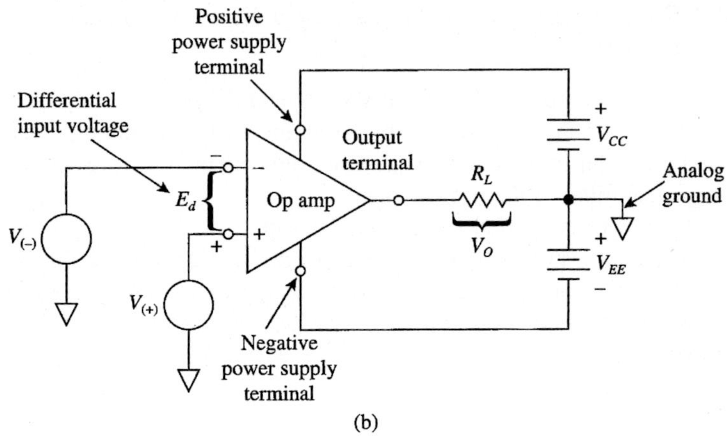

FIGURE 1-3(b)

1-2.3 Input Stage—Differential Amplifier

The input stage of the op amp in Fig. 1-3(a) is called a differential amplifier. It has very high input impedance as well as a large voltage gain. When input signals $V_{(+)}$ and $V_{(-)}$ are applied, the difference voltage, E_d, is amplified by this stage and appears as the output voltage V_1. (Examples of how E_d is calculated are given in Chapter 2.)

1-2.4 Intermediate Stage—Level Shifter

Signal voltage V_1 at the output of the differential amplifier is directly coupled to the input of the intermediate level shifter stage. This stage performs two functions. First, it shifts the dc voltage level at the output of the differential amplifier to a value required to bias the output stage. Second, this stage allows input signal V_1 to pass nearly unaltered and become the input signal V_2 for the output stage.

1-2.5 Output Stage—Push-Pull

The signal voltage V_2 at the output of the intermediate stage is coupled directly into the output stage. The most common output stage is a *pnp-npn* push-pull transistor configuration. Using a push-pull circuit as the final stage allows the op amp to have a very low output resistance. As shown in Figs. 1-3(a) and (b), load resistor R_L is connected between the output terminal and common to develop output voltage V_o.

This simplified model of the op amp in Fig. 1-3(a) presents the basic information on its internal architecture. The actual circuitry is more complicated, but the functions are similar.

1-3 PACKAGING AND PINOUTS

1-3.1 Packaging

The op amp is fabricated on a tiny silicon chip and packaged in a suitable case. Fine-gage wires connect the chip to external leads extending from a metal, plastic, or ceramic package. Common op amp packages are shown in Figs. 1-4(a) to (d).

The metal can package shown in Fig. 1-4(a) is available with 3, 5, 8, 10, and 12 leads. The silicon chip is bonded to the bottom metal sealing plane to expedite the dissipation of heat. In Fig. 1-4(a) the tab identifies pin 8, and the pins are numbered counterclockwise when you view the metal can from the top.

The popular 14-pin and 8-pin dual-in-line packages (DIPs) are shown in Figs. 1-4(b) and (c). Either plastic or ceramic cases are available. As viewed from the top, a notch or dot identifies pin 1 and terminals are numbered counterclockwise.

Complex integrated circuits involving many op amps and other ICs can now be fabricated on a single large chip or by interconnecting many chips and placing them in a single package. For ease of manufacture and assembly, pads replace the leads. The resulting

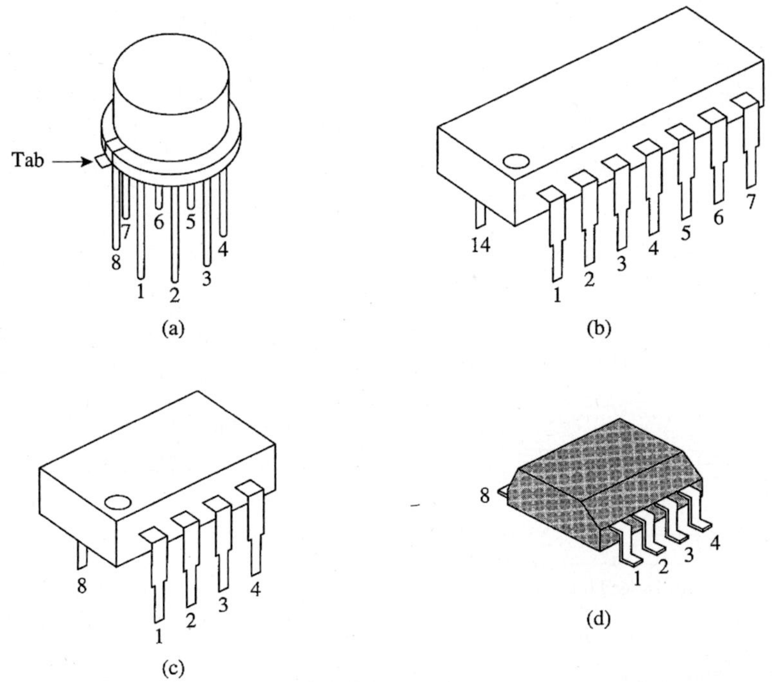

FIGURE 1-4 The three most popular op amp packages are the metal can in (a) and the 14- and 8-pin dual-in-line packages in (b) and (c), respectively. For systems requiring high density, surface-mounted technology (SMT) packages are used as shown in (d).

structure is called surface-mounted technology (SMT), shown in Fig. 1-4(d). These packages provide a higher circuit density for a package of a given size. Additionally, SMTs have lower noise and improved frequency–response characteristics. SMT components are available in (1) *plastic lead chip carriers* (PLCCs), (2) *small outline integrated circuits* (SOICs), and (3) *leadless ceramic chip carriers* (LCCCs).

1-3.2 Combining Symbol and Pinout

Manufacturers are now combining the circuit symbol for an op amp together with the package view into a single drawing. For example, the four most common types of packages that house a 741 chip are shown in Fig. 1-4. Compare Figs. 1-5(a) and (d) to see that

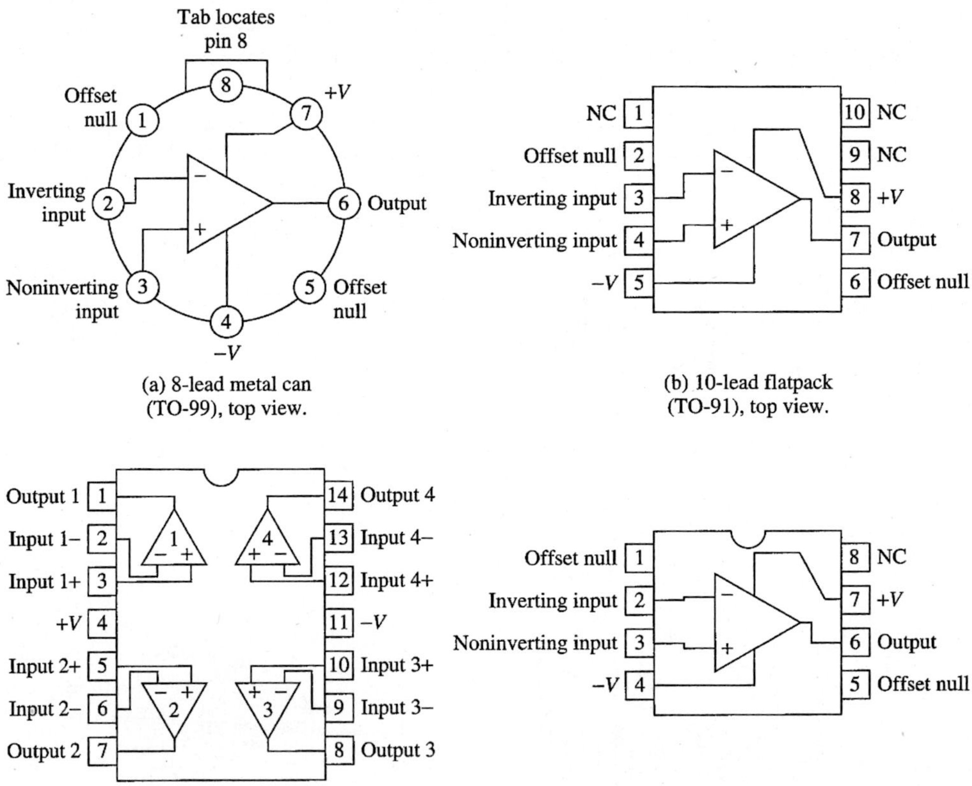

(a) 8-lead metal can (TO-99), top view.

(b) 10-lead flatpack (TO-91), top view.

(c) 14-lead dual-in-line package (DIP, TO-116), top view.

(d) 8-lead mini-DIP top view.

FIGURE 1-5 Connection diagrams for typical op amp packages. The abbreviation NC stands for "no connection." That is, these pins have no internal connection, and the op amp's terminals can be used for spare junction terminals. Diagram (c) shows how four op amps can be configured in a single package. Not shown in (c) are the internal connections for $+V$ and $-V$.

the numbering schemes are identical for an 8-pin can and an 8-pin DIP. A notch or dot identifies pin 1 on the DIPs, and a tab identifies pin 8 on the TO-5 (or the similar TO-99) package. From a top view, the pin count proceeds counterclockwise.

The final tasks in this chapter are to learn how to buy a specific type of op amp and to present advice on basic breadboarding techniques.

1-4 HOW TO IDENTIFY OR ORDER AN OP AMP

1-4.1 The Identification Code

Each type of op amp has a letter–number identification code. This code answers four questions:

1. What type of op amp is it? (Example: 741.)
2. Who made it? (Example: Analog Devices.)
3. How good is it? (Example: the guaranteed temperature range for operation.)
4. What kind of package houses the op amp chip? (Example: plastic DIP.)

Not all manufacturers use precisely the same code, but most use an identification code that consists of four parts written in the following order: (1) letter prefix, (2) circuit designator, (3) letter suffix, and (4) military specification code.

Letter prefix. The letter prefix code usually consists of two or three letters that identify the manufacturer. The following examples list some of the codes used by a manufacturer. You may wish to visit their Web site to obtain data sheets and application notes about a particular product. Their main Web site address is given.

Letter prefix	Manufacturer	Manufacturer's Web Site
AD/OP	Analog Devices	www.analog.com
INA/OPA	Burr-Brown	www.burr-brown.com
CD	Cirrus Logic	www.cirrus.com
LF/LT/LTC	Linear Technology	www.linear-tech.com
MAX	Maxim	www.maxim-ic.com
MC	Motorola	www.motorola.com
LF/LM/LMC/LMV	National Semiconductor	www.national.com
TL/TLC/TH/TM	Texas Instruments	www.ti.com

Circuit designation. The circuit designator consists of three to seven numbers and letters. They identify the type of op amp and its temperature range. For example:

324C

Part number ↗ ↖ "C" identifies commercial temperature range

The three *temperature-range codes* are as follows:

1. C: commercial, 0 to 70°C
2. I: industrial, −25 to 85°C
3. M: military, −55 to 125°C

Letter suffix. A one- or two-letter suffix identifies the package style that houses the op amp chip. You need the package style to get the correct pin connections from the data sheet (see Appendix 1). Three of the most common package suffix codes are

Package code	Description
D	Plastic dual-in-line for surface mounting on a pc board
J	Ceramic dual-in-line
N, P	Plastic dual-in-line for insertion into sockets. (Leads extend through the top surface of a pc board and are soldered to the bottom surface.)

Military specification code. The military specification code is used only when the part is for high-reliability applications.

1-4.2 Order Number Example

A 741 general-purpose op amp would be completely identified in the following way:

Prefix	Designator	Suffix
↓	↓	↓
LM	741C	N (8-pin)
↓	↓	↓
National	General-purpose op amp, commercial temperature range	Plastic package

1-5 SECOND SOURCES

Some op amps are so widely used that they are made by more than one manufacturer. This is called *second sourcing.* The company (Fairchild) who designed and made the original 741 contracted for licenses with other manufacturers to make 741s in exchange for a license to make op amps or other devices.

As time went on, the original 741 design was modified and improved by all manufacturers. The present 741 has evolved over several generations. Thus, if you order a 741 8-pin DIP from a supplier, it may have been built by Texas Instruments (TL741), Analog Devices (AD741), National Semiconductor (LM741), or others. Therefore, always check the manufacturer's data sheets that correspond to the device you have. You will then have information on its exact performance and a key to the identification codes on the device.

1-6 BREADBOARDING OP AMP CIRCUITS

1-6.1 The Power Supply

Power supplies for general-purpose op amps are bipolar. As shown in Fig. 1-6(a), the typical commercially available power supply outputs ±15 V. The common point between the +15 V supply and −15 V is called the *power supply common.* It is shown with a common symbol for two reasons. First, *all* voltage measurements are made with respect to this point. Second, the power supply common is usually wired to the third wire of the line cord that extends ground (usually from a water pipe in the basement) to the chassis containing the supply.

The schematic drawing of a portable supply is shown in Fig. 1-6(b). This is offered to reinforce the idea that a bipolar supply contains two separate power supplies connected in series aiding.

1-6.2 Breadboarding Suggestions

It should be possible to breadboard and test the performance of all circuits presented in this text. A few circuits require printed circuit board construction. Before we proceed to learn how to use an op amp, it is prudent to give some time-tested advice on breadboarding a circuit:

1. Do all wiring with power off.
2. Keep wiring and component leads as short as possible.
3. Wire the $+V$ and $-V$ supply leads *first* to the op amp. It is surprising how often this vital step is omitted.
4. Try to wire all ground leads to one tie point, the power supply common. This type of connection is called *star grounding.* Do not use a ground bus, because you may create a ground loop, thereby generating unwanted noise voltages.
5. Recheck the wiring before applying power to the op amp.

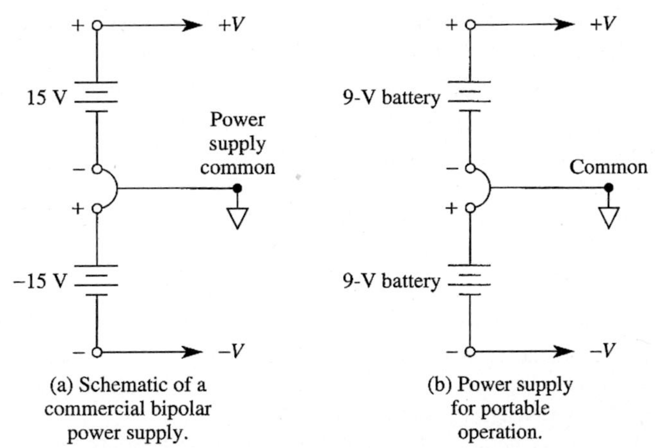

FIGURE 1-6 Power supplies for general-purpose op amps must be bipolar.

6. Connect signal voltages to the circuit only after the op amp is powered.
7. Take all measurements with respect to common. For example, if a resistor is connected between two terminals of an IC, do not connect either a meter or an oscilloscope across the resistor; instead, measure the voltage on one side of the resistor with respect to common, then the voltage on the other side, and calculate the voltage across the resistor.
8. Avoid using ammeters, if possible. Measure the voltage as in step 7 and calculate current.
9. Disconnect the input signal before the dc power is removed. Otherwise, the IC may be destroyed.
10. These ICs will stand much abuse. But *never:*
    a. Reverse the polarity of the power supplies,
    b. Drive the op amp's input pins above or below the potentials at the $+V$ and $-V$ terminal, or
    c. Leave an input signal connected with no power on the IC.
11. If unwanted oscillations appear at the output and the circuit connections seem correct:
    a. Connect a 0.1-μF capacitor between the op amp's $+V$ pin and ground and another 0.1-μF capacitor between the op amp's $-V$ pin and ground.
    b. Shorten your leads, and
    c. Check the test instrument, signal generator, load, and power supply ground leads. They should come together at one point.
12. The same principles apply to all other linear ICs.

We now proceed to our first experience with an op amp.

PROBLEMS

1-1. In the term *operational amplifier*, what does the word *operational* stand for?

1-2. Is the LM324 op amp a single op amp housed in one package, a dual op amp in one package, or a quad op amp in one package?

1-3. With respect to an op amp, what does the abbreviation PIN stand for?

1-4. Does the letter prefix of a PIN identify the manufacturer or the package style?

1-5. Does the letter suffix of a PIN identify the manufacturer or the package style?

1-6. Which manufacturer makes the AD741CN?

1-7. Does the tab on a metal can package identify pin 1 or pin 8?

1-8. Which pin is identified by the dot on an 8-pin mini-DIP?

1-9. **(a)** How do you identify power supply common on a circuit schematic?
(b) Why do you need to do so?

1-10. When breadboarding an op amp circuit, should you use a ground bus or star grounding?

1-11. Search a manufacturer's Web site and download a 741 data sheet.
(a) What is the manufacturer's identification code?
(b) What package styles are available?
(c) List three applications that the 741 op amp can be used in.

CHAPTER 2

First Experiences with an Op Amp

LEARNING OBJECTIVES

Upon completing this chapter on first experiences with an op amp, you will be able to:

- Briefly describe the task performed by the power supply and input and output terminals of an op amp.
- Show how the single-ended output voltage of an op amp depends on its open-loop gain and differential input voltage.
- Calculate the differential input voltage E_d, and the resulting output voltage V_o.
- Draw the circuit schematic for an inverting or noninverting zero-crossing detector.
- Draw the output voltage waveshape of a zero-crossing detector if you are given the input voltage waveshape.
- Draw the output–input voltage characteristics of a zero-crossing detector.
- Sketch the schematic of a noninverting or inverting voltage-level detector.

- Describe at least two practical applications of voltage-level detectors.
- Analyze the action of a pulse-width modulator and tell how it can interface an analog signal with a microcomputer.
- Use voltage reference ICs to design precise voltage-level detectors.
- Use SPICE to analyze a basic comparator circuit.

2-0 INTRODUCTION

The name *operational amplifier* was originally given to early high-gain vacuum-tube amplifiers designed to perform mathematical operations of addition, subtraction, multiplication, division, differentiation, and integration. They could also be interconnected to solve differential equations.

The modern successor of those amplifiers is the *linear integrated-circuit op amp.* It inherits the name, works at lower voltages, and is available in a variety of specialized forms. Today's op amp is so low in cost that millions are now used annually. Their low cost, versatility, and dependability have expanded their use far beyond applications envisioned by early designers. Some present-day uses for op amps are in the fields of signal conditioning, process control, communications, computers, power and signal sources, displays, and testing or measuring systems. The op amp is still basically a very good high-gain dc amplifier.

One's first experience with a linear IC op amp should concentrate on its most important and fundamental properties. Accordingly, our objectives in this chapter will be to identify each terminal of the op amp and to learn its purpose, some of its electrical limitations, and how to apply it usefully.

2-1 OP AMP TERMINALS

Remember from Fig. 1-2 that the circuit symbol for an op amp is an arrowhead that symbolizes high gain and points from input to output in the direction of signal flow. Op amps have five basic terminals: two for supply power, two for input signals, and one for output. Internally they are complex, as was shown by the schematic diagram in Fig. 1-3(a). It is not necessary to know much about the internal operation of the op amp in order to use it. We will refer to certain internal circuitry, when appropriate. The people who

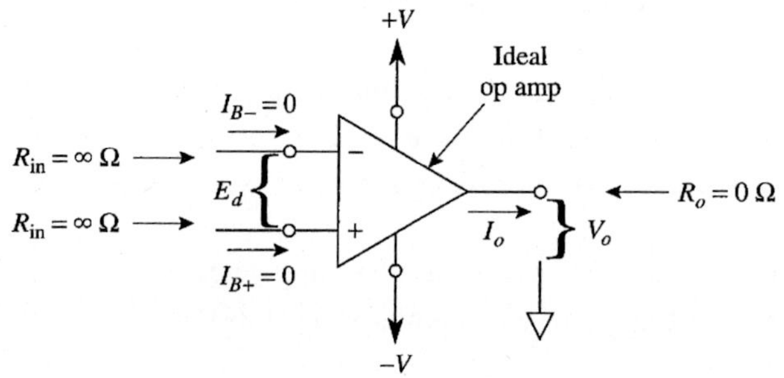

FIGURE 2-1 The ideal op amp has infinite gain and input resistances plus zero output resistance.

design and build op amps have done such an outstanding job that external components connected to the op amp determine what the overall system will do.

The ideal op amp of Fig. 2-1 has infinite gain and infinite frequency response. The input terminals draw no signal or bias currents and exhibit infinite input resistance. Output impedance is zero ohms, and the power supply voltages are without limit. We now examine the function of each op amp terminal to learn something about the limitations of a real op amp.

2-1.1 Power Supply Terminals

Op amp terminals labeled $+V$ and $-V$ identify those op amp terminals that must be connected to the power supply (see Fig. 2-2 and Appendices 1 and 2). Note that the power supply has *three* terminals: positive, negative, and power supply common. The power supply common terminal may or may not be wired to earth ground via the third wire of line cord. *All voltage measurements are made with respect to power supply common.*

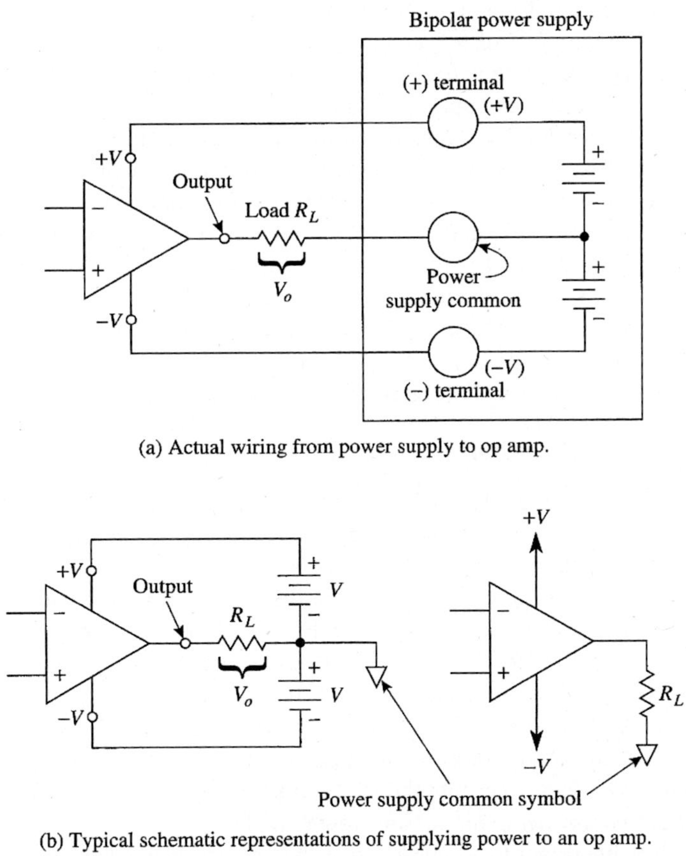

FIGURE 2-2 Wiring power and load to an op amp.

The power supply in Fig. 2-2 is called a bipolar or split supply and has typical values of ±15 V. Some op amps are now designed to operate from a single-polarity supply such as +5 or +15 V and ground. Note that the common is *not* wired to the op amp in Fig. 2-2. Currents returning to the supply from the op amp must return through external circuit elements such as the load resistor R_L. The maximum supply voltage that can be applied between $+V$ and $-V$ is typically 36 V or ±18 V.

2-1.2 Output Terminal

In Fig. 2-2 the op amp's output terminal is connected to one side of the load resistor R_L. The other side of R_L is wired to ground. Output voltage V_o is measured with respect to ground. Since there is only one output terminal in an op amp, it is called a *single-ended output.* There is a limit to the current that can be drawn from the output terminal of an op amp, usually on the order of 5 to 10 mA. There are also limits on the output terminal's voltage levels; these limits are set by the supply voltages and by the op amp's output transistors (see also Appendix 1, "Output Voltage Swing as a Function of Supply Voltage"). These output transistors need about 1 to 2 V from collector to emitter to ensure that they are acting as amplifiers and not as switches. Thus the output terminal can rise approximately to within 1 V of $+V$ and drop to within 2 V of $-V$. The upper limit of V_o is called the *positive saturation voltage,* $+V_{sat}$, and the lower limit is called the *negative saturation voltage,* $-V_{sat}$. For example, with a supply voltage of ±15 V, $+V_{sat} = +14$ V and $-V_{sat} = -13$ V. Therefore, V_o is restricted to a symmetrical peak-to-peak swing of ±13 V. Both current and voltage limits place a *minimum* value on the load resistance R_L of 2 kΩ. However, op amps are now available especially for applications that operate from low supply voltages (+3.3 V) and have MOS rather than bipolar output transistors. The output of these op amps can be brought to within millivolts of either $+V$ or $-V$.

Most op amps, like the 741, have internal circuitry that automatically limits current drawn from the output terminal. Even with a short circuit for R_L, output current is limited to about 25 mA, as noted in Appendix 1. This feature prevents destruction of the op amp in the event of a short circuit.

2-1.3 Input Terminals

In Fig. 2-3 there are two input terminals, labeled − and +. They are called *differential input terminals* because output voltage V_o depends on the *difference* in voltage between them, E_d, and the gain of the amplifier, A_{OL}. As shown in Fig. 2-3(a), the output terminal is positive with respect to ground when the (+) input is positive with respect to, or above, the (−) input. When E_d is reversed in Fig. 2-3(b) to make the (+) input negative with respect to, or below, the (−) input, V_o becomes negative with respect to ground.

We conclude from Fig. 2-3 that the polarity of the output terminal is the same as the polarity of (+) input terminal with respect to the (−) input terminal. Moreover, the polarity of the output terminal is opposite or inverted from the polarity of the (−) input terminal. For these reasons, the (−) input is designated the *inverting input* and the (+) input the *noninverting input* (see Appendix 1).

It is important to emphasize that the polarity of V_o depends only on the *difference* in voltage between inverting and noninverting inputs. This difference voltage can be found by

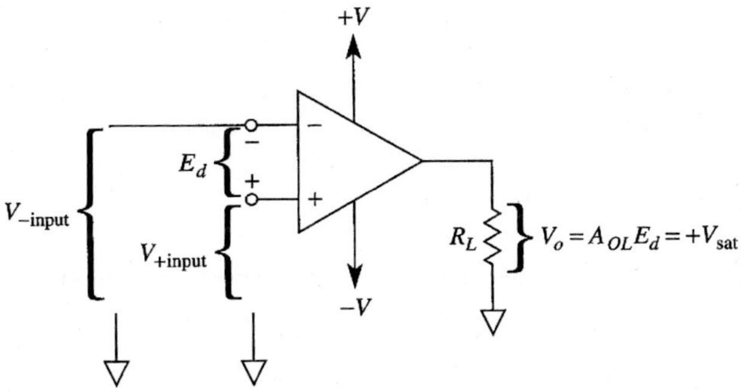

(a) V_o goes positive when the (+) input is more positive than (above) the (−) input, $E_d = (+)$.

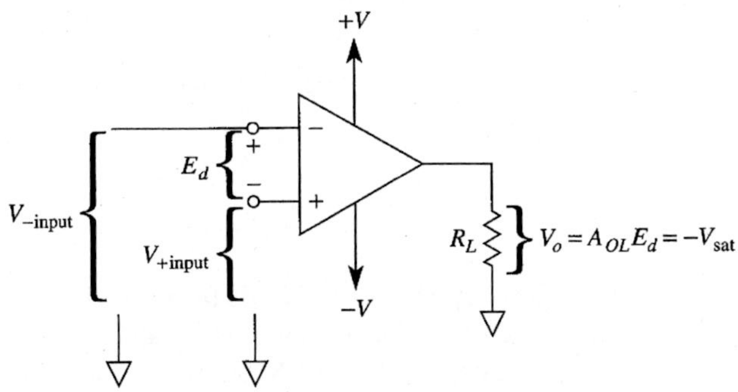

(b) V_o goes negative when the (+) input is less positive than (below) the (−) input, $E_d = (-)$.

FIGURE 2-3 Polarity of *single-ended* output voltage V_o depends on the polarity of *differential* input voltage E_d. If the (+) input is *above* the (−) input, E_d is positive and V_o is *above* ground at $+V_{sat}$. If the (+) input is *below* the (−) input, E_d is negative and V_o is *below* ground at $-V_{sat}$.

$$E_d = \text{voltage at the } (+) \text{ input} - \text{voltage at the } (-) \text{ input} \tag{2-1}$$

Both input voltages are *measured with respect to ground.* The sign of E_d tells us (1) the polarity of the (+) input with respect to the (−) input and (2) the polarity of the output terminal with respect to ground. This equation holds if the inverting input is grounded, if the noninverting input is grounded, and even if both inputs are above or below ground potential. Thus, if the polarity of E_d matches the op amp's symbol, the output voltage goes to $+V_{sat}$. When the polarity of E_d is opposite the op amp's symbol, the output voltage goes to $-V_{sat}$.

Review. We have chosen the words in Fig. 2-3 very carefully. They simplify analysis of open-loop operation (no connection from output to either input). Another memory aid is this: If the (+) input is *above* the (−) input, the output is *above* ground and at $+V_{sat}$. If the (+) input is *below* the (−) input, the output is *below* ground at $-V_{sat}$.

2-1.4 Input Bias Currents and Offset Voltage

The input terminals of real op amps draw tiny bias currents and signal currents to activate the internal transistors. The input terminals also have a small imbalance called *input offset voltage*, V_{io}. It is modeled as a voltage source V_{io} in series with the (+) input. In Chapter 9, the effects of V_{io} are explained in detail.

We must learn much more about op amp circuit operation, particularly involving negative feedback, before we can measure bias currents and offset voltage. For this reason, in these introductory chapters we will assume that both are negligible.

2-2 OPEN-LOOP VOLTAGE GAIN

2-2.1 Definition

Refer to Fig. 2-3. Output voltage V_o is determined by E_d and the *open-loop voltage gain*, A_{OL}. A_{OL} is called open-loop voltage gain because possible feedback connections from output terminal to input terminals are left open. Accordingly, V_o is expressed by the relationship

$$\text{output voltage} = \text{differential input voltage} \times \text{open-loop gain} \tag{2-2}$$

$$V_o = E_d \times A_{OL}$$

2-2.2 Differential Input Voltage, E_d

The value of A_{OL} is extremely large, often 200,000 or more. Recall from Section 2-1.2 that V_o can never exceed the positive or negative saturation voltages $+V_{sat}$ and $-V_{sat}$. For a ±15-V supply, the saturation voltages are approximately ±13 V. Thus, for the op amp to act as an amplifier, E_d must be limited to a maximum voltage of ±65 μV. This conclusion is reached by rearranging Eq. (2-2).

$$E_{d\,\max} = \frac{+V_{sat}}{A_{OL}} = \frac{13\text{ V}}{200{,}000} = 65\ \mu\text{V}$$

$$-E_{d\,\max} = \frac{-V_{sat}}{A_{OL}} = \frac{-13\text{ V}}{200{,}000} = -65\ \mu\text{V}$$

In the laboratory or shop it is difficult to measure 65 μV, because induced noise, 60-Hz hum, and leakage currents on the typical test setup can easily generate a millivolt (1000 μV). Furthermore, it is difficult and inconvenient to measure very high gains. The op amp also has tiny internal unbalances that act as a small voltage that may exceed E_d. As mentioned in Section 2-1.4, this small voltage is called an *offset voltage* and is discussed in Chapter 9.

2-2.3 Conclusions

There are three conclusions to be drawn from these brief comments. First, V_o in the circuit of Fig. 2-3 either will be at one of the limits $+V_{sat}$ or $-V_{sat}$ or will be oscillating between these limits. Don't be disturbed, because this behavior is what a high-gain amplifier usually does. Second, to maintain V_o between these limits we must go to a feedback type of circuit that forces V_o to depend on stable, precision elements such as resistors and capacitors. Feedback circuits are introduced in Chapter 3.

Without learning any more about the op amp, it is possible to understand basic comparator applications. In a comparator application, the op amp performs not as an amplifier but as a device that tells when an unknown voltage is below, above, or just equal to a known reference voltage. Before introducing the op amp as a comparator in the next section, Example 2-1 is given to illustrate ideas presented thus far.

Example 2-1

In Fig. 2-3, $+V = 15$ V, $-V = -15$ V, $+V_{sat} = +13$ V, $-V_{sat} = -13$ V, and gain $A_{OL} =$ 200,000. Assuming ideal conditions, find the magnitude and polarity of V_o for each of the following input voltages. These input voltages are given with respect to ground.

	Voltage at (−) input	Voltage at (+) input
(a)	−10 μV	−15 μV
(b)	−10 μV	+15 μV
(c)	−10 μV	−5 μV
(d)	+1.000001 V	+1.000000 V
(e)	+5 mV	0 V
(f)	0 V	+5 mV

Solution The polarity of V_o is the same as the polarity of the (+) input with respect to the (−) input. The (+) input is more negative than the (−) input in (a), (d), and (e). This is shown by Eq. (2-1), and therefore V_o will go negative. From Eq. (2-2), the magnitude of V_o is A_{OL} times the difference, E_d, between voltages at the (+) and (−) inputs, but if $A_{OL} \times E_d$ exceeds $+V$ or $-V$, then V_o must stop at $+V_{sat}$ or $-V_{sat}$. Calculations are summarized as follows:

	E_d [using Eq. (2-1)]	Polarity of (+) input with respect to (−) input	Theoretical V_o [from Eq. (2-2)]	Actual V_o
(a)	−5 μV	−	−5 μV × 200,000 = −1.0 V	−13 V
(b)	25 μV	+	25 μV × 200,000 = 5.0 V	+13 V
(c)	5 μV	+	5 μV × 200,000 = 1.0 V	+13 V
(d)	−1 μV	−	−1 μV × 200,000 = −0.2 V	−13 V
(e)	−5 mV	−	−5 mV × 200,000 = −1000 V	−13 V
(f)	5 mV	+	5 mV × 200,000 = 1000 V	+13 V

2-3 ZERO-CROSSING DETECTORS

2-3.1 Noninverting Zero-Crossing Detector

The op amp in Fig. 2-4(a) operates as a comparator. Its (+) input compares voltage E_i with a reference voltage of 0 V ($V_{ref} = 0$ V). When E_i is above V_{ref}, V_o equals $+V_{sat}$. This is because the voltage at the (+) input is more positive than the voltage at the (−) input. Therefore, the sign of E_d in Eq. (2-1) is positive. Consequently, V_o is positive, from Eq. (2-2).

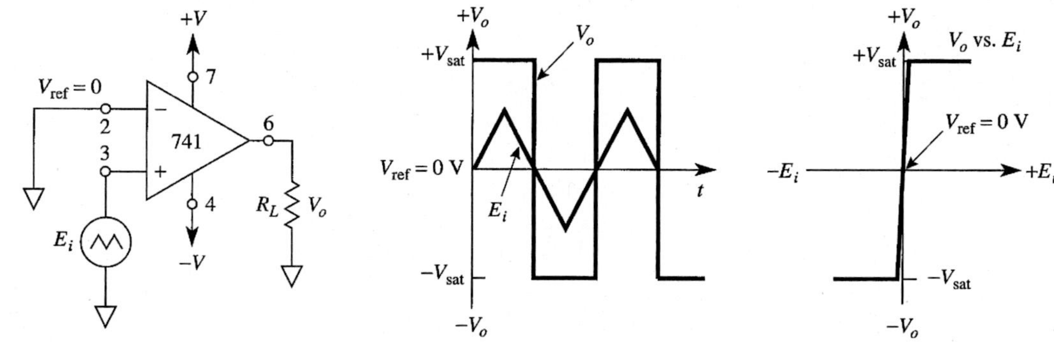

(a) Noninverting: When E_i is above V_{ref}, $V_o = +V_{sat}$.

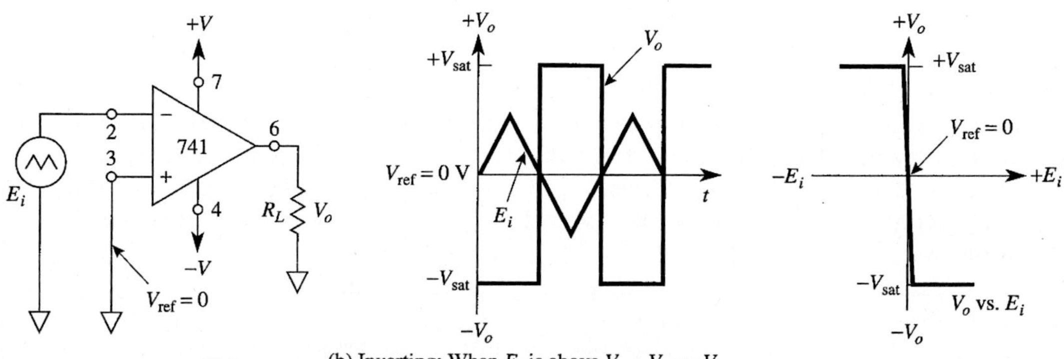

(b) Inverting: When E_i is above V_{ref}, $V_o = -V_{sat}$.

FIGURE 2-4 Zero-crossing detectors, noninverting in (a) and inverting in (b). If the signal E_i is applied to the (+) input, the circuit action is noninverting. If the signal E_i is applied to the (−) input, the circuit action is inverting.

The polarity of V_o "tells" if E_i is above or below V_{ref}. The *transition* of V_o tells *when* E_i crossed the reference and in what direction. For example, when V_o makes a positive-going transition from $-V_{sat}$ to $+V_{sat}$, it indicates that E_i just crossed 0 in the positive direction. The circuit of Fig. 2-4(a) is a noninverting zero-crossing detector.

2-3.2 Inverting Zero-Crossing Detector

The op amp's (−) input in Fig. 2-4(b) compares E_i with a reference voltage of 0 V (V_{ref} = 0 V). This circuit is an *inverting zero-crossing detector.* The waveshapes of V_o versus time and V_o versus E_i in Fig. 2-4(b) can be explained by the following summary:

1. If E_i is more positive than V_{ref}, then V_o equals $-V_{sat}$.
2. Where E_i crosses the reference going positive, V_o makes a negative-going transition from $+V_{sat}$ to $-V_{sat}$.

Summary. If the signal or voltage to be monitored is connected to the (+) input, a noninverting comparator results. If the signal or voltage to be monitored is connected to the (−) input, an inverting comparator results.

When $V_o = +V_{sat}$, the signal is *above* (more positive than) V_{ref} in a noninverting comparator and *below* (more negative than) V_{ref} in an inverting comparator.

2-4 POSITIVE- AND NEGATIVE-VOLTAGE-LEVEL DETECTORS

2-4.1 Positive-Level Detectors

In Fig. 2-5 a positive reference voltage V_{ref} is applied to one of the op amp's inputs. This means that the op amp is set up as a comparator to detect a positive voltage. If the voltage to be sensed, E_i, is applied to the op amp's (+) input, the result is a *noninverting positive-level detector.* Its operation is shown by the waveshapes in Fig. 2-5(a). When E_i is above V_{ref}, V_o equals $+V_{sat}$. When E_i is below V_{ref}, V_o equals $-V_{sat}$.

If E_i is applied to the inverting input as in Fig. 2-5(b), the circuit is an inverting positive-level detector. Its operation can be summarized by the statement: When E_i is above V_{ref}, V_o equals $-V_{sat}$. This circuit action can be seen more clearly by observing the plot of E_i and V_{ref} versus time in Fig. 2-5(b).

2-4.2 Negative-Level Detectors

Figure 2-6(a) is a *noninverting negative-level detector.* This circuit detects when input signal E_i crosses the negative voltage $-V_{ref}$. When E_i is above $-V_{ref}$, V_o equals $+V_{sat}$. When E_i is below $-V_{ref}$, $V_o = -V_{sat}$. The circuit of Fig. 2-6(b) is an *inverting negative-level detector.* When E_i is above $-V_{ref}$, V_o equals $-V_{sat}$, and when E_i is below $-V_{ref}$, V_o equals $+V_{sat}$.

2-5 TYPICAL APPLICATIONS OF VOLTAGE-LEVEL DETECTORS

2-5.1 Adjustable Reference Voltage

ICs are available to set precise voltage references. These reference chips will be introduced in the next section. In this section a resistive divider network is used to set V_{ref}. Figure 2-7 shows how to make an adjustable reference voltage. Two 10-kΩ resistors and

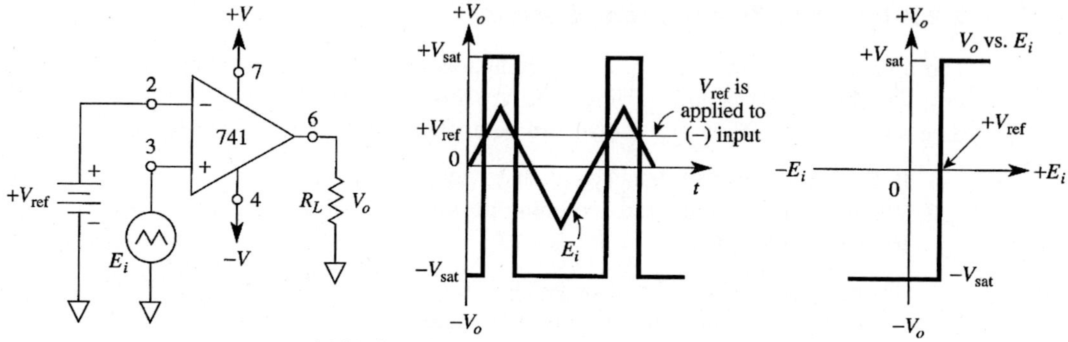

(a) Noninverting: When E_i is above V_{ref}, $V_o = +V_{sat}$.

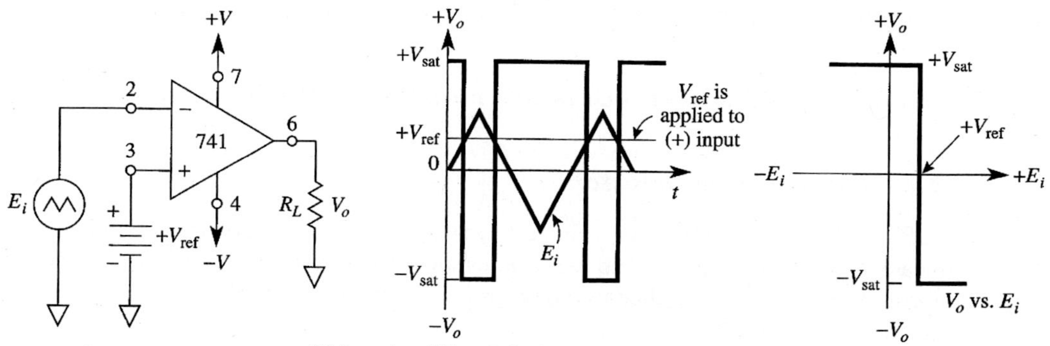

(b) Inverting: When E_i is above V_{ref}, $V_o = -V_{sat}$.

FIGURE 2-5 Positive-voltage-level detector, noninverting in (a) and inverting in (b). If the signal E_i is applied to the (+) input, the circuit action is noninverting. If the signal E_i is applied to the (−) input, the circuit action is inverting.

a 10-kΩ potentiometer are connected in series to make a 1-mA voltage divider. Each kilohm of resistance corresponds to a voltage drop of 1 V. V_{ref} can be set to any value between −5 and +5 V. Remove the $-V$ connection to the bottom 10-kΩ resistor and substitute a ground. You now have a 0.5-mA divider, and V_{ref} can be adjusted from 5 to 10 V.

2-5.2 Sound-Activated Switch

Figure 2-8 first shows how to make an adjustable reference voltage of 0 to 100 mV. Pick a 10-kΩ pot, 5-kΩ resistor, and +15-V supply to generate a convenient large adjustable voltage of 0 to 10 V. Next connect a 100:1 (approximately) voltage divider that divides the 0-to-10-V adjustment down to the desired 0-to-100-mV adjustable reference voltage. (*Note:* Pick the large 100-kΩ divider resistor to be 10 times the potentiometer resistance; this avoids loading down the 0-to-10-V adjustment.)

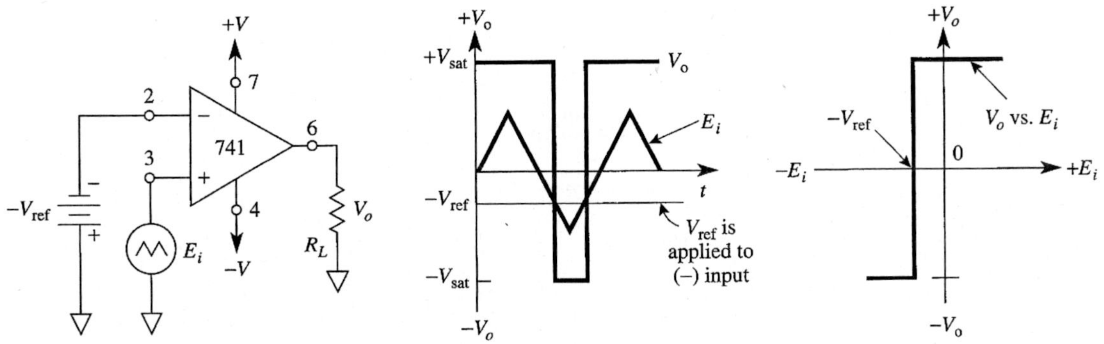

(a) Noninverting: When E_i is above V_{ref}, $V_o = +V_{sat}$.

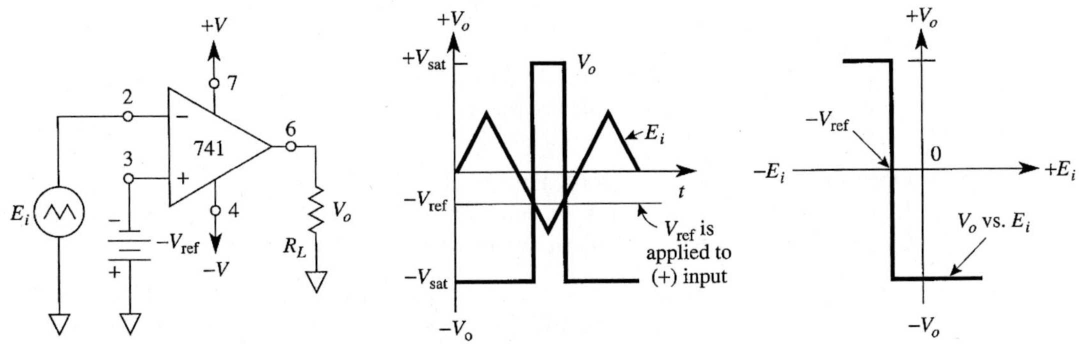

(b) Inverting: When E_i is above V_{ref}, $V_o = -V_{sat}$.

FIGURE 2-6 Negative-voltage-level detector, noninverting in (a) and inverting in (b).

A practical application that uses a positive level detector is the sound-activated switch shown in Fig. 2-8. Signal source E_i is a microphone, and an alarm circuit is connected to the output. The procedure to arm the sound switch is as follows:

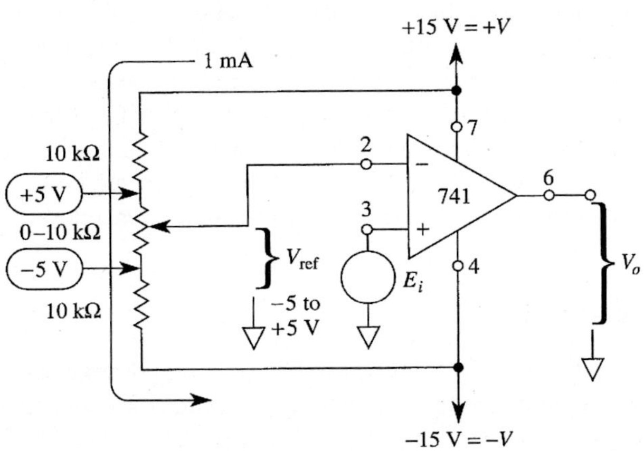

FIGURE 2-7 A variable reference voltage can be obtained by using the op amp's bipolar supply along with a voltage-divider network. (*Note:* Any fluctuations in the power supply result in a change in V_{ref}; hence, the need to use a stable or precise voltage may be a requirement in your system.)

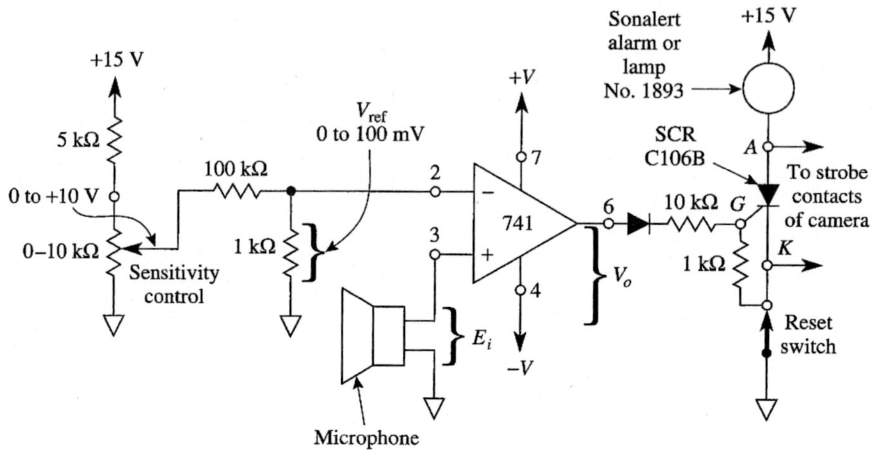

FIGURE 2-8 A sound-activated switch is made by connecting the output of a noninverting voltage-level detector to an alarm circuit.

1. Open the reset switch to turn off both SCR and alarm.
2. In a quiet environment, adjust the sensitivity control until V_o just swings to $-V_{sat}$.
3. Close the reset switch. The alarm should remain off.

Any noise signal will now generate an ac voltage and be picked up by the microphone as an input. The first positive swing of E_i above V_{ref} drives V_o to $+V_{sat}$. The diode now conducts a current pulse of about 1 mA into the gate, *G*, of the silicon-controlled rectifier (SCR). Normally, the SCR's anode, *A*, and cathode, *K*, terminals act like an open switch. However, the gate current pulse makes the SCR turn on, and now the anode and cathode terminals act like a closed switch. The audible or visual alarm is now activated. Furthermore, the alarm stays on because once an SCR has been turned on, it stays on until its anode–cathode circuit is opened.

The circuit of Fig. 2-8 can be modified to photograph high-speed events such as a bullet penetrating a glass bulb. Some cameras have mechanical switch contacts that close to activate a stroboscopic flash. To build this sound-activated flash circuit, remove the alarm and connect anode and cathode terminals to the strobe input in place of the camera switch. Turn off the room lights. Open the camera shutter and fire the rifle at the glass bulb. The rifle's sound activates the switch. The strobe does the work of apparently stopping the bullet in midair. Close the shutter. The position of the bullet in relation to the bulb in the picture can be adjusted experimentally by moving the microphone closer to or farther away from the rifle.

2-5.3 Light-Column Voltmeter

A light-column voltmeter displays a column of light whose height is proportional to voltage. Manufacturers of audio and medical equipment may replace analog meter panels with light-column voltmeters because they are easier to read at a distance.

A light-column voltmeter is constructed in the circuit of Fig. 2-9. R_{cal} is adjusted so that 1 mA flows through the equal resistor divider network R_1 to R_{10}. Ten separate reference voltages are established in 1-V steps from 1 V to 10 V.

When E_i = 0 V or is less than 1 V, the outputs of all op amps are at $-V_{sat}$. The silicon diodes protect the light-emitting diodes against excessive reverse bias voltage. When

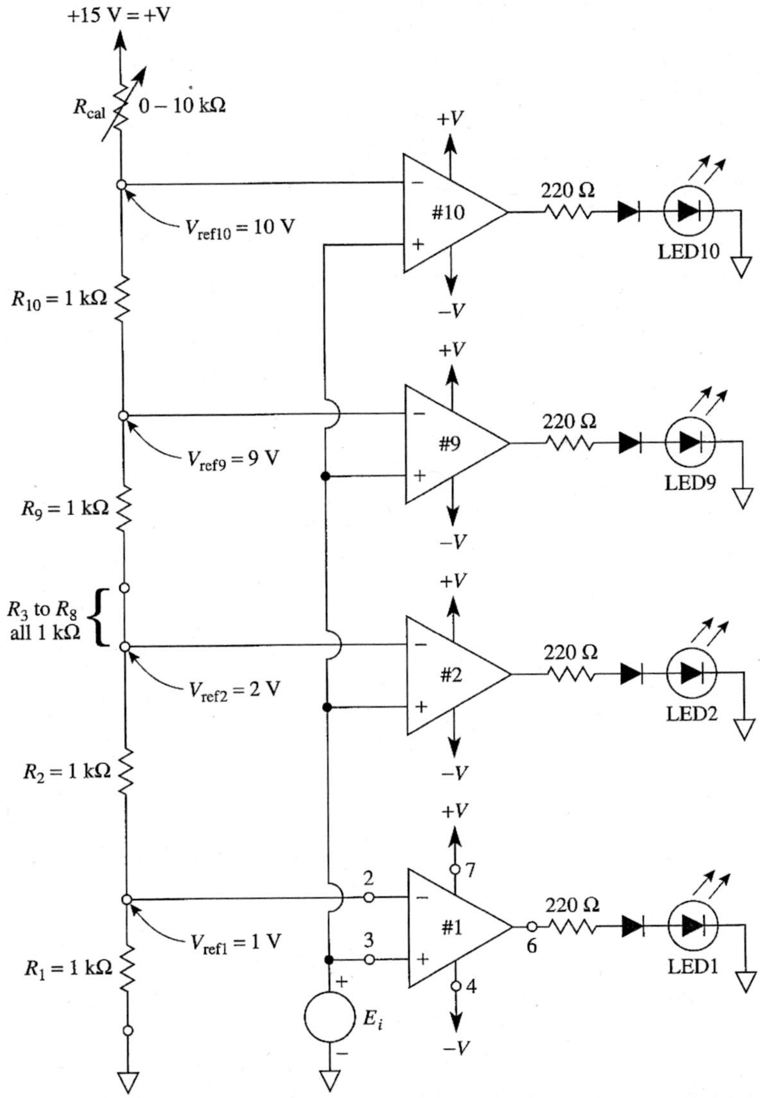

FIGURE 2-9 Light-column voltmeter. Reference voltages to each op amp are in steps of 1 V. As E_i is increased from 1 V to 10 V, LED1 through LED10 light in sequence. R_1 to R_{10} are 1% resistors. The op amps are 741 8-pin mini-DIPS.

E_i is increased to a value between 1 and 2 V, only the output of op amp 1 goes positive to light LED1. Note that the op amp's output current is automatically limited by the op amp to its short-circuit value approximately 20 to 25 mA. The 220-Ω output resistors divert heat away from the op amp.

As E_i is increased, the LEDs light in numerical order. This circuit can also be built using two-and-one-half LM324 quad op amps; some manufacturers have designed IC packages for this particular application, such as National Semiconductor's LM3914.

2-5.4 Smoke Detector

Another practical application of a voltage-level detector is a smoke or dust particle detector, as shown in Fig. 2-10. The lamp and photoconductive cell are mounted in an enclosed chamber that admits smoke or dust but not external light. The photoconductor is a light-sensitive resistor. In the absence of smoke or dust, very little light strikes the photoconductor and its resistance stays at some high value, typically several hundred kilohms. The 10-kΩ sensitivity control is adjusted until the alarm turns off.

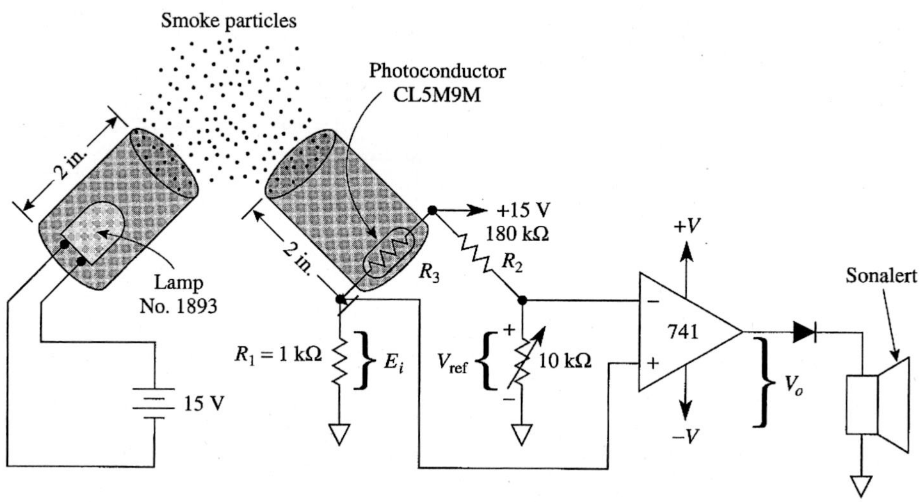

FIGURE 2-10 With no smoke or dust present the 10-kΩ sensitivity control is adjusted until the alarm stops. Light reflected off any smoke or dust particles causes the alarm to sound.

Any particles entering the chamber cause light to reflect off the particles and strike the photoconductor. This, in turn, causes the photoconductor's resistance to decrease and the voltage across R_1 to increase. As E_i increases above V_{ref}, V_o switches from $-V_{sat}$ to $+V_{sat}$, causing the alarm to sound. The alarm circuit of Fig. 2-10 does not include an SCR. Therefore, when the particles leave the chamber, the photoconductor's resistance increases and the alarm turns off. If you want the alarm to stay on, use the SCR alarm circuit shown in Fig. 2-8. The lamp and photoresistor must be mounted in a flat black, lightproof box that admits smoke. Ambient (room) light prevents proper operation. The resistive network at the input of the op amp forms a Wheatstone bridge. This circuit can be used to monitor the level of dust particles in a clean room environment.

2-6 VOLTAGE REFERENCE ICs

2-6.1 Introduction

Voltage reference ICs are used to provide a precise voltage for circuit and system designers, especially when setting the reference voltage for comparator circuits as well as A/D or D/A converters. Any fluctuation on the reference pin(s) of converter devices produces an inaccuracy in the conversion. Fluctuations on the reference input to a comparator can result in data being lost or erroneous data being sent to a computer system. In Figs. 2-7 and 2-8, we needed a reference voltage, V_{ref}, and used a resistor divider network connected between the supply voltages or the positive supply and ground. Although this circuit may be adequate for some quick-testing or low-cost designs, a better solution is to use a precision voltage reference chip. Many of these chips are inexpensive (less than a dollar), set a constant output voltage independent of temperature, and can be operated from a wide range of input power supply voltages. Variations in power supply voltages do not affect their output reference voltage. Some of these chips use the bandgap diode principle to produce a constant voltage of 1.2 V. This temperature-independent voltage is followed by an amplifier and buffer (amplifiers and buffers are topics covered in Chapter 3) to provide standard output voltages such as 2.5 V, 5 V, or 10 V. Other reference chips use a Zener diode as the reference followed by a buffer and amplifier to provide output voltages such as +5 V and −10 V, as well as ±5 V and ±10 V tracking outputs. Some of the most commonly used IC voltage reference chips are the REF-01 (+10 V), REF-02 (5 V), and REF-03 (2.5 V). We'll use the REF-02 IC as an introduction to precision voltage reference devices.

2-6.2 REF-02

The REF-02 IC outputs a stable +5.0 V, which can be adjusted by ±6% (±300 mV) using one external potentiometer as shown in Fig. 2-11(a). The 10-kΩ potentiometer allows the actual output voltage to be adjusted from 4.7 to 5.3 V. Thus for an 8-bit A/D converter the reference voltage can be set to 5.12 V, creating a resolution of 20 mV/bit. (Resolution of A/D converters is discussed in detail in Chapter 14.) The REF-02 can operate from an input supply voltage of from 7 to 40 V, making it an ideal voltage reference device for a wide range of applications. Two common package styles are shown in Figs. 2-11(b) and (c).

2-6.3 REF-02/Voltage Level Detector Applications

Figure 2-11(d) shows how the REF-02 can be connected to an op amp comparator to set the reference voltage at 5.0 V. In this circuit, the adjustment potentiometer is not used and the REF-02 is used in its basic configuration. In this application, V_{ref} for the comparator

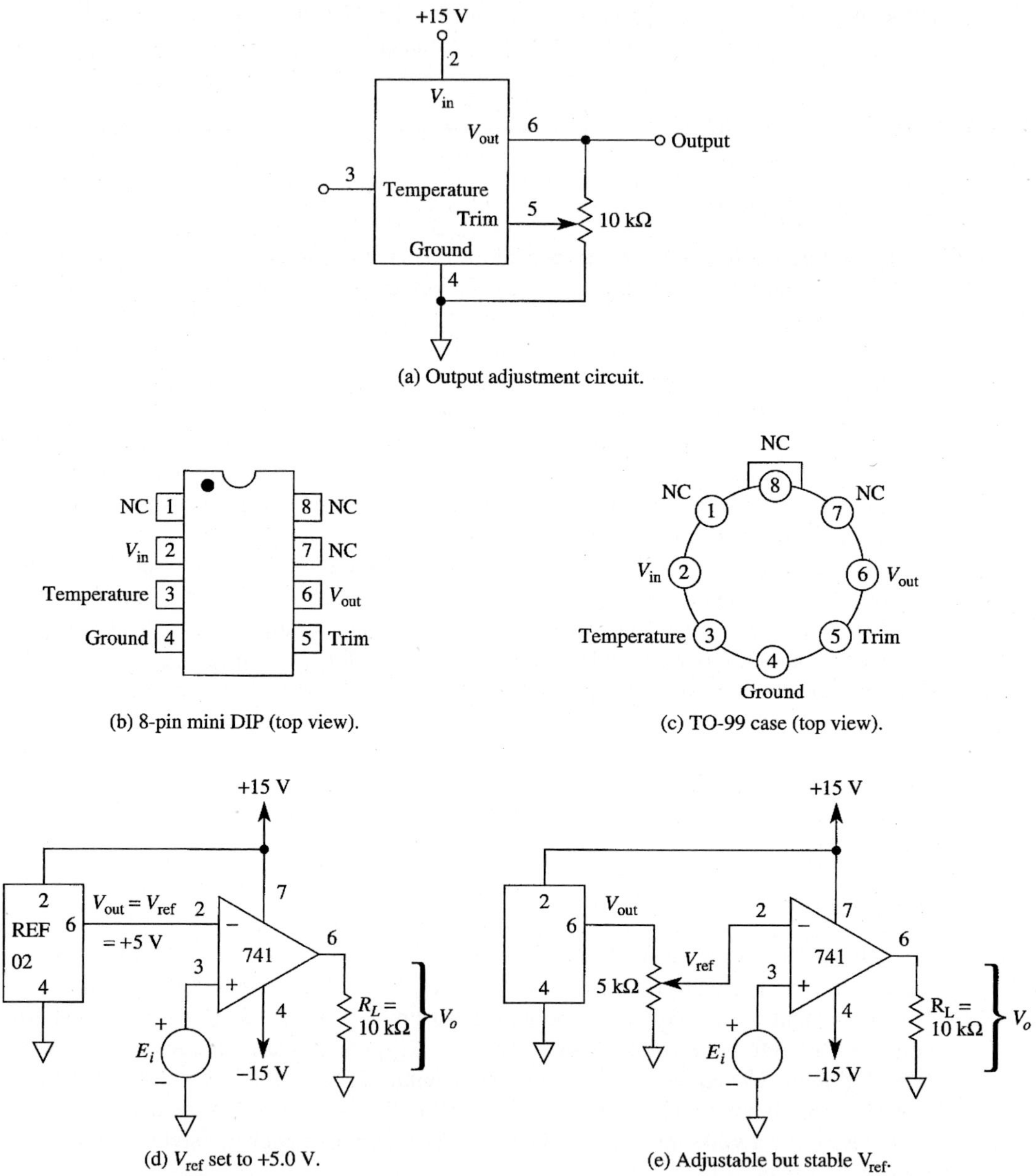

(a) Output adjustment circuit.

(b) 8-pin mini DIP (top view).

(c) TO-99 case (top view).

(d) V_{ref} set to +5.0 V.

(e) Adjustable but stable V_{ref}.

FIGURE 2-11 REF-02 pin assignments, package styles, and use with an op amp comparator.

should be within ±15 mV because the manufacturer's specification for output voltage of the REF-02 is 0.3% of 5.0 V. If your design requires closer tolerance of V_{ref}, use the output adjustment circuit shown in Fig. 2-11(a).

If your application requires a stable but variable reference voltage of 0 to 5.0 V you still may use the REF-02 with a potentiometer connected between the REF-02's output terminal and common, as shown in Fig. 2-11(e). (*Note:* The 5-kΩ potentiometer in this figure allows us to vary V_{ref} for the comparator.) This potentiometer is not being used to adjust the output voltage of the REF-02 but rather the input reference to the comparator so that V_{ref} can be varied from 0 to 5.0 V. You still may use the circuit of Fig. 2-11(a) if you need to set the REF-02's maximum output voltage to 5.0 V ±300 mV. The temperature pin (pin 3) is used if the REF-02 is being used as a temperature sensor. For an example, refer to Analog Devices' Web Site, specifically the REF-02's data sheet, to see this device as a sensor in a temperature controller application.

2-7 SIGNAL PROCESSING WITH VOLTAGE-LEVEL DETECTORS

2-7.1 Introduction

Armed with only the knowledge gained thus far, we will make a sine-to-square wave converter, an analog-to-digital converter, and a pulse-width modulator out of the versatile op amp. These open-loop comparator (or voltage-level detector) applications are offered to show how easy it is to use op amps.

2-7.2 Sine-to-Square Wave Converter

The zero-crossing detector of Fig. 2-4 will convert the output of a sine-wave from a function generator into a variable-frequency square wave. If E_i is a sine wave, triangular wave, or a wave of any other shape that is symmetrical around zero, the zero-crossing detector's output will be square. The frequency of E_i should be below 100 Hz, for reasons that are explained in Chapter 10.

2-7.3 Sawtooth-to-Pulse Wave Converter

Zener Diode Method: The voltage level detector circuits of Section 2-4 can be used to convert a sawtooth wave to a pulse wave provided that the output of the op amp is modified to create only a positive pulse. This modification is shown in Fig. 2.12(a) and consists of a silicon diode, a resistor, and a zener diode in series. When the output voltage of the op amp is at $+V_{sat}$, the resistor limits the current to approximately 5 mA, enough current to cause zener breakdown. For this condition the output voltage of the circuit, V_o, equals the zener voltage. V_z 5 4.7 V in Fig. 2.12(a). When the op amp's output voltage is at $-V_{sat}$, diode D1 is reversed biased and the op amp's output current is zero, hence $V_o = 0$ V. This circuit is a quick way of converting a sawtooth-to-pulse wave that is TTL compatible. The input and output waveforms are shown in Figs. 2.12(b) and (c), respectively. For this application, however, a better method is to use an integrated circuit called a comparator because we can get V_o to swing between 0 and 5 V without the external diodes.

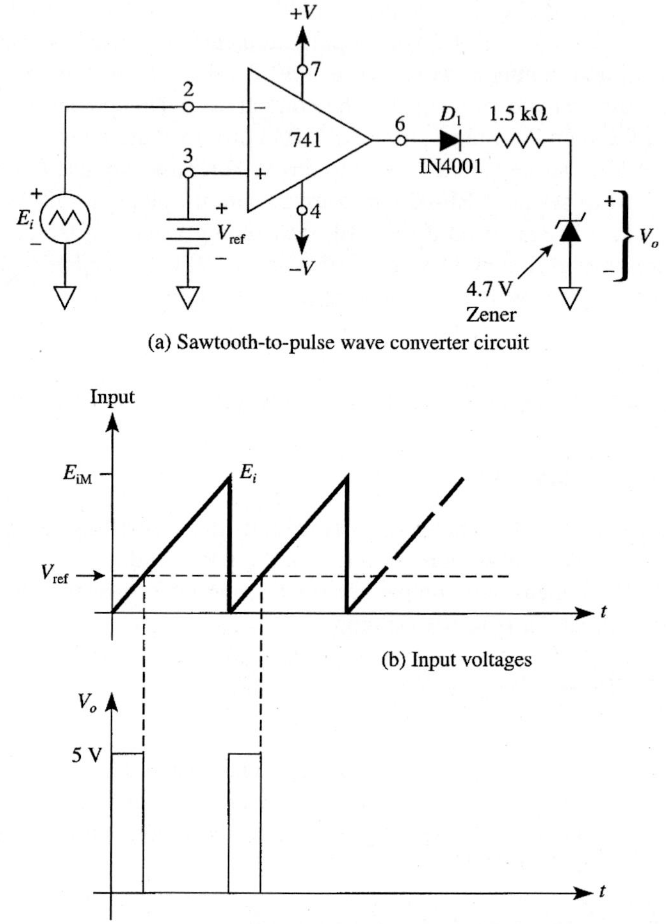

(a) Sawtooth-to-pulse wave converter circuit

(b) Input voltages

(c) Pulse output wave

FIGURE 2-12 A modification to a basic voltage-level detector converts a sawtooth wave to a pulse wave.

Comparator Method: Although comparator circuits are covered in detail in Chapter 4, comparator integrated circuit LM339 is introduced here so that a computer interfacing application using voltage-level detectors may be studied in the next section.

2-7.4 Quad Voltage Comparator, LM339

The pinouts and operation of a specialized op amp, the LM339, are shown in Figs. 2-13 and 2-14. The LM339 houses four independent op amps that have been specially designed to be flexible voltage comparators. We examine its operation by analyzing the role played by each terminal.

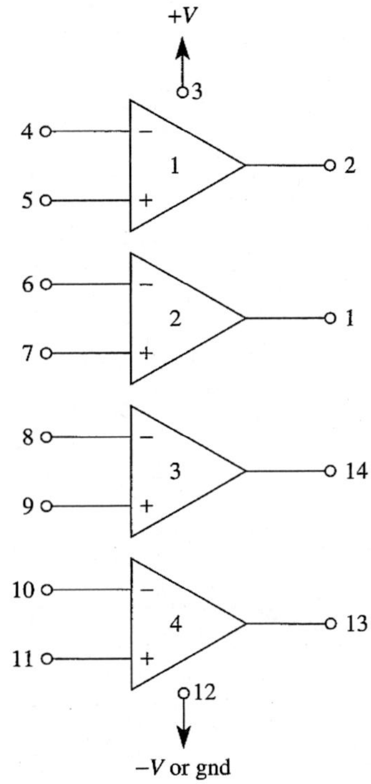

FIGURE 2-13 Connection diagram for the LM339 quad comparator. Four voltage comparators are contained in one 14-pin dual-in-line package.

Power supply terminals. Pins 3 and 12 are positive and negative supply voltage terminals, respectively, for all four comparators. Maximum supply voltage between pins 3 and 12 is ±18 V. In most applications, the negative supply terminal, pin 12, is grounded. Then pin 3 can be any voltage from 2 to 36 V dc. The LM339 is used primarily for single-supply operation.

Output terminals. The output terminal of each op amp is an open-collector *npn* transistor. Each transistor collector is connected to the respective output terminals 2, 1, 14, and 13. All emitters are connected together and then to pin 12. If pin 12 is grounded, the output terminal acts like a switch. A closed switch extends the ground from pin 12 to the output terminal [see Fig. 2-14(b)].

If you want the output to go high when the switch is open, you must install a pull-up resistor and an external voltage source. As shown in Fig. 2-14(a), this feature allows easy interfacing between a ±15-V analog system and a 5-V digital system. The output terminal should not sink more than 16 mA.

Input terminals. The input terminals are differential. Use Eq. (2-1) to determine the sign for E_d. If E_d is *positive,* the output switch is *open,* as in Fig. 2-14(a). If E_d is *negative,* the output switch is *closed,* as in Fig. 2-14(b). Unlike many other op amps, the input terminals can be brought down to ground potential when pin 12 is grounded.

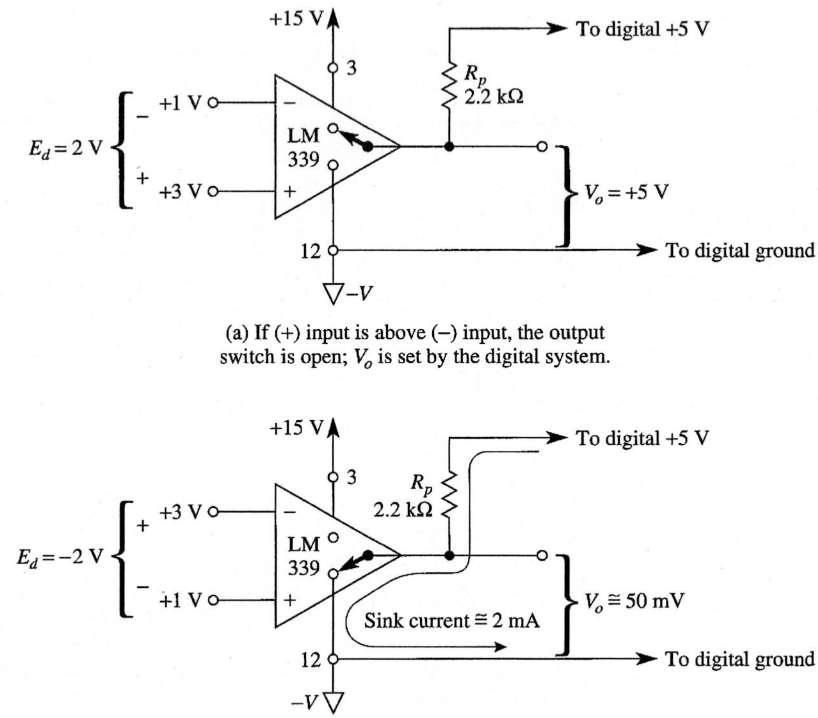

(a) If (+) input is above (−) input, the output switch is open; V_o is set by the digital system.

(b) If (+) input is below (−) input, the output switch is closed and sinks 2 mA of current from pull-up resistor R_p and the 5 V supply.

FIGURE 2-14 Operation of an LM339 (open collector output) comparator. When E_d is positive in (a), V_o goes high. V_o is determined by an external positive supply, pull-up resistor R_p and any external load resistor. If E_d is negative as in (b), the output goes low to essentially ground potential.

Summary. If the (+) input of an LM339 is *above* its (−) input, the output is pulled *high* by the pull-up resistor. If the (+) input is *below* the (−) input, the output is pulled *down* to the ground potential at pin 12. We now have information to analyze a pulse-width modulator application.

2-8 COMPUTER INTERFACING WITH VOLTAGE-LEVEL DETECTORS

2-8.1 Introduction

There are many characteristics of our environment or manufacturing processes that change very slowly. Examples are room temperature or the temperature of a large acid bath. A transducer can convert temperature changes to resistance or current changes. In Chapters 5 and 8 we show how you can convert these resistance or current changes into voltage changes quite easily with an op amp and a few parts.

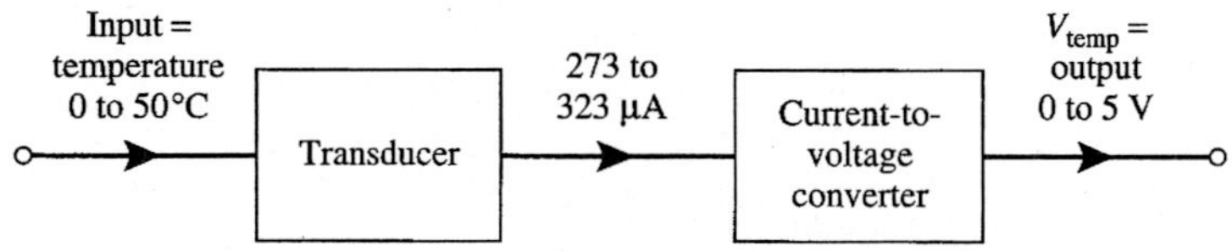

(a) Block diagram of a temperature-to-voltage converter.

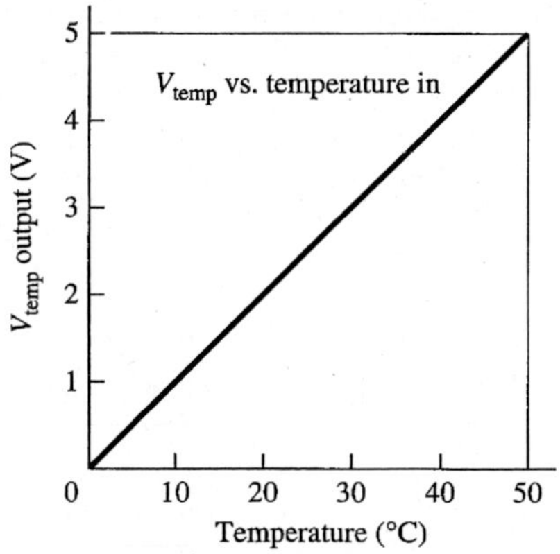

(b) Input–output characteristic of a temperature-to-voltage converter.

FIGURE 2-15 An example of how room or a process temperature is measured electronically.

Assume that you have available a circuit that gives 0 to 5 V out for a room-temperature change of 0° to 50°C (see Fig. 2-15). The output, V_{temp}, can now be used as a measurement of temperature, or it can be used to control temperature. Suppose that you want to send this temperature information to a computer so that the computer could monitor, control, or change room temperature. A voltage-level detector can accomplish this task. To understand how this can be done, we present a *pulse-width modulator* using the LM339 comparator.

2-8.2 Pulse-Width Modulator, Noninverting

The LM339 comparator in Fig. 2-16(a) compares two input voltages, E_c and V_{temp}. [Figure 2-16(b) is similar to Fig. 2-12(a).] A sawtooth wave, E_c, with constant frequency is connected to the (−) input. It is called a *carrier wave.* V_{temp} is a temperature-controlled voltage. Its rate of change must be much less than that of E_c. In this design, V_{temp} is the signal from the temperature transducer. It can be treated as a variable reference voltage when Fig. 2-16(a) is compared to Fig. 2-12(a) or Fig. 2-5(b).

In this circuit the *input signal is defined as* V_{temp}. The *output is defined as the high time,* T_H of V_o. In Fig. 2-16(b), the output stays high for 2 ms when $V_{temp} = 1$ V. If V_{temp} increases to 4 V, high time T_H increases to 8 ms as in Fig. 2-16(c).

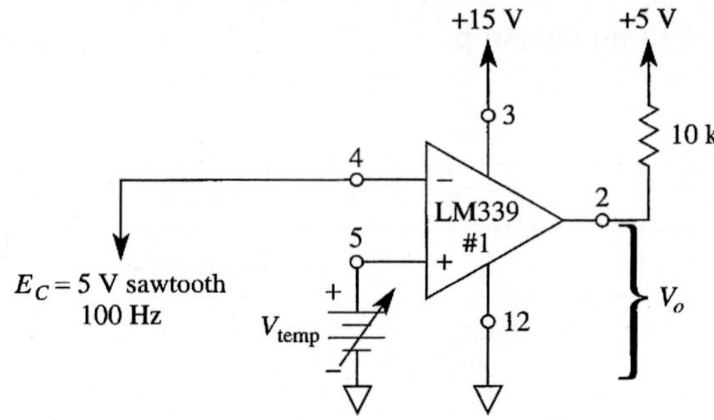

(a) Noninverting pulse-width-modulator circuit.

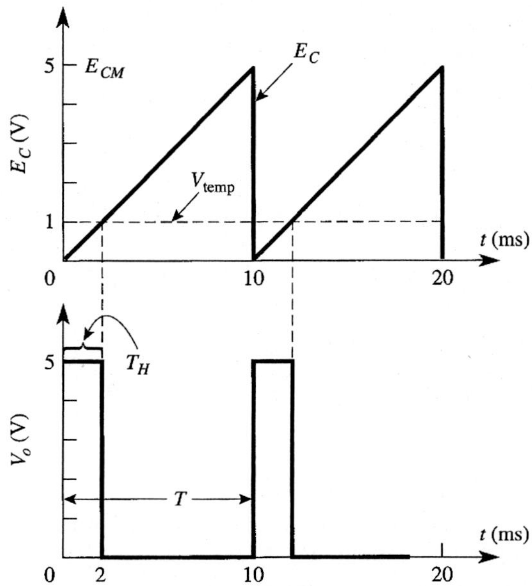

(b) $V_{temp} = 1$ V, $T_H = 2$ ms.

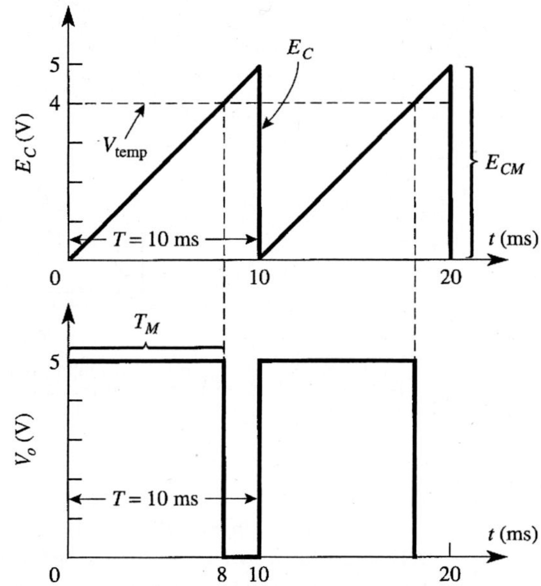

(c) $V_{temp} = 4$ V, $T_H = 8$ ms.

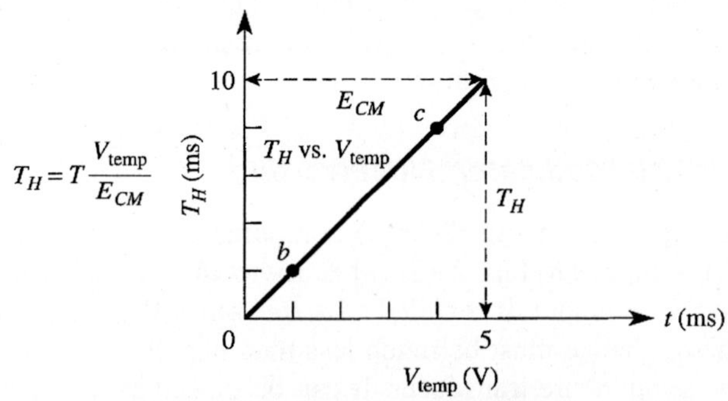

(d) Input voltage V_{temp} vs. output high time T_H.

FIGURE 2-16 V_{temp} is defined as the input signal in (a). As V_{temp} *increases* from 0 to 5 V, the high time of output voltage V_o *increases* from 0 to 10 ms. The circuit is called a *noninverting pulse-width modulator.*

Operation of the circuit is summarized by the input–output characteristics in Fig. 2-16(d). The width of output pulse T_H is changed (modulated) by V_{temp}. The constant period of the output wave is set by E_c. *Thus E_c carries the information contained in V_{temp}.* V_o is then said to be a pulse-width-modulated wave. The input–output equation is

$$\text{output } T_H = (V_{temp}) \frac{T}{E_{CM}} \tag{2-3}$$

where T = period of sawtooth carrier wave
E_{CM} = maximum peak voltage of a sawtooth carrier

Example 2-2 shows that the pulse-width modulator can also be called a *duty-cycle controller.*

Example 2-2

A 10-V, 50-Hz sawtooth wave is pulse-width modulated by a 4-V signal. Find the output's (a) high time; (b) duty cycle.

Solution Period T is found from the reciprocal of the frequency:

$$T = \frac{1}{f} = \frac{1}{50\text{ Hz}} = 20\text{ ms}$$

(a) From Eq. (2-3),

$$T_H = (4\text{ V})\frac{20\text{ ms}}{10\text{ V}} = 8\text{ ms}$$

(b) Duty cycle is defined as the ratio of high time to the period and is expressed in percent:

$$\text{duty cycle} = \frac{T_H}{T} \times 100 \tag{2-4}$$

$$= \frac{8\text{ ms}}{20\text{ ms}} \times 100$$

Thus the output stays high for 40% of each signal.

2-8.3 Inverting and Noninverting Pulse-Width Modulators

Figure 2-17 shows the difference between noninverting and inverting pulse-width modulators. If signal V_{temp} is applied to the (+) input, the circuit is defined as noninverting [see Figs. 2-17(a), (b), and (c)]. The *slope* of T_H versus V_{temp} rises to the right and is *positive* or *noninverting.*

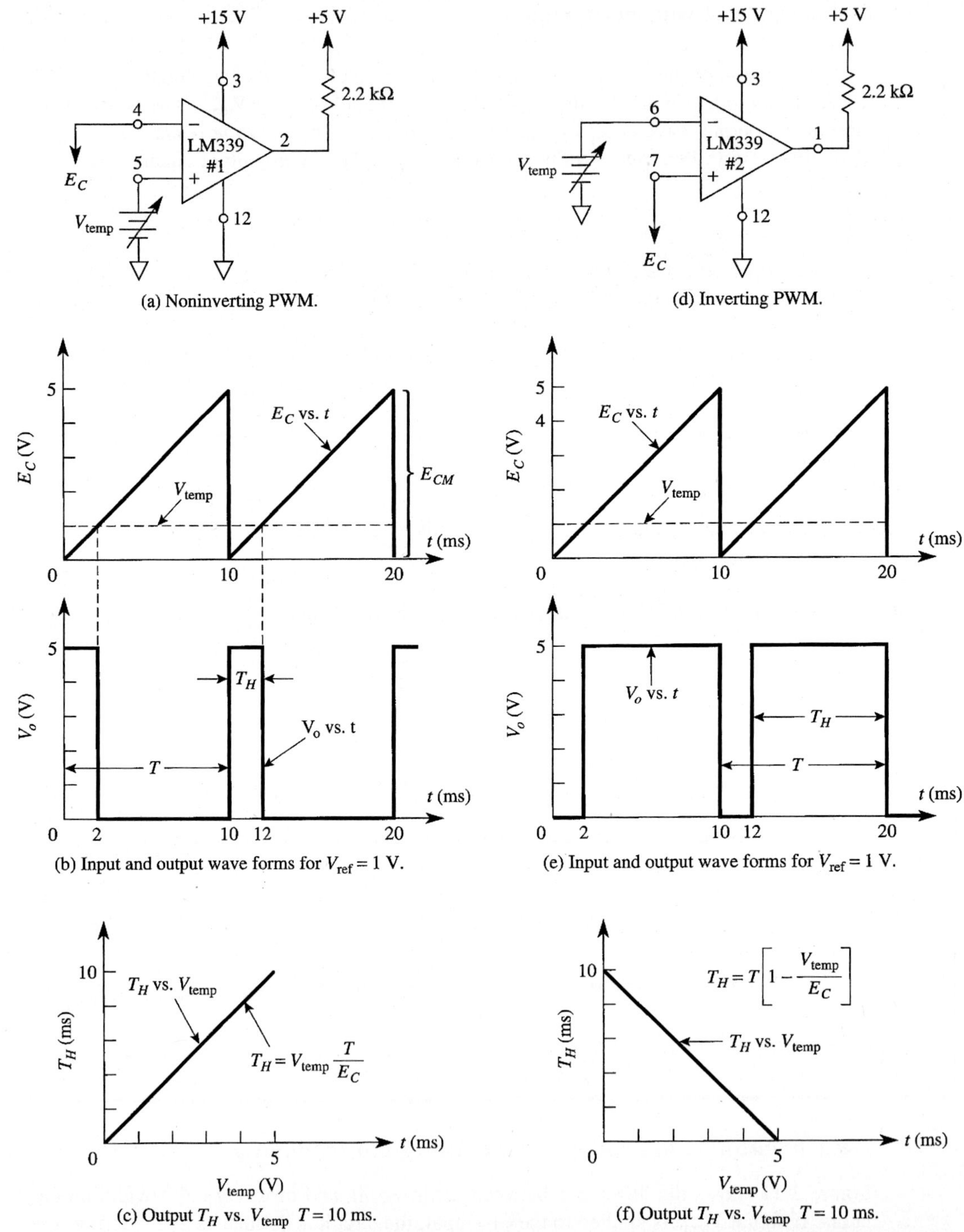

(a) Noninverting PWM.

(d) Inverting PWM.

(b) Input and output wave forms for $V_{ref} = 1$ V.

(e) Input and output wave forms for $V_{ref} = 1$ V.

(c) Output T_H vs. V_{temp} $T = 10$ ms.

(f) Output T_H vs. V_{temp} $T = 10$ ms.

FIGURE 2-17 Output high time *increases* as input V_{temp} *increases* in a noninverting pulse-width modulator [see (a), (b), and (c)]. Output high time *decreases* as V_{temp} *increases* in an *inverting* pulse-width modulator.

V_{temp} is applied to the (−) input in Fig. 2-17(d). As V_{temp} increases, T_H decreases. The *slope* of T_H versus V_{temp} is shown in Fig. 2-17(f) and is *negative.* The inverting performance equation is

$$T_H = T\left(1 - \frac{V_{temp}}{E_{CM}}\right) \tag{2-5}$$

Example 2-3

Calculate the output high time if $V_{temp} = 4$ V in Fig. 2-17(d).

Solution From Eq. (2-5),

$$T_H = 10 \text{ ms}\left(1 - \frac{4 \text{ V}}{5 \text{ V}}\right) = 2 \text{ ms}$$

2-9 A PULSE-WIDTH MODULATOR INTERFACE TO A MICROCONTROLLER

Either circuit of Fig. 2-17 can be used to transmit temperature information as a pulse-width modulated signal to a computer. The advantage of such an analog interface circuit is to eliminate a voltage drop over distances of several hundred feet. Thus the pulse-width modulator can interface an analog signal with an input port of a microcontroller (see Fig. 2-18). The temperature is first converted to a voltage by the sensor. A noninverting pulse-width modulator then converts this analog voltage to an output that is digital in nature; that is, its output is either high or low and the high time is directly proportional to temperature.

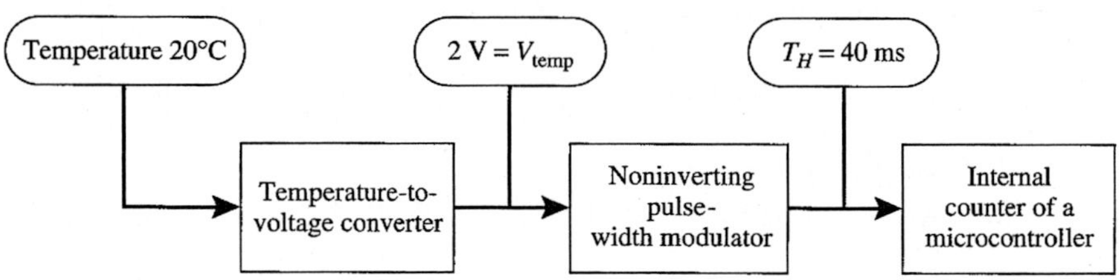

FIGURE 2-18 Block diagram of a computerized temperature measurement.

The computer programmer can perform the analog-to-digital conversion of the high time to a digital code. This may be done by using a 1-ms timing loop and counting the number of times that the loop is executed. Another and more efficient method is to use the internal counter designed into most microcontrollers. The 0-to-5-V transition of V_o is used to start the microcontroller's counter and the 5-to-0-V transition stops the counter. The count, which is automatically stored in one of the microcontroller's internal registers, is directly proportional to the temperature.

2-10 OP AMP COMPARATOR CIRCUIT SIMULATION

2-10.1 Introduction

PSpice is a software package for analog and digital design analysis. Students who are studying op amps usually have already used PSpice in previous courses, so all the PSpice fundamentals are not introduced; however, enough introduction steps are included throughout this text to allow first-time users to create and analyze their circuits.

2-10.2 Creating, Initializing, and Simulating a Circuit

Let us create and analyze the noninverting positive-level detector circuit shown in Fig. 2-5(a). We will use a sine wave for the input signal because it is easy to obtain from the basic parts list. (*Note:* The parts list does not contain a triangular waveform although one can be created, which we will do in a later chapter.) To create and simulate Fig. 2-5(a), open a new worksheet either by clicking on **File** => **New,** or if the PSpice window is not open, double-click on the Schematics icon in the window. If necessary, enlarge the work area to fill the entire screen.

The basic parts list browser may be obtained by clicking **Draw** from the Menu bar and then clicking **Get New Part** from the drop-down menu. These steps will be represented by

Draw => **Get New Part**

A shortcut for obtaining the parts list is to click the icon on the toolbar. (*Note*: The icon symbol is different for different versions of PSpice.) Either method produces a pop-up menu that contains the **Parts Browser's** basic list. Click **Advanced** >> and the basic menu expands to include a window to show you the part before you place it in the work area. Other libraries of parts can be obtained by clicking on the **Libraries** button.

The general guidelines for creating and simulating a circuit in PSpice are:

1. Open a new work area.
2. Obtain each part from the parts list and place it in the work area. Then close the parts list.
3. Arrange the parts the way they appear in the circuit schematic.
4. Interconnect the parts.
5. Change any attribute value(s) for a part if necessary.
6. Initialize setup parameters—**Analysis** => **Setup.**
7. Initialize probe setup if you want a plot—**Analysis** => **Probe Setup**
8. Save the schematic as a file with the .SCH extension.
9. Ensure there are no wiring errors—**Analysis** => **Create Netlist.**
10. Execute the program to observe the results—**Analysis** => **Simulate.**

Let us create the noninverting positive-level detector circuit of Fig. 2-5(a) by calling up the following parts and placing them in the work area. It is easier if you get all the parts at once and place them in the right section of the work area, close the parts list, and then arrange the parts as they appear in the circuit schematic. For this application, we will use three dc supplies for $+V$, $-V$, and V_{ref}.

Draw => Get New Part

Part	Number	Version 6.2 Library
=> **UA741**	pins 1 and 5 are shown but not used	**eval. slb**
=> **VDC**	place three for $+V$, $-V$, and V_{ref}	**source.slb**
=> **VSIN**	sine wave	**source.slb**
=> **GLOBAL**	place six	**port.slb**
=> **AGND**	analog ground, place five	**port.slb**
=> **R**	resistor for RL	**analog.slb**

Close the parts list and arrange the parts as in Fig. 2-5(a). (*Note:* The op amp PSpice model comes from the parts list with the inverting terminal at the bottom and the noninverting terminal at the top of the diagram. For now we will leave it with this orientation. The terminals can be switched if the op amp is rotated twice and then flipped. In this new orientation, however, $+V$ is at the bottom and $-V$ is at the top.) To interconnect the parts, click **Draw** => **Wire** or click the thin Pencil icon in the toolbar. Figure 2-19 shows how the parts can be interconnected.

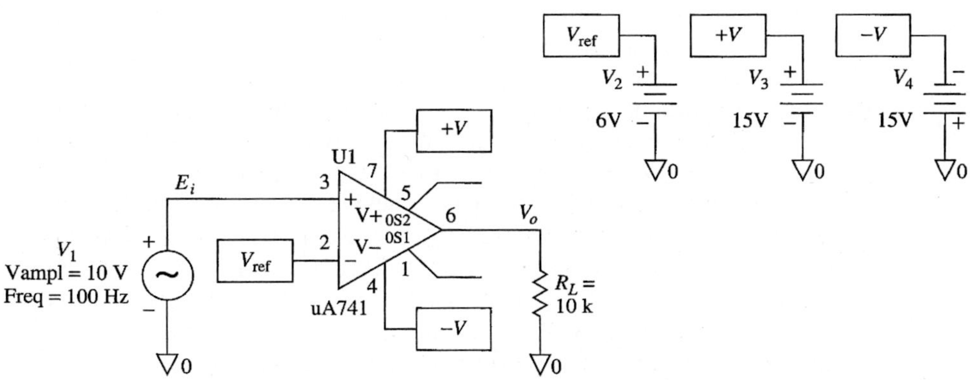

FIGURE 2-19 PSpice model of a noninverting comparator circuit.

The parts in this circuit that require setting new values (attributes) are the three dc supply voltages; the six globals; the sine wave's amplitude, frequency, and offset; and the value of R_L. Changing a part's attributes is done by first double-clicking on the part or value to be changed and then entering the new value. Double-clicking highlights the part or value in red and then opens an attribute box that allows you to enter the new value.

One at a time, double-click on 0 V and set the voltages at the supplies as:

+V = 15 V
−V = 15 V (Note the orientation of this supply.)
Vref = 6 V

One at a time, set the six GLOBAL labels as:

+V—at pin 7 of the op amp and that connected to +15 V
−V—at pin 4 of the op amp and that connected to −15 V
Vref—at pin 2 of the op amp and the source used for **Vref**.

Similarly the label of the resistor and its value can be changed to R_L and 10 kΩ, respectively. To change the attributes of the input sine-wave signal, double-click the symbol and a VSIN attribute box appears. One at a time, change each attribute by double-clicking the attribute and setting the new value in the window. For this circuit, amplitude, frequency, and offset values have to be changed as shown:

AMPL = to 10 V => **Save Attr** => **Change Display** => **Both name and value**
FREQ = to 100 Hz => **Save Attr** => **Change Display** => **Both name and value**
VOFF = to O V => **Save Attr** (not necessary to change display for this application)

In this application, we want a plot of E_i, V_{ref}, and V_o versus time similar to what is shown in Fig. 2-5(a). In order to do this, we first must add the location of E_i and V_o to the op amp's inverting and output terminals, respectively. The location of V_{ref} is already shown on the circuit diagram. This step is done by double-clicking the "wire" connection at the point of interest and entering the label in the window of the pop-up box. Figure 2-19 shows the completed schematic ready for analysis. To obtain these plots, open **Analysis** => **Probe Setup** and click Automatically Run Probe After Simulation. Now open **Analysis** => **Setup** and click the box next to **Transient.** An *x* appears indicating it has been selected. Now click on **Transient** and set Print Step to 0.05 ms and Final Time to 20 ms. This will allow Probe to display two complete cycles of a 100-Hz sine wave.

Save the file by **File** => **Save** or by clicking the Disk icon in the toolbar. You may use any file name but be sure to use extension .SCH. A check of the wiring corrections is done by creating a netlist—**Analysis** => **Create Netlist.** A warning appears if there are any wiring errors. Click OK and a list of the error location(s) is obtained. If there are no errors, then the circuit is ready to run the simulation program. This step is done by **Analysis** => **Simulate** or using the hot key F11. The Probe window (a black screen) appears. To plot the graphs, use **Trace** => **Add** and click **V[Ei], V[Vref],** and **V[Vo]** and then OK. The three waveforms should now be plotted as shown in Fig. 2-20. To add labels to the graphs, use **Tools** => **Label** => **Text** and a text box appears. Type in the label you wish to place on the graph and then click OK. Use the mouse to place the label where you want it and repeat the procedure for any new labels. To add arrows, use **Tools** => **Label** => **Arrow.** Use the mouse to place the tail of the arrow at the starting point and draw out the arrow. Click the left mouse button to stop and the completed arrow is drawn.

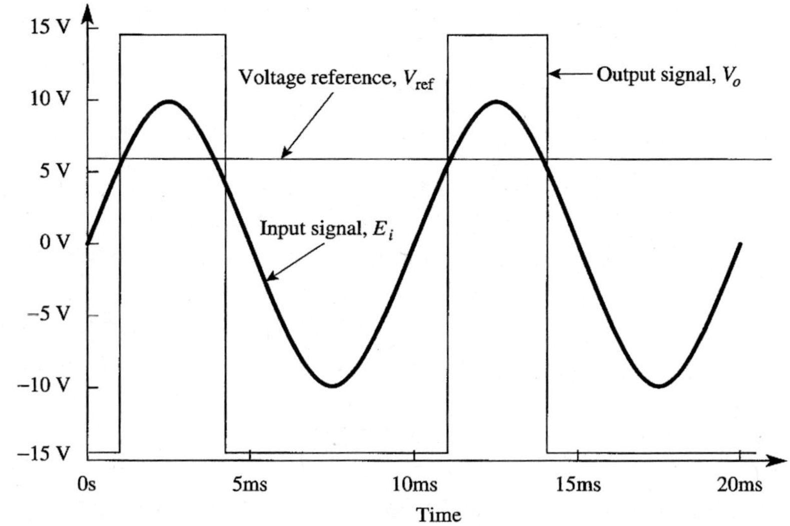

FIGURE 2-20 PSpice output display for a 10-V sine wave input and a reference level of 6 V.

PROBLEMS

2-1. Name the five basic terminals of an op amp.

2-2. Name the manufacturer of an AD741 op amp.

2-3. A 741 op amp is manufactured in an 8-pin dual-in-line package. What are the terminal numbers for the **(a)** inverting input; **(b)** noninverting input; **(c)** output?

2-4. A 741 op amp is connected to a ±15-V supply. What are the output terminal's operating limits under normal conditions with respect to **(a)** output voltage; **(b)** output current?

2-5. When the load resistor of an op amp is short-circuited, what is the op amp's **(a)** output voltage; **(b)** approximate output current?

2-6. Both op amps of Fig. P2-6 are in 14-pin dual-in-line packages. **(a)** Number the terminals. **(b)** Calculate E_d. **(c)** Find V_o.

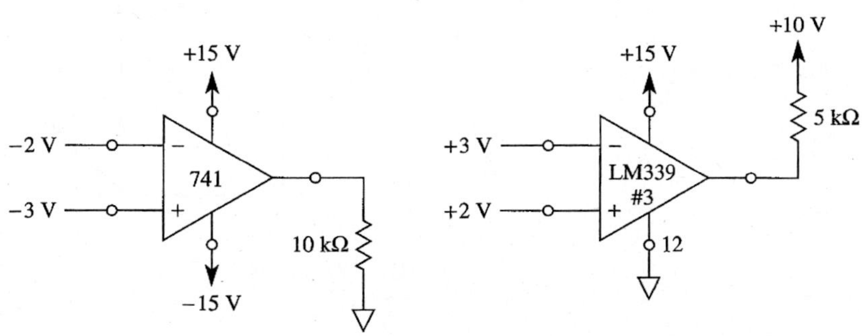

FIGURE P2-6

2-7. E_i is applied to the (−) input and ground to the (+) input of a 741 in Fig. P2-7. Sketch accurately **(a)** V_o vs. t and **(b)** V_o vs. E_i.

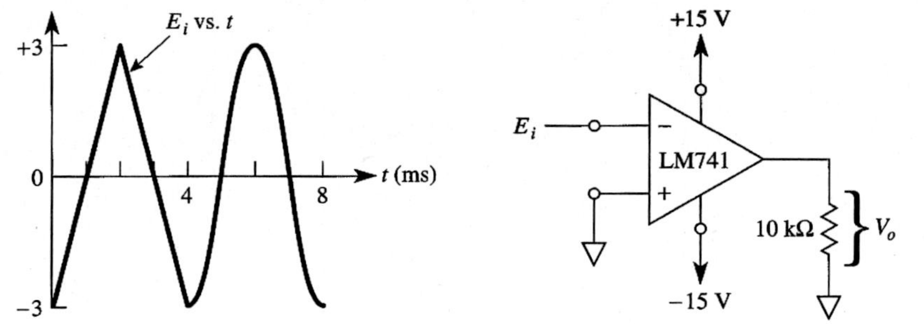

FIGURE P2-7

2-8. Swap the input connections to E_i and ground in Fig. P2-7. Sketch **(a)** V_o vs. t and **(b)** V_o vs. E_i.

2-9. Refer to Problems 2-7 and 2-8. Which circuit is the noninverting zero-crossing detector, and which is the inverting zero-crossing detector?

2-10. To which input would you connect a reference voltage to make an inverting level detector?

2-11. You need a 741 noninverting voltage-level detector. **(a)** Will the output be at $+V_{sat}$ or $-V_{sat}$ when the signal voltage is above the reference voltage? **(b)** To which input do you connect the signal?

2-12. Design a reference voltage that can be varied from 0 to −5 V. Assume that the negative supply voltage is −15 V.

2-13. Design a 0 to +50 mV adjustable reference voltage. Derive it from the +15-V supply.

2-14. The frequency of carrier wave E_c is constant at 50 Hz in Fig. P2-14. If $V_{temp} = 5$ V, **(a)** calculate high time T_H; **(b)** plot V_o vs. time.

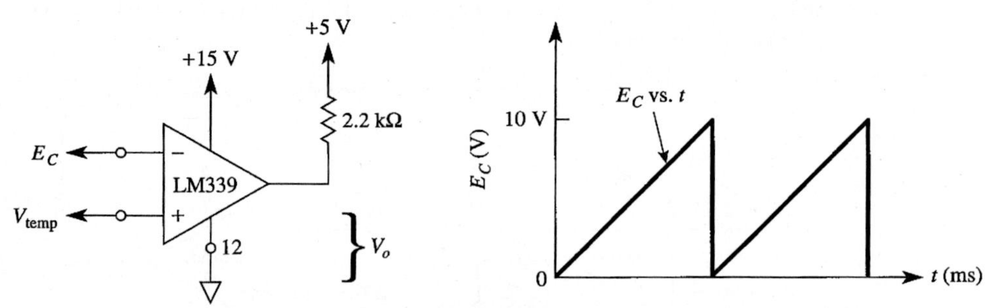

FIGURE P2-14

2-15. Assume that V_{temp} is varied from 0 V to +10 V in Problem 2-14. Plot T_H vs. V_{temp}.

2-16. In Fig. P2-16, E_{in} is a triangle wave. The amplitude is -5 V to $+5$ V and the frequency is 100 Hz. Sketch accurately the graphs of **(a)** V_o vs. E_{in}; **(b)** V_o vs. t.

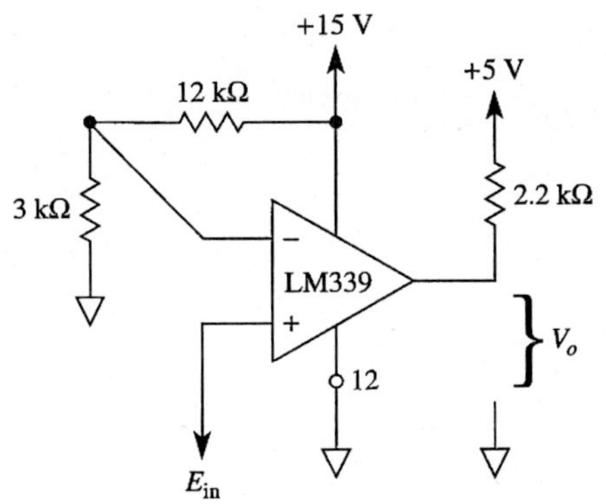

FIGURE P2-16

2-17. Draw the schematic of a circuit whose output voltage will go positive to $+V_{sat}$, when the input signal crosses $+5$ V in the positive direction.

2-18. Is the solution of Problem 2-17 classified as an inverting or noninverting comparator?

2-19. Draw a circuit whose output goes to $+V_{sat}$ when the input signal is below -4 V. The output should be at $-V_{sat}$ when the input is above -4 V.

2-20. Does the solution circuit for Problem 2-19 represent an **(a)** inverting or noninverting, **(b)** positive- or negative-voltage-level detector?

CHAPTER 3

Inverting and Noninverting Amplifiers

LEARNING OBJECTIVES

Upon completing this chapter on inverting and noninverting amplifiers, you will be able to:

- Draw the circuit for an inverting amplifier and calculate all voltages and currents for a given input signal.
- Draw the circuit for a noninverting amplifier and calculate all voltages and currents.
- Plot the output voltage waveshape and output–input characteristics of either an inverting or a noninverting amplifier for any input voltage waveshape.
- Design an amplifier to meet a gain and input resistance specification.
- Build an inverting or noninverting adder and audio mixer.
- Use a voltage follower to make an ideal voltage source.
- Create a negative output voltage from a positive reference voltage.
- Add a dc offset voltage to an ac signal voltage.

- Measure the average value of several signals.
- Design with single-supply op amps.
- Build a subtractor.
- Design a signal conditioning circuit for a temperature sensor.
- Analyze inverting and noninverting amplifier circuits using PSpice.

3-0 INTRODUCTION

This chapter uses the op amp in one of its most important applications—making an amplifier. An *amplifier* is a circuit that receives a signal at its input and delivers an undistorted larger version of the signal at its output. All circuits in this chapter have one feature in common: An external feedback resistor is connected between the output terminal and (−) input terminal. This type of circuit is called a *negative feedback circuit.*

There are many advantages obtained with negative feedback, all based on the fact that circuit performance no longer depends on the open-loop gain of the op amp, A_{OL}. By adding the feedback resistor, we form a loop from output to (−) input. The resulting circuit now has a *closed-loop gain* or *amplifier gain,* A_{CL}, which is independent of A_{OL} (provided that A_{OL} is much larger than A_{CL}).

As will be shown, the closed-loop gain, A_{CL}, depends only on external resistors. For best results 1% resistors should be used, and A_{CL} will be known within 1%. Note that adding external resistors does not change the open-loop gain A_{OL}. A_{OL} still varies from op amp to op amp, so adding negative feedback will allow us to ignore changes in A_{OL} as long as A_{OL} is large. We begin with the inverting amplifier to show that A_{CL} depends simply on the ratio of two resistors.

3-1 THE INVERTING AMPLIFIER

3-1.1 Introduction

The circuit of Fig. 3-1 is one of the most widely used op amp circuits. It is an amplifier whose closed-loop gain from E_i to V_o is set by R_f and R_i. It can amplify ac or dc signals. To understand how this circuit operates, we make two realistic simplifying assumptions that were introduced in Chapter 2.

1. The voltage E_d between the (+) and (−) inputs is essentially 0 if V_o is not in saturation.
2. The current drawn by either the (+) or the (−) input terminal is negligible.

3-1.2 Positive Voltage Applied to the Inverting Input

In Fig. 3-1, positive voltage E_i is applied through input resistor R_i to the op amp's (−) input. Negative feedback is provided by feedback resistor R_f. The voltage between the (+) and (−) inputs is essentially equal to 0 V. Therefore, the (−) input terminal is also at

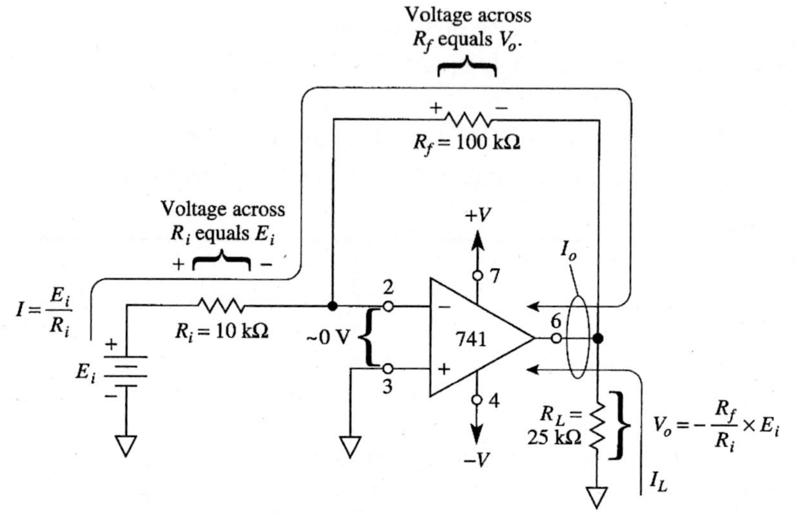

FIGURE 3-1 A positive input voltage is applied to the (−) input of an inverting amplifier. R_i converts this voltage to a current, I; R_f converts I back into an amplified version of E_i.

0 V, so ground potential is at the (−) input. For this reason, the (−) input is said to be at *virtual* ground.

Since one side of R_i is at E_i and the other is at 0 V, the voltage drop across R_i is E_i. The current I through R_i is found from Ohm's law:

$$I = \frac{E_i}{R_i} \tag{3-1a}$$

R_i includes the resistance of the signal generator. This point is discussed further in Section 3-5.2.

All of the input current I flows through R_f, since a negligible amount is drawn by the (−) input terminal. Note that the current through R_f is set by R_i and E_i; *not* by R_f, V_o, or the op amp.

The voltage drop across R_f is simply $I\ (R_f)$, or

$$V_{R_f} = I \times R_f = \frac{E_i}{R_i} R_f \tag{3-1b}$$

As shown in Fig. 3-1, one side of R_f and one side of load R_L are connected. The voltage from this connection to ground is V_o. The other sides of R_f and of R_L are at ground potential. Therefore, V_o equals V_{R_f} (the voltage across R_f). To obtain the polarity of V_o, note that the left side of R_f is at ground potential. The current direction established by E_i forces the right side of R_f to go negative. Therefore, V_o is negative when E_i is positive. Equating V_o with V_{R_f} and adding a minus sign to signify that V_o goes negative when E_i goes positive, we have

$$V_o = -E_i\frac{R_f}{R_i} \tag{3-2a}$$

Now, introducing the definition that the closed-loop gain of the amplifier is A_{CL}, we rewrite Eq. (3-2a) as

$$A_{CL} = \frac{V_o}{E_i} = -\frac{R_f}{R_i} \tag{3-2b}$$

The minus sign in Eq. (3-2b) shows that the polarity of the output V_o is inverted with respect to E_i. For this reason, the circuit of Fig. 3-1 is called an *inverting amplifier.*

3-1.3 Load and Output Currents

The load current I_L that flows through R_L is determined only by R_L and V_o and is furnished from the op amp's output terminal. Thus $I_L = V_o/R_L$. The current I through R_f must also be furnished by the output terminal. Therefore, the op amp output current I_o is

$$I_o = I + I_L \tag{3-3}$$

Example 3-1

For Fig. 3-1, let $R_f = 100\ \text{k}\Omega$, $R_i = 10\ \text{k}\Omega$, and $E_i = 1$ V. Calculate (a) I; (b) V_o; (c) A_{CL}.

Solution (a) From Eq. (3-1a),

$$I = \frac{E_i}{R_i} = \frac{1\ \text{V}}{10\ \text{k}\Omega} = 0.1\ \text{mA}$$

(b) From Eq. (3-2a),

$$V_o = -\frac{R_f}{R_i} \times E_i = -\frac{100\ \text{k}\Omega}{10\ \text{k}\Omega}(1\ \text{V}) = -10\ \text{V}$$

(c) Using Eq. (3-2b), we obtain

$$A_{CL} = -\frac{R_f}{R_i} = -\frac{100\ \text{k}\Omega}{10\ \text{k}\Omega} = -10$$

This answer may be checked by taking the ratio of V_o to E_i:

$$A_{CL} = \frac{V_o}{E_i} = \frac{-10\ \text{V}}{1\ \text{V}} = -10$$

Example 3-2

Using the values given in Example 3-1 and $R_L = 25\ \text{k}\Omega$, determine (a) I_L; (b) the total current into the output pin of the op amp.

Solution (a) Using the value of V_o calculated in Example 3-1, we obtain

$$I_L = \frac{V_o}{R_L} = \frac{10 \text{ V}}{25 \text{ k}\Omega} = 0.4 \text{ mA}$$

The direction of current is shown in Fig. 3-1.
(b) Using Eq. (3-3) and the value of I from Example 3-1, we obtain

$$I_o = I + I_L = 0.1 \text{ mA} + 0.4 \text{ mA} = 0.5 \text{ mA}$$

The input resistance seen by E_i is R_i. In order to keep input resistance of the circuit high, R_i should be equal to or greater than 10 kΩ.

Note: The maximum value of I_o is usually between 5 and 10 mA.

3-1.4 Negative Voltage Applied to the Inverting Input

Figure 3-2 shows a negative voltage, E_i, applied via R_i to the inverting input. All the principles and equations of Sections 3-1.1 to 3-1.3 still apply. The only difference between Figs. 3-1 and 3-2 is the direction of the currents. Reversing the polarity of the input voltage, E_i, reverses the direction of all currents and the voltage polarities. Now the output of the amplifier will go positive when E_i goes negative.

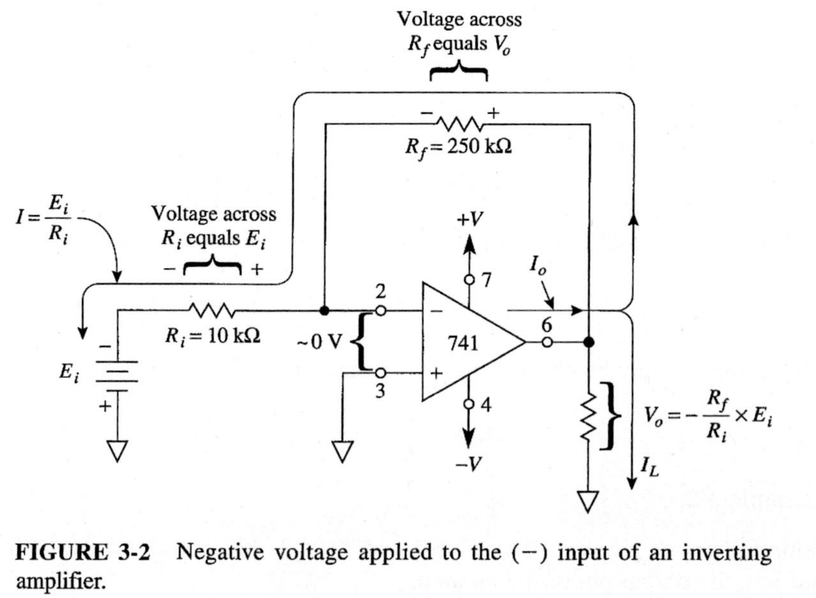

FIGURE 3-2 Negative voltage applied to the (−) input of an inverting amplifier.

Example 3-3

For Fig. 3-2, let $R_f = 250\ \text{k}\Omega$, $R_i = 10\ \text{k}\Omega$, and $E_i = -0.5$ V. Calculate (a) I; (b) the voltage across R_f; (c) V_o.

Solution (a) From Eq. (3-1a),

$$I = \frac{E_i}{R_i} = \frac{0.5\ \text{V}}{10\ \text{k}\Omega} = 50\ \mu\text{A} = 0.05\ \text{mA}$$

(b) From Eq. (3-1b),

$$\begin{aligned} V_{R_f} &= I \times R_f \\ &= (50\ \mu\text{A})(250\ \text{k}\Omega) \\ &= 12.5\ \text{V} \end{aligned}$$

(c) From Eq. (3-2a),

$$V_o = -\frac{R_f}{R_i} \times E_i = -\frac{250\ \text{k}\Omega}{10\ \text{k}\Omega}(-0.5\ \text{V}) = +12.5\ \text{V}$$

Thus the magnitude of the output voltage does equal the voltage across R_f, and $A_{CL} = -25$.

Example 3-4

Using the values in Example 3-3, determine (a) R_L for a load current of 2 mA; (b) I_o; (c) the circuit's input resistance.

Solution (a) Using Ohm's law and V_o from Example 3-3,

$$R_L = \frac{V_o}{I_L} = \frac{12.5\ \text{V}}{2\ \text{mA}} = 6.25\ \text{k}\Omega$$

(b) From Eq. (3-3) and Example 3-3,

$$I_o = I + I_L = 0.05\ \text{mA} + 2\ \text{mA} = 2.05\ \text{mA}$$

(c) The circuit's input resistance, or the resistance seen by E_i, is $R_i = 10\ \text{k}\Omega$.
A PSpice model and simulation results are given in Section 3-13.

3-1.5 Voltage Applied to the Inverting Input

Figure 3-3(a) shows an ac signal voltage E_i applied via R_i to the inverting input. For the positive half-cycle, the voltage polarities and the direction of currents are the same as in Fig. 3-1. For the negative half-cycle voltage, the polarities and direction of currents are the same as in Fig. 3-2. The output waveform is the negative (or 180° out of phase) of the input wave as shown in Fig. 3-3(b). That is, when E_i is positive, V_o is negative; and vice versa. The equations developed in Section 3-1.2 are applicable to Fig. 3-3 for ac voltages.

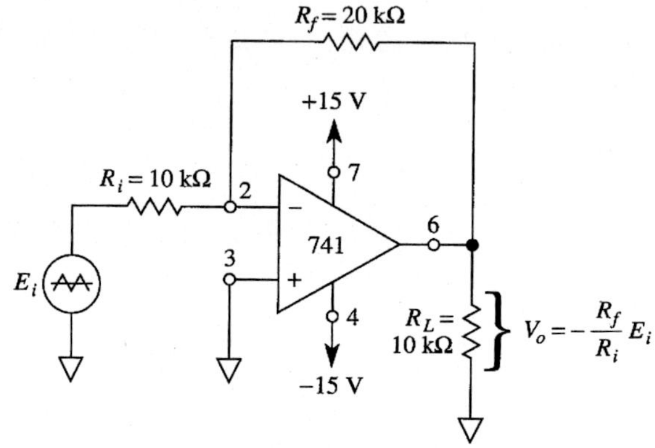

(a) Ac input voltage E_i is amplified by −2.

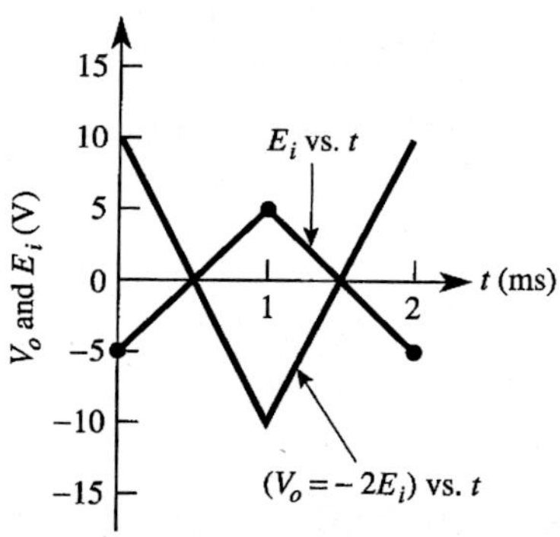

(b) Oscilloscope display of E_i vs. t and V_o vs. t.

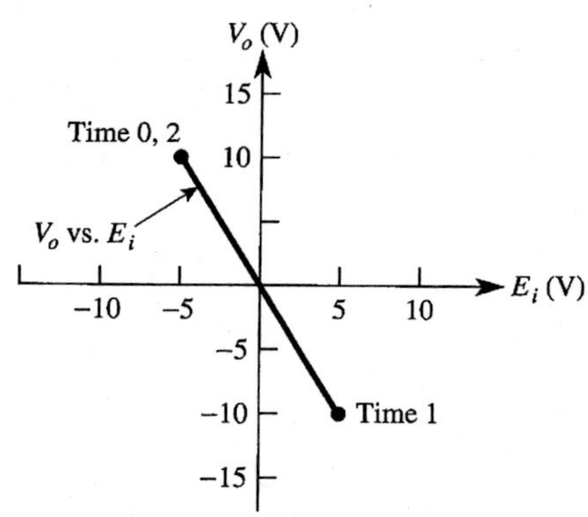

(c) xy scope display of V_o vs. E_i.

FIGURE 3-3 The inverting amplifier circuit in (a) has an ac input signal and a gain of −2. Time plots are shown in (b) and the output–input characteristic in (c). Note that the slope of V_o versus E_i in (c) is the closed loop gain A_{CL} (rise/run = V_o/E_i).

Example 3-5

For the circuit of Fig. 3-3, $R_f = 20$ kΩ and $R_i = 10$ kΩ, calculate the voltage gain A_{CL}.

Solution From Eq. (3-2b),

$$A_{CL} = -\frac{R_f}{R_i} = \frac{-20\text{ k}\Omega}{10\text{ k}\Omega} = -2$$

Example 3-6

If the input voltage in Example 3-5 is −5 V, determine the output voltage.

Solution Using Eq. (3-2a) or rearranging Eq. (3-2b), we obtain

$$V_o = -\frac{R_f}{R_i} \times E_i = A_{CL}\, E_i = (-2)(-5\text{ V}) = 10\text{ V}$$

See time 0 in Figs. 3-3(b) and (c). The frequency of the output and input signals is the same.

A PSpice model and simulation are given in Section 3-13. The simulation uses a 5-V-peak sine wave with a frequency set at 500 Hz as the input signal.

3-1.6 Design Procedure

Following is an example of the design procedure for an inverting amplifier.

Design Example 3-7

Design an amplifier with a gain of −25. The input resistance R_{in} should equal or exceed 10 kΩ.

Design Procedure

1. Choose the circuit type illustrated in Figs. 3-1 to 3-3.
2. Pick R_i = 10 kΩ (safe, prudent choice).
3. Calculate R_f from R_f = (gain)(R_i). (For this calculation, use the magnitude of gain.)

3-1.7 Analysis Procedure

You are interviewing for a job in the electronics field. The technical interviewer asks you to analyze the circuit. Assume that you recognize the circuit as that of an inverting amplifier. Then,

1. Look at R_i. State that the input resistance of the circuit equals the resistance of R_i.
2. Divide the value of R_f by the value of R_i. State that the magnitude of gain equals R_f/R_i. Also, the output voltage will be negative when the input voltage is positive.

3-2 INVERTING ADDER AND AUDIO MIXER

3-2.1 Inverting Adder

In the circuit of Fig. 3-4, V_o equals the sum of the input voltages with polarity reversed. Expressed mathematically,

$$V_o = -(E_1 + E_2 + E_3) \tag{3-4}$$

Circuit operation is explained by noting that the summing point S and the $(-)$ input are at ground potential. Current I_1 is set by E_1 and R, I_2 by E_2 and R, and I_3 by E_3 and R. Expressed mathematically,

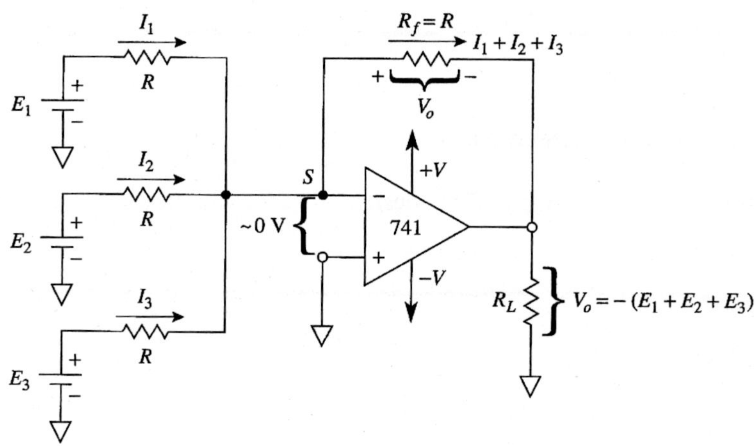

FIGURE 3-4 Inverting adder, $R = 10\ \text{k}\Omega$.

$$I_1 = \frac{E_1}{R}, \qquad I_2 = \frac{E_2}{R}, \qquad I_3 = \frac{E_3}{R} \tag{3-5}$$

Since the $(-)$ input draws negligible current, I_1, I_2, and I_3 all flow through R_f. That is, the sum of the input currents flows through R_f and sets up a voltage drop across R_f equal to V_o, or

$$V_o = -(I_1 + I_2 + I_3)R_f$$

Substituting for the currents from Eq. (3-5) and substituting R for R_f, we obtain Eq. (3-4):

$$V_o = -\left(\frac{E_1}{R} + \frac{E_2}{R} + \frac{E_3}{R}\right)R = -(E_1 + E_2 + E_3)$$

Example 3-8

In Fig. 3-4, $E_1 = 2$ V, $E_2 = 3$ V, $E_3 = 1$ V, and all resistors are 10 kΩ. Evaluate V_o.

Solution From Eq. (3-4), $V_o = -(2\text{ V} + 3\text{ V} + 1\text{ V}) = -6$ V. The PSpice model for this circuit is shown in Fig. 3-24.

Example 3-9

If the polarity of E_3 is reversed in Fig. 3-4 but the values are the same as in Example 3-8, find V_o.

Solution From Eq. (3-4), $V_o = -(2\text{ V} + 3\text{ V} - 1\text{ V}) = -4$ V.

If only two input signals, E_1 and E_2, are needed, simply replace E_3 with a short circuit to ground. If four signals must be added, simply add another equal resistor R between the fourth signal and the summing point S. Equation (3-4) can be changed to include any number of input voltages.

3-2.2 Audio Mixer

In the adder of Fig. 3-4, all the input currents flow through feedback resistor R_f. This means that I_1 does not affect I_2 or I_3. More generally, the input currents do *not* affect one another because each "sees" ground potential at the summing node. Therefore, the input currents—and consequently the input voltages E_1, E_2, and E_3—do *not* interact.

This feature is especially desirable in an audio mixer. For example, let E_1, E_2, and E_3 be replaced by microphones. The ac voltages from each microphone will be added or mixed at every instant. Then if one microphone is carrying guitar music, it will not come out of a second microphone facing the singer. If a 100-kΩ volume control is installed between each microphone and associated input resistor, their relative volumes can be adjusted and added. A weak singer can then be heard above a very loud guitar.

3-2.3 DC Offsetting an AC Signal

Some applications require that you add a dc offset voltage or current to an ac signal. Suppose that you must transmit an audio signal via an infrared emitting diode (IRED) or light-emitting diode. It is first necessary to bias the IRED on with a dc current. Then the audio signal can be superimposed as an ac current that rides on or modulates the dc current. The result is a light or infrared beam whose intensity changes directly with the audio signal. We illustrate this principle by an example.

Example 3-10

Design a circuit that allows you to add a dc voltage to a triangle wave.

Solution Select a two-channel adder circuit as in Fig. 3-5(a). A variable dc offset voltage E_{dc}, is connected to one channel. The ac signal, E_{ac}, is connected to the other.

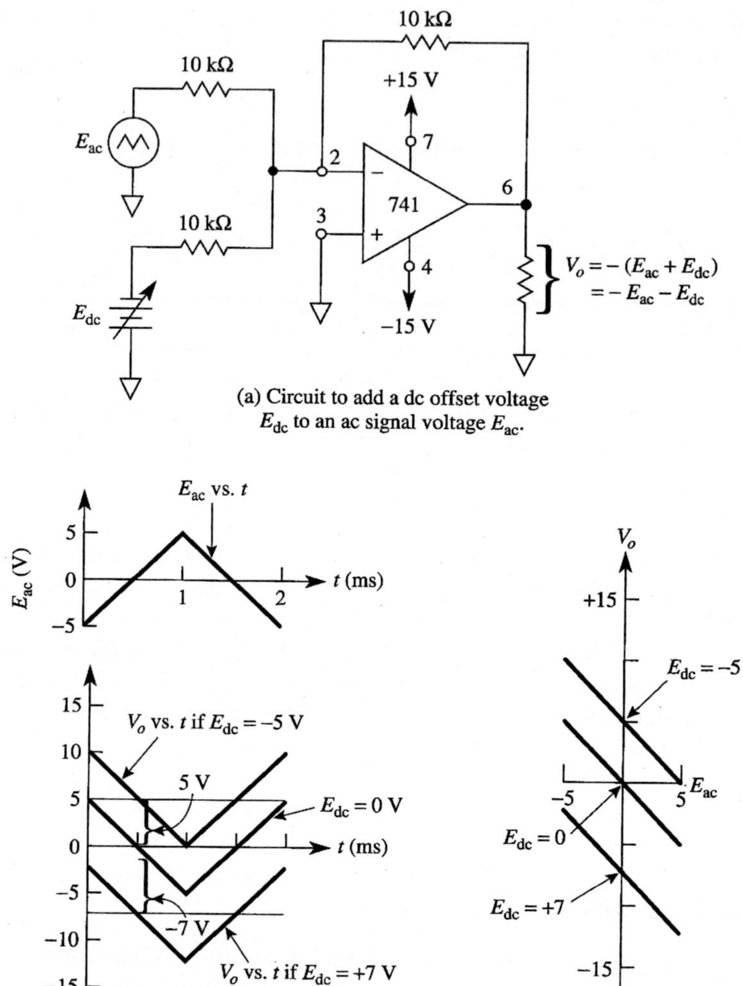

(a) Circuit to add a dc offset voltage E_{dc} to an ac signal voltage E_{ac}.

(b) Waveshapes of V_o for, $E_{dc} = 0$, −5 V, or +7 V.

(c) Output–input characteristic.

FIGURE 3-5 E_{ac} is transmitted with a gain of -1. If E_{dc} is *positive*, the average (dc) value of V_o is shifted *negative* by the same value.

Circuit analysis. If E_{dc} is 0 V, E_{ac} appears inverted at V_o (gain is -1) [see Figs. 3-5(b) and (c)]. If E_{dc} is -5 V, it appears at the output as a +5-V dc offset voltage upon which rides the inverted E_{ac}. If E_{dc} is +7 V, then E_{ac} is shifted *down* by 7 V. Most function generators contain this type of circuit. We will return to this circuit in Section 3-12 to design a signal conditioning circuit that interfaces between a temperature sensor and a microcontroller.

3-3 MULTICHANNEL AMPLIFIER

3-3.1 The Need for a Multichannel Amplifier

Suppose you had low-, medium-, and high-level signal sources. You need to combine them and make their relative amplitudes reasonably equal. You can use a three-input adder circuit to combine the signals. The versatile adder circuit will also allow you to equalize the signal amplitudes at its output. Simply design the required gain for each input channel by the selection of R_f and input resistors R_1, R_2, and R_3 as shown in Fig. 3-6.

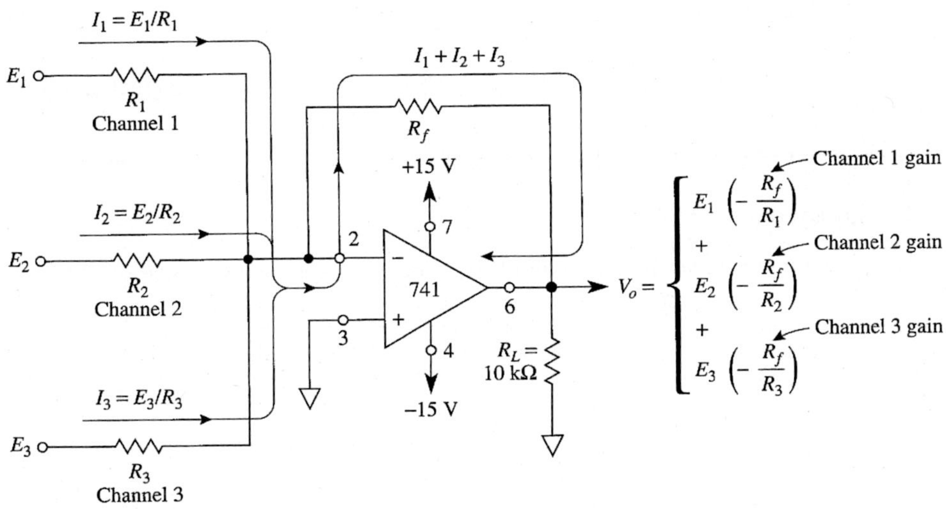

FIGURE 3-6 Multichannel amplifier. The inverting voltage gain of each channel depends on the values of its input resistor and R_f.

3-3.2 Circuit Analysis

As shown in Fig. 3-6, each channel input signal sees its associated input resistor connected to a virtual ground at the op amp's (−) input. Therefore, the input resistance of each channel is equal to the corresponding value selected for R_1, R_2, or R_3.

Input currents I_1, I_2, and I_3 are added in feedback resistor R_f and then converted back to a voltage V_{R_f}.

$$V_{R_f} = (I_1 + I_2 + I_3)R_f \tag{3-6a}$$

where

$$I_1 = \frac{E_1}{R_1}, \qquad I_2 = \frac{E_2}{R_2}, \qquad I_3 = \frac{E_3}{R_3} \tag{3-6b}$$

As was shown in Section 3-2.1, output voltage $V_o = -V_{R_f}$. Therefore,

$$V_o = -\left(E_1\frac{R_f}{R_1} + E_2\frac{R_f}{R_2} + E_3\frac{R_f}{R_3}\right) \quad \text{(3-7a)}$$

Equation (3-7a) shows that the gain of each channel can be changed independently of the others by simply changing its input resistor.

$$A_{CL_1} = -\frac{R_f}{R_1}, \qquad A_{CL_2} = -\frac{R_f}{R_2}, \qquad A_{CL_3} = -\frac{R_f}{R_3} \quad \text{(3-7b)}$$

or

$$V_o = E_1A_{CL_1} + E_2A_{CL_2} + E_3A_{CL_3}$$

3-3.3 Design Procedure

Following is an example of the design procedure for a multichannel amplifier.

Design Example 3-11

Design a three-channel inverting amplifier. The gains for each channel will be

Channel number	Voltage gain
1	−10
2	−5
3	−2

Design Procedure

1. Select a 10-kΩ resistor for the input resistance of the channel with the *highest* gain. Choose $R_1 = 10$ kΩ since A_{CL_1} is the largest.
2. Calculate feedback resistor R_f from Eq. (3-7b):

$$A_{CL_1} = -\frac{R_f}{R_1}, \qquad -10 = -\frac{R_f}{10\text{ k}\Omega}, \qquad R_f = 100\text{ k}\Omega$$

3. Calculate the remaining input resistors from Eq. (3-7b) to get $R_2 = 20$ kΩ and $R_3 = 50$ kΩ.

3-4 INVERTING AVERAGING AMPLIFIER

Suppose that you had to measure the average temperature at three locations in a dwelling. First make three temperature-to-voltage converters (shown in Section 5-14). Then connect their outputs to an *averaging* amplifier. An *averaging* amplifier gives an output voltage proportional to the average of all the input voltages. If there are three input voltages, the averager should add the input voltages and divide the sum by 3. The averager is the same circuit arrangement as the inverting adder in Fig. 3-4 or the inverting adder with gain in Fig. 3-6. The difference is that the input resistors are made equal to some convenient value R and the feedback resistor is made equal to R divided by the number of inputs. Let n equal the number of inputs. Then for a three-input averager, $n = 3$ and $R_f = R/3$. Proof is found by substituting into Eq. (3-7a), for $R_f = R/3$ and $R_1 = R_2 = R_3 = R$ to show that

$$V_o = -\left(\frac{E_1 + E_2 + E_3}{n}\right) \tag{3-8}$$

Example 3-12

In Fig. 3-4, $R_1 = R_2 = R_3 = R = 100\ \text{k}\Omega$ and $R_f = 100\ \text{k}\Omega/3 \approx 33\ \text{k}\Omega$. If $E_1 = +5$ V, $E_2 = +5$ V, and $E_3 = -1$ V, find V_o.

Solution Since $R_f = R/3$, the amplifier is an averager, and from Eq. (3-8) with $n = 3$, we have

$$V_o = -\left[\frac{5\text{ V} + 5\text{ V} + (-1\text{ V})}{3}\right] = -\frac{9\text{ V}}{3} = -3\text{ V}$$

Up to now we have dealt with amplifiers whose input signals were applied via R_i to the op amp's inverting input. We turn our attention next to amplifiers in which E_i is applied directly to the op amp's noninverting input.

3-5 NONINVERTING AMPLIFIER

3-5.1 Circuit Analysis

Figure 3-7 is a noninverting amplifier; that is, the output voltage, V_o, is the same polarity as the input voltage, E_i. The input resistance of the inverting amplifier (Section 3-1) is R_i, but the input resistance of the noninverting amplifier is extremely large, typically exceeding 100 MΩ. Since there is practically 0 voltage between the (+) and (−) pins of the op amp, both pins are at the same potential E_i. Therefore, E_i appears across R_1. E_i causes current I to flow as given by

$$I = \frac{E_i}{R_1} \tag{3-9a}$$

The direction of I depends on the polarity of E_i. Compare Figs. 3-7(a) and (b). The input current to the op amp's (−) terminal is negligible. Therefore, I flows through R_f and the voltage drop across R_f is represented by V_{R_f} and expressed as

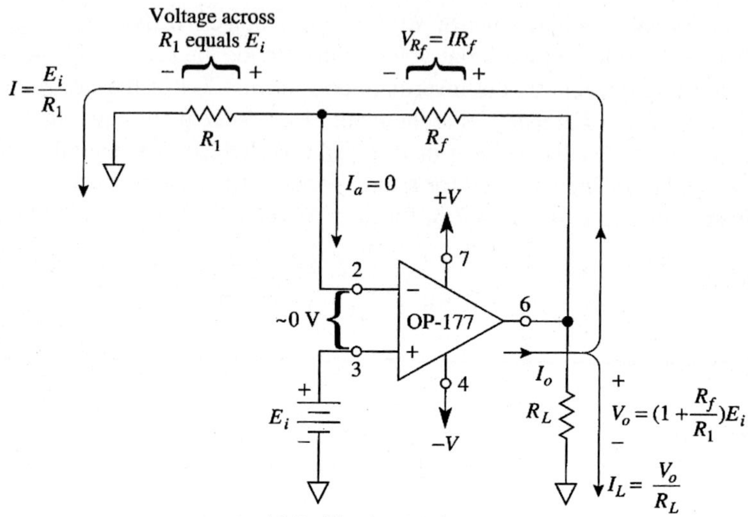

(a) Positive input voltages.

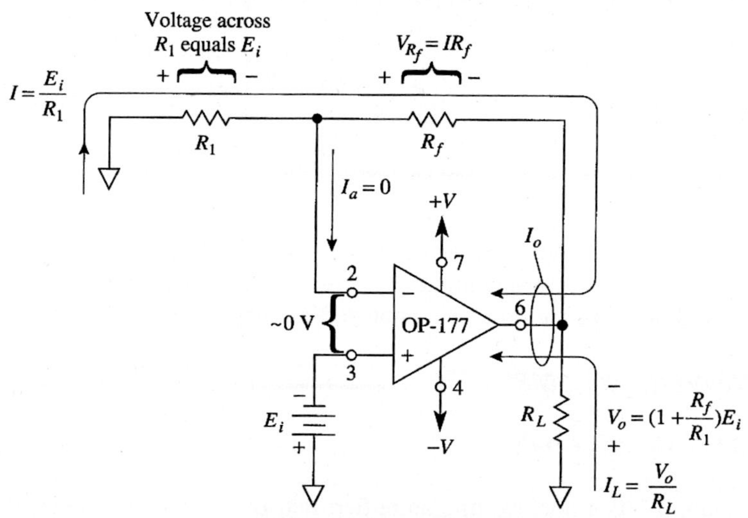

(b) Negative input voltages.

FIGURE 3-7 Voltage polarities and direction of currents for noninverting amplifiers.

$$V_{R_f} = I\,(R_f) = \frac{R_f}{R_1}\,E_i \tag{3-9b}$$

Equations (3-9a) and (3-9b) are similar to Eqs. (3-1a) and (3-1b).

The output voltage V_o is found by adding the voltage drop across R_1, which is E_i, to the voltage across R_f, which is V_{R_f}:

$$V_o = E_i + \frac{R_f}{R_1} E_i$$

or

$$V_o = \left(1 + \frac{R_f}{R_1}\right) E_i \tag{3-10a}$$

Rearranging Eq. (3-10a) to express voltage gain, we get

$$A_{CL} = \frac{V_o}{E_i} = 1 + \frac{R_f}{R_1} = \frac{R_f + R_1}{R_1} \tag{3-10b}$$

Equation (3-10b) shows that the voltage gain of a noninverting amplifier is always greater than 1.

The load current I_L is given by V_o/R_L and therefore depends only on V_o and R_L. I_o is the op amp's output current and is given by Eq. (3-3).

Example 3-13

(a) Find the voltage gain for the noninverting amplifier of Fig. 3-8. If E_i is a 100-Hz triangle wave with a 2-V peak, plot (b) V_o vs. t; (c) V_o vs. E_i.

Solution (a) From Eq. (3-10b),

$$A_{CL} = \frac{R_f + R_1}{R_1} = \frac{(40 + 10)\ \text{k}\Omega}{10\ \text{k}\Omega} = 5$$

(b) See Fig. 3-8(b). These are the waveshapes that would be seen on a dc-coupled, dual-trace oscilloscope.
(c) See Fig. 3-8(c). Set an oscilloscope for an x–y display, vertical 5 V/div, horizontal 1 V/div. Note that the slope rises to the right and is positive. Rise over run gives you the gain magnitude of +5.

3-5.2 Design Procedure

Following is an example of the design procedure for a noninverting amplifier.

Design Example 3-14

Design an amplifier with a gain of +10.

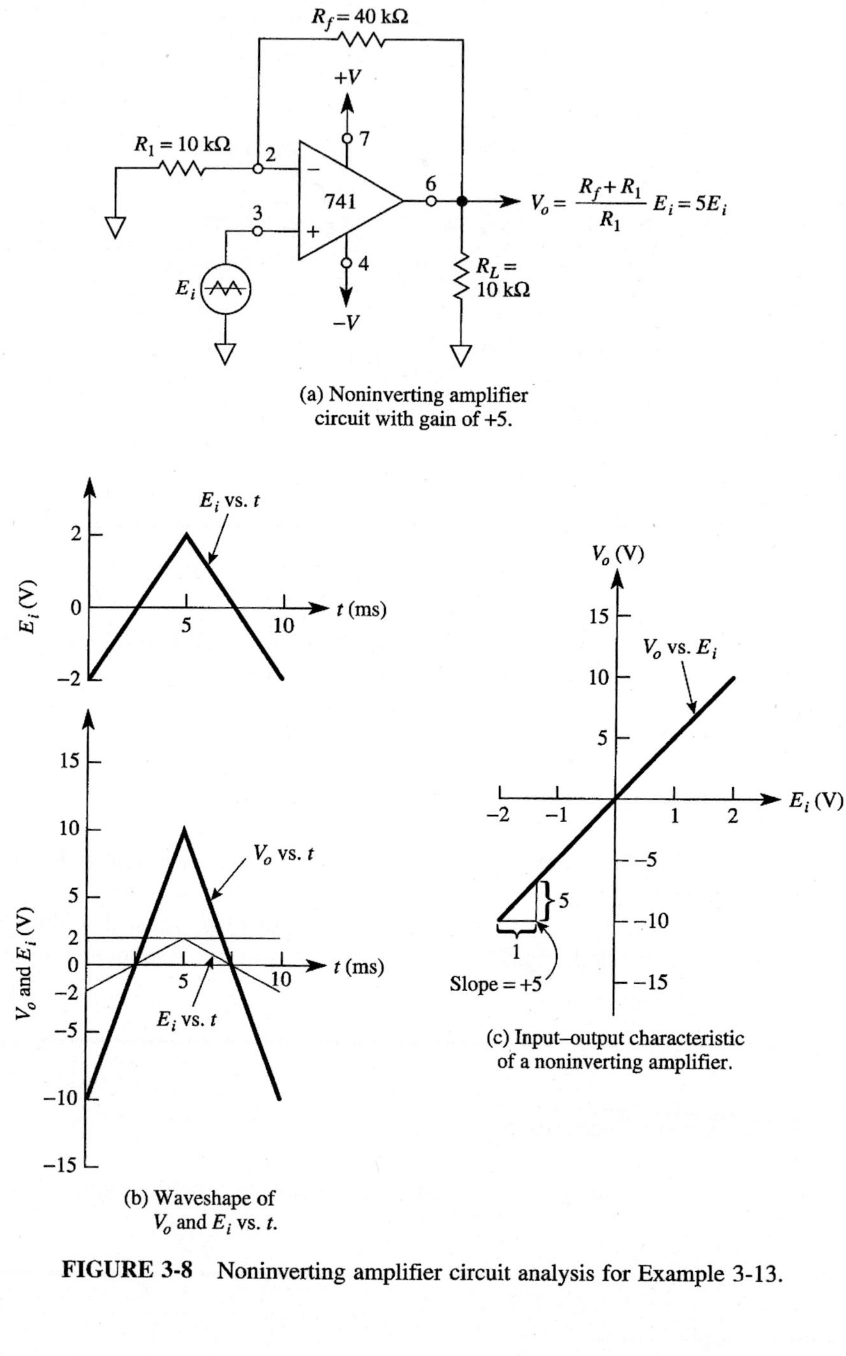

FIGURE 3-8 Noninverting amplifier circuit analysis for Example 3-13.

Design Procedure

1. Since the gain is positive, select a noninverting amplifier. That is, we apply E_i to the op amp's (+) input.
2. Choose $R_1 = 10\ \text{k}\Omega$.
3. Calculate R_f from Eq. (3-10b).

$$A_{CL} = 1 + \frac{R_f}{R_1}, \qquad 10 = 1 + \frac{R_f}{10\ \text{k}\Omega}, \qquad R_f = 9(10\ \text{k}\Omega) = 90\ \text{k}\Omega$$

3-6 VOLTAGE FOLLOWER

3-6.1 Introduction

The circuit of Fig. 3-9 is called a *voltage follower,* but it is also referred to as a *source follower, unity-gain amplifier, buffer amplifier,* or *isolation amplifier.* It is a special case of the noninverting amplifier. The input voltage, E_i, is applied directly to the (+) input. Since the voltage between (+) and (−) pins of the op amp can be considered 0,

$$V_o = E_i \tag{3-11a}$$

Note that the output voltage equals the input voltage in both magnitude and sign. Therefore, as the name of the circuit implies, the output voltage *follows* the input or source voltage. The voltage gain is 1 (or unity), as shown by

$$A_{CL} = \frac{V_o}{E_i} = 1 \tag{3-11b}$$

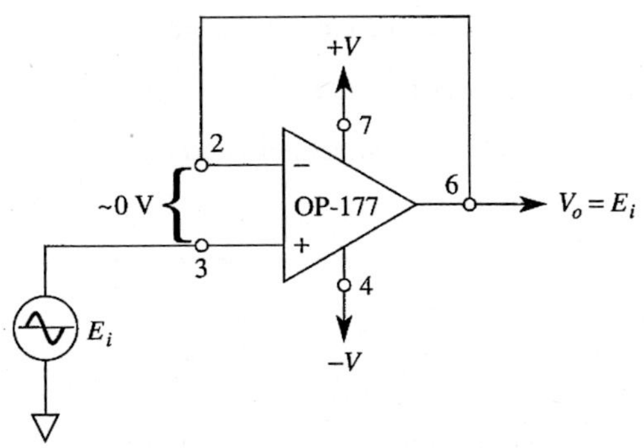

FIGURE 3-9 Voltage follower.

Example 3-15

For Fig. 3-10(a), determine (a) V_o; (b) I_L; (c) I_o.

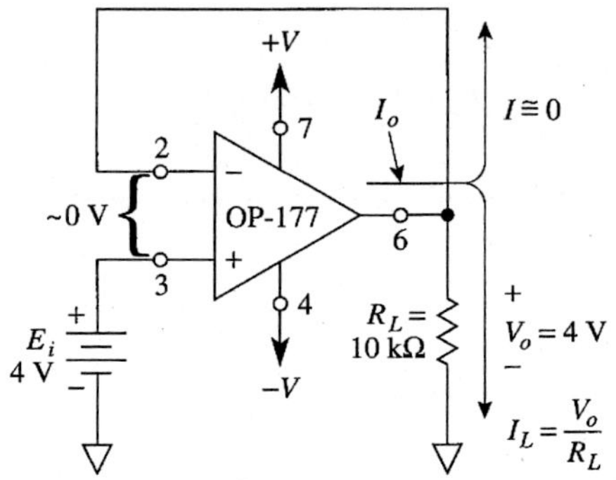

(a) Voltage follower for a positive input voltage.

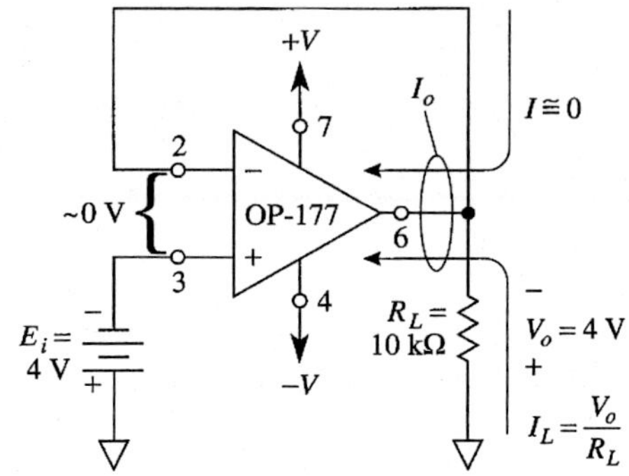

(b) Voltage follower for a negative input voltage.

FIGURE 3-10 Circuits for Example 3-15.

Solution (a) From Eq. (3-11a),

$$V_o = E_i = 4 \text{ V}$$

(b) From Ohm's law,

$$I_L = \frac{V_o}{R_L} = \frac{4 \text{ V}}{10 \text{ k}\Omega} = 0.4 \text{ mA}$$

(c) From Eq. (3-3),

$$I_o = I + I_L$$

This circuit is still a negative-feedback amplifier because there is a connection between output and (−) input. Remember that it is negative feedback that forces E_d to be 0 V. Also $I \approx 0$, since input terminals of op amps draw negligible current; therefore,

$$I_o = 0 + 0.4 \text{ mA} = 0.4 \text{ mA}$$

If E_i were reversed, the polarity of V_o, the direction of currents would be reversed, as shown in Fig. 3-10(b).

3-6.2 Using the Voltage Follower

A question that arises quite often is "Why bother to use an amplifier with a gain of 1?" The answer is best seen if we compare a voltage follower with an inverting amplifier. In this example, we are not primarily concerned with the polarity of voltage gain but rather with the input loading effect.

The voltage follower is used because its input resistance is high (many megohms). Therefore, it draws negligible current from a signal source. For example, in Fig. 3-11(a) the signal source has an open circuit or generator voltage, E_{gen}, of 1.0 V. The generator's internal resistance is 90 kΩ. Since the input terminal of the op amp draws negligible current, the voltage drop across R_{int} is 0 V. The terminal voltage E_i of the signal source becomes the input voltage to the amplifier and equals E_{gen}. Thus

$$V_o = E_i = E_{gen}$$

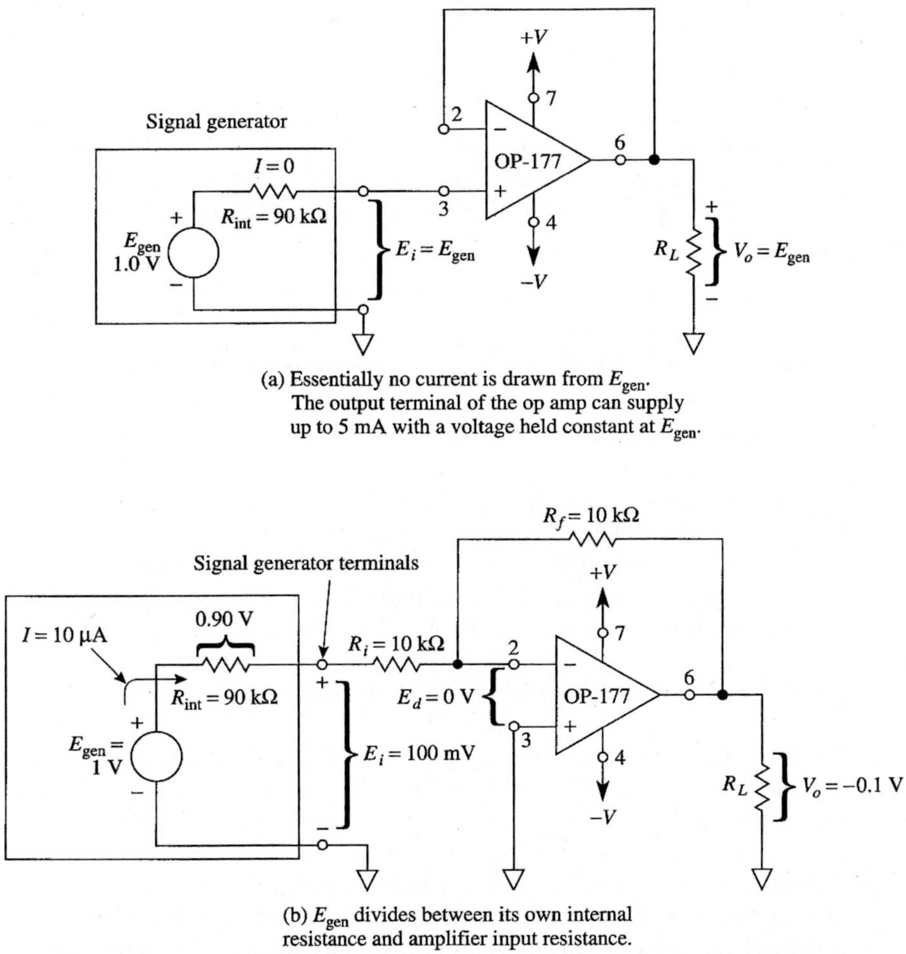

(a) Essentially no current is drawn from E_{gen}. The output terminal of the op amp can supply up to 5 mA with a voltage held constant at E_{gen}.

(b) E_{gen} divides between its own internal resistance and amplifier input resistance.

FIGURE 3-11 Comparison of loading effect between an inverting amplifier and a voltage follower on a high-resistance source.

Now let us consider the same signal source connected to an inverting amplifier whose gain is -1 [see Fig. 3-11(b)]. As stated in Section 3-1.3, the input resistance to an inverting amplifier is R_i. This causes the generator voltage E_{gen} to divide between R_{int} and R_i. Using the voltage division law to find the generator terminal voltage E_i yields

$$E_i = \frac{R_i}{R_{int} + R_i} \times E_{gen} = \frac{10\ \text{k}\Omega}{10\ \text{k}\Omega + 90\ \text{k}\Omega} \times (1.0\ \text{V}) = 0.1\ \text{V}$$

Thus it is this 0.1 V that becomes the input voltage to the inverting amplifier. If the inverting amplifier has a gain of only -1, the output voltage V_o is -0.1 V.

In conclusion, if a high-impedance source is connected to an inverting amplifier, the voltage gain from V_o to E_{gen} is not set by R_f and R_i as given in Eq. (3-2b). The actual gain must include R_{int}, as

$$\frac{V_o}{E_{gen}} = -\frac{R_f}{R_i + R_{int}} = -\frac{10\ \text{k}\Omega}{100\ \text{k}\Omega} = -0.1$$

If you must amplify and invert a signal from a high-impedance source and wish to draw no signal current, first *buffer* the source with a voltage follower. Then feed the follower's output into an inverter. If you need buffering and do not want to invert the input signal, use the noninverting amplifier.

3-7 THE "IDEAL" VOLTAGE SOURCE

3-7.1 Definition and Awareness

The ideal voltage source is first encountered in textbooks concerned with fundamentals. By definition, the voltage does not vary regardless of how much current is drawn from it. You may not be aware of the fact that you create a perfect voltage source when you measure the frequency response of an amplifier or filter. We explain how this apparently perfect performance comes about in the next section.

3-7.2 The Unrecognized Ideal Voltage Source

The lab or field procedure typically goes like this: Set the input signal amplitude at 0.2 V rms and frequency at the lowest limit. Measure output V_o. Hold E_{in} at 0.2 V rms for each measurement. Plot V_o or V_o/E_{in} versus frequency. As you dial higher frequencies, E_{in} begins to decrease (because of input capacitance loading). You automatically increase the function generator's volume control to hold E_{in} at 0.2 V. You have just, by definition, created an "ideal" voltage source. E_{in} never varied throughout the test sequence no matter how much current was drawn from it. This is an example of the unrecognized ideal voltage source.

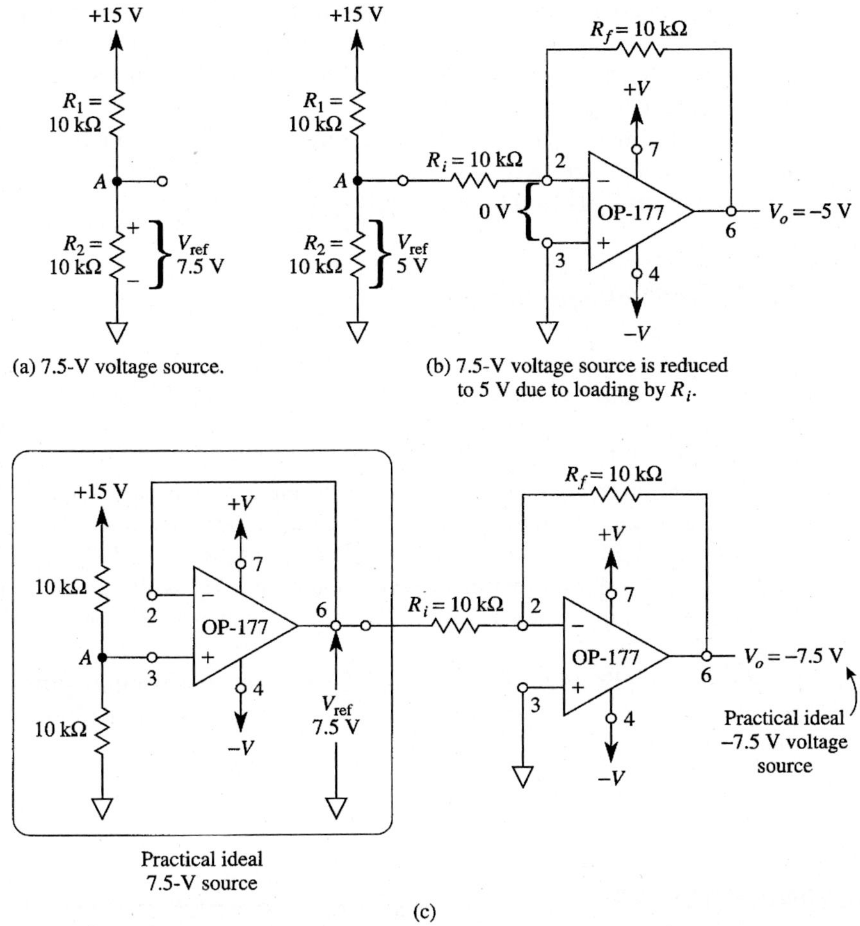

FIGURE 3-12 A voltage divider and (+) supply voltage gives a 7.5-V test or reference voltage in (a). V_{ref} drops to 5 V in (b) when connected to an inverter. A voltage follower converts the voltage divider into an ideal voltage source in (c).

3-7.3 The Practical Ideal Voltage Source

A circuit schematic shows a battery symbol labeled −7.5 V. Your job is to make one. The convenient +15-V supply voltage is available, and a simple voltage divider network gives 7.5 V as shown in Fig. 3-12(a). This 7.5-V source is fine as long as you never use it by connecting a load.

As shown in Fig. 3-12(b), R_i of the inverter appears in parallel with R_2 to form an equivalent resistance of 10 kΩ || 10 kΩ = 5 kΩ. The 15-V supply divides between R_1 = 10 kΩ and 5 kΩ, and V_{ref} drops to 5 V.

To preserve the value of any reference voltage, simply *buffer* it with a voltage follower. The 7.5-V reference voltage is connected to a voltage follower in Fig. 3-12(c). The output of the follower equals V_{ref}. You can extract up to 5 mA from the follower's output with no change in V_{ref}.

The buffer makes an excellent clandestine bug. You can monitor what is going on at any circuit point. Since a follower has a high input impedance, it draws negligible current from the circuit. Therefore, it is nearly impossible to detect.

3-7.4 Precise Voltage Sources

Section 2-6 introduced precision voltage reference ICs such as the REF-02 (a precision +5-V reference chip). You can use these voltage reference chips with an inverting amplifier to create precise negative voltages as well as positive and negative voltages. The circuit of Figure 3-13(a) shows how a negative voltage of −5 V can be created using the REF-02 and an inverting amplifier. This circuit has a lower parts count and more precision than the circuit of Figure 3-12(c). The parts count is obvious, and the precision is obtained by the REF-02 in place of R_i and R_f. Consider R_i and R_f to be 1% resistors. Then there is a possibility that one resistor could be +1% while the other is −1%. This will produce an output voltage with a 2% error, which may not be acceptable for your design.

Another application using the REF-02 with an inverting amplifier is shown in Figure 3-13(b). This circuit creates a ±5-V source from a single REF-02 chip and an inverting amplifier.

3-8 NONINVERTING ADDER

A three-input noninverting adder is constructed with a passive averager and noninverting amplifier as shown in Fig. 3-14(a). The passive averager circuit consists of three equal resistors R_A and the three voltages to be added. The output of the passive averager is E_{in}, where E_{in} is the average of E_1, E_2, and E_3, or $E_{in} = (E_1 + E_2 + E_3)/3$. Connect a voltage follower to E_{in} if you need a noninverting averager (in contrast with Sec. 3-4).

Output V_o results from amplifying E_{in} by a gain equal to the number of inputs n. In Fig. 3-14(a), $n = 3$. Design the amplifier by choosing a convenient value for resistor R. Then find R_f from

$$R_f = R(n - 1) \tag{3-12}$$

As shown in Fig. 3-14(a), the value for R_f should be $R_f = 10\ \text{k}\Omega(3 - 1) = 20\ \text{k}\Omega$. If E_1, E_2, and E_3 are not ideal voltage sources, such as a battery or output of an op amp, buffer them with followers as in Fig. 3-14(b).

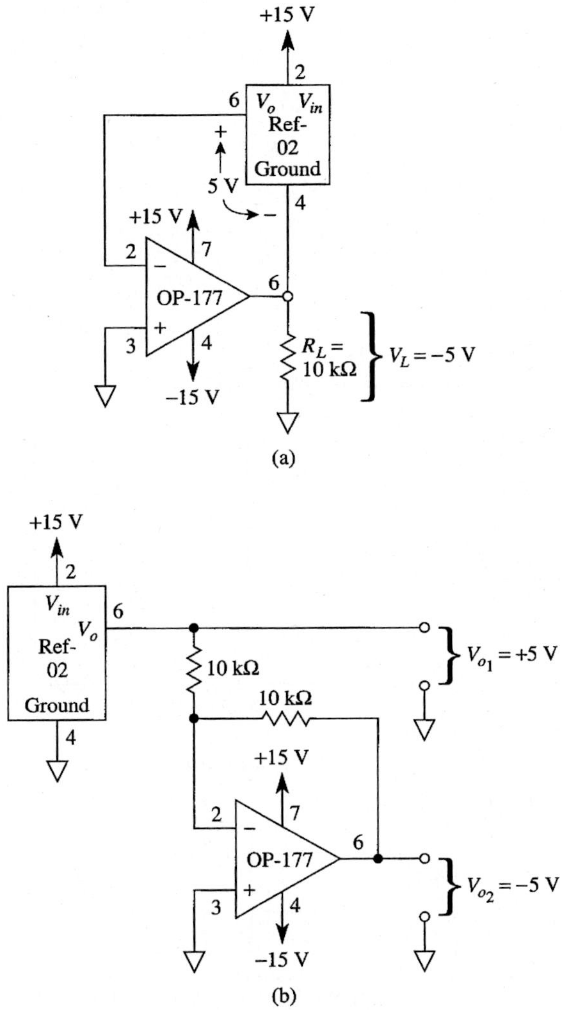

FIGURE 3-13 (a) A −5 V precision output voltage; (b) a ±5 V output voltage.

3-9 SINGLE-SUPPLY OPERATION

Some amplifier designs require battery-powered operation and output voltage swings to within millivolts of the supply voltages. Thus, you want an op amp described by manufacturers as a device capable of single-supply and rail-to-rail operation. Two such devices are the AD820 and the OP-90 from Analog Devices. These op amps can operate from either a single or dual supply. For example, the AD820 can operate from a dual supply

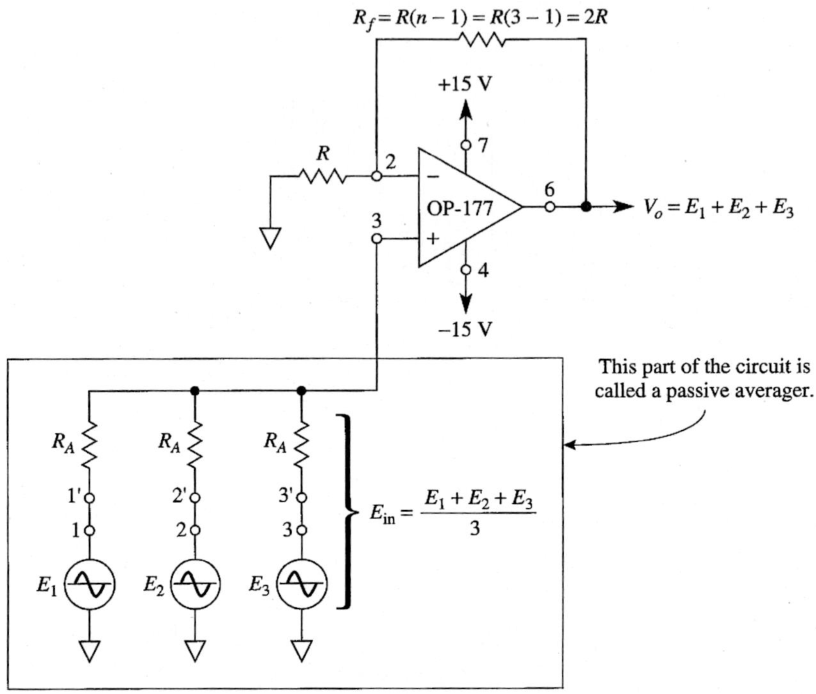

(a) Noninverting adder.

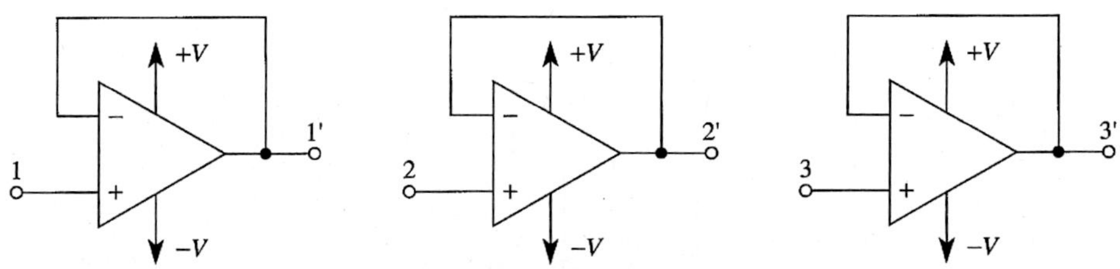

(b) If E_1, E_2, and E_3 are not ideal voltage sources, simply buffer each one with a voltage follower.

FIGURE 3-14 All resistors of an n-input noninverting adder are equal except the feedback resistor; choose $R = 10\ \text{k}\Omega$ and $R_A = 10\ \text{k}\Omega$. Then R_f equals R times the number of inputs minus one: $R_f = R(n - 1)$.

(±1.5 V to ±18 V) or from a single supply (+3 V to +36 V). The input signal applied to these devices can be brought to ground, and the output can swing to within 10 mV of either supply voltage. They are available in an 8-pin mini-DIP package with the same pinouts as the 741 or OP-177 op amps. Single-supply op amps are often used in battery-powered applications, portable instruments, medical instrumentation, and data acquisition units. Often they are used to amplify positive signals coming from sensors such as strain

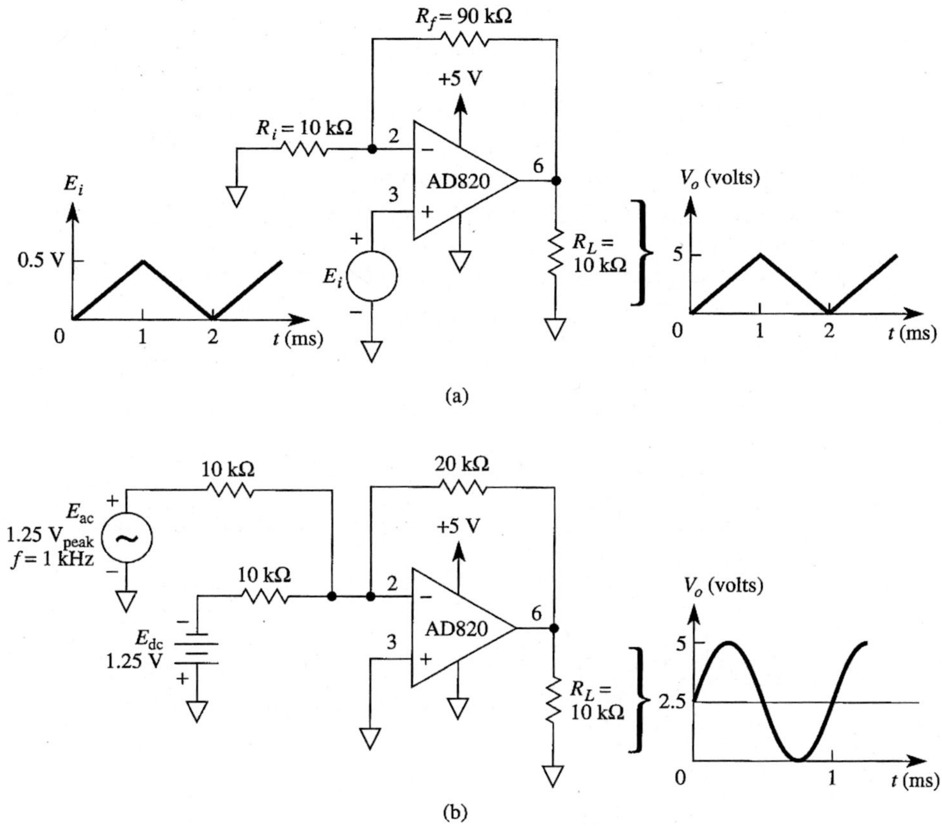

FIGURE 3-15 The AD820 can operate for single-supply applications as shown in (a) and (b).

gages or thermocouples. Figure 3-15(a) shows the AD820 wired as a noninverting amplifier with a gain of 10. If you are operating the AD820 from a single supply and want to amplify an ac signal, then the input ac signal has to have a dc offset or be combined with a dc voltage as shown in Figure 3-15(b). (*Note:* This circuit is similar to the inverting adder studied in Figure 3-5(a) but now operated from a single supply.)

3-10 DIFFERENCE AMPLIFIERS

The differential amplifier and its more powerful relative, the instrumentation amplifier, will be studied in Chapter 8. However, as other applications of inverting and noninverting amplifiers, we offer two examples of the *difference* amplifier in this section and the design of a signal conditioning circuit for a temperature sensor in the next.

3-10.1 The Subtractor

A circuit that takes the difference between two signals is called a *subtractor* [see Fig. 3-16(a)]. It is made by connecting an inverting amplifier to a two-input inverting averager. To analyze this circuit, note that E_i is transmitted through op amp A with a gain of -1 and appears as $V_{o_1} = -E_1$. V_{o_1} is then inverted (times -1) by the top channel of the inverting amplifier B. Thus E_1 is inverted once by op amp A and again by op amp B to appear at V_o as E_1.

E_2 is inverted by the bottom channel of op amp B and drives V_o to $-E_2$. Thus V_o responds to the difference between E_1 and E_2, or

$$V_o = E_1 - E_2 \tag{3-13a}$$

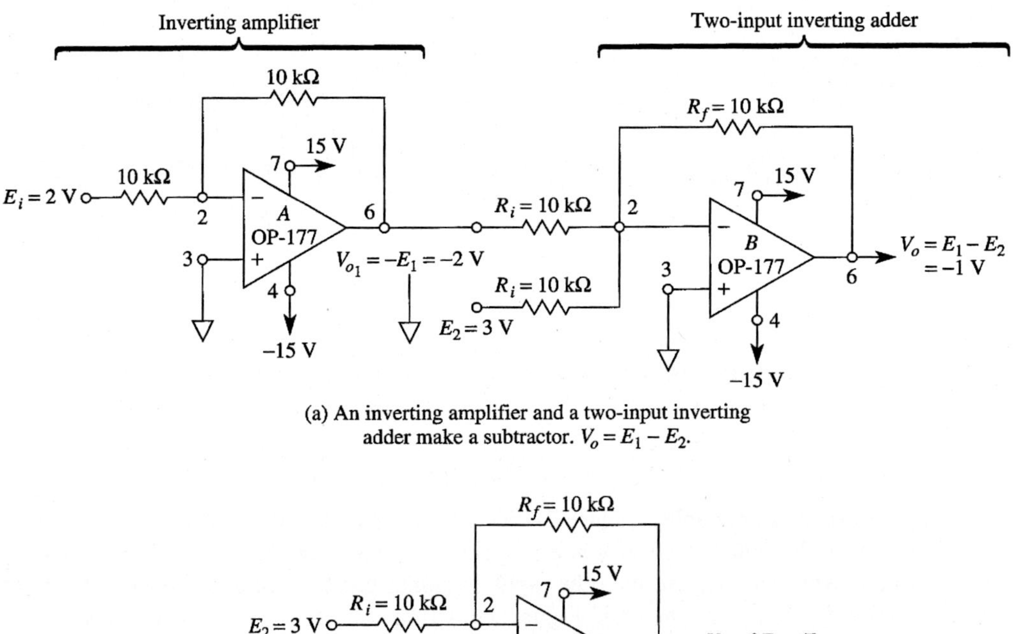

(a) An inverting amplifier and a two-input inverting adder make a subtractor. $V_o = E_1 - E_2$.

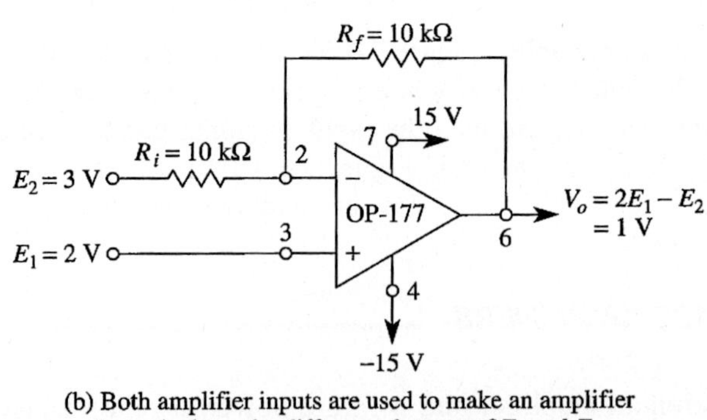

(b) Both amplifier inputs are used to make an amplifier that calculates the difference between $2E_1$ and E_2.

FIGURE 3-16 Two examples of difference amplifiers are the subtractor in (a) and using the op amp as both an inverting and a noninverting amplifier in (b).

As shown in Fig. 3-16(a), for $E_1 = 2$ V and $E_2 = 3$ V, $V_o = 2 - 3 = -1$ V. If the value of R_f is made larger than R_i, the subtractor will have gain

$$V_o = \frac{R_f}{R_i}(E_1 - E_2) \tag{3-13b}$$

3-10.2 Inverting–Noninverting Amplifier

In Fig. 3-16(b), signal E_1 is applied to the amplifier's noninverting input and signal E_2 is applied to the inverting input. We will use superposition to analyze this circuit. First remove E_2 and replace it by a ground. E_1 sees a noninverting amplifier with a gain of $(R_f + R_i)/R_i$, or 2. Thus E_1 alone drives V_o to $2E_1$. Next reconnect E_2 and replace E_1 by a ground. E_2 sees an inverting amplifier with a gain of -1. E_2 drives V_o to $-E_2$. When *both* E_1 and E_2 are connected, V_o is given by

$$V_o = 2E_1 - E_2 \tag{3-14}$$

As shown in Fig. 3-16(b), $V_o = 1$ V when $E_1 = 2$ V and $E_2 = 3$ V.

We will now show how to design the subtractor circuit of Fig. 3-16(a) to be the analog interface circuitry connected between a temperature sensor and the analog–digital converter of a microcontroller. This analog interface circuit is also known as a signal conditioning circuit.

3-11 DESIGNING A SIGNAL CONDITIONING CIRCUIT*

Another way of viewing the circuit in Fig. 3-5(a) (redrawn in Figure 3-17 for convenience) is that it allows us to design a signal conditioning circuit (SCC) for a microcontroller application that satisfies the equation of a straight line, $y = mx + b$. This equation occurs quite often when designing SCCs. Comparing the equation of $y = mx + b$ to the circuit of Fig. 3-17, the y term is the output voltage, V_o; the x term is the input signal voltage, E_i; the m term is the gain of the circuit, R_f/R_1; and the b term is R_f/R_2 times E_{dc}. Therefore, if your application uses a sensor that generates an output signal measured with respect to ground, which must be amplified and offset, then an SCC similar to Fig. 3-17 may be used. (*Note:* The outputs of some sensors generate a differential output, and these devices require an SCC capable of measuring a differential voltage. Such circuits are studied in Chapter 8.) The design of any SCC unit requires obtaining the equation of the circuit. This equation is obtained from what you've *got* (the output conditions of the sensor) to what you *want* (the input conditions of the microcontroller's A/D converter). Let us study this topic.

Statement of the problem. Design a signal conditioning circuit to interface between a temperature sensor and the A/D converter of a microcontroller. The tempera-

* For more examples of the linear circuit design procedure refer to *Data Acquisition and Process Control with the M68HC11 Microcontroller,* 2nd Ed. by F. Driscoll, R. Coughin, and R. Villanucci, Prentice Hall (2000).

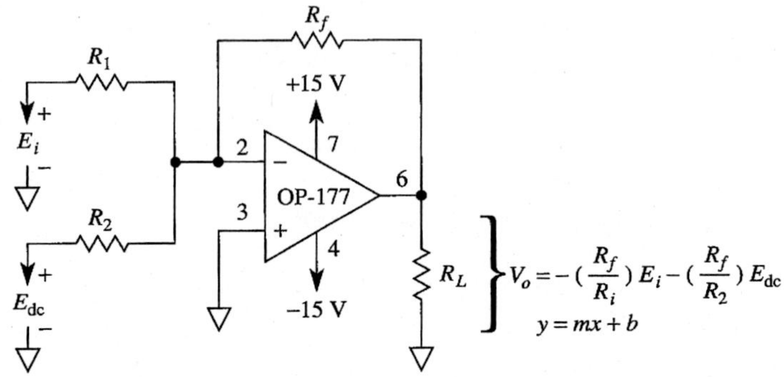

FIGURE 3-17 The inverting summer can be designed to satisfy the equation of a straight line, $y = mx + b$.

ture range to be measured is 0° to 50°C, and the range of the A/D converter is from 0 to 5 V. You want the output of the SCC to be linear; that is, when the sensor is measuring 0°C, the output of the SCC is 0 V; when the sensor is measuring 10°C, the SCC's output is to be 1 V; and so forth up to 50°C, at which temperature the SCC outputs 5 V.

Solution. Although our goal is to design the SCC, which is an op amp circuit, our starting point is the sensor and writing an equation for it, because the output of the sensor is the input to the SCC. Therefore, once the sensor is picked, this is what you've *got.* What you *want* is the output of the SCC to fit the range of the microcontroller's A/D converter. Therefore, the SCC design is being squeezed between what you've *got* and what you *want.* Let's first learn about one type of temperature sensor and how to write the equation for it.

Introduction to a temperature sensor. For this application, we shall choose the LM335, which is a solid-state temperature sensor that belongs to a family of devices that has a sensitivity of 10 mV/°K. It is used in applications that require measuring temperatures from −10° to 100°C and is modeled as a two-terminal zener. The package style and model are shown in Figures 3-18(a) and (b), respectively. This device is capable of operating over a current range of from 400 μA to 5 mA. The data sheet for the LM335 gives the device's sensitivity as 10 mV/°K. However, our application is to measure degrees Celsius. The relationship between degrees Kelvin and degrees Celsius is: A 1-degree rise in Kelvin equals a 1-degree rise in Celsius, and the freezing point of water is 0°C, which equals 273°K.

Writing an equation that describes the sensor. A plot of the output voltage of the LM335 versus temperature is given in Figure 3-18(c). The slope of the line is the device's sensitivity—10 mV/°K. Therefore, in terms of °K the output voltage is

$$V_T = (10\ \text{mV/°K})\ (T_{\text{in°K}}) \tag{3-15}$$

where T is the temperature in °K. At 273°K (0°C), the sensor's output voltage is

$$V_T = (10\ \text{mV/°K})\ (273\text{°K}) = 2.73\ \text{V}$$

as shown in Figure 3-18(c).

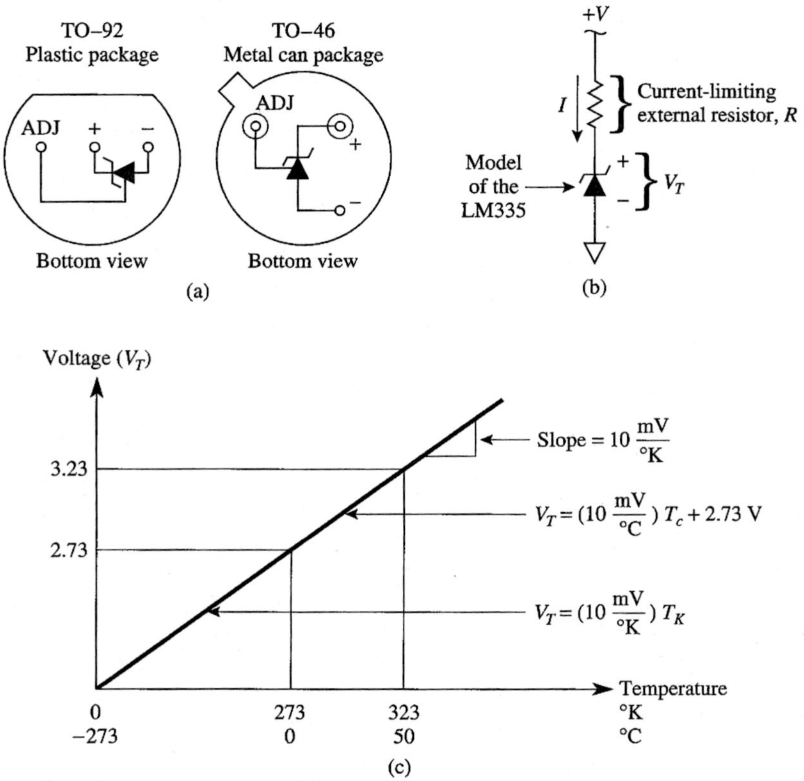

FIGURE 3-18 The LM335 (a) package styles, (b) model, and (c) voltage versus temperature characteristic.

Now the sensor's output voltage can be written in terms of degrees Celsius as

$$V_T = (10\ \text{mV/°C})\ (T_{\text{in°C}}) + 2.73\ \text{V} \tag{3-16}$$

where T is the temperature in degrees Celsius. For our application, at 0°C $V_T = 2.73$ V and at 50°C $V_T = 3.23$ V. This is the input voltage range for the SCC. The output range of the SCC is the input range of the A/D converter, which is 0 V to 5 V. Figure 3-19 shows a block diagram of this data acquisition system for measuring temperature.

Writing an equation that describes the SCC. From the information we know about the sensor and the A/D converter, we can plot the output/input characteristics of the SCC. Figure 3-19 includes such a plot. The output values of the SCC are plotted on the y-axis. Remember, these values are the voltage range of the A/D converter—0 V to 5 V. The input values to the SCC are plotted on the x-axis. These values are the voltage range of the sensor—2.73 V to 3.23 V for this application.

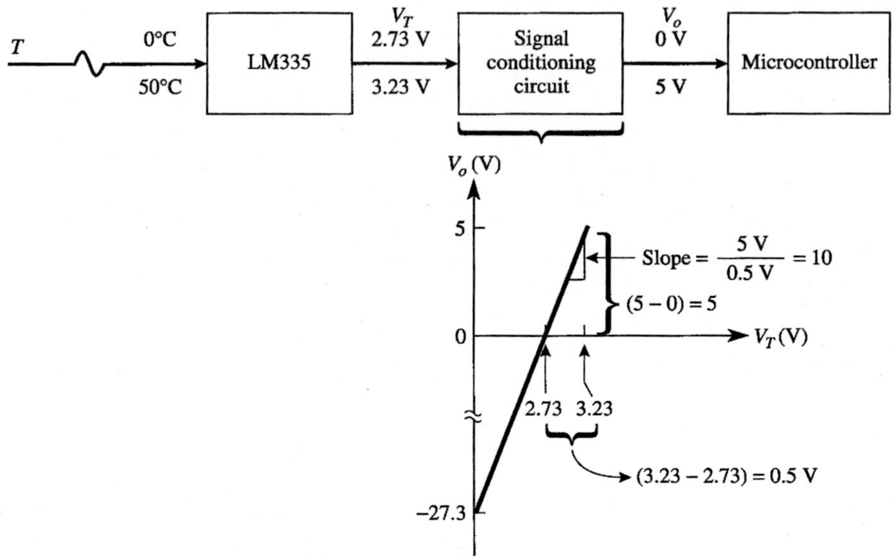

FIGURE 3-19 Block diagram of a temperature measuring system and the desired output–input characteristics of the SCC.

The slope of the line is

$$m = \frac{(5 - 0)\ \text{V}}{(3.23 - 2.73)\ \text{V}} = 10 = \frac{\Delta V_o}{\Delta V_T} \tag{3-17}$$

This value of 10 is the gain that V_T must be multiplied by. The dc offset is found from choosing a point on the line and substituting into the equation of a straight line—$y = mx + b$. Choosing the coordinate pair (2.73, 0), we obtain

$$0 = (10)\,(2.73) + b$$

Solving for b yields

$$b = -27.3\ \text{V}$$

Thus, the equation of the SCC's output voltage is

$$V_o = (10)\,(V_T) - 27.3\ \text{V} \tag{3-18}$$

Note: Although the dc offset is −27.3 V, the output voltage, V_o, never goes to this value because the range of V_T is from 2.73 V to 3.23 V. This range of V_T limits V_o from 0 to 5 V.

Designing the signal conditioning circuit. Now that we know the equation for the SCC and it is in the form of $y = mx + b$, we want a circuit in which the gain of 10 and the offset of −27.3 V can be set independently. A noninverting summer is not the answer because the gain and offset cannot be set independently. What is required is an op amp circuit as shown in Fig. 3-20—an inverting amplifier with a gain of −1 followed by an inverting adder. This circuit is similar to Fig. 3-16(a). The general equation for the output voltage of the adder is

$$V_o = -(R_f/R_1)\,(-V_T) - (R_f/R_2)\,E_{dc} \tag{3-19a}$$

$$y = mx + b$$

or

$$V_o = (R_f/R_1)\,V_T - (R_f/R_2)\,E_{dc} \tag{3-19b}$$

Matching the coefficients of V_T in Eq. (3-18) and Eq. (3-19b) yields

$$\frac{R_f}{R_1} = 10$$

choosing $R_1 = 10\ \text{k}\Omega$, then $R_f = 100\ \text{k}\Omega$. Matching the dc offset terms in Eq. (3-18) and Eq. (3-19b) yields

$$\frac{-R_f}{R_2}\,E_{dc} = -27.3\ \text{V}$$

Let E_{dc} be wired to the +15-V supply. Since $R_f = 100\ \text{k}\Omega$, then $R_2 = 54.9\ \text{k}\Omega$. Note that E_{dc} is a positive voltage and V_T is a negative voltage at the input of inverting adder. Since the LM335 generates a positive voltage, V_T, the inverting amplifier with a gain of −1 is used to generate $-V_T$ as shown in the complete circuit of Fig. 3-20.

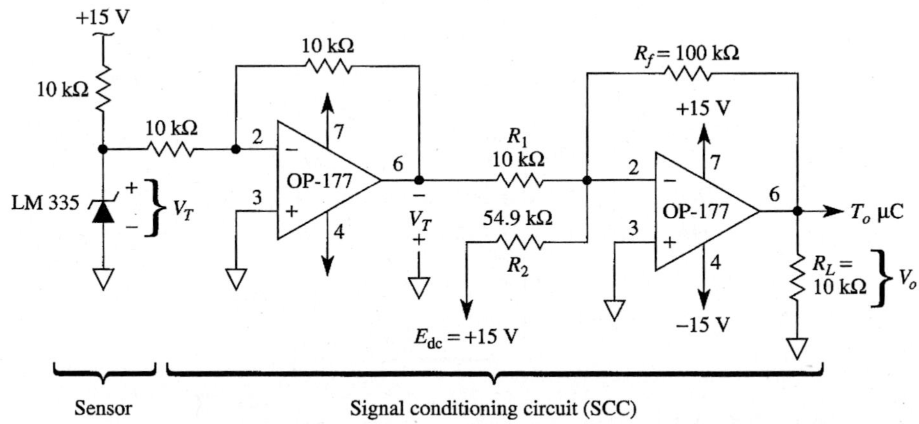

FIGURE 3-20 Design of the SCC for the temperature measuring system of Fig. 3-19. The SCC is an interface between the sensor and microcontroller.

3-12 PSPICE SIMULATION

In this section, we will use PSpice and simulate the performance of four circuits studied in this chapter.

3-12.1 Inverting Amplifier—DC Input

Create the PSpice model of the inverting amplifier of Fig. 3-2. Set the input voltage, E_i, to 0.5 V. Use **IPROBE** to measure currents, I, I_L, and I_o, and **VIEWPOINT** to measure, with respect to ground, the voltage at the inverting input and the op amp's output voltage. Use the procedure outlined in Chapter 2 to obtain and place the following parts on the right side of the work area.

Draw => Get New Part

Part	Number	Library
=> **uA741**	1	**eval.slb**
=> **VDC**	3	**source.slb**
=> **R**	3	**analog.slb**
=> **GLOBAL**	4	**port.slb**
=> **AGND**	5	**port.slb**
=> **IPROBE**	3	**special.slb**
=> **VIEWPOINT**	2	**special.slb**

Arrange the parts as shown in the schematic of Fig. 3-2 but include the **IPROBE**s and **VIEWPOINT**s. Save the file with an **.SCH** extension and click **Analysis => Simulate.** The completed circuit with current and voltage values is shown in Fig. 3-21.

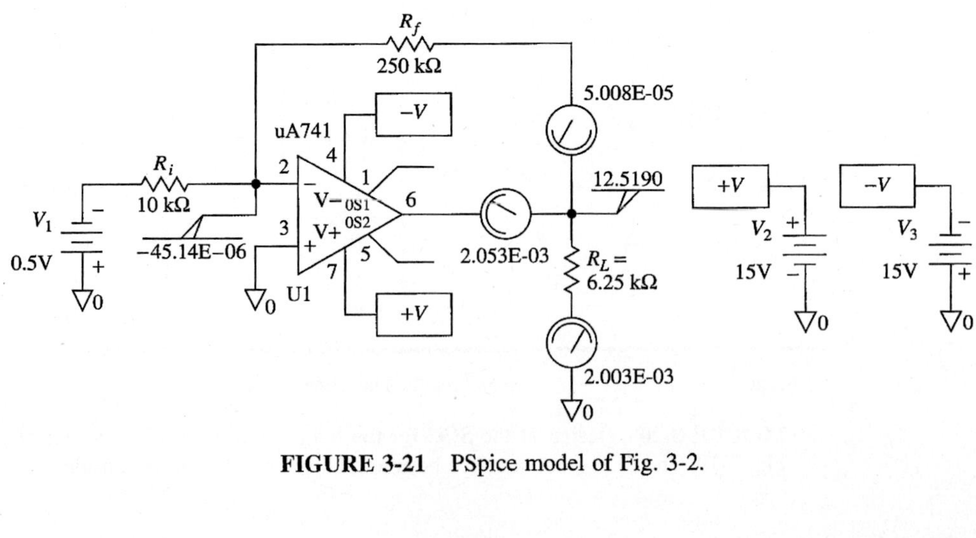

FIGURE 3-21 PSpice model of Fig. 3-2.

3-12.2 Inverting Amplifier—AC Input

Refer to Fig. 3-3 and create the PSpice model of the circuit. Set the input voltage to a sine wave with a peak value of 5 V and a frequency of 500 Hz. Obtain a plot of E_i and V_o versus time. To begin, place the following parts in the work area.

Draw = > Get New Part

Part	Number	Library
=> **uA741**	1	**eval.slb**
=> **VSIN**	1	**source.slb**
=> **VDC**	2	**source.slb**
=> **R**	3	**analog.slb**
=> **GLOBAL**	4	**port.slb**
=> **AGND**	5	**port.slb**

Note that we are using a sine wave as the input signal instead of a triangular wave as shown in Fig. 3-3. Arrange the parts as shown in Fig. 3-3. Change the attributes of the parts as given in Fig. 3-3. Set up the sine-wave attributes by double-clicking on the symbol. In the pop-up window change **VOFF**, **VAMPL**, and **FREQ**.

VOFF = > 0 = > Save Attr
VAMPL = > 5V = > Save Attr = > Change Display = > Both name and value
FREQ = > 500Hz = > Save Attr = > Change Display = > Both name and value

Double-click on the lead from the sine wave generator to **Ri** and label it **Ei**. Double-click on the lead from the output terminal of the op amp and label it **Vo**. See Fig. 3-22.

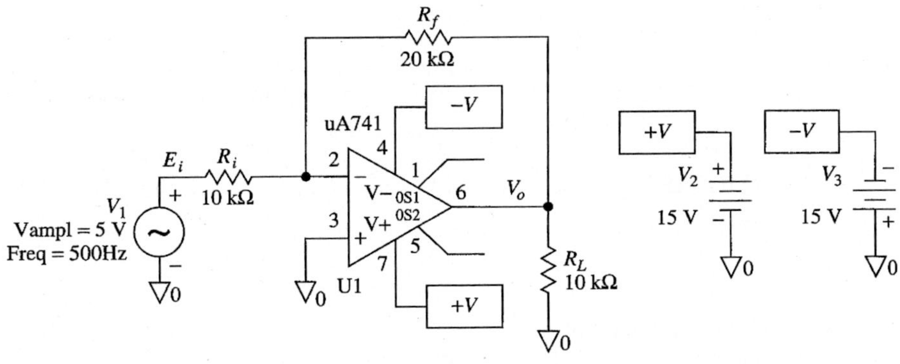

FIGURE 3-22 PSpice model for Fig. 3-3.

In order to obtain a plot of E_i and V_o versus time, we must initialize the Transient menu.

Select

Analysis = > Setup = > Transient
Click **Transient = > Print Step: = > 20μs**
= > Final Time: = > 4ms

Save the circuit as a file with the **.SCH** extension. Run the simulation

Analysis => Simulate

In the Probe window, select

Trace => Add = > V[Ei]
Trace => Add => V[Vo]

Label the plots and obtain a printout as shown in Fig. 3-23.

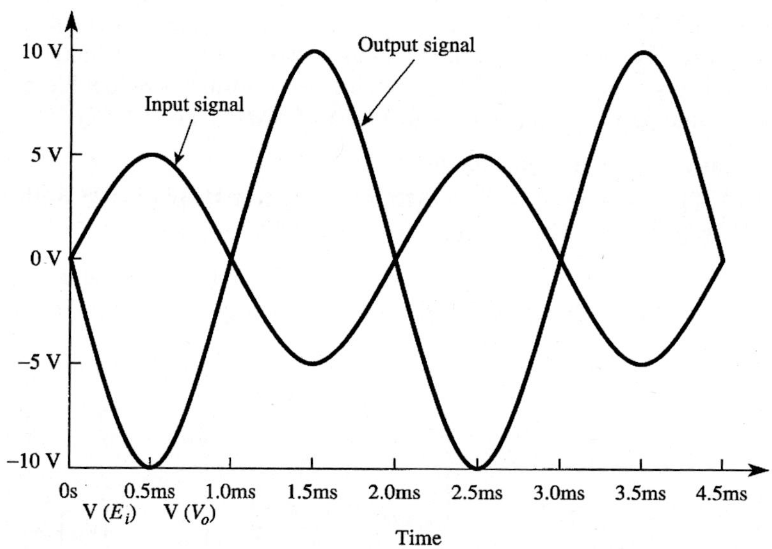

FIGURE 3-23 Plot of E_i and V_o versus time for the circuit of Figs. 3-22 and 3-3.

3-12.3 Inverting Adder

Create the PSpice model and simulate the inverting adder shown in Fig. 3-4. Use **IPROBE** to show $I_1 + I_2 + I_3 = I_f$. Measure the voltage at the summing node, $V(-)$, and V_L using **VIEWPOINT**.

Draw => Get New Part

Part	Number	Library
=> **uA741**	1	**eval.slb**
=> **VDC**	5	**source.slb**
=> **R**	5	**analog.slb**
=> **GLOBAL**	4	**port.slb**
=> **AGND**	7	**port.slb**
=> **IPROBE**	4	**special.slb**
=> **VIEWPOINT**	2	**special.slb**

Arrange and wire the parts as show in Fig. 3-4. Place the **IPROBE**s to measure I_1, I_2, I_3, and I_f, and place the **VIEWPOINT**s to measure the voltage at the summing point, *S*, and V_o. Change the parts attributes to correspond to Fig. 3-4. Save the circuit in a file and run the simulation **Analysis => Simulate.** The results are shown in Fig. 3-24.

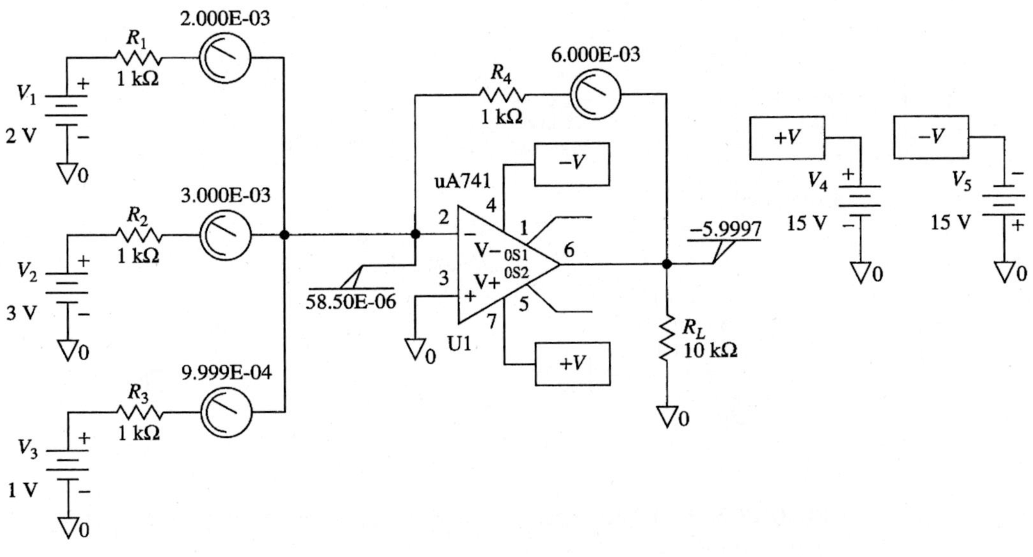

FIGURE 3-24 PSpice model of Fig. 3-4—the inverting adder.

3-12.4 Noninverting Amplifier

Create a PSpice model of the noninverting amplifier shown in Fig. 3-7(a) with $R_f =$ 20 kΩ, $R_1 = 10$ kΩ, and $E_i = 2$ V. Use **IPROBE** to measure I_o, I_f, I_L, and $I(-)$. Use **VIEWPOINT** to measure V_L and the voltage at the (−) input. If you are using the evaluation software package, build the circuit with a 741 op amp instead of the OP-177 as shown in Fig. 3-7(a). Place the following parts in the work area.

Draw => Get New Part

Part	Number	Library
=> **uA741**	1	**eval.slb**
=> **VDC**	3	**source.slb**
=> **R**	3	**analog.slb**
=> **GLOBAL**	4	**port.slb**
=> **AGND**	5	**port.slb**
=> **IPROBE**	4	**special.slb**
=> **VIEWPOINT**	2	**special.slb**

Arrange the parts and include the **IPROBE**s and **VIEWPOINT**s. Save the file and run the simulation **Analysis => Simulate.** The results are shown in Fig. 3-25.

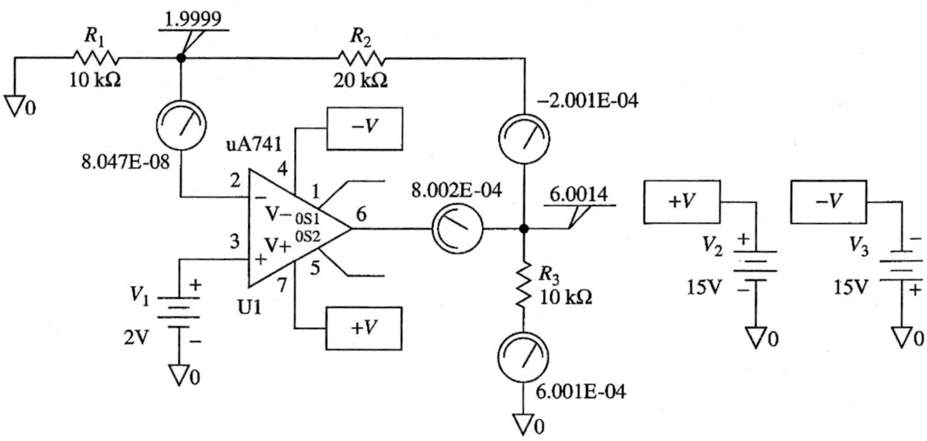

FIGURE 3-25 PSpice model for the noninverting amplifier of Fig. 3-7(a).

PROBLEMS

3-1. What type of feedback is applied to an op amp when an external component is connected between the output terminal and the inverting input?

3-2. If the open-loop gain is very large, does the closed-loop gain depend on the external components or the op amp?

3-3. What two assumptions have been used to analyze the circuits in this chapter?

3-4. Identify the circuit in Fig. P3-4.

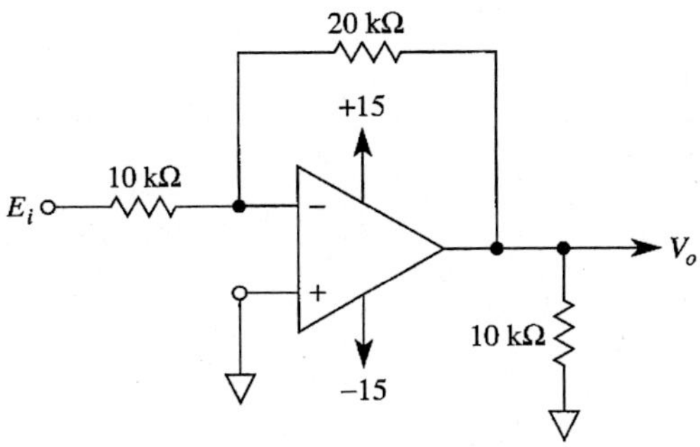

FIGURE P3-4

3-5. Calculate V_o and the op amp's output current in Fig. P3-4 if E_1 equals **(a)** +5 V; **(b)** −2 V. For each situation, state if the op amp sources or sinks current.

3-6. Calculate E_i in Fig. P3-4 if V_o equals **(a)** +5 V; **(b)** −2 V.

3-7. Let E_i be a triangle wave with a frequency of 100 Hz and a peak value of 5 V in Fig. P3-4. **(a)** Plot E_i and V_o vs. time; **(b)** V_o vs. E_i.

3-8. Repeat Problem 3-7 but let E_i be increased in amplitude to 8 V. (Assume $\pm V_{sat} = \pm 15$ V for ease of plotting.)

3-9. Identify the circuit in Fig. P3-9 and calculate V_o if E_i equals **(a)** +5 V; **(b)** −2 V. Compare your results with Problem 3-5.

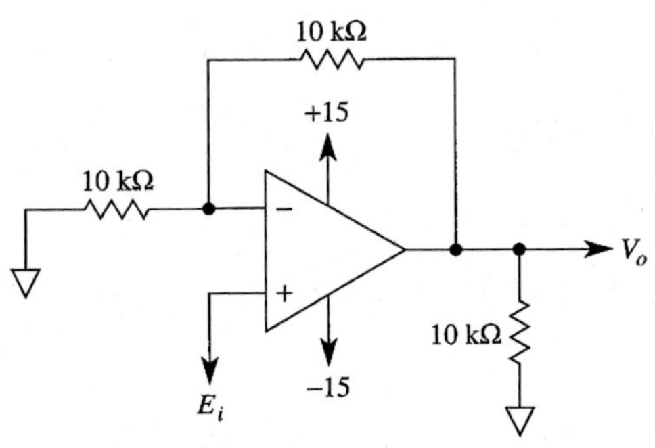

FIGURE P3-9

3-10. Repeat Problem 3-7 except apply it to Fig. P3-9. Compare solutions of both problems to distinguish between inverting and noninverting operation.

3-11. Design an inverting amplifier with a gain of −5 and an input resistance of 10 kΩ.

3-12. Design a noninverting amplifier with a gain of 5.

3-13. Input–output characteristics are shown for three different circuits in Fig. P3-13. Design circuits to re-create plots *A*, *B*, and *C*.

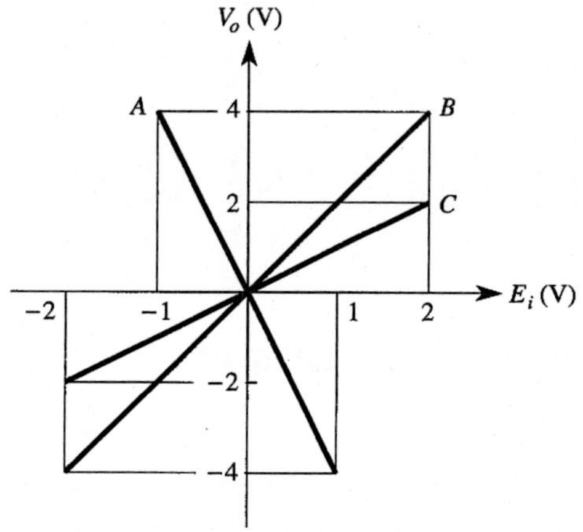

FIGURE P3-13

3-14. The circuit of Fig. P3-14 is called a "subtractor." Is E_1 subtracted from E_2, or vice versa?

3-15. A 5-V peak-to-peak sine wave, E_2, is applied to the summing node in Fig. P3-14. Plot V_o vs. E_1 if voltage E_1 is **(a)** +5 V; **(b)** −5 V.

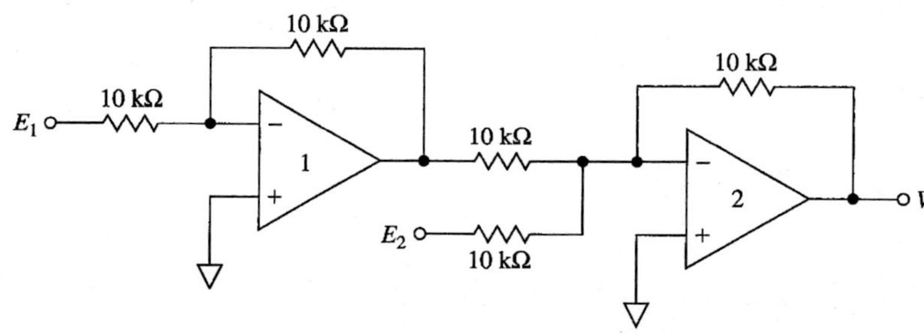

FIGURE P3-14

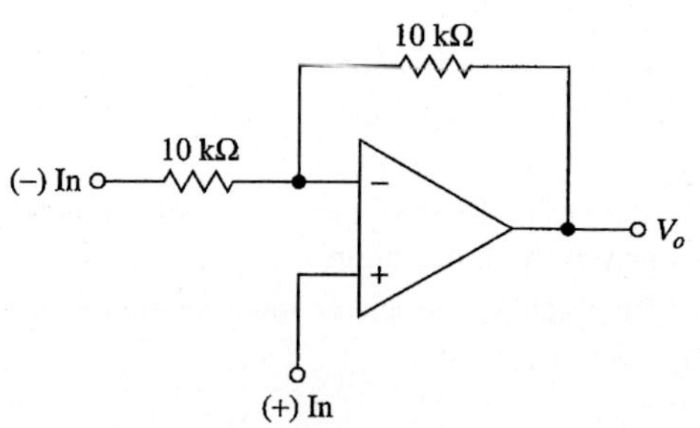

FIGURE P3-15

3-16. A 5-V peak-to-peak sine wave, E_i, is applied to (+) In of Fig. P3-15. Plot V_o vs. E_i if the voltage of (−) In is **(a)** +5 V; **(b)** −5 V. (Assume that $\pm V_{sat} = \pm 15$ V.)

3-17. Design a three-channel inverting amplifier. Gains are to be −1 for channel 1, −3 for channel 2, and −5 for channel 3 (refer to Section 3-3.2).

3-18. Design a two op amp circuit to subtract 1 V from 3 V. Show the output voltage present at each op amp.

3-19. Design a circuit to amplify the difference between E_1 and E_2 by 5. The inputs E_1 and E_2 should be buffered.

3-20. Redesign the system of Fig. 3-20 to measure a temperature range from 0° to 100°C.

3-21. Obtain the data sheet of the LM135/335 temperature sensor from National Semiconductor's Web site and determine

(a) temperature range (continuous) of the LM135.
(b) temperature range (continuous) of the LM335.
(c) package styles of the LM335.
(d) operating current range.
(e) LM335 temperature accuracy (typical) at 25°C.

CHAPTER 4

Comparators and Controls

LEARNING OBJECTIVES

Upon completion of this chapter on comparators and controls, you will be able to:

- Draw the circuit for a zero-crossing detector and plot its output–input characteristic.
- Identify the upper and lower threshold voltages on an output–input characteristic.
- Calculate hysteresis voltage if you know the threshold voltages.
- Explain how hysteresis gives a measure of noise immunity to comparator circuits.
- Explain why hysteresis must be present in all on–off control circuitry, using the familiar wall thermostat as an example.
- Make a battery-charger control circuit.
- Build and calibrate an independently adjustable setpoint controller.
- Describe the operation of an LM311 precision comparator.
- Know how to control the strobe terminal of a comparator with a microcontroller signal.

- Use a comparator circuit in a biomedical application.
- Connect two LM311 comparators to make a window detector.
- Give the definition of *propagation delay* and know how to measure it.
- Simulate comparator circuit operation using PSpice.

4-0 INTRODUCTION

A comparator compares a signal voltage on one input with a reference voltage on the other input. Voltage-level-detector circuits are also comparators and were introduced in Chapter 2 to show how easy it is to use op amps to solve some types of signal comparison applications without the need to know much about the op amp itself. The general-purpose op amp was used as a substitute for ICs designed only for comparator applications.

Unfortunately, the general-purpose op amp's output voltage does not change very rapidly. Also, its output changes between limits fixed by the saturation voltages $+V_{sat}$ and $-V_{sat}$ that are typically about ±13 V. Therefore, the output cannot drive devices that require voltage levels between 0 and +5, such as TTL digital logic ICs or input port lines of a microcontroller. These disadvantages are eliminated by ICs that have been specifically designed to act as comparators. One such device is the 311 comparator. It will be introduced at the end of this chapter.

Neither the general-purpose op amp nor the comparator can operate properly if noise is present at either input. To solve this problem, we will learn how the addition of *positive feedback* overcomes the noise problem. Note that positive feedback does not eliminate the noise but makes the op amp less responsive to it. These circuits will show how to make better voltage-level detectors and also build a foundation to understand square-wave generators (multivibrators) and single-pulse generators (one-shots), which are covered in Chapter 6.

4-1 EFFECT OF NOISE ON COMPARATOR CIRCUITS

Input signal E_i is applied to the (−) input of a 741 op amp in Fig. 4-1 (the 741 is a general-purpose op amp). If no noise is present, the circuit operates as an inverting zero-crossing detector because $V_{ref} = 0$.

Noise voltage is shown, for simplicity, as a square wave in series with E_i. To show the effect of noise voltage, the op amp's input signal voltage is drawn both with and without noise in Fig. 4-2. The waveshape of V_o vs. time shows clearly how the addition of noise causes false output signals. V_o should indicate only the crossings of E_i, *not* the crossings of E_i plus noise voltage.

If E_i approaches V_{ref} very slowly or actually hovers close to V_{ref}, V_o can either follow all the noise voltage oscillations or burst into high-frequency oscillation. These false crossings can be eliminated by *positive feedback.*

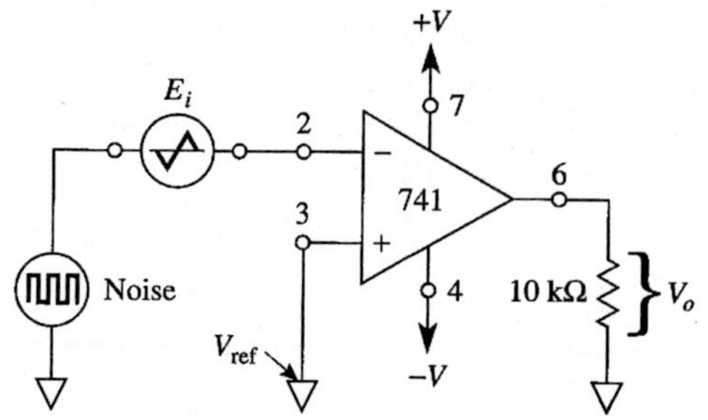

FIGURE 4-1 Inverting zero-crossing detector.

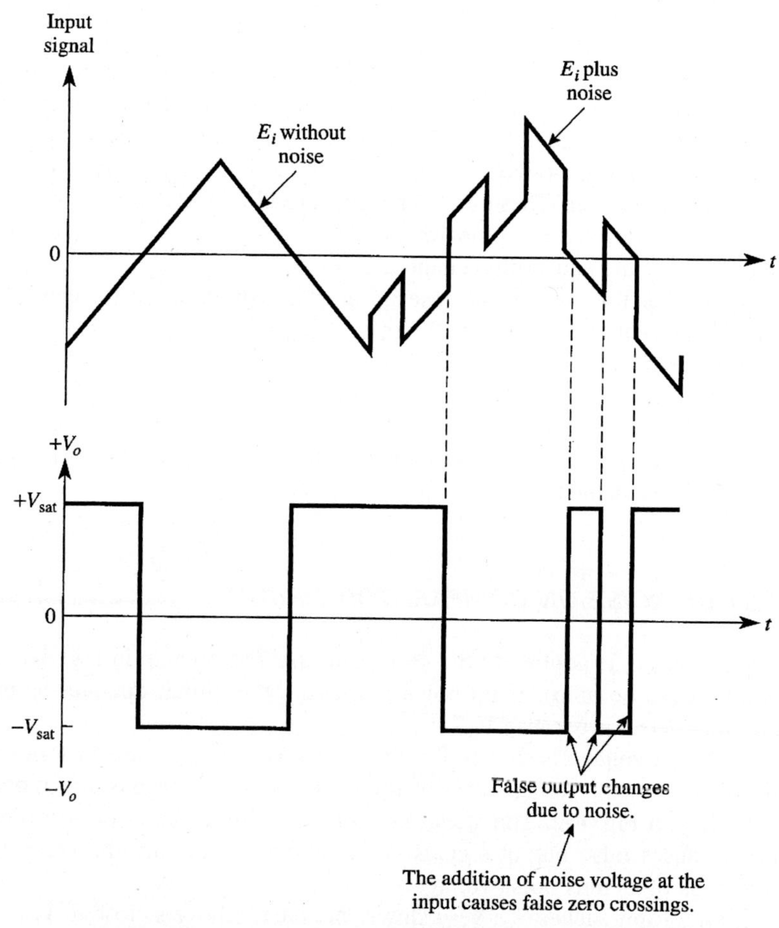

FIGURE 4-2 Effect of noise on a zero-crossing detector.

4-2 POSITIVE FEEDBACK

4-2.1 Introduction

Positive feedback is accomplished by taking a fraction of the output voltage V_o and applying it to the (+) input. In Fig. 4-3(a), output voltage V_o divides between R_1 and R_2. A fraction of V_o is fed back to the (+) input and creates a reference voltage that depends on V_o. The idea of a reference voltage was introduced in Chapter 2. We will now study positive feedback and how it can be used to eliminate false output changes due to noise.

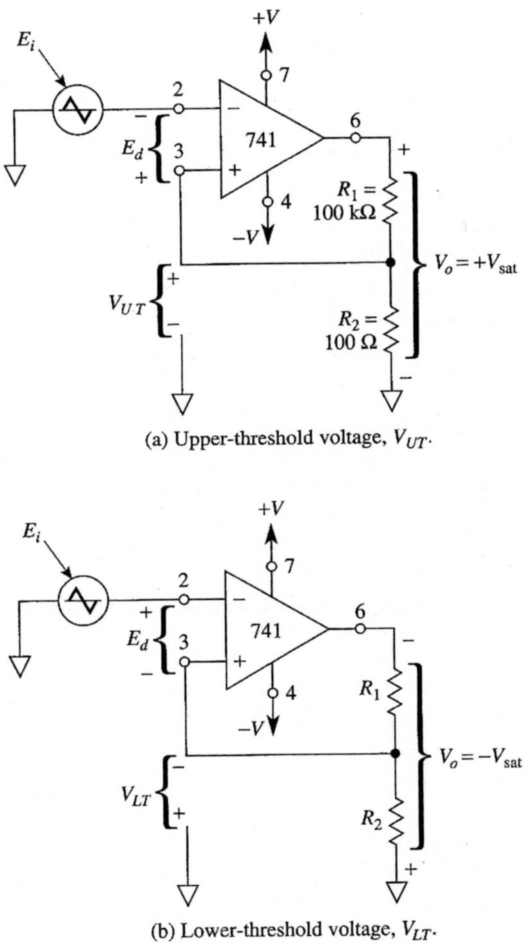

FIGURE 4-3 R_1 and R_2 feed back a reference voltage from the output to the (+) input terminal.

4-2.2 Upper-Threshold Voltage

In Fig. 4-3(a), output voltage V_o divides between R_1 and R_2. A fraction of V_o is fed back to the (+) input. When $V_o = +V_{sat}$, the fed-back voltage is called the *upper-threshold voltage*, V_{UT}. V_{UT} is expressed from the voltage divider as

$$V_{UT} = \frac{R_2}{R_1 + R_2}(+V_{sat}) \tag{4-1}$$

For E_i values below V_{UT}, the voltage at the (+) input is above the voltage at the (−) input. Therefore, V_o is locked at $+V_{sat}$.

If E_i is made slightly more positive than V_{UT}, the polarity of E_d, as shown, reverses and V_o begins to drop in value. Now the fraction of V_o fed back to the positive input is smaller, so E_d becomes larger. V_o then drops even faster and is driven quickly to $-V_{sat}$. The circuit is then stable at the condition shown in Fig. 4-3(b).

4-2.3 Lower-Threshold Voltage

When V_o is at $-V_{sat}$, the voltage fed back to the (+) input is called *lower-threshold voltage*, V_{LT}, and is given by

$$V_{LT} = \frac{R_2}{R_1 + R_2}(-V_{sat}) \tag{4-2}$$

Note that V_{LT} is negative with respect to ground. Therefore, V_o will stay at $-V_{sat}$ as long as E_i is above, or positive with respect to, V_{LT}. V_o will switch back to $+V_{sat}$ if E_i goes more negative than, or below, V_{LT}.

We conclude that positive feedback induces a snap action to switch V_o faster from one limit to the other. Once V_o begins to change, it causes a regenerative action that makes V_o change even faster. If the threshold voltages are larger than the peak noise voltages, positive feedback will eliminate false output transitions. This principle is investigated in the following examples.

Example 4-1

If $+V_{sat} = 14$ V in Fig. 4-3(a), find V_{UT}.

Solution By Eq. (4-1),

$$V_{UT} = \frac{100\ \Omega}{100{,}100\ \Omega}(14\ \text{V}) \approx 14\ \text{mV}$$

Example 4-2

If $-V_{sat} = -13$ V in Fig. 4-3(b), find V_{LT}.

Solution By Eq. (4-2),

$$V_{LT} = \frac{100\ \Omega}{100{,}100\ \Omega}(-13\ \text{V}) \approx -13\ \text{mV}$$

Example 4-3

In Fig. 4-4, E_i is a triangular wave applied to the (−) input in Fig. 4-3(a). Find the resultant output voltage.

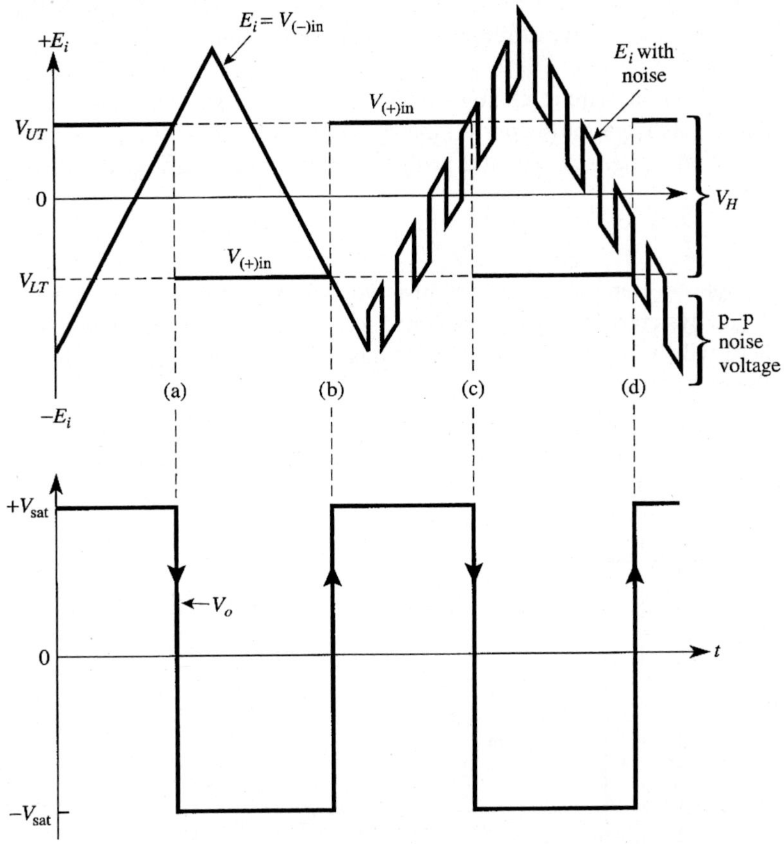

FIGURE 4-4 Solution to Example 4-3. When E_i goes above V_{UT} at time (c), V_o goes to $-V_{sat}$. The peak-to-peak noise voltage would have to equal or exceed V_H to pull E_i below V_{LT} and generate a false crossing. Thus V_H tells us the margin against peak-to-peak noise voltage.

Solution The dashed lines drawn on E_i in Fig. 4-4 locate V_{UT} and V_{LT}. At time $t = 0$, E_i is below V_{LT}, so V_o is at $+V_{sat}$ (as in Fig. 4-4). When E_i goes above V_{UT}, at times (a) and (c), V_o switches quickly to $-V_{sat}$. When E_i again goes below V_{LT}, at times (b) and (d), V_o switches quickly to $+V_{sat}$. Observe how positive feedback has eliminated the false crossings.

4-3 ZERO-CROSSING DETECTOR WITH HYSTERESIS

4-3.1 Defining Hysteresis

There is a standard technique of showing comparator performance on one graph instead of two graphs, as in Fig. 4-4. By plotting E_i on the horizontal axis and V_o on the vertical axis, we obtain the output–input voltage characteristic, as in Fig. 4-5. For E_i less than V_{LT}, $V_o = +V_{\text{sat}}$. The vertical line (a) shows V_o going from $+V_{\text{sat}}$ to $-V_{\text{sat}}$ as E_i becomes greater than V_{UT}. Vertical line (b) shows V_o changing from $-V_{\text{sat}}$ to $+V_{\text{sat}}$ when E_i becomes less than V_{LT}. The difference in voltage between V_{UT} and V_{LT} is called the *hysteresis voltage,* V_H.

Whenever any circuit changes from one state to a second state at some input signal and then reverts from the second to the first state at a *different* input signal, the circuit is said to exhibit *hysteresis.* For the positive-feedback comparator, the difference in input signals is

$$V_H = V_{UT} - V_{LT} \tag{4-3}$$

For Examples 4-1 and 4-2, the hysteresis voltage is 14 mV − (−13 mV) = 27 mV.

If the hysteresis voltage is designed to be greater than the *peak-to-peak* noise voltage, there will be no false output crossings. Thus V_H tells us how much peak-to-peak noise the circuit can withstand.

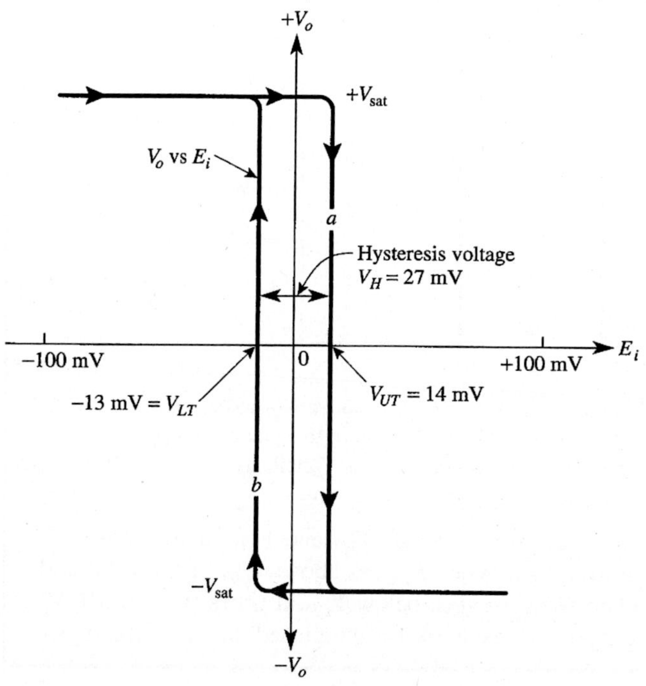

FIGURE 4-5 Plot of V_o vs. E_i illustrates the amount of hysteresis voltage in a comparator circuit.

4-3.2 Zero-Crossing Detector with Hysteresis as a Memory Element

If E_i has a value that lies between V_{LT} and V_{UT}, it is impossible to predict the value of V_o unless *you already know* the value of V_o. For example, suppose that you substitute ground for E_i ($E_i = 0$ V) in Fig. 4-3 and turn on the power. The op amp will go to *either* $+V_{\text{sat}}$ or $-V_{\text{sat}}$, depending on the inevitable presence of noise. If the op amp goes to $+V_{\text{sat}}$, E_i must then go above V_{UT} in order to change the output. If V_o had gone to $-V_{\text{sat}}$, then E_i would have to go below V_{LT} to change V_o.

Thus the comparator with hysteresis exhibits the property of *memory*. That is, if E_i lies between V_{UT} and V_{LT} (within the hysteresis voltage), the op amp remembers whether the last switching value of E_i was above V_{UT} or below V_{LT}.

4-4 VOLTAGE-LEVEL DETECTORS WITH HYSTERESIS

4-4.1 Introduction

In the zero-crossing detectors of Sections 4-2 and 4-3, the hysteresis voltage V_H is centered on the zero reference voltage V_{ref}. It is also desirable to have a collection of circuits that exhibit hysteresis about a center voltage that is either positive or negative. For example, an application may require a positive output, V_o, when an input E_i goes above an upper-threshold voltage of $V_{UT} = 12$ V. Also, we may wish V_o to go negative when E_i goes below a lower threshold voltage of, for example, $V_{LT} = 8$ V. These requirements are summarized on the plot of V_o vs. E_i in Fig. 4-6. V_H is evaluated from Eq. (4-3) as

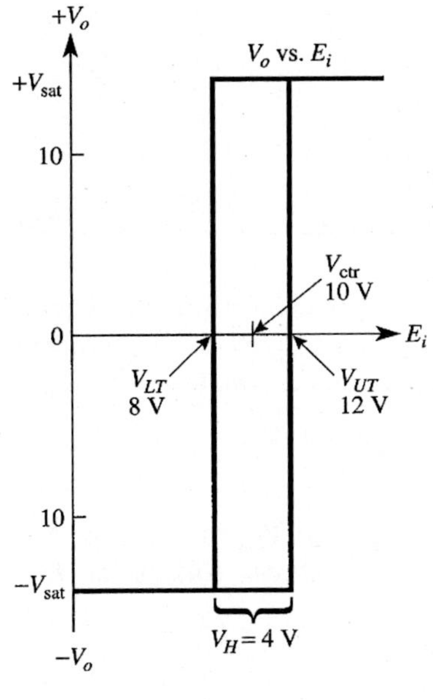

FIGURE 4-6 Positive-voltage-level detector. Hysteresis voltage V_H is symmetrical about the center voltage V_{ctr}. This voltage-level detector is a noninverting type because V_o goes positive when E_i goes above V_{UT}.

$$V_H = V_{UT} - V_{LT} = 12\text{ V} - 8\text{ V} = 4\text{ V}$$

The hysteresis voltage V_H should be centered on the average of V_{UT} and V_{LT}. This average is called center voltage, V_{ctr}, where

$$V_{ctr} = \frac{V_{UT} + V_{LT}}{2} = \frac{12\text{ V} + 8\text{ V}}{2} = 10\text{ V}$$

When we try to build this type of voltage-level detector, it is desirable to have four features: (1) an adjustable resistor to set the value of V_H; (2) a separate adjustable resistor to set the value of V_{ctr}; (3) the setting of V_H and V_{ctr} should *not* interact; and (4) the center voltage V_{ctr} should equal or be simply related to an external reference voltage V_{ref}. For the lowest possible parts count, the op amp's regulated supply voltage and a resistor network can be used for selecting V_{ref}.

Sections 4-4.2 and 4-4.3 deal with circuits that do not have all these features but are low in parts count and consequently cost. Section 4-5 presents a circuit that has all four features but at the cost of a higher parts count.

4-4.2 Noninverting Voltage-Level Detector with Hysteresis

The positive feedback resistor from output to (+) input indicates the presence of hysteresis in the circuit of Fig. 4-7. E_i is applied via R to the (+) input, so the circuit is noninverting. (Note that E_i must be a low-impedance source or the output of either a voltage follower or op amp amplifier.) The reference voltage V_{ref} is applied to the op amp's (−) input.

The upper- and lower-threshold voltages can be found from the following equations:

$$V_{UT} = V_{ref}\left(1 + \frac{1}{n}\right) - \frac{-V_{sat}}{n} \tag{4-4a}$$

$$V_{LT} = V_{ref}\left(1 + \frac{1}{n}\right) - \frac{+V_{sat}}{n} \tag{4-4b}$$

Hysteresis voltage V_H is expressed by

$$V_H = V_{UT} - V_{LT} = \frac{(+V_{sat}) - (-V_{sat})}{n} \tag{4-5}$$

In zero-crossing detectors, V_H is centered on the zero-volts reference. For the circuit of Fig. 4-7, V_H is *not* centered on V_{ref} but is symmetrical about the *average* value of V_{UT} and V_{LT}. This average value is called *center voltage* V_{ctr} and is found from

$$V_{ctr} = \frac{V_{UT} + V_{LT}}{2} = V_{ref}\left(1 + \frac{1}{n}\right) \tag{4-6}$$

Compare the locations of V_{ctr} and V_{ref} in Fig. 4-6. Also compare Eqs. (4-5) and (4-6) to see that n appears in *both* equations. This means that any adjustment in resistor nR affects *both* V_{ctr} and V_H.

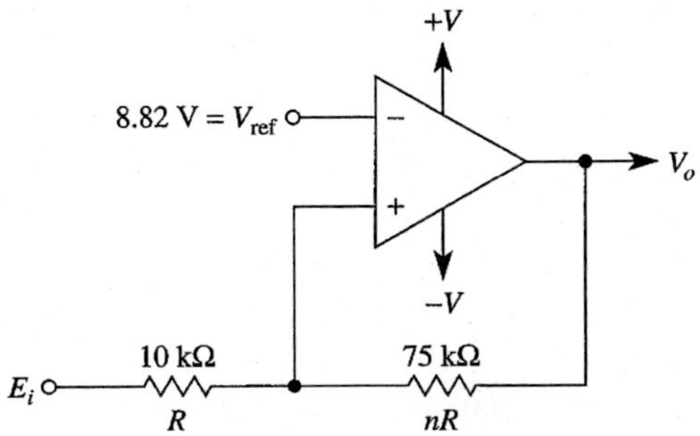

(a) The ratio of nR to R or n and V_{ref} determines V_{UT}, V_{LT}, V_H, and V_{ctr}.

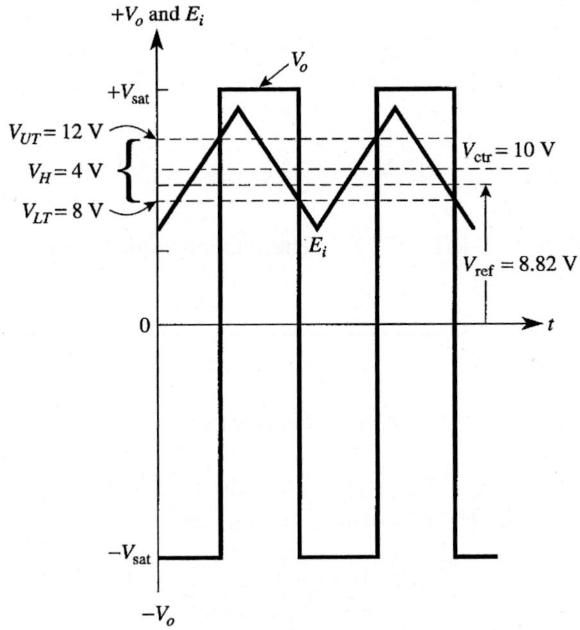

(b) V_o and E_i vs. time.

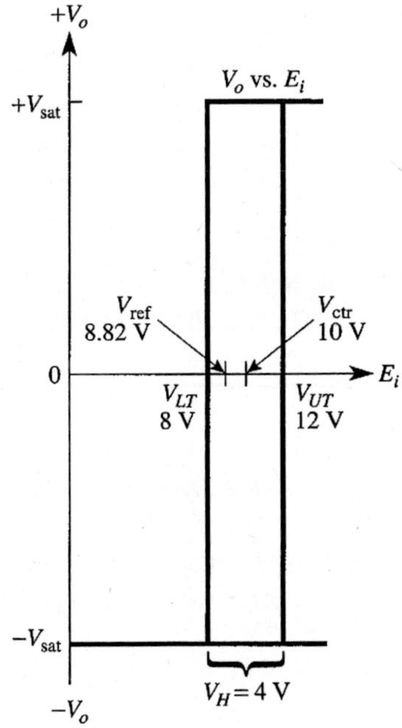

(c) V_o vs. E_i.

FIGURE 4-7 Noninverting voltage-level detector with hysteresis. Center voltage V_{ctr} and hysteresis voltage V_H cannot be adjusted independently since both depend on the ratio n.

Design Example 4-4

Design the circuit of Fig. 4-7 to have $V_{UT} = 12$ V and $V_{LT} = 8$ V. Assume that $\pm V_{sat} = \pm 15$ V.

Design Procedure

1. From Eqs. (4-5) and (4-6), calculate V_H and V_{ctr}:

$$V_H = 12\text{ V} - 8\text{ V} = 4\text{ V}, \qquad V_{ctr} = \frac{12\text{ V} + 8\text{ V}}{2} = 10\text{ V}$$

2. Find n from Eq. (4-5):

$$n = \frac{+V_{sat} - (-V_{sat})}{V_H} = \frac{+15\text{ V} - (-15\text{ V})}{4} = 7.5$$

3. Find V_{ref} from Eq. (4-6):

$$V_{ref} = \frac{V_{ctr}}{1 + 1/n} = \frac{10\text{ V}}{1 + 1/7.5} = 8.82\text{ V}$$

4. Select $R = 10\text{ k}\Omega$ and $nR = 7.5 \times 10\text{ k}\Omega = 75\text{ k}\Omega$. The relationships between E_i and V_o are shown in Figs. 4-7(b) and (c).

4-4.3 Inverting Voltage-Level Detector with Hysteresis

If E_i and V_{ref} are interchanged in Fig. 4-7(a), the result is the inverting voltage-level detector with hysteresis (see Fig. 4-8). The expressions for V_{UT} and V_{LT} are

$$V_{UT} = \frac{n}{n+1}(V_{ref}) + \frac{+V_{sat}}{n+1} \tag{4-7a}$$

$$V_{LT} = \frac{n}{n+1}(V_{ref}) + \frac{-V_{sat}}{n+1} \tag{4-7b}$$

V_{ctr} and V_H are then found to be

$$V_{ctr} = \frac{V_{UT} + V_{LT}}{2} = \left(\frac{n}{n+1}\right)V_{ref} \tag{4-8}$$

$$V_H = V_{UT} - V_{LT} = \frac{(+V_{sat}) - (-V_{sat})}{n+1} \tag{4-9}$$

Note that V_{ctr} and V_H both depend on n and therefore are *not* independently adjustable.

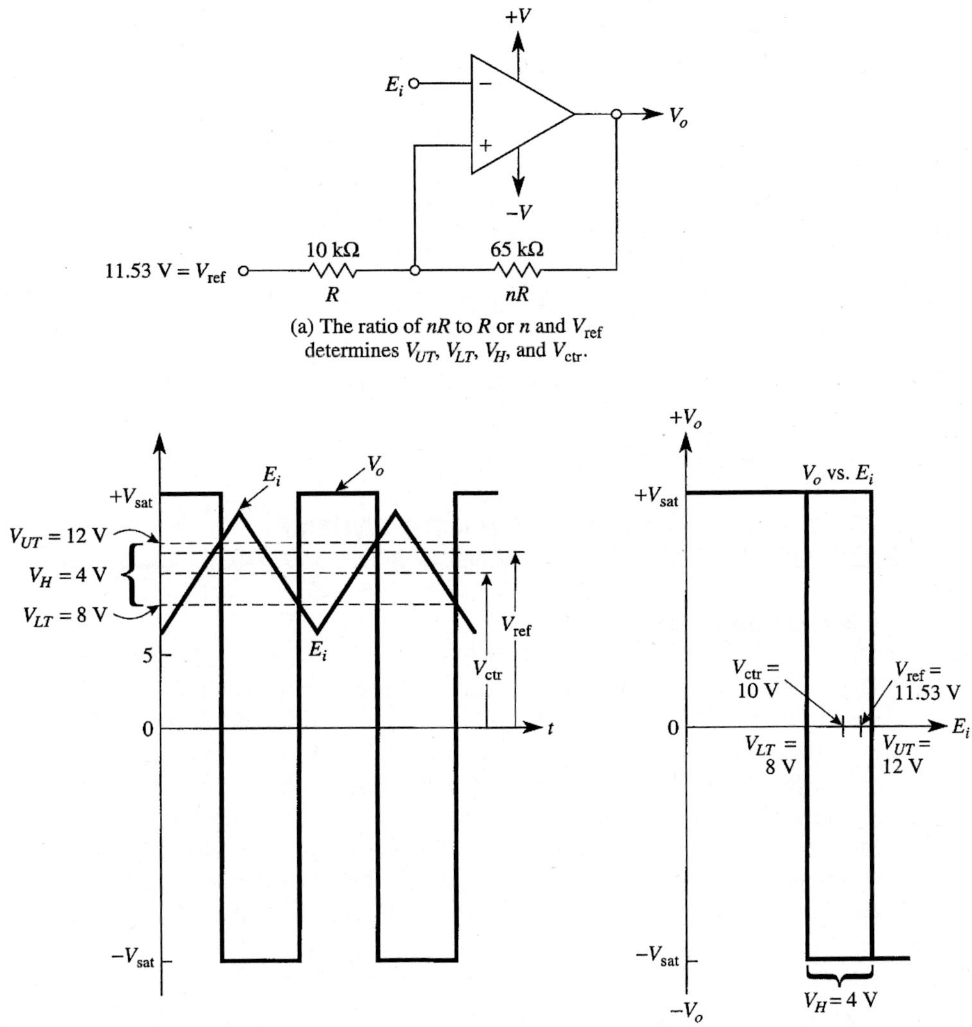

(a) The ratio of nR to R or n and V_{ref} determines V_{UT}, V_{LT}, V_H, and V_{ctr}.

(b) V_o and E_i vs. time. (c) V_o vs. E_i.

FIGURE 4-8 Inverting voltage-level detector with hysteresis. Center voltage V_{ctr} and V_H cannot be adjusted independently since both depend on n.

Design Example 4-5

Complete a design for Fig. 4-8 that has $V_{UT} = 12$ V and $V_{LT} = 8$ V. To make this example comparable with Example 4-4, assume that $\pm V_{sat} = \pm 15$ V. Therefore, $V_{ctr} = 10$ V and $V_H = 4$ V.

Design Procedure

1. Find n from Eq. (4-9):

$$n = \frac{(+V_{\text{sat}}) - (-V_{\text{sat}})}{V_H} - 1 = \frac{15\text{ V} - (-15\text{ V})}{4\text{ V}} - 1 = 6.5$$

2. Find V_{ref} from Eq. (4-8):

$$V_{\text{ref}} = \frac{n+1}{n}(V_{\text{ctr}}) = \frac{6.5+1}{6.5}(10) = 11.53\text{ V}$$

3. Choose $R = 10\text{ k}\Omega$; therefore, resistor nR will be $6.5 \times 10\text{ k}\Omega = 65\text{ k}\Omega$. These circuit values and waveshapes are shown in Fig. 4-8.

4-5 VOLTAGE-LEVEL DETECTOR WITH INDEPENDENT ADJUSTMENT OF HYSTERESIS AND CENTER VOLTAGE

4-5.1 Introduction

The circuit of Fig. 4-9 is a noninverting voltage-level detector with *independent adjustment of hysteresis and center voltage.* In this circuit, the center voltage V_{ctr} is determined by both resistor mR and the reference voltage V_{ref}. V_{ref} can be either supply voltage +V or −V. Remember that the op amp's supply voltage is being used for a lower parts count. Hysteresis voltage V_H is determined by resistor nR. If resistor nR is adjustable, then V_H can be adjusted *independently of* V_{ctr}. Adjusting resistor mR adjusts V_{ctr} without affecting V_H. Note that the signal source, E_i, must be a low-impedance source.* The key voltages are shown in Fig. 4-9 and are designed or evaluated from the following equations:

$$V_{UT} = -\frac{-V_{\text{sat}}}{n} - \frac{V_{\text{ref}}}{m} \tag{4-10a}$$

$$V_{LT} = \frac{-V_{\text{ref}}}{m} - \frac{+V_{\text{sat}}}{n} \tag{4-10b}$$

$$V_H = V_{UT} - V_{LT} = \frac{(+V_{\text{sat}}) - (-V_{\text{sat}})}{n} \tag{4-11}$$

$$V_{\text{ctr}} = \frac{V_{UT} + V_{LT}}{2} = -\frac{V_{\text{ref}}}{m} - \frac{+V_{sat} + (-V_{sat})}{2n} \tag{4-12a}$$

* If not, buffer E_i with a voltage follower as shown in Section 3-6.

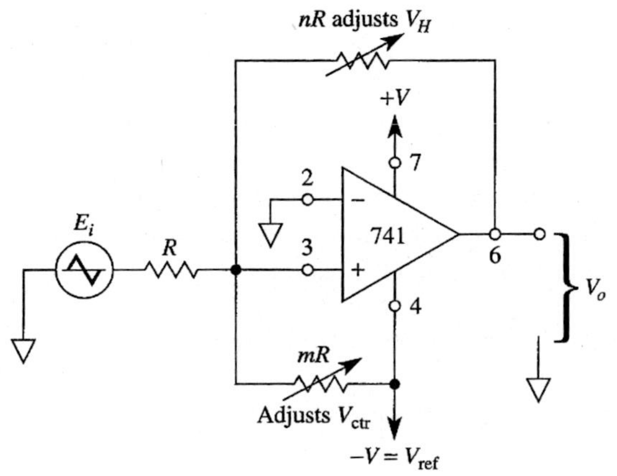

(a) Comparator with independent adjustments for hysteresis and reference voltage.

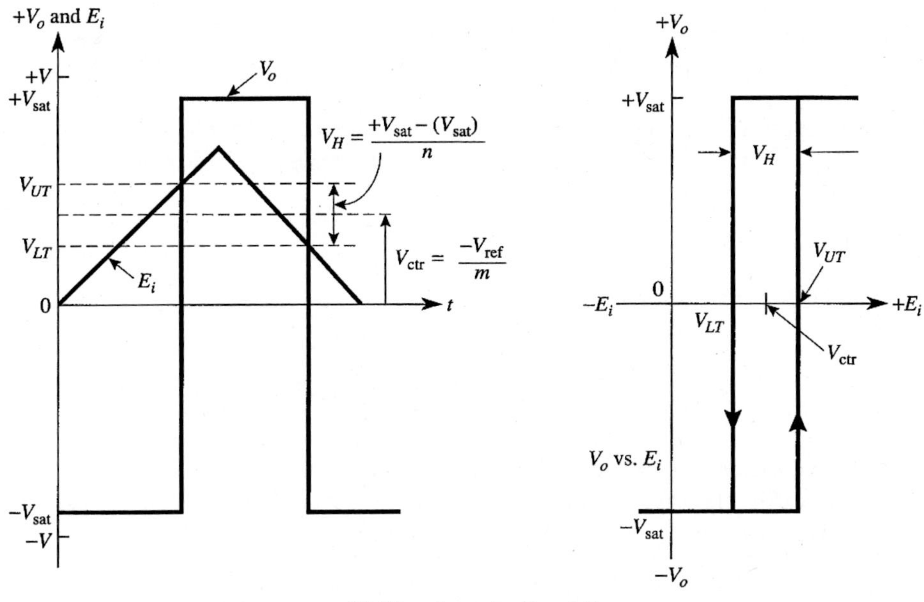

(b) Waveshapes for V_o and E_i.

FIGURE 4-9 Resistor mR and supply voltage $-V$ establish the center voltage V_{ctr}. Resistor nR allows independent adjustment of the hysteresis voltage V_H symmetrically around V_{ctr}.

The general equation for V_{ctr} seems complex. However, if the magnitudes of $+V_{sat}$ and $-V_{sat}$ are nearly equal, then V_{ctr} is expressed simply by

$$V_{ctr} = -\frac{V_{ref}}{m} \tag{4-12b}$$

So V_{ctr} depends only on m and V_H depends only on n.

The following example shows how easy it is to design a battery-charger control circuit using the principles studied in this section.

4-5.2 Battery-Charger Control Circuit

Following is an example of the design procedure for a battery-charger control circuit.

Design Example 4-6

Assume that you want to monitor a 12-V battery. When the battery's voltage drops below 10.5 V, you want to connect it to a charger. When the battery voltage reaches 13.5 V, you want the charger to be disconnected. Therefore, $V_{LT} = 10.5$ V and $V_{UT} = 13.5$ V. Let us use the $-V$ supply voltage for V_{ref} and assume that it equals $-$ 15.0 V. Further, let us assume that $\pm V_{sat} = \pm 13.0$ V. Find (a) V_H and V_{ctr}; (b) resistor mR; (c) resistor nR.

Design Procedure

1. From Eqs. (4-11) and (4-12a), find V_H and V_{ctr}.

$$V_H = V_{UT} - V_{LT} = 13.5\text{ V} - 10.5\text{ V} = 3.0\text{ V}$$

$$V_{ctr} = \frac{V_{UT} + V_{LT}}{2} = \frac{13.5\text{ V} + 10.5\text{ V}}{2} = 12.0\text{ V}$$

Note that the center voltage is the battery's nominal voltage.

2. Arbitrarily choose resistor R to be a readily available value of 10 kΩ. From Eq. (4-12b), choose V_{ref} as -15 V to make the sign of m positive:

$$m = -\left(\frac{V_{ref}}{V_{ctr}}\right) = -\left(\frac{-15\text{ V}}{12\text{ V}}\right) = 1.25$$

Therefore, $mR = 1.25 \times 10\text{ k}\Omega = 12.5\text{ k}\Omega$.

3. From Eq. (4-11), find n.

$$n = \frac{+V_{sat} - (-V_{sat})}{V_H} = \frac{13\text{ V} - (-13\text{ V})}{3} = 8.66$$

Therefore, $nR = 86.6\text{ k}\Omega$.

The final circuit is shown in Fig. 4-10. When E_i drops below 10.5 V, V_o goes negative, releasing the relay to its normally closed position. The relay's normally closed (NC) contacts connect the charger to battery E_i. Diode D_1 protects the transistor against excessive reverse bias when $V_o = -V_{sat}$. When the battery charges to 13.5 V, V_o switches to $+V_{sat}$, which turns on the transistor and operates the relay, whose NC contacts open to disconnect the charger. Diode D_2 protects both op amp and transistor against transients developed by the relay's collapsing magnetic field.

One final note. Suppose that the application requires an inverting voltage-level detector with hysteresis. That is, V_o must go low when E_i goes above V_{UT} and V_o must go high when E_i drops below V_{LT}. For this application, do not change the circuit or design procedure for the noninverting voltage-level detectors, simply add an inverting amplifier, or inverting comparator, to the output V_o.

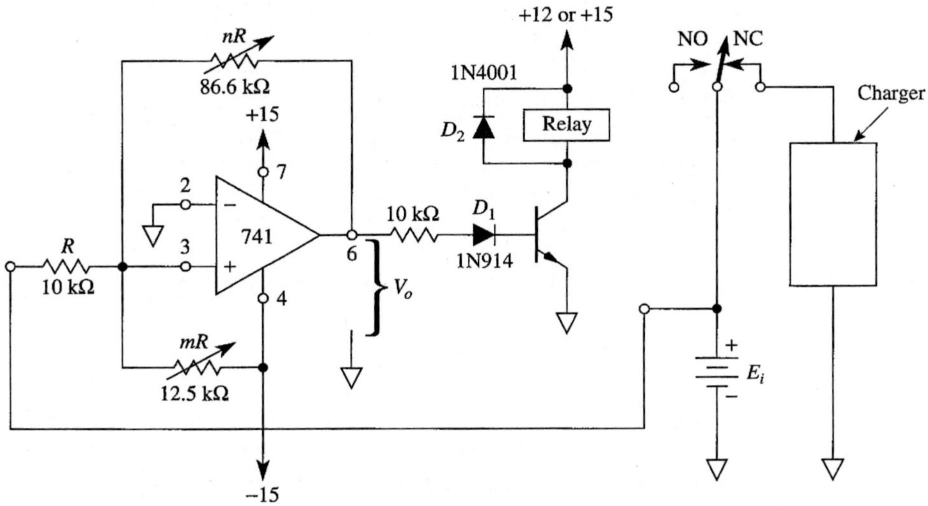

FIGURE 4-10 Battery-charge control for solution to Example 4-6. Adjust mR for $V_{ctr} = 12$ V in the test circuit of Fig. 4-9 and adjust nR for $V_H = 3$ V centered on V_{ctr}.

4-6 ON–OFF CONTROL PRINCIPLES

4-6.1 Comparators in Process Control

Design Example 4-6 illustrates one of the most important applications of positive feedback circuits with hysteresis. They make excellent low-cost on–off controls. The circuit of Fig. 4-10 turns a charger on when battery voltage is below 10.5 V. It also turns the charger off when battery voltage exceeds 13.5 V. Note that the 10.5-to-13.5-V area is the *memory* or hysteresis range. If the battery voltage is 12.0 V, it can be either in the process of charging or discharging, depending on the last command.

4-6.2 The Room Thermostat as a Comparator

The temperature control in a room is a familiar example of on–off control. Let's draw an analogy to our comparator circuits. You adjust the temperature pointer to 65°F. This corresponds to V_{ctr}. The manufacturer builds in the hysteresis. Turn heat *on* if the temperature is below 63°F. Turn heat *off* if the temperature is above 67°F. If the temperature is in the memory range (63° to 67°F), remember the last command. The memory range corresponds to V_H.

4-6.3 Selection/Design Guideline

Controls that are to be operated by the general public all share common characteristics with the comparators of Sections 4-4 and 4-5. The customer can only adjust V_{ctr}. V_H is present or, in some applications, the V_H control is brought out for the customer. These types of control circuits are fail safe. If temperature is set to 50°F, the room simply gets colder. They do not contain the possibility of catastrophic failure. That is, they won't mix up V_{UT} and V_{LT}.

The control circuit of the next section gives high-precision adjustments for the upper and lower setpoints for process control applications. It contains the possibility of runaway operation. Controls for V_{UT} and V_{LT} must be available only to a knowledgeable person and never to the general public.

4-7 AN INDEPENDENTLY ADJUSTABLE SETPOINT CONTROLLER

4-7.1 Principle of Operation

The circuit to be presented will allow both upper setpoint voltage V_{UT} and lower setpoint voltage V_{LT} to be adjusted independently and with precision. The principle of operation is straightforward. Set up a voltage with one fixed resistor and two adjustable resistors as in Fig. 4-11.

When a single-pole double-throw (SPDT) switch is thrown to one position, it grounds the upper setpoint adjust pot. Adjust R_{UT} so that the upper setpoint voltage V_{UT} appears at the setpoint voltage output, V_{set} [see Fig. 4-11(a)]. Then throw the switch to its remaining position as in Fig. 4-11(b). Adjust R_{LT} so that V_{LT} appears on the V_{set} line.

Before we design and analyze the hardware to make an electronically controlled SPDT switch, let us define the required output–input characteristics.

4-7.2 Output–Input Characteristics of an Independently Adjustable Setpoint Controller

Two outputs are required from our basic controller. The first is shown in Fig. 4-12(a). The setpoint voltage output V_{set} is needed for two reasons. First, it will be applied to one input of the in–out comparator (number 1 in Fig. 4-13). Second, it must be available for system test or calibration to trained test persons (not the public).

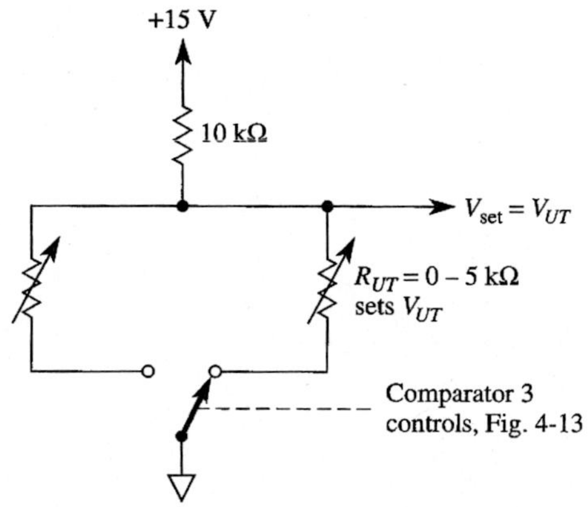

(a) If $E_i > V_{UT}$, the switch is thrown to the right and grounds the upper setpoint adjust pot. If $V_{LT} < E_i < V_{UT}$, stay there (memory).

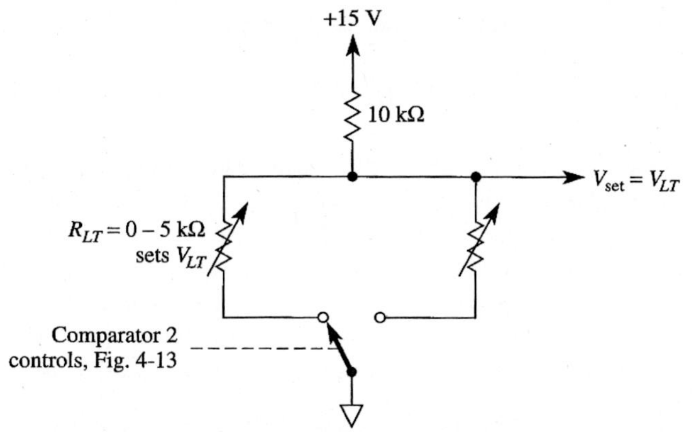

(b) If $E_i < V_{LT}$, the switch is thrown to the left and grounds the lower setpoint adjust pot. If $V_{LT} < E_i < V_{UT}$, stay there (remember last command).

FIGURE 4-11 Separately adjustable setpoint or threshold voltages are designed with an electronically controlled SPDT switch, two pots, one resistor, and a power supply. For the resistor values shown, setpoint voltages are independently adjusted for any value between 0 and 5 V.

The second output voltage required is shown in Fig. 4-12(b). It is the input–output characteristic of the control system (number 1 in Fig. 4-13). Both output characteristics exhibit hysteresis.

4-7.3 Choice of Setpoint Voltages

In Fig. 4-11, the resistor components were chosen to give a range of 0 to 5 V for the setpoint voltage adjustments. In Fig. 4-12, V_{UT} was chosen to be 2.0 V and V_{LT} was chosen to be 0.5 V.

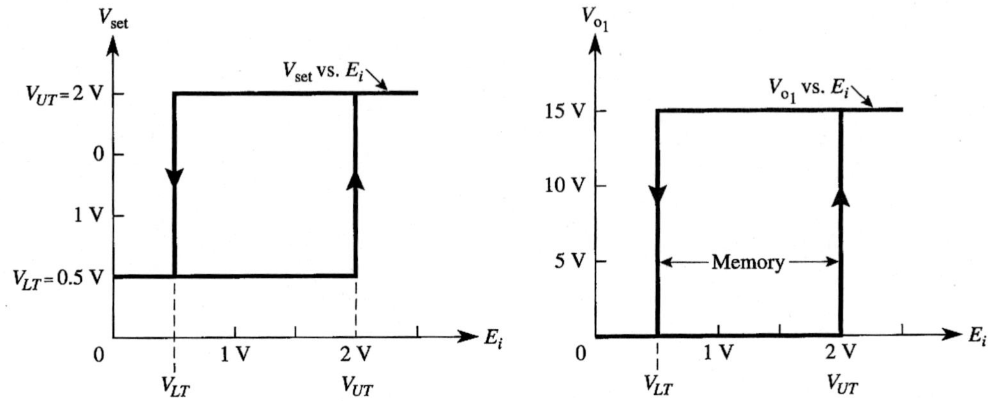

(a) Controller setpoint output vs. E_i.

(b) Controller input–output characteristic.

FIGURE 4-12 An on–off controller with independent setpoint voltages, V_{UT} and V_{LT}, has two distinct characteristics. The setpoint voltage depends on the input voltage as in (a). The input–output characteristic exhibits hysteresis as in (b).

4-7.4 Circuit for Independently Adjustable Setpoint Voltage

We only have to add two more parts to the basic voltage divider of Fig. 4-11. One is a 10-kΩ resistor and the other is three-fourths of an LM339 open-collector comparator.

Recall from Chapter 2 that the LM339 usually works single supply with a ground at pin 12 on its negative rail. If differential input voltage E_d is negative for any one of its four comparators, the corresponding output switch is closed. This grounds the comparator's output terminal (and, incidentally, a setpoint pot). If E_d is positive [(+) In above (−) In] the output switch is open (and, incidentally, disconnects an adjustment pot).

The control circuit is finally presented in Fig. 4-13(a). The alignment procedure is presented as a flowchart showing the cause-and-effect sequence in Fig. 4-13(b). Study the alignment procedure carefully. It explains how the circuit works.

You will probably draw the following conclusions from your study.

1. Comparators 2 and 3 form a single-pole, double-throw switch to ground the bottom terminal and activate either R_{UT} or R_{LT} (see Fig. 4-11).
2. Comparator 1 is the output–input control. V_{set} is either at V_{UT} or at V_{LT} and is applied to 1's (−) input. Since E_i is applied to the (+) input, the circuit is inherently noninverting [see Fig. 4-12(b)].
3. This circuit makes an excellent laboratory experiment since it has a parts count of only five.

In practice, V_{o_1} will usually drive a relay or optocoupler with triac output. Therefore, V_{o_1} should be buffered or use the remaining LM339 comparator to (a) invert the output–input characteristic, or (b) avoid loading R_{PU} below either setpoint voltages.

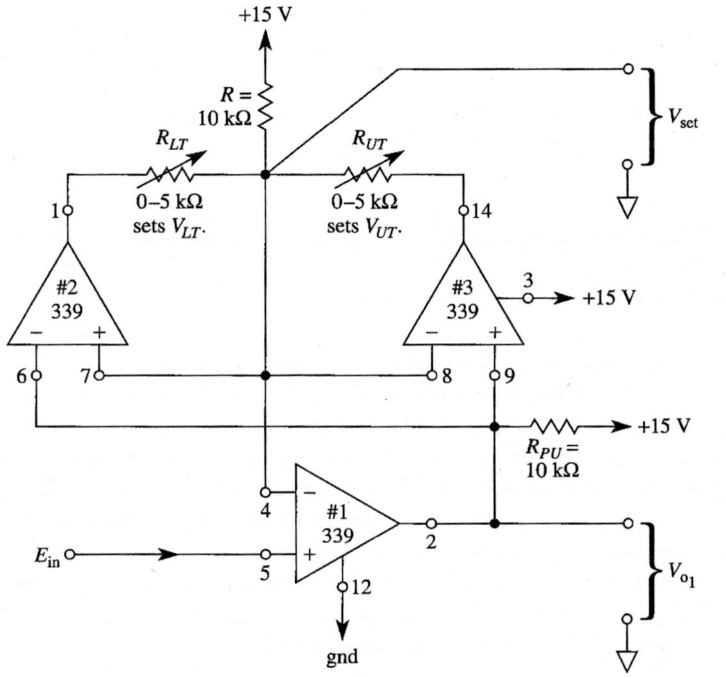

(a) Circuit for independent settings of upper and lower setpoint voltages V_{UT} and V_{LT}.

Alignment procedure:

1. Adjust R_{LT} and R_{UT} for maximum resistance.
2. Ground E_{in}, → E_d of No. 1 goes negative.

↓

No. 1 output switch closes, V_{o_1}, ≅ 0.1 V

↓

No. 2 E_d = positive, switch opens → R_{LT} disconnected

No. 3 E_d = negative, switch closes → Pin 14 grounded, set R_{UT} for $V_{ref} = V_{UT}$

3. Set E_{in} greater than V_{UT}, → E_d of No. 1 goes positive,

↓

No. 2 E_d = negative, switch closes → Pin 1 grounded, set R_{LT} for $V_{ref} = V_{LT}$

No. 3 E_d = positive, switch opens → R_{UT} disconnected

(b) Alignment procedure and operating sequence.

FIGURE 4-13 Circuit, alignment procedure, and operating sequence for a control module that allows independent adjustment of upper and lower setpoint voltages.

4-7.5 Precautions

One last time, suppose that V_{UT} is adjusted to a value below V_{LT}. Upon power up, one of two events will occur: (1) the system will not start up, or (2) the system will turn on and remain on until destruction occurs.

4-8 IC PRECISION COMPARATOR, 111/311

4-8.1 Introduction

The 111 (military) or 311 (commercial) comparator is an IC that has been designed and optimized for superior performance in voltage-level-detector applications. A comparator should be fast. That is, its output should respond quickly to changes at its inputs. The 311 comparator is much faster than the 741 op amp but not as fast as many of today's high-speed comparators. The subject of speed is discussed in Section 4-11, "Propagation Delay."

The 311 is an excellent choice for a low-cost comparator because of its versatility. Its output is designed *not* to bounce between $\pm V_{sat}$ but can be changed quite easily. As a matter of fact, if you are interfacing to a system with a different supply voltage, you simply connect the output to the new supply voltage via a pull-up resistor. We begin by examining the operation of the output terminal.

4-8.2 Output Terminal Operation

A simplified model of the 311 in Fig. 4-14(a) shows that its output behaves like a switch connected between output pin 7 and pin 1. Pin 7 can be wired to any voltage V^{++} with magnitudes up to 40 V more positive than the $-V$ supply terminal (pin 4). When (+) input pin 2 is more positive than (−) input pin 3, the 311's equivalent output switch is open. V_o is then determined by V^{++} and is +5 V.

When the (+) input is less positive than (below) the (−) input, the 311's equivalent output switch closes and extends the ground at pin 1 to output pin 7. Here is one important difference between the 311 and the 339. The 339 has no equivalent to pin 1. There is no separate switch return terminal on the 339 as there is on the 311.

R_f and R_i add about 50 mV of hysteresis to minimize noise effects so that pin 2 is essentially at 0 V. Waveshapes for V_o and E_i are shown in Fig. 4-14(b). V_o is 0 V (switch closed) for positive half-cycles of E_i. V_o is +5 V (switch open) for negative half-cycles of E_i. This is a typical interface circuit; that is, voltages may vary between levels of +15 V and −15 V, but V_o is restrained between +5 V and 0 V, which are typical digital signal levels, so the 311 can be used for converting analog voltage levels to digital voltage levels for interfacing applications.

4-8.3 Strobe Terminal Operation

The strobe terminal of the 311 is pin 6 (see also Appendix 3). This strobe feature allows the comparator output either to respond to input signals or to be independent of input signals. Fig. 4-15 uses the 311 comparator as a zero-crossing detector. A 10-kΩ resistor is

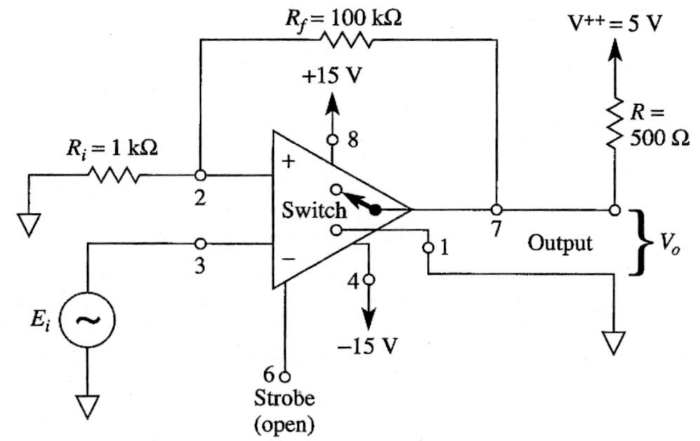

(a) Zero-crossing detector with hysteresis.

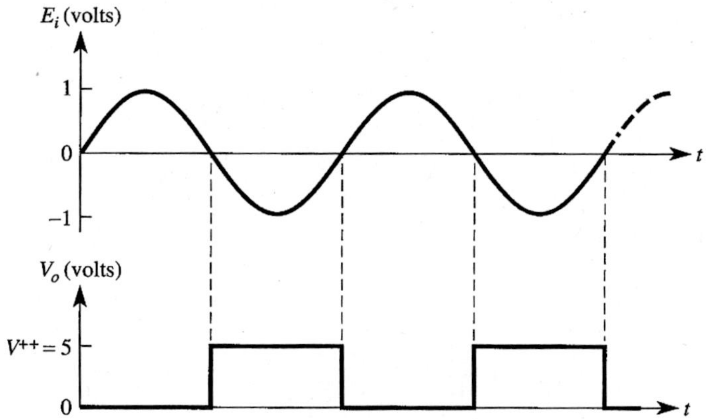

(b) Output and input voltages.

FIGURE 4-14 Simplified model of the 311 comparator with input and output voltage waveforms.

connected to the strobe terminal. The other side of the resistor is connected to a switch. With the strobe switch open, the 311 operates normally. That is, the output voltage is at V^{++} for negative values of E_i and at 0 for positive values of E_i. When the strobe switch is closed (connecting the 10 kΩ to ground), the output voltage goes to V^{++} regardless of the input signal. V_o will stay at V^{++} as long as the strobe switch is closed [see Fig. 4-15(b)]. The output is then independent of the inputs until the strobe switch is again opened.

The strobe feature is useful when the comparator is used with a microcontroller for handshake (control) signals. If the microcontroller outputs a logic 0 to pin 1 of the 311 comparator, the strobe switch is closed. Thus no extraneous signal will be connected to the microcontroller. When the programmer wishes the microcontroller to receive data from the 311, a logic 1 is applied to pin 1. Current from the strobe terminal should be

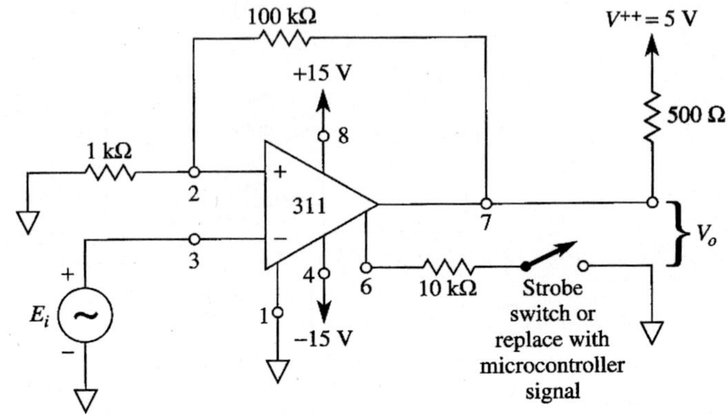

(a) 311 with strobe control.

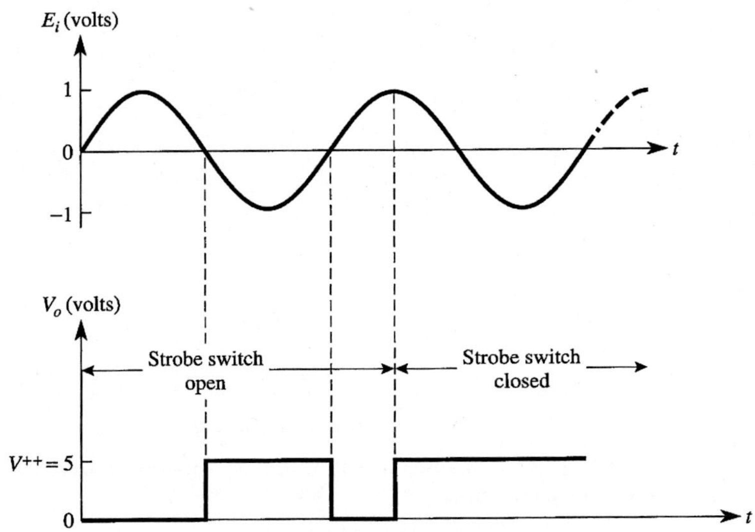

(b) When the strobe switch is closed, $V_o = V^{++}$.

FIGURE 4-15 Operation of the strobe terminal.

limited to about 3 mA. If the strobe feature is not used, the strobe terminal is left open or wired to $+V$ (see Appendix 3).

4-9 BIOMEDICAL APPLICATION

As previously stated, the circuit of Fig. 4-14(a) can be used to convert analog signals to digital signals with TTL levels. Consider the electrocardiogram (EKG or ECG) waveform shown in Fig. 4-16. It is to be signal conditioned for a microcontroller's counter input pin

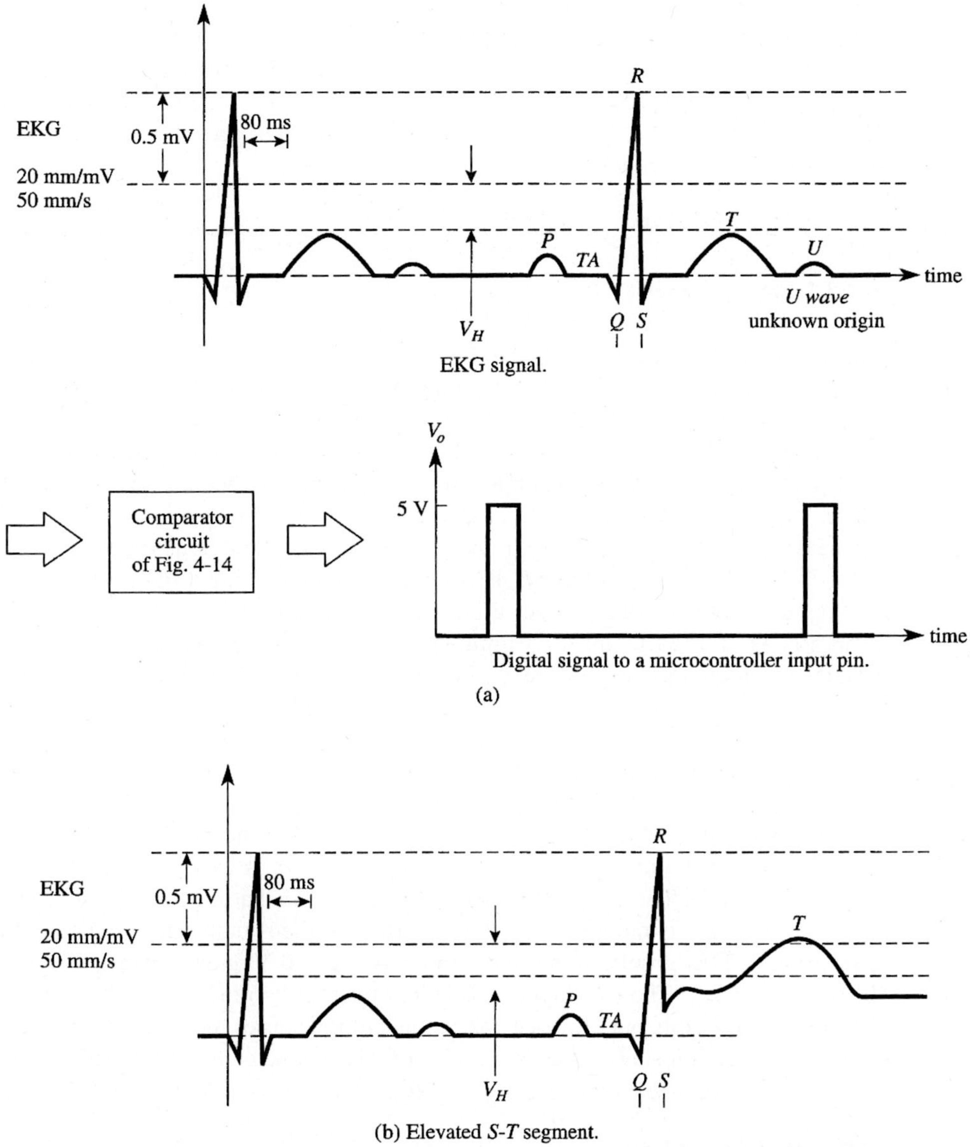

FIGURE 4-16 A 311 comparator circuit signal conditions a biomedical signal for a computer input.

or its input capture pin. The 311 with hysteresis can be used for this application. The hysteresis level can be set so that the 311's output goes to a logic 1 only on the *R* wave and above a specified voltage level. The comparator's output remains at a logic 0 for normal *P* and *T* waves. If a patient has an elevated *S-T* segment, then this level can also cause the output of the 311 to change state and be recorded by the microcontroller. In a hospital or a doctor's office the electrocardiogram waveform is obtained from a person through sensors and passes through an isolation/instrumentation amplifier and then to the analog-to-digital signal conditioning circuit. Thus the patient is always isolated from any ac power source. In the laboratory, function/arbitrary waveform generators such as Hewlett-Packard's HP33120A output an EKG waveform so that you may design and test biomedical equipment without having to monitor a patient.

4-10 WINDOW DETECTOR

4-10.1 Introduction

The circuit of Fig. 4-17 is designed to monitor an input voltage and indicate when this voltage goes either above or below prescribed limits. For example, IC logic power supplies for TTL must be regulated to 5.0 V. If the supply voltage should exceed 5.5 V, the logic may be damaged, and if the supply voltage should drop below 4.5 V, the logic may exhibit marginal operation. Therefore, the limits for TTL power supplies are 4.5 V and 5.5 V. The power supply should be looking through a window whose limits are 4.5 V and 5.5 V, hence the name *window detector*. This circuit is sometimes called a *double-ended limit detector*.

In Fig. 4-17 input voltage E_i is connected to the $(-)$ input of comparator A and the $(+)$ input of comparator B. Upper limit V_{UT} is applied to the $(+)$ input of A, while lower limit V_{LT} is applied to the $(-)$ input of B. When E_i lies between V_{LT} and V_{UT}, the light/alarm is off, but when E_i drops below V_{LT} or goes above V_{UT}, the light/alarm goes on to signify that E_i is not between the prescribed limits.

4-10.2 Circuit Operation

Circuit operation is as follows. Assume that $E_i = 5$ V. Since E_i is greater than V_{LT} and less than V_{UT}, the output voltage of both comparators is at V^{++} because both output switches are open. The lamp/alarm is off. Next, assume that $E_i = 6.0$ V or $E_i > V_{UT}$. The input at pin 3 of A is more positive than at pin 2, so the A output is at the potential of pin 1 or ground. This ground lights the lamp, and $V_o = 0$ V. Now assume that E_i drops to 4.0 V or $E_i < V_{LT}$. The $(+)$ input of B is less than its $(-)$ input, so the B output goes to 0 V (the voltage at its pin 1). Once again, this ground causes the lamp/alarm to light. Note that this application shows that output pins of 311 can be connected together and the output is at V^{++} only when the output of each comparator is at V^{++}.

4-11 PROPAGATION DELAY

4-11.1 Definition

Suppose that a signal E_i is applied to the input of a comparator as in Fig. 4-18. There will be a measurable time interval for the signal to propagate through all the transistors within the comparator. After this time interval the output begins to change. This time interval is called *response time, transit time,* or *propagation delay.*

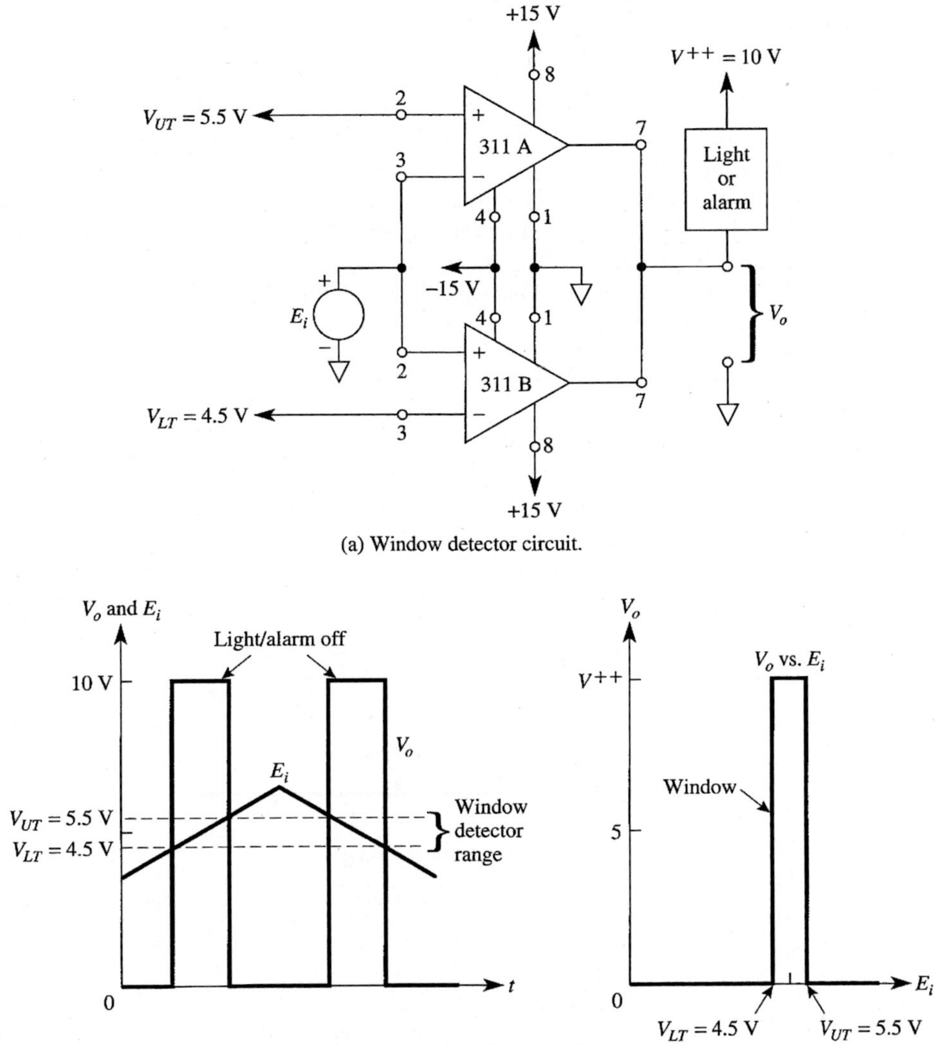

(a) Window detector circuit.

(b) Waveshapes for window detector.

FIGURE 4-17 Upper- and lower-threshold voltages are independently adjustable in the window detector circuit.

Before the signal is applied, the comparator is in saturation. This means that some of its internal transistors contain an excess amount of charge. It is the time required to clean out these charges that is primarily responsible for propagation delay.

4-11.2 Measurement of Propagation Delay

The comparator's depth of saturation depends directly on the amount of differential input voltage. The conventional method of comparing performance of one comparator with another is to first connect a +100-mV reference voltage to one input. In Fig. 4-18(a) the reference voltage is connected to the (−) input. The other (+) input is connected to 0 V. This forces all comparators, under test for propagation delay, into the same initial state of saturation. In Fig. 4-18(b) the outputs are shown at about 0.4 V before time 0.

A fast-rising signal voltage E_i is then applied to the (+) input at time $t = 0$ in Fig. 4-18. If E_i is brought up to 100 mV, the comparator will be on the verge of switching but

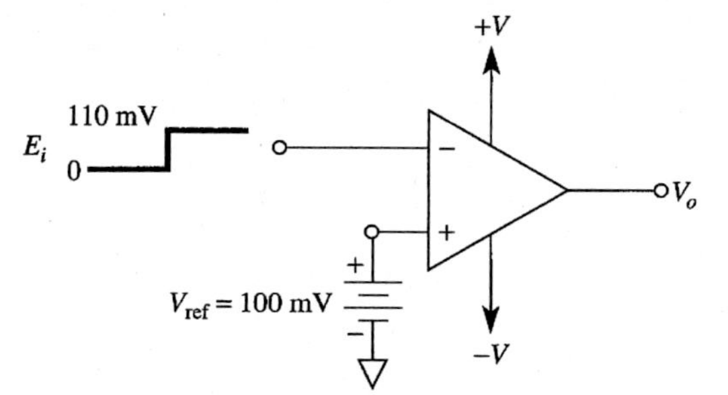

(a) Test circuit for propagation delay.

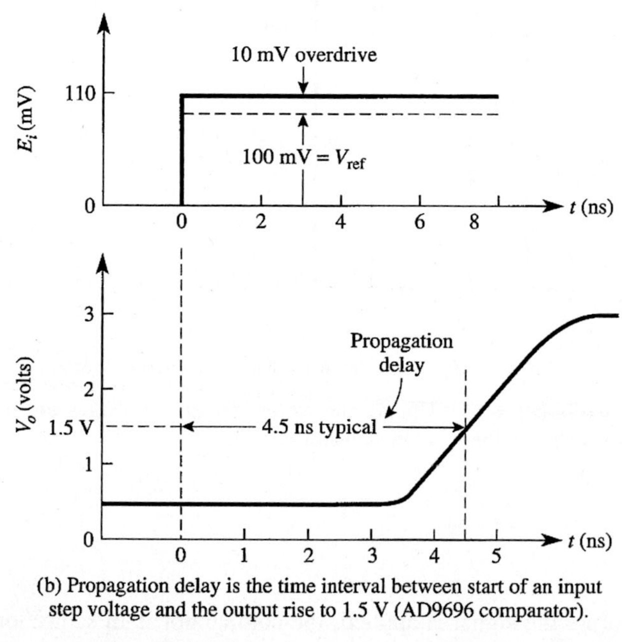

(b) Propagation delay is the time interval between start of an input step voltage and the output rise to 1.5 V (AD9696 comparator).

FIGURE 4-18 Propagation delay is measured by the test circuit in (a) and defined by the waveshapes in (b).

will *not* switch. However, if E_i is brought up quickly to 100 mV plus a small amount of *overdrive,* the overdrive signal will propagate through the comparator. After a propagation delay the output comes out of saturation and rises to a specified voltage. This voltage is typically 1.5 V.

As shown in Fig. 4-18(b), a 5-mV overdrive results in a propagation delay of 4.5 ns for an AD9696 comparator. Typical response times for the 311, 522, and 710 comparators and the 301 general-purpose op amp are

Comparator	Response time for 5-mV overdrive (ns)	Response time for 20-mV overdrive (ns)
311[a]	170	100
522	17	15
710	40	20
301	>10,000	>10,000

[a] $V^{++} = 5$ V with a 500-Ω pull-up resistor.

4-12 USING PSPICE TO MODEL AND SIMULATE COMPARATOR CIRCUITS

In this section, we will use PSpice and simulate the performance of two comparator circuits—the zero-crossing detector with hysteresis and the window detector. In the laboratory, you can test the performance of each and compare it to the simulation so that you can draw conclusions about both. Your version of PSpice may not have the 301 op amp or the 311 comparator listed, so we will use the 741 op amp and the LM111 ICs in their place. The input signal to these circuits will be a triangle wave, and you will learn how to set the attributes of a pulse wave to create a triangle wave as well as how to plot V_o versus E_i.

4-12.1 Simulation of the Zero-Crossing Detector with Hysteresis

Refer to Fig. 4-3 and create the PSpice model of the circuit. We will set the input voltage of the triangular wave to a peak value of 1 V and a frequency of 500 Hz. We wish to obtain a plot of V_o versus E_i. To begin, place the following parts in the work area.

Draw => Get New Part

Part	Number	Library
=> **uA741**	1	**eval.slb**
=> **VPULSE**	1	**source.slb**
=> **VDC**	2	**source.slb**
=> **R**	2	**analog.slb**
=> **GLOBAL**	4	**port.slb**
=> **AGND**	5	**port.slb**

Arrange the parts as shown in Fig. 4-3. Change the attributes of the parts as given in Fig. 4-3. Now we are going to set the attributes of a pulse wave to generate a triangular wave. Set up the triangular-wave attributes by double-clicking the pulse symbol. In the pop-up box, we need to set values for **DC** (dc offset value), **V1** (minimum input voltage), **V2** (maximum input voltage), **TD** (time delay), **TR** (rise time), **TF** (fall time), **PW** (pulse width), and **PER** (period) as given:

DC => 0 => Save Attr
V1 => −1V => Save Attr => Change Display => Both name and value
V2 => 1V => Save Attr => Change Display => Both name and value
TD => 0 => Save Attr
TR => 1ms => Save Attr
TF => 1ms => Save Attr
PW => 1ns => Save Attr
PER => 2ms => Save Attr

Close the attribute box. Double-click on the lead from the triangle wave to the op amp and label it **Ei**. Double-click on the lead from the output terminal of the op amp and label it ***Vo***. See Figure 4-19.

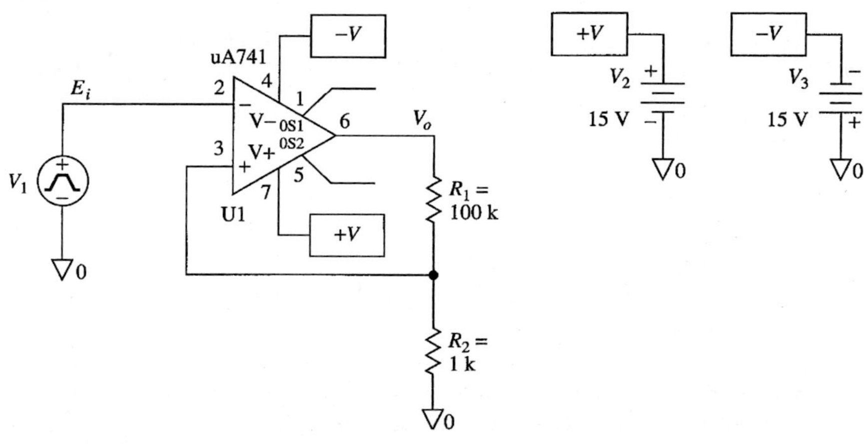

FIGURE 4-19 PSpice model of Fig. 4-3.

Initialize the Transient menu.
Select

Analysis => Setup = > Transient
Click **Transient => Print Step: => 1μs**
=> Final Time: => 4ms

Save the circuit as a file with the **.SCH** extension. Run the simulation

Analysis => Simulate

In the Probe window, select

Plot => X Axis Settings => User Defined => −1V to 1V => Axis Variable => V[Ei]
Plot => Y Axis Settings => User Defined => −15V to 15V

Now select

Trace => Add => V[Vo]

Obtain a printout as shown in Fig. 4-20.

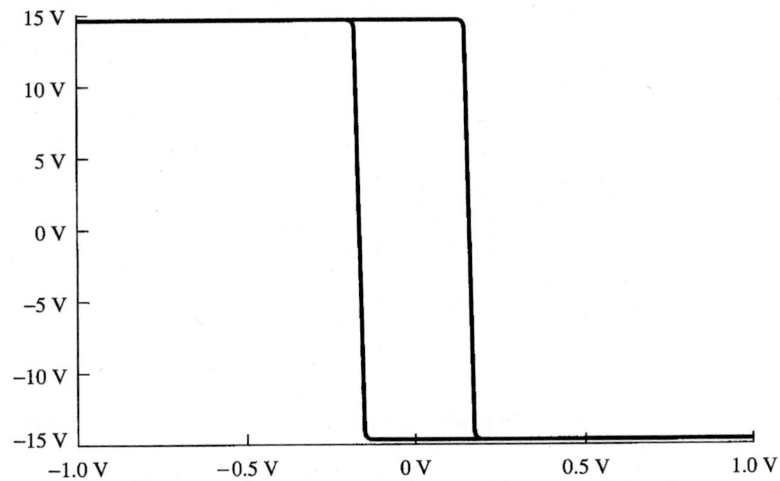

FIGURE 4-20 Plot of V_o vs E_i from PSpice simulation.

4-12.2 Window Detector

In this section, we wish to model and simulate the operation of the window detector circuit shown in Fig. 4-17(a). The input signal is a triangular wave with a peak value of 10 V and a frequency of 500 Hz. Obtain a plot of V_o and E_i versus time as shown in Fig. 4-17(b). Refer to Fig. 4-17(a) and create the PSpice model of the circuit. To begin, place the following parts in the work area.

Draw => Get New Part

Part	Number	Library
=> **LM111**	2	**eval.slb**
=> **VPULSE**	1	**source.slb**
=> **VDC**	5	**source.slb**
=> **R**	1	**analog.slb**
=> **GLOBAL**	7	**port.slb**
=> **AGND**	7	**port.slb**

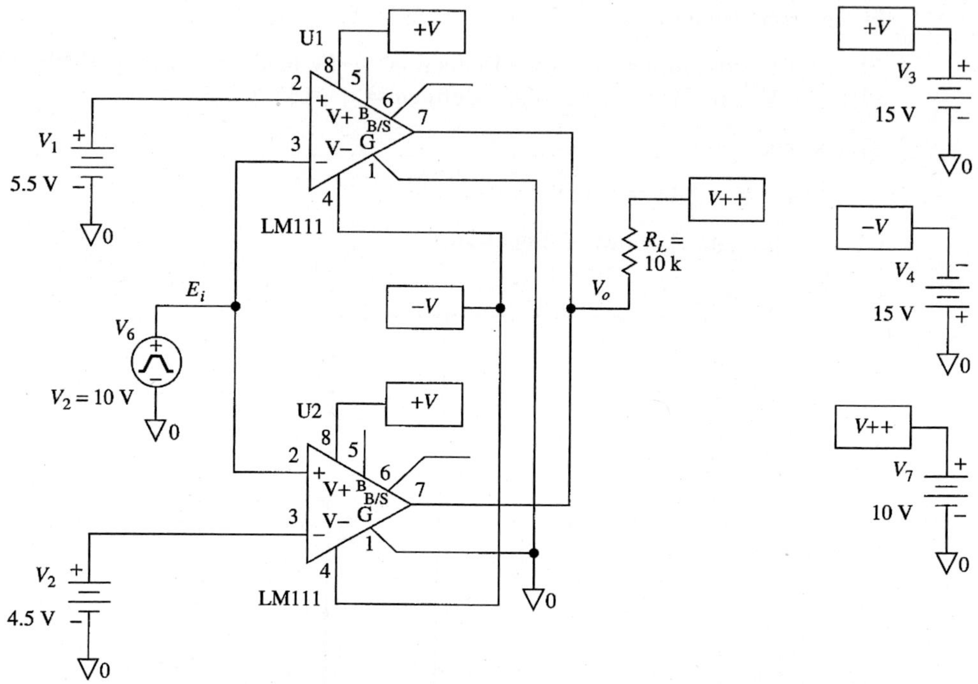

FIGURE 4-21 PSpice model of the window detector circuit of Fig. 4-17(a).

Arrange the parts as shown in Fig. 4-21. Change the attributes of the parts as given in Fig. 4-17(a) with a load resistor set at 10 kΩ. Now we are going to set the attributes of a pulse wave to generate a positive triangular wave. The triangular-wave attributes are set by double-clicking the pulse symbol. In the pop-up box, the attributes **DC** (dc offset value), **V1** (minimum input voltage), **V2** (maximum input voltage), **TD** (time delay), **TR** (rise time), **TF** (fall time), **PW** (pulse width), and **PER** (period) must be set.

DC => 0V => Save Attr
V1 => 0V => Save Attr => Change Display => Both name and value
V2 => 10V => Save Attr => Change Display => Both name and value
TD => 0 => Save Attr
TR => 1ms => Save Attr
TF => 1ms => Save Attr
PW => 1ns => Save Attr
PER => 2ms => Save Attr

Close the attribute box. Double-click on the lead from the input source to the circuit and label it **Ei**. Double-click on the lead from the output terminal of the window detector and label it **Vo**. See Fig. 4-21.

Initialize the Transient menu.

Select

Analysis => Setup => Transient
Click **Transient => Print Step: => 1μs**
=> Final Time: => 2ms

Save the circuit as a file with the **.SCH** extension. Run the simulation

Analysis => Simulate

In the Probe window, select

Trace => Add => V[Ei] and V[Vo]

Obtain a printout as shown in Fig. 4-22.

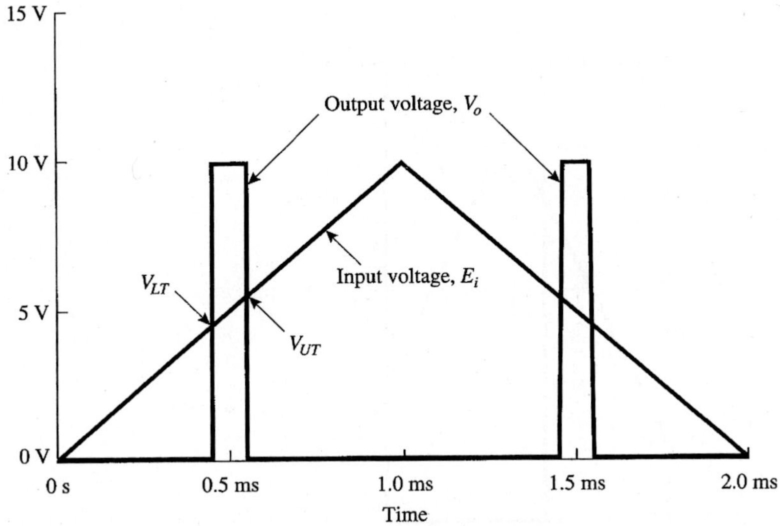

FIGURE 4-22 Input and output waveforms of the window detector using PSpice.

PROBLEMS

4-1. How do you recognize when positive feedback is present in the schematic of an op amp circuit?

4-2. In Fig. P4-2, $R_1 = 25\text{ k}\Omega$ and $R_2 = 5\text{ k}\Omega$. Assume for simplicity that $\pm V_{sat} = \pm 15$ V. Calculate **(a)** V_{UT}; **(b)** V_{LT}; **(c)** V_H.

4-3. For the values given in Problem 4-2, plot **(a)** E_i vs. t; **(b)** V_o vs. t; **(c)** V_o vs. E_i. Let E_i be a 100-Hz triangular wave with peak values of ± 10 V.

4-4. Label V_{UT}, V_{LT}, and V_H on your sketches from Problem 4-3.

4-5. Given the waveshapes of E_i vs. t and V_o vs. t in Fig. P4-5, identify **(a)** the frequency of E_i; **(b)** the peak amplitude of E_i; **(c)** the value of V_{UT}; **(d)** the value of V_{LT}; **(e)** V_H.

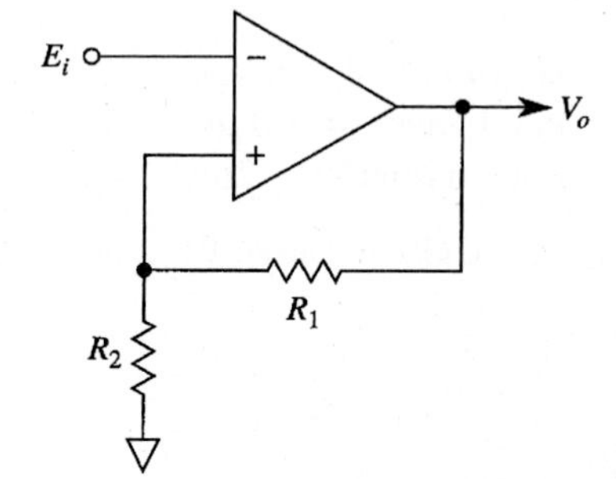

FIGURE P4-2

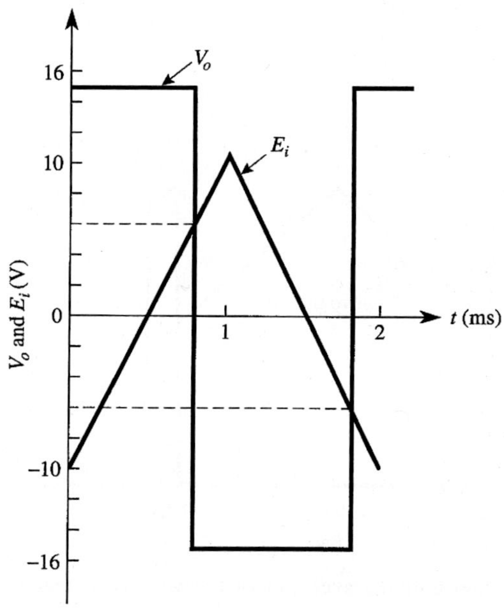

FIGURE P4-5

4-6. Use Fig. 4-7 and Design Example 4-4 for guidance. Design a noninverting voltage-level detector with $V_{UT} = 2.0$ V and $V_{LT} = 0.5$ V.

4-7. To see how negative threshold voltages are handled, redesign the voltage-level detector of Problem 4-6 for $V_{UT} = -0.5$ V and $V_{LT} = -2.0$ V. (Note that $V_H = 1.5$ V in both problems.)

4-8. Refer to Fig. 4-8 and Design Example 4-5 for guidance. Design **(a)** a circuit whose output is at $(+)V_{sat}$ when its input is *below* $V_{LT} = 0.5$ V; **(b)** a circuit whose output is at $(-)V_{sat}$ when its input is above $V_{UT} = 2.0$ V.

4-9. Redesign the circuit of Fig. 4-9 for $V_{UT} = 2.0$ V and $V_{LT} = 0.5$ V (see Design Example 4-6).

4-10. For the circuit of Fig. P4-10 calculate **(a)** V_{ctr}; **(b)** V_H; **(c)** V_{UT}; **(d)** V_{LT}. Assume that $\pm V_{sat} = \pm 15$ V.

4-11. If E_i is grounded in Fig. 4-14, calculate V_o.

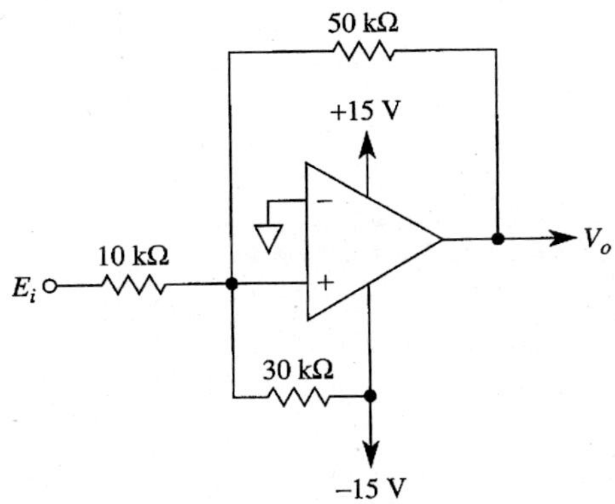

FIGURE P4-10

4-12. Refer to the 311 circuit in Fig. 4-14. The strobe terminal is wired to +15 V. Find the value of V_o when **(a)** $E_i = 1$ V; **(b)** $E_i = -1$ V.

4-13. Repeat Problem 4-12 but with the strobe terminal wired to ground via a 10-kΩ resistor.

4-14. Design a window detector circuit whose output is high when the input voltage is between +2 and +0.5 V.

4-15. Which comparator has a faster response time, the 311 or the 301?

CHAPTER 5

Selected Applications of Op Amps

LEARNING OBJECTIVES

Upon completing this chapter on selected applications of op amps, you will be able to:

- Appreciate how one or two op amps plus a few components can give an inexpensive solution to a number of practical applications.
- Make a universal high-resistance voltmeter.
- Test diodes, LEDs, IREDs, and low-voltage zeners with a constant current.
- Build a circuit that interfaces a 4-to-20 mA transmitter to a microcomputer.
- Measure the power received by a solar cell, photodiode, or photoresistor.
- Explain how to measure solar energy.
- Shift the phase angle of a fixed-frequency sine wave by a precise amount and independent of its amplitude.

- Show how to design a signal conditioning circuit for a Celsius or Fahrenheit thermometer with an AD590 temperature transducer and a current-to-voltage converter.
- Use integrating and differentiating amplifier circuits.
- Use PSpice to simulate circuit performance.

5-0 INTRODUCTION

Why is the op amp such a popular device? This chapter attempts to answer that question by presenting a wide selection of applications. They were selected to show that the op amp can perform as a very nearly ideal device. Moreover, the diversity of operations that the op amp can perform is almost without limit. In fact, applications that are normally very difficult, such as measuring short-circuit current, are rendered easy by the op amp. Together with a few resistors and a power supply, the op amp can, for example, measure the output from photodetectors, give audio tone control, equalize tones of different amplitudes, control high currents, and allow matching of semiconductor device characteristics. We begin with selecting an op amp circuit to make a high-resistance dc and ac voltmeter.

5-1 HIGH-RESISTANCE DC VOLTMETER

5-1.1 Basic Voltage-Measuring Circuit

Figure 5-1 shows a simple but very effective high-input-resistance dc voltmeter. The voltage to be measured, E_i, is applied to the (+) input terminal. Since the differential input voltage is 0 V, E_i is developed across R_i. The meter current I_m is set by E_i and R_i just as in the noninverting amplifier.

$$I_m = \frac{E_i}{R_i} \tag{5-1}$$

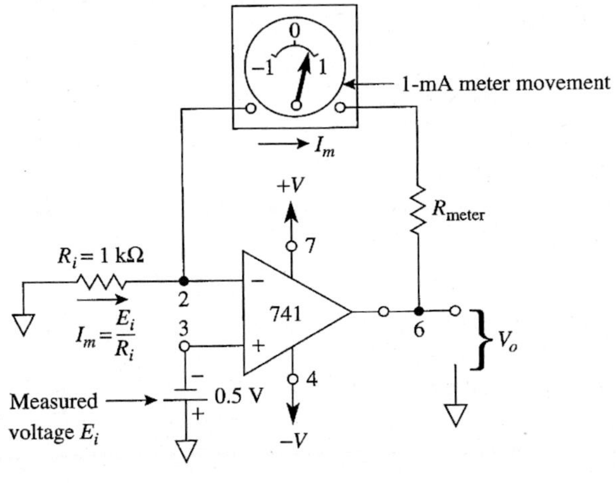

FIGURE 5-1 High-input-resistance dc voltmeter.

If R_i is 1 kΩ, then 1 mA of meter current will flow for $E_i = 1$ V dc. Therefore, the milliammeter can be calibrated directly in volts. As shown, this circuit can measure any dc voltage from −1 V to +1 V.

Example 5-1

Find I_m in Fig. 5-1.

Solution From Eq. (5-1), $I_m = 0.5$ V/1 kΩ $= 0.5$ mA. The needle is deflected halfway between 0 and +1 mA.

One advantage of Fig. 5-1 is that E_i sees the very high input impedance of the (+) input. Since the (+) input draws negligible current, it will not load down or change the voltage being measured. Another advantage of placing the meter in the feedback loop is that if the meter resistance should vary, it will have no effect on meter current. Even if we added a resistor in series with the meter, within the feedback loop, it would not affect I_m, because I_m is set only by E_i and R_i. The output voltage will change if meter resistance changes, but in this circuit we are not concerned with V_o. This circuit is sometimes called a *voltage-to-current* converter.

5-1.2 Voltmeter Scale Changing

Since the input voltage in Fig. 5-1 must be less than the power supply voltages (±15 V), a convenient maximum limit to impose on E_i is ±10 V. The simplest way to convert Fig. 5-1 from a ±1-V voltmeter to a ±10-V voltmeter is to change R_i to 10 kΩ. In other words, pick R_i so that the full-scale input voltage E_{FS} equals R_i times the full-scale meter current I_{FS} or

$$R_i = \frac{E_{FS}}{I_{FS}} \tag{5-2}$$

Example 5-2

A microammeter with 50 μA = I_{FS} is to be used in Fig. 5-1. Calculate R_i for $E_{FS} = 5$ V.

Solution By Eq. (5-2), $R_i = 5$ V/50 μA $= 100$ kΩ.

To measure higher input voltages, use a voltage-divider circuit. The output of the divider is applied to the (+) input.

5-2 UNIVERSAL HIGH-RESISTANCE VOLTMETER

5-2.1 Circuit Operation

The voltage-to-current converter of Fig. 5-2 can be used as a universal voltmeter. That is, it can be used to measure positive or negative dc voltage or the rms, peak, or peak-to-peak (p-p) value of a *sine wave.* To change from one type of voltmeter to another, it is necessary to change only a single resistor. The voltage to be measured, E_i, is applied to the op amp's (+) input. Therefore, the meter circuit has a high input resistance.

When E_i is positive, current flows through the meter movement and diodes D_3 and D_4. When E_i is negative, current flows in the *same* direction through the meter and diodes D_1 and D_2. Thus meter current direction is the same whether E_i is positive or negative.

A dc meter movement measures the *average* value of current. Suppose that a basic meter movement is rated to give full-scale deflection when conducting a current of 50 μA. A voltmeter circuit containing the basic meter movement is to indicate at full scale when E_i is a sine wave with a peak voltage of 5 V. The meter *face* should be calibrated linearly from 0 to +5 V instead of 0 to 50 μA. The *circuit* and meter movement would then be called a *peak reading voltmeter* (for sine waves only) with a full-scale deflection for E_{ip} = 5 V. The following section shows how easy it is to design a universal voltmeter.

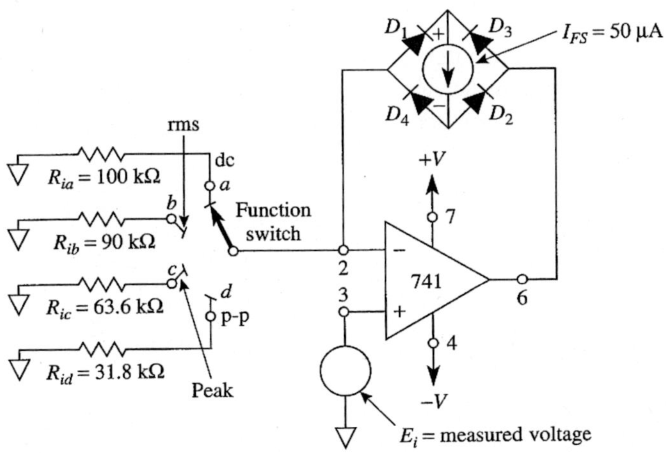

FIGURE 5-2 Basic high-resistance universal voltmeter circuit. The meaning of a full-scale meter deflection depends on the function switch position as follows: 5 V dc on position *a,* 5 V ac rms on position *b,* 5 V peak ac on position *c,* and 5 V ac p-p on position *d.*

5-2.2 Design Procedure

The design procedure is as follows: Calculate R_i according to the application from one of the following equations:

1. Dc voltmeter:

$$R_i = \frac{\text{full-scale } E_{\text{dc}}}{I_{FS}} \tag{5-3a}$$

2. Rms ac voltmeter (sine wave only):

$$R_i = 0.90 \frac{\text{full-scale } E_{\text{rms}}}{I_{FS}} \tag{5-3b}$$

3. Peak reading voltmeter (sine wave only)

$$R_i = 0.636 \frac{\text{full-scale } E_{\text{peak}}}{I_{FS}} \tag{5-3c}$$

4. Peak-to-peak ac voltmeter (sine wave only)

$$R_i = 0.318 \frac{\text{full-scale } E_{\text{p-p}}}{I_{FS}} \tag{5-3d}$$

where I_{FS} is the meter's full-scale current rating in amperes. The design procedure is illustrated by an example.

Design Example 5-3

A basic meter movement (such as the Simpson 260) is rated at 50 μA for full-scale deflection (with a meter resistance of 5 kΩ). Design a simple switching arrangement and select resistors to indicate full-scale deflection when the voltage to be measured is (a) 5 V dc; (b) 5 V rms; (c) 5 V peak; (d) 5 V p-p.

Design Procedure From Eqs. (5-3a) to (5-3d):

(a) $R_{ia} = \dfrac{5\text{ V}}{50\ \mu\text{A}} = 100\text{ k}\Omega$ (b) $R_{ib} = 0.9 \dfrac{5\text{ V}}{50\ \mu\text{A}} = 90\text{ k}\Omega$

(c) $R_{ic} = 0.636 \dfrac{5\text{ V}}{50\ \mu\text{A}} = 63.6\text{ k}\Omega$ (d) $R_{id} = 0.318 \dfrac{5\text{ V}}{50\ \mu\text{A}} = 31.8\text{ k}\Omega$

The resulting circuit is shown in Fig. 5-2.

It must be emphasized that neither meter resistance nor diode voltage drops affect meter current. Only R_i and E_i determine average or dc meter current.

5-3 VOLTAGE-TO-CURRENT CONVERTERS: FLOATING LOADS

5-3.1 Voltage Control of Load Current

From Sections 5-1 and 5-2 we learned not just how to make a voltmeter but that current in the feedback loop depends on the input voltage and R_i. There are applications where we need to pass a constant current through a load and hold it constant despite any changes in load resistance or load voltage. If the load does not have to be grounded, we simply place the load in the feedback loop and control both input and load current by the principle developed in Section 5-1.

5-3.2 Zener Diode Tester

Suppose that we have to test the breakdown voltage of a number of zener diodes at a current of precisely 5 mA. If we connect the zener in the feedback loop as in Fig. 5-3(a), our voltmeter circuit of Fig. 5-1 becomes a zener diode tester. That is, E_i and R_i set the load or zener current at a constant value. E_i forces V_o to go negative until the zener breaks down and clamps the zener voltage at V_Z. R_i converts E_i to a current, and as long as R_i and E_i are constant, the load current will be constant regardless of the value of the zener voltage. Zener breakdown voltage can be calculated from V_o and E_i as $V_Z = V_o - E_i$.

Example 5-4

In the circuit of Fig. 5-3(a), $V_o = 10.3$ V, $E_i = 5$ V, and $R_i = 1$ kΩ. Find (a) the zener current; (b) the zener voltage.

Solution (a) From Eq. (3-1), $I = E_i/R_i$ or $I = 5\text{ V}/1\text{ k}\Omega = 5$ mA. (b) From Fig. 5-3(a), rewrite the equation for V_o.

$$V_Z = V_o - E_i = 10.3\text{ V} - 5\text{ V} = 5.3\text{ V}$$

5-3.3 Diode Tester

Suppose that we needed to select diodes from a production batch and find pairs with matching voltage drops at a particular value of diode current. Place the diode in the feedback loop as shown in Fig. 5-3(b). E_i and R_i will set the value of I. The (−) input draws negligible current, so I passes through the diode. As long as E_i and R_i are constant, current through any diode I will be constant at $I = E_i/R_i$. V_o will equal the diode voltage for the same reasons that V_o was equal to V_{R_f} in the inverting amplifier (see Section 3-1).

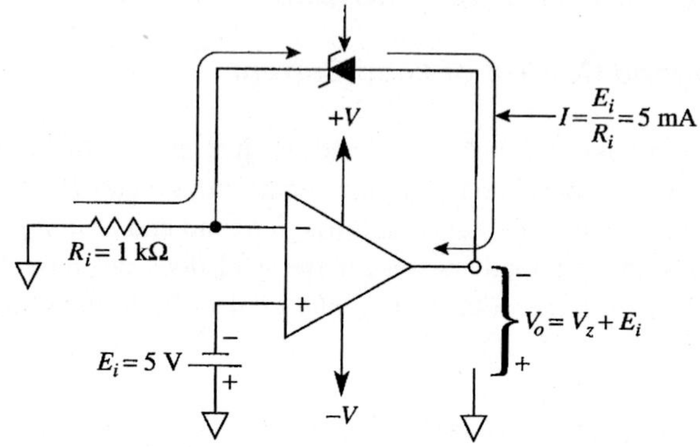

(a) Negligible current drawn from E_i, load current furnished by op amp.

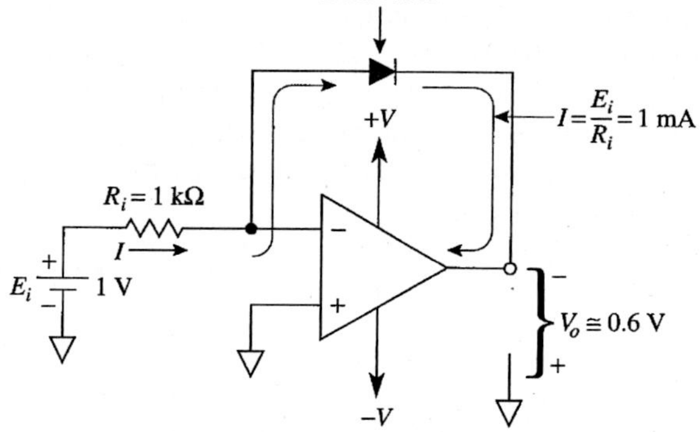

(b) Load current equals input current.

FIGURE 5-3 Voltage-controlled load currents with loads in the feedback loop.

Example 5-5

$E_i = 1$ V, $R_i = 1$ kΩ, and $V_o = 0.6$ V in Fig. 5-3(b). Find (a) the diode current; (b) the voltage drop across the diode.

Solution (a) $I = E_i/R_i = 1$ V/1 kΩ = 1 mA. (b) $V_{\text{diode}} = V_o = 0.6$ V (with polarity as shown).

There is one disadvantage with the circuit of Fig. 5-3(b): E_i must be able to furnish the current. Both circuits in Fig. 5-3 can only furnish currents up to 10 mA because of the op amp's output current limitation. Higher-load currents can be furnished from the power supply terminal and a current boost transistor as shown in Fig. 5-4.

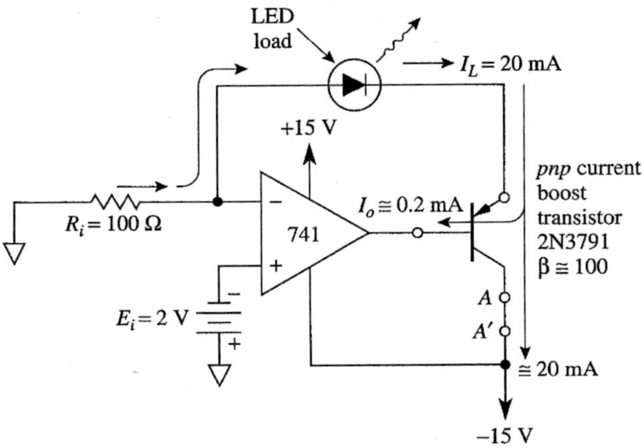

FIGURE 5-4 Voltage to high-current converter.

5-4 LIGHT-EMITTING-DIODE TESTER

The circuit of Fig. 5-4 converts E_i to a 20-mA load current based on the same principles discussed in Sections 5-1 to 5-3. Since the 741's output terminal can only supply about 5 to 10 mA, we cannot use the circuits of Figs. 5-1 to 5-3 for higher load currents, but if we add a transistor as in Fig. 5-4, load current is furnished from the negative supply voltage. The op amp's output terminal is required to furnish only base current, which is typically 1/100 of the load current. The factor 1/100 comes from assuming that the transistor's beta equals 100. Since the op amp can furnish an output current of up to 5 mA into the transistor's base, this circuit can supply a maximum load current of 5 mA $\times$ 100 = 0.5 A.

A light-emitting diode such as the MLED50 is specified to have a typical brightness of 750 fL (foot lamberts) provided that the forward diode current is 20 mA. E_i and R_i will set the diode current I_L equal to $E_i/R_i = 2\text{ V}/100\ \Omega = 20$ mA. Now brightness of LEDs can be measured easily one after another for test or matching purposes, because the current through each diode will be exactly 20 mA regardless of the LED's forward voltage.

It is worthwhile to note that a load of two LEDs can be connected in series with the feedback loop and both would conduct 20 mA. The load could also be connected in Fig. 5-4 between points *AA′* (which is in series with the transistor's collector) and still conduct about 20 mA. This is because the collector and emitter currents of a transistor are essentially equal. A load in the feedback loop is called a *floating load.* If one side of the load is grounded, it is a *grounded load.* To supply a constant current to a grounded load, another type of circuit must be selected, as shown in Section 5-5.

5-5 FURNISHING A CONSTANT CURRENT TO A GROUNDED LOAD

5-5.1 Differential Voltage-to-Current Converter

The circuit of Fig. 5-5 can be called a differential voltage-to-current converter because the load current I_L depends on the *difference* between input voltages E_1 and E_2 and resistor R. I_L does *not* depend on load resistor R_L. Therefore, if E_1 and E_2 are constant, the grounded load is driven by a constant current. Load current can flow in either direction, so this circuit can either source or sink current.

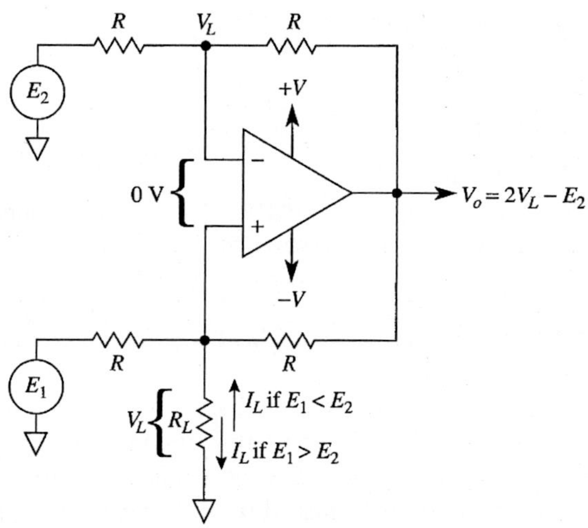

FIGURE 5-5 Differential voltage-to-current converter or constant-current source with grounded load.

Load current I_L is determined by

$$I_L = \frac{E_1 - E_2}{R} \tag{5-4}$$

A positive value for I_L signifies that it flows downward in Fig. 5-5 and V_L is positive with respect to ground. A negative value of I_L means that V_L is negative with respect to ground and current flows upward.

Load voltage V_L (not I_L) depends on load resistor R_L from

$$V_L = I_L R_L \tag{5-5}$$

To ensure that the op amp does not saturate, V_o must be known and can be calculated from

$$V_o = 2V_L - E_2 \tag{5-6}$$

Circuit operation is illustrated by the following examples.

Example 5-6

In Fig. 5-5, $R = 10\ \text{k}\Omega$, $E_2 = 0$, $R_L = 5\ \text{k}\Omega$, and $E_1 = 5$ V. Find (a) I_L; (b) V_L; (c) V_o.

Solution (a) From Eq. (5-4),

$$I_L = \frac{5\ \text{V} - 0}{10\ \text{k}\Omega} = 0.5\ \text{mA}$$

(b) From Eq. (5-5),

$$V_L = 0.5\ \text{mA} \times 5\ \text{k}\Omega = 2.5\ \text{V}$$

(c) From Eq. (5-6),

$$V_o = 2 \times 2.5\ \text{V} = 5\ \text{V}$$

Reversing the polarity of E_1 reverses I_L and the polarity of V_o and V_L.

Example 5-7

In Fig. 5-5, $R = 10\ \text{k}\Omega$, $E_2 = 5$ V, $R_L = 5\ \text{k}\Omega$, and $E_1 = 0$. Find (a) I_L; (b) V_L; (c) V_o. Compare this example with Example 5-6.

Solution (a) From Eq. (5-4),

$$I_L = \frac{0 - 5\ \text{V}}{10\ \text{k}\Omega} = -0.5\ \text{mA}$$

(b) From Eq. (5-5),

$$V_L = -0.5\ \text{mA} \times 5\ \text{k}\Omega = -2.5\ \text{V}$$

(c) From Eq. (5-6),

$$V_o = 2(-2.5\ \text{V}) - 5\ \text{V} = -10\ \text{V}$$

Note: V_L and I_L are reversed in polarity and direction, respectively, from Example 5-6. If the polarity of E_2 is reversed, I_L and V_L change sign but *not* magnitude.

5-5.2 Constant-High-Current Source, Grounded Load

In certain applications, such as electroplating, it is desirable to furnish a high current, of constant value, to a grounded load. The circuit of Fig. 5-6 will furnish constant currents above 500 mA provided that the transistor is heat-sinked properly (above 5 W) and has a high beta ($\beta > 100$). The circuit operates as follows. The zener diode voltage is applied to one end of current sense resistor R_s and the op amp's positive input. Since the differ-

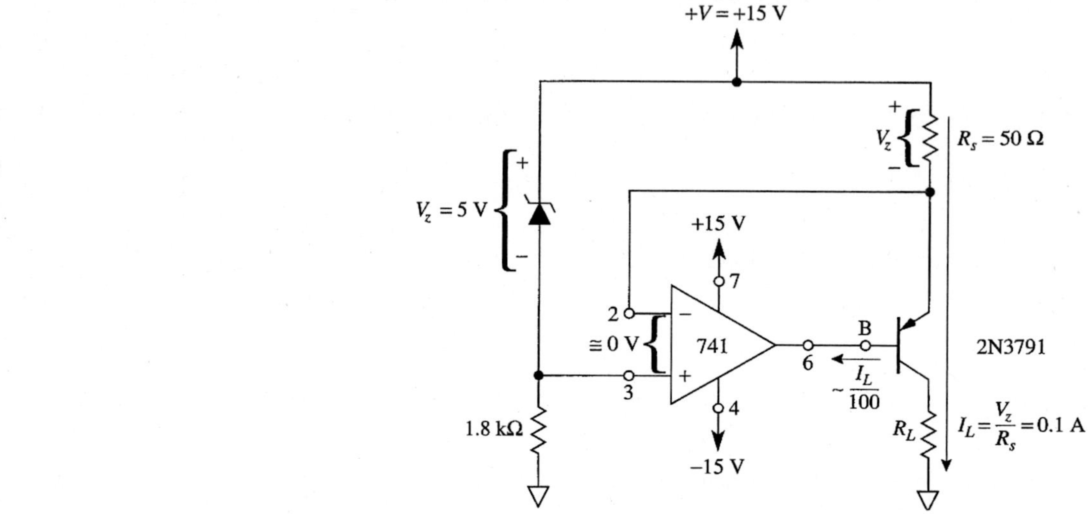

FIGURE 5-6 Constant-high-current source.

ential input voltage is 0 V, the zener voltage is developed across R_s. R_s and V_z set the emitter current, I_E, constant at V_z/R_s. The emitter and collector currents of a bipolar junction transistor are essentially equal. Since the collector current is load current I_L and $I_L \approx I_E$, the load current I_L is set by V_z and R_s.

If the op amp can furnish a base current drive of over 5 mA and if the beta of the transistor is greater than 100, then I_L can exceed 5 mA × 100 = 500 mA. The voltage across the load must not exceed the difference between the supply and the zener voltage; otherwise, the transistor and the op amp will go into saturation. (If oscillations occur, install a 100-Ω resistor between terminals 6 and B.)

5-5.3 Interfacing a Microcontroller Output to a 4-to-20-mA Transmitter

A TTL digital circuit, microprocessor, microcontroller, or microcomputer communicates to the outside world in a binary language that has only two symbols, 0 and 1. Their corresponding electrical voltages are low (<0.8 V) and high (>2.4 V). Many valves, actuators, and other devices commonly used in process control need a serial pulse train of either 20- or 4-mA current pulses. The 20-mA current pulse may be used to energize a selector magnet, and the 4-mA current pulse releases the magnet thus opening and closing a solenoid. Although 4-to-20-mA transmitter devices can be purchased as single ICs, such as the AD694, you may have an application where it is necessary to quickly test a system or only a single interface circuit is needed. The principles of operation for an op amp circuit that will interface between a microcontroller and a 4-to-20-mA current loop are shown in Fig. 5-7 and analyzed in Section 5-5.4.

5-5.4 Digitally Controlled 4-to-20-mA Current Source

In the circuit of Fig. 5-7(a), resistors R_1 and R_2 form an unloaded voltage divider. Since E_d of the op amp is zero volts, the 2-V drop always appears between the positive rail and emitter of the current boost transistor Q_B. Operation of the circuit is summarized in Fig. 5-7(b).

If the 400-Ω resistor is *not* shorted out, the current through emitter resistor R_E (and collector or loop current I) equals 2 V/500 Ω = 4 mA. If the 400-Ω resistor *is* shorted out by Q_2, the loop current I equals 2 V/100 Ω = 20 mA.

The selection of either 4 or 20 mA is determined by E_{in}. E_{in} can be (1) a TTL or (other logic family) open-collector gate circuit output, (2) an output port from a microcontroller, or (3) a discrete (Q_1) bipolar junction transistor. (For a stand-alone circuit add a 2.2-kΩ resistor in series with the base.)

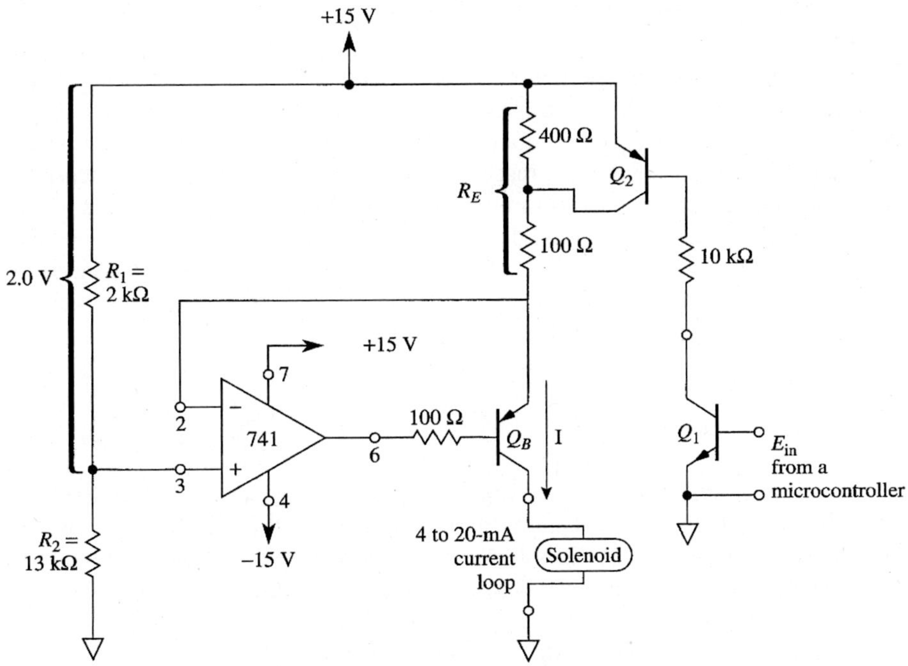

(a)

E_{in}	Q_1	Q_2	I_{out}
High = 1	On	On	2 V/100 Ω = 20 mA
Low = 0	Off	Off	2 V/500 Ω = 4 mA

(b)

FIGURE 5-7 Digitally controlled 4-to-20-mA current source.

When E_{in} is high, Q_1 saturates and I_{CE_1} equals about 1.4 mA. The collector current of Q_1 is the base current of Q_2, and Q_2 saturates. When Q_2 saturates, it effectively shorts out the 400-Ω resistor and fixes Q_B's emitter current and thus the loop current to 20 mA.

When E_{in} is low, Q_1 is cut off, which in turn cuts off Q_2. Q_2 then appears as an open circuit to the 400-Ω resistor, and the loop current, I, is clamped at 2 V/500 Ω = 4 mA. The compliance voltage of this circuit is 12 V. Allow 12 V for V_{R1} plus 1 V to keep Q_B out of saturation [15 − (2 + 1) = 12 V].

5-6 SHORT-CIRCUIT CURRENT MEASUREMENT AND CURRENT-TO-VOLTAGE CONVERSION

5-6.1 Introduction

Transducers such as microphone pickups and solar cells convert some physical quantity into electrical signals. For convenience, the transducers may be modeled by a signal generator as in Fig. 5-8(a). It is often desirable to measure their maximum output current under short-circuit conditions; that is, we should place a short circuit across the output terminals and measure current through the short circuit. This technique is particularly suited to signal sources with very high internal resistance. For example, in Fig. 5-8(a), the short-circuit current I_{SC} should be 2.5 V/50 kΩ = 50 μA. However, if we place a microammeter across the output terminals of the generator, we no longer have a short circuit but a 5000-Ω resistance. The meter indication is

$$\frac{2.5\text{ V}}{50\text{ k}\Omega + 5\text{ k}\Omega} \cong 45\ \mu\text{A}$$

High-resistance sources are better modeled by an equivalent Norton circuit. This model is simply the ideal short-circuit current, I_{SC}, in parallel with its own internal resistance as in Fig. 5-8(b). This figure shows how I_{SC} splits between its internal resistance and the meter resistance. To eliminate this current split, we will use the op amp.

5-6.2 Using the Op Amp to Measure Short-Circuit Current

The op amp circuit of Fig. 5-8(c) effectively places a short circuit around the current source. The (−) input is at virtual ground because the differential input voltage is almost 0 V. The current source "sees" ground potential at both of its terminals, or the equivalent of a short circuit. *All* of I_{SC} flows toward the (−) input and on through R_f. R_f converts I_{SC} to an output voltage, revealing the basic nature of this circuit to be a *current-to-voltage converter.*

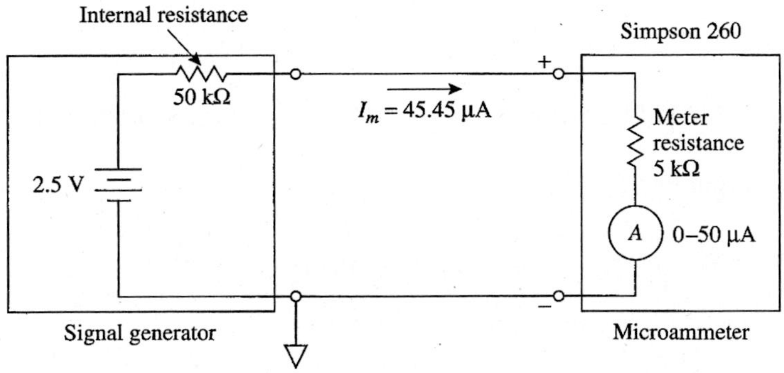

(a) Ammeter resistance reduces short-circuit current from the signal generator.

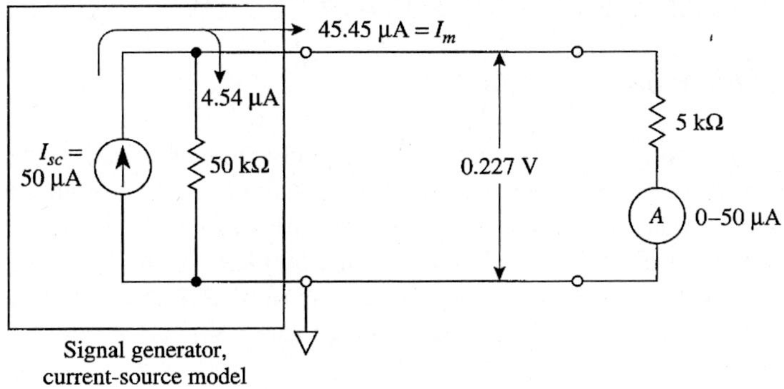

(b) Current-source model of signal generator in (a).

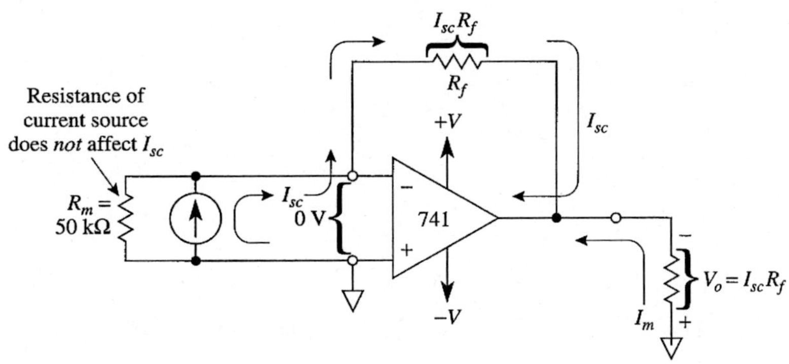

(c) Current-to-voltage converter.

FIGURE 5-8 Current-measuring circuits.

Example 5-8

V_o measures 5 V in Fig. 5-8(c), and $R_f = 100\ \text{k}\Omega$. Find the short-circuit current I_{SC}.

Solution From Fig. 5-8(c),

$$I_{SC} = \frac{V_o}{R_f} = \frac{5\ \text{V}}{100\ \text{k}\Omega} = 50\ \mu\text{A}$$

The resistance R_m is the resistance of the source.

5-7 MEASURING CURRENT FROM PHOTODETECTORS

5-7.1 Photoconductive Cell

With the switch at position 1 in Fig. 5-9, a photoconductive cell, sometimes called a light-sensitive resistor (LSR), is connected in series with the (−) input and E_i. The resistance of a photoconductive cell is very high in darkness and much lower when illuminated. Typically, its dark resistance is greater than 500 kΩ and its light resistance in bright sun is approximately 5 kΩ. If $E_i = 5$ V, then current through the photoconductive cell, I, would be 5 V/500 kΩ = 10 μA in darkness and 5 V/5 kΩ = 1 mA in sunlight.

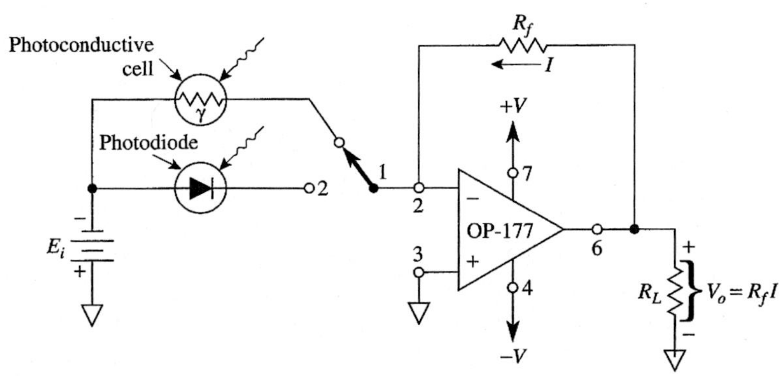

FIGURE 5-9 Using the op amp to measure output current from photodetectors.

Example 5-9

In Fig. 5-9 the switch is in position 1 and $R_f = 10\ \text{k}\Omega$. If the current through the photoconductive cell is 10 μA in darkness and 1 mA in sunlight, find V_o for (a) the dark condition; (b) the light condition.

Solution From Fig. 5-9 $V_o = R_f I$. (a) V_o = 10 kΩ × 10 μA = 0.1 V; (b) V_o = 10 kΩ × 1 mA = 10 V. Thus the circuit of Fig. 5-9 converts the output current from the photoconductive cell into an output voltage (a current-to-voltage converter).

5-7.2 Photodiode

When the switch is in position 2 in Fig. 5-9, E_i is on one side of the photodiode and virtual ground on the other. The photodiode is reverse biased, as it must be for normal operation. In darkness the photodiode conducts a small leakage current on the order of nanoamperes, but depending on the radiant energy striking the diode, it will conduct 50 μA or more. Therefore, current I depends only on the energy striking the photodiode and not on E_i. This current is converted to a voltage by R_f.

Example 5-10

With the switch in position 2 in Fig. 5-9 and R_f = 100 kΩ, find V_o as the light changes photodiode current from (a) 1 μA to (b) 50 μA.

Solution From $V_o = R_f I_L$, (a) V_o = 100 kΩ × 1 μA = 0.1 V; (b) V_o = 100 kΩ × 50 μA = 5.0 V.

5-8 CURRENT AMPLIFIER

Characteristics of high-resistance signal sources were introduced in Section 5-6.1. There is no point in converting a current into an equal current, but a circuit that converts a small current into a large current can be very useful. The circuit of Fig. 5-10 is a current multiplier or current amplifier (technically, a current-to-current converter). The signal current source I_{SC} is effectively short-circuited by the input terminals of the op amp. All of I_{SC} flows through resistor mR, and the voltage across it is mRI_{SC}. (Resistor mR is known as a multiplying resistor and m as the multiplier.) Since R and mR are in parallel, the voltage across R is also mRI_{SC}. Therefore, the current through R must be mI_{SC}. Both currents add to form load current I_L. I_L is an amplified version of I_{SC} and is found from

$$I_L = (1 + m)I_{SC} \tag{5-7}$$

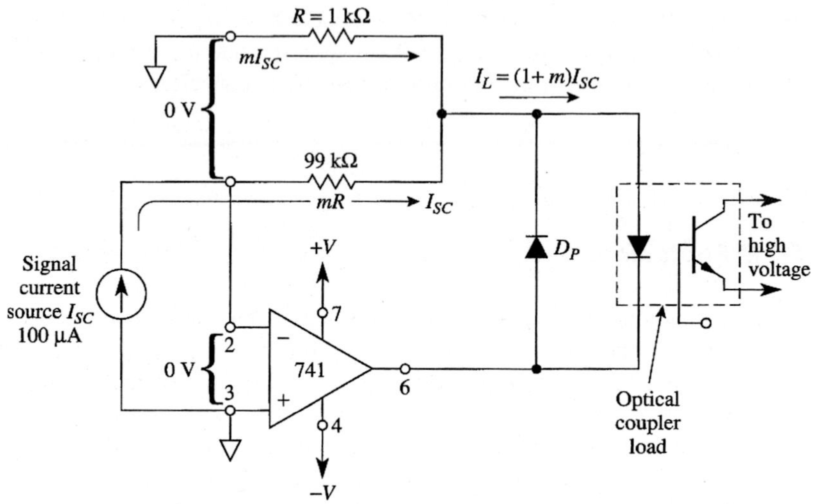

FIGURE 5-10 Current amplifier with optical coupler load.

Analysis Example 5-11

In Fig. 5-10, $R = 1\ \text{k}\Omega$ and $mR = 99\ \text{k}\Omega$. Therefore, $m = 99\ \text{k}\Omega/1\ \text{k}\Omega = 99$. Find the current I_L through the emitting diode of the optical coupler.

Solution By Eq. (5-7), $I_L = (1 + 99)(10\ \mu\text{A}) = 1.0\ \text{mA}$.

It is important to note that the load does not determine load current. Only the multiplier m and I_{SC} determine load current. For variable current gain, mR and R can be replaced by a single 100-kΩ potentiometer. The wiper goes to the emitting diode, one end to ground and the other end to the (−) input. The optical coupler isolates the op amp circuit from any high-voltage load. D_P is an ordinary silicon diode that protects the emitting diode against a reverse bias voltage.

5-9 SOLAR CELL ENERGY MEASUREMENTS

5-9.1 Introduction to the Problems

A solar cell (also called a photovoltaic cell) is a device that converts light energy directly into electrical energy. The best way to record the amount of power received by the solar cell is to measure its short-circuit current. For example, one type of solar cell furnishes a short-circuit current I_{SC} that ranges from 0 to 0.5 A as sunlight varies from complete darkness to maximum brightness.

One problem facing users of these devices is to convert the 0 to 0.5-A solar cell output current from 0 to 10 V so that its performance can be monitored with a strip-chart recorder. Another problem is to measure 1/2 A of current with a low-current meter movement (0 to 0.1 mA). To solve this problem, I_{SC} must be divided so that it can be measured on site with an inexpensive basic meter movement. The final problem is that the value of I_{SC} is too large to be used with the op amp circuits studied thus far in this chapter.

5-9.2 Converting Solar Cell Short-Circuit Current to a Voltage

The circuit of Fig. 5-11 solves several problems. First, the solar cell sees the (−) input of the op amp as a virtual ground. Therefore, it can deliver its short-circuit current I_{SC}. A second problem is solved when I_{SC} is converted by R_f to a voltage V_o. To obtain a 0- to 10-V output for a 0- to 0.5-A input, R_f should have a value of

$$R_f = \frac{V_o \text{ (full scale)}}{I_{SC} \text{ (max)}} = \frac{10 \text{ V}}{0.5 \text{ A}} = 20.0 \ \Omega$$

V_o should be buffered by a voltage follower. The solar cell current of 0.5 A is too large to be handled by the op amp. This problem is solved by adding an *npn* current boost transistor.

The solar cell current flows through the emitter and collector of the boost transistor to + *V.* Current gain of the transistor should exceed $\beta = 100$ to ensure that the op amp has to furnish no more than 0.5 A/100 = 5 mA, when $I_{SC} = 0.5$ A.

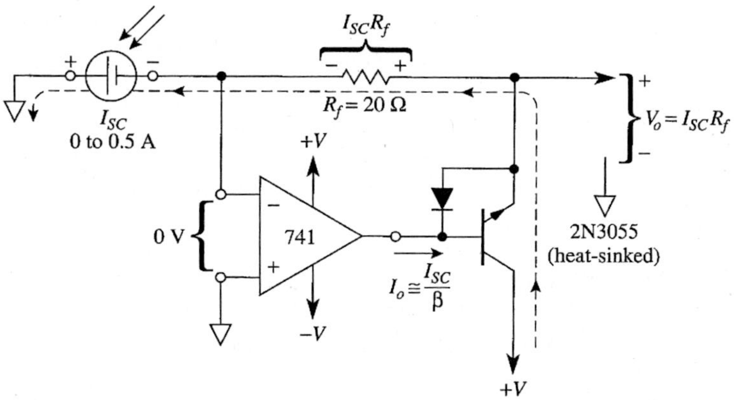

FIGURE 5-11 This circuit forces the solar cell to deliver a short-circuit current I_{SC}. I_{SC} is converted to a voltage by R_f. Current boost is furnished by the *npn* transistor. The diode protects the base-emitter junction of the 2N3055 against accidental excessive reverse bias.

5-9.3 Current-Divider Circuit (Current-to-Current Converter)

Only a slight addition to the circuit of Fig. 5-11 allows us to measure I_{SC} with a low-current milliammeter or microammeter. The current-divider resistance dR_f is shown in Fig. 5-12. Resistance dR_f is made up of the meter resistance R_m plus the scale resistor R_{scale}.

The short-circuit current develops a voltage drop across R_f equal to V_o. V_o is also equal to the voltage across resistance dR_f. Thus the current divider d can be found by equating the voltage drops across dR_f and R_f.

$$V_o = I_{SC}\, R_f = I_m dR_f \tag{5-8a}$$

so

$$d = \frac{I_{SC}}{I_m} \tag{5-8b}$$

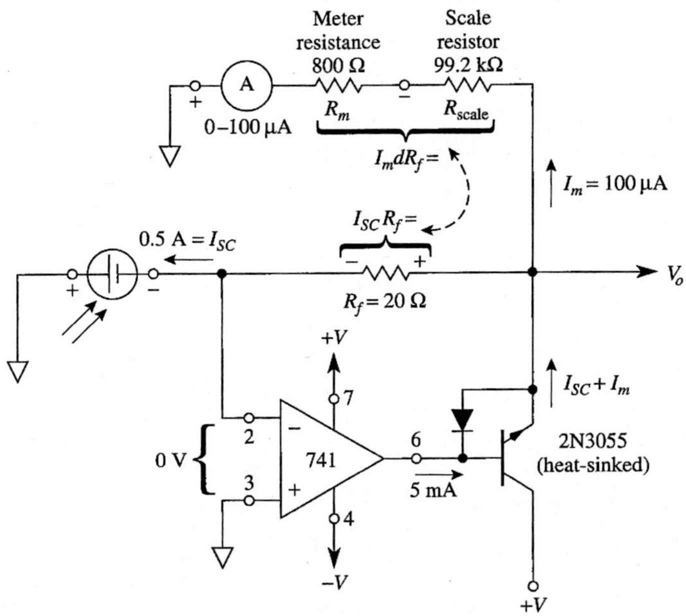

FIGURE 5-12 Current-to-current converter. Divider resistance dR_f equals the sum of meter resistance R_m and scale resistance R_{scale}. Short-circuit current $I_{SC} = 0.5$ A is converted by d down to 100 μA for measurement by a low-current meter.

Design Example 5-12

If the meter in Fig. 5-12 is to indicate full scale at $I_m = 100$ μA when $I_{SC} = 0.5$ A, find resistance dR_f and R_{scale}.

Design Procedure From Eq. (5-8b), $d = 0.5\ \text{A}/100\ \mu\text{A} = 5000\ \Omega$, and $dR_f = 5000 \times 20\ \Omega = 100\ \text{k}\Omega$. Then $R_{\text{scale}} = dR_f - R_m = 100\ \text{k}\Omega - 0.8\ \text{k}\Omega = 99.2\ \text{k}\Omega$.

5-10 PHASE SHIFTER

5-10.1 Introduction

An ideal phase-shifting circuit should transmit a wave without changing its amplitude but should change its phase angle by a preset amount. For example, a sine wave E_i with a frequency of 1 kHz and peak value of 1 V is the input of the phase shifter in Fig. 5-13(a).

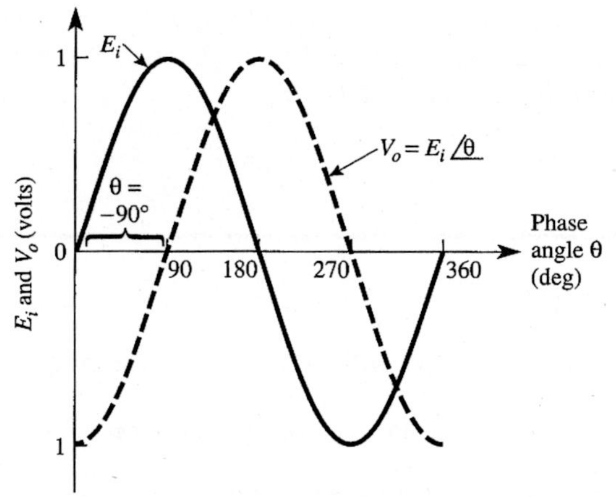

(a) Input and output voltages for θ = −90°.

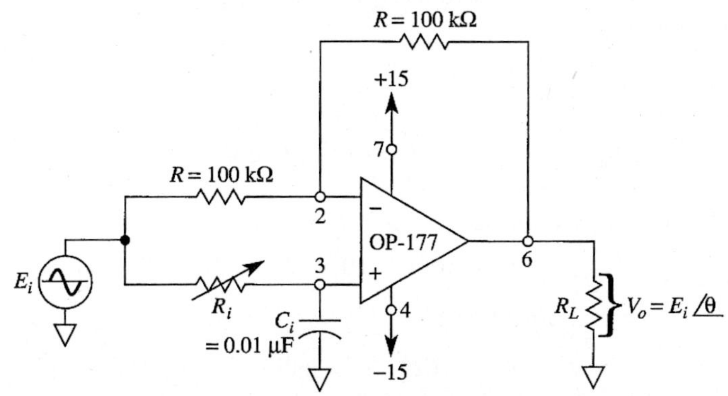

(b) Phase shifter circuit.

FIGURE 5-13 Phase shifter.

The output V_o has the same frequency and amplitude but lags E_i by 90°. That is, V_o goes through 0 V $\underline{/90°}$ *after* E_i goes through 0 V. Mathematically, V_o can be expressed by $V_o = E_i \underline{/-90°}$. A general expression for the output voltage of the phase-shifter circuit in Fig. 5-13(b) is given by

$$V_o = E_i \underline{/-\theta} \tag{5-9}$$

where θ is the phase angle and will be found from Eq. (5-10a).

5-10.2 Phase-Shifter Circuit

As shown in Fig. 5-13(b) one op amp, three resistors, and one capacitor are all that is required to make an excellent phase shifter. The resistors R must be equal, and any convenient value from 10 to 220 kΩ may be used. Phase angle θ depends only on R_i, C_i, and the frequency f of E_i. The relationship is

$$\theta = 2 \arctan 2\pi f R_i C_i \tag{5-10a}$$

where θ is in degrees, f in hertz, R_i in ohms, and C_i in farads. Equation (5-10a) is useful to find the phase angle if f, R_i, and C_i are known. If the desired phase angle is known, choose a value for C_i and solve for R_i:

$$R_i = \frac{\tan(\theta/2)}{2\pi f C_i} \tag{5-10b}$$

Design Example 5-13

Find R_i in Fig. 5-13(b) so that V_o will lag E_i by 90°. The frequency of E_i is 1 kHz.

Design Procedure Since $\theta = 90°$, tan (90°/2) = tan (45°) = 1; from Eq. (5-10b),

$$R_i = \frac{1}{2\pi \times 1000 \times 0.01 \times 10^{-6}} = 15.9 \text{ k}\Omega$$

With $R_i = 15.9$ kΩ, V_o will have the phase angle shown in Fig. 5-13(a). This waveform is a negative cosine wave.

Analysis Example 5-14

If $R_i = 100$ kΩ in Fig. 5-13(b), find the phase angle θ.

Solution From Eq. (5-10a),

$$\begin{aligned}\theta &= 2 \arctan (2\pi)(1 \times 10^3)(100 \times 10^3)(0.01 \times 10^{-6}) \\ &= 2 \arctan 6.28 \\ &= 2 \times 81° = 162° \qquad \text{and} \qquad V_o = E_i \underline{/-162°}\end{aligned}$$

It can be shown from Eq. (5-10a) that $\theta = -90°$ when R_i equals the reactance of C_i, or $1/(2\pi f C_i)$. As R_i is varied from 1 kΩ to 100 kΩ, θ varies from approximately $-12°$ to $-168°$. Thus the phase shifter can shift phase angles over a range approaching 180°. If R_i and C_i are interchanged in Fig. 5-12(b), the phase angle is positive, and the circuit becomes a leading phase-angle shifter. The magnitude of θ is found from Eq. (5-10a), but the output is given by $V_o = E_i \angle 180° - \theta$.

5-11 TEMPERATURE-TO-VOLTAGE CONVERTERS

5-11.1 AD590 Temperature Transducer

An electronic thermometer can be made from a temperature transducer, an op amp, and resistors. We shall select the AD590, manufactured by Analog Devices, as the temperature transducer. The AD590 converts its ambient temperature in degrees Kelvin into an output current, I_T, that is 1 μA for every degree Kelvin. In terms of Celsius temperatures, $I_T = 273\ \mu$A at 0°C (273°K) and 373 μA at 100°C (373°K). In terms of Fahrenheit temperatures, $I_T = 255\ \mu$A at 0°F and 310 μA at 100°F. Thus the AD590 acts as a current

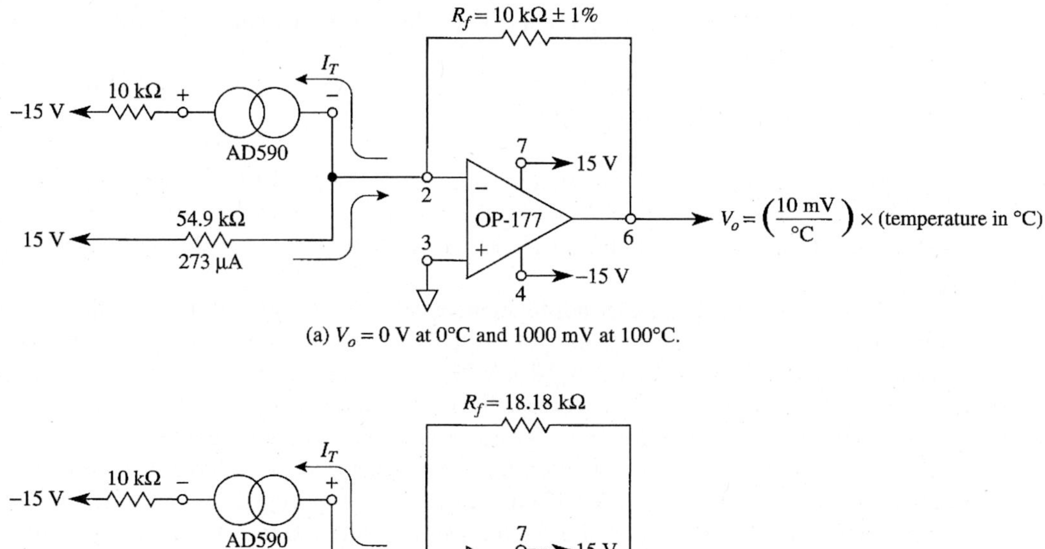

(a) V_o = 0 V at 0°C and 1000 mV at 100°C.

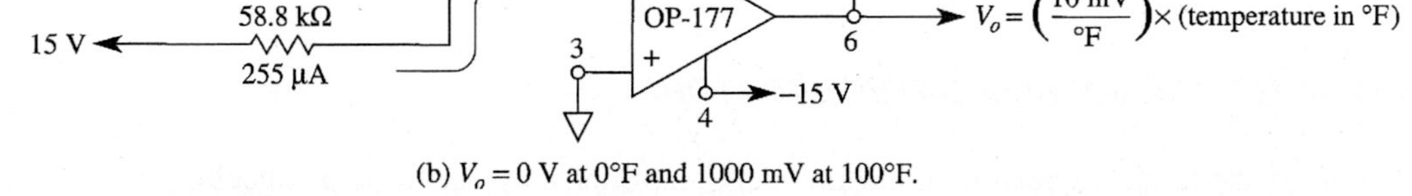

(b) V_o = 0 V at 0°F and 1000 mV at 100°F.

FIGURE 5-14 Temperature-to-voltage converters for (a) Celsius degrees and (b) Fahrenheit degrees.

source that depends on temperature. If, however, we need a voltage reading to indicate temperature, such as 10 mV/°C or 10 mV/°F, a current-to-voltage converter circuit is required.

The circuit symbol for an AD590 is the same as a current source, as shown in Figs. 5-14(a) and (b). Also, the AD590 requires a supply voltage exceeding 4 V to bias its internal transistor circuitry. Let's use this device to build a Celsius or Fahrenheit thermometer.

5-11.2 Celsius Thermometer

In the Celsius thermometer of Fig. 5-14(a), all of the AD590's current is steered into the virtual ground at pin 2 and flows through the 10-kΩ feedback resistor, producing a voltage drop equal to V_o. Each microampere of current thus causes V_o to go positive by 1 μA $\times$ 10 kΩ = 10 mV. A change of 1°C causes I_T to change by 1 μA and consequently produces a change in V_o of 10 mV. The temperature-to-voltage converter thus has a conversion gain of 10 mV/°C.

At 0°C, $I_T = 273$ μA. But we want V_o to equal zero volts. For this reason, an equal and opposite current of 273 μA through the 15-V supply and 54.9-kΩ resistor is required. This results in the net current through R_f to be zero and thus V_o is zero volts. For every increase of 1 μA/°C above 0°C, the net current through R_f increases by 1 μA and V_o increases by 10 mV.

5-11.3 Fahrenheit Thermometer

A circuit for a Fahrenheit thermometer is shown in Fig. 5-14(b). At 0°F we want V_o to equal 0 V. Since $I_T = 255$ μA at 0°F, it must be nulled out by an equal but opposite current through R_f. This current is generated by the 15-V supply and the 58.8-kΩ resistor.

An increase of 1°F corresponds to an increase of 5/9°C or 0.555°C. Thus the AD590 increases its output current by 0.555 μA/°F. This increase is converted by R_f into a voltage of 0.555 μA/°F $\times$ 18.18 kΩ = 10 mV/°F. In conclusion, for every temperature increase of 1°F above 0°F, V_o will increase by 10 mV above 0 V.

Both op amp circuits of Fig. 5-14 are classified as signal conditioning circuits (SCCs). They are designed to condition the input (what you *got*—0 to 100°C or F) to an output (what you *want*—0 to 1 V). If your application requires the output of the SCC to be an input to an A/D converter of a microcontroller with a 0-to-5-V range, then R_f must be changed. (See Problem 5-22.)

5-12 INTEGRATORS AND DIFFERENTIATORS

If the feedback resistor in an inverting amplifier is replaced by a capacitor as shown in Fig. 5.15(a), the new op amp circuit is known as an integrator. Integrators are used in the

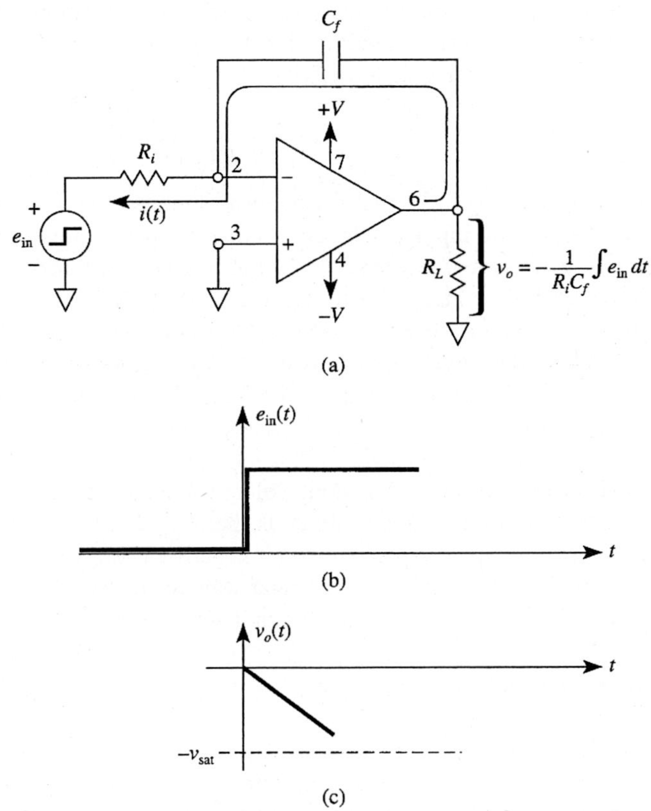

FIGURE 5-15 (a) An integrator circuit. The output voltage ramps negative in (c) for a positive input step function in (b).

design of signal generators and signal processing circuits. Chapter 6 describes signal generating circuits, but the basics of an integrator and a signal processing application are described here. In Fig. 5-15(a), the current that charges the capacitor is set by e_{in} and R_i and is given by

$$i(t) = \frac{e_{in}}{R_i} \tag{5-11}$$

Remember that Eq. (5-11) assumes that the differential voltage of the op amp, E_d, is 0 V. The equation for voltage across a capacitor is

$$v_c = \frac{1}{C}\int i\,dt \tag{5-12}$$

From Chapter 3, we learned that for an inverting circuit the output voltage is negative of the input voltage. Therefore, substituting Eq. (5-11) into Eq. (5-12) and knowing that the output is negative of the input yields

$$v_o = -\frac{1}{R_i C_f} \int e_{\text{in}} dt \tag{5-13}$$

Equation (5-13) states that the ouput voltage is directly proportional to the negative of the integral of the input voltage and indirectly proportional to the time constant $R_i C_f$. If the input wave is a sine wave, the output is a negative cosine wave. The period of the input wave should be greater than the time constant $R_i C_f$.

If e_{in} is constant, then the current $i(t)$ is constant and C charges at a constant rate. In Fig. 5-15(b), the input voltage is a positive step function. For this input signal, V_o aims toward $-V_{\text{sat}}$ as C charges. See Fig. 5-15(c). If a negative step function is applied, C charges in the opposite direction and the output voltage, V_o, aims toward $+V_{\text{sat}}$.

The conclusion is: If the input voltage is a step function, the output voltage is a ramp function. Thus the ouput voltage is the integral of the input voltage. In Chapter 6, we will alternately apply a positive and a negative voltage at the input of an integrator to design a circuit that generates a triangular waveform. We will now use the integrator followed by an inverting amplifier to design a servoamplifier.

5-12.2 Servoamplifier

Introduction

A servoamplifier is a circuit whose output is a delayed response to its input. An application may be ground-following radar signals by a cruise missile or the positioning of an *xy* table in a manufacturing process. In both applications, we may need a circuit that delays the output response due to clutter or noise at the input. Such a circuit is a servoamplifier as shown in Fig. 5-16. We proceed to answer two questions about this circuit's operation.

1. If E_1 is in equilibrium, what is V_o in equilibrium?
2. How long will it take for V_o to change from one equilibrium to another in response to a change in E_i.

Servoamplifier Circuit Analysis

We analyze the circuit behavior of Fig. 5-16 at *equilibrium* as follows:

1. Assume that $E_i = 2$ V, the capacitor C is charged, no current flows through R_i, and its voltage drop is zero.
2. The voltage at pin 2 equals E_i (since R_i current equals 0) and negative feedback makes voltages equal at pins 2 and 3 of op amp A.
3. Therefore, $V_F = E_i$.

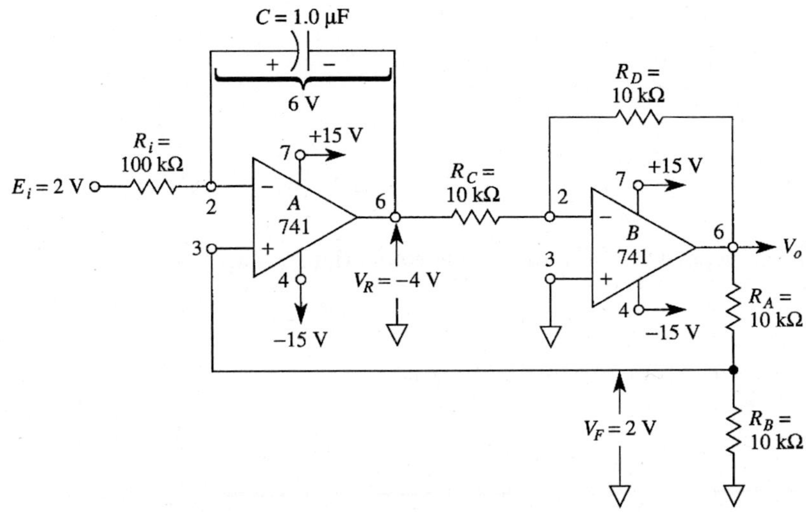

FIGURE 5-16 V_o exhibits a delayed response to a change of E_i in this servoamplifier circuit.

4. V_F causes a current through R_B of $I = V_F/R_B$. This current flows through R_A. V_o is set by I flowing through *both* R_A and R_B.

$$V_o = I\,(R_A + R_B) = \frac{V_F}{R_B}\,(R_A + R_B)$$

For $R_A = R_B = 10\ \text{k}\Omega$, $V_o = 2V_F$.

5. Since op amp B has a gain of -1, $V_o = -V_R$, or rather, $V_R = -V_o$.
6. Capacitor voltage V_{cap} is an equilibrium at $E_i - V_R$.

Summary

$$V_o = 2V_F = 2E_i = -V_R \tag{5-14a}$$

$$V_{\text{cap}} = E_i - V_R = 3E_i \tag{5-14b}$$

We have answered question 1. Next we use two examples to prepare an answer for question 2.

Example 5-15

Calculate the equilibrium voltages for the servoamplifier in Fig. 5-16.

Solution From Eqs. (5-14a) and (5-14b):

1. $E_i = 2$ V, forcing V_F to 2 V.
2. V_F forces V_o to $2V_F = 4$ V.
3. V_o forces V_R to -4 V.
4. V_{cap} stabilizes at $3E_i = 6$ V.

Example 5-16

If E_i is abruptly stepped to 4 V, find the new equilibrium voltages.

Solution

1. $E_i = 4$ V forces V_F to 4 V, forcing V_o to 8 V.
2. V_R decreases toward -8 V.
3. V_{cap} must charge to 12 V.

Delay Action

Examples 5-15 and 5-16 show that V_o must servo from 4 to 8 V when E_i is stepped from 2 to 4 V. A delay will occur (as V_o servos toward 8 V) because the capacitor must charge from 6 to 12 V. Any noise (clutter) as E_i changes from 2 to 4 V will be zeroed out. The time constant for the capacitor charge is

$$T = 3R_iC \tag{5-15a}$$

Assume that we need 5 time constants for the capacitor to fully charge. Thus, equilibrium will be achieved in

$$\text{equilibrium time} = 5T \tag{5-15b}$$

Example 5-17

How long does it take for V_o to reach equilibrium in the servoamplifier of Fig. 5-16?

Solution From Eqs. (5-15a) and (5-15b):

$$T = 3R_iC = 3(1 \times 10^5\ \Omega)(1 \times 10^{-6}\ \text{F}) = 0.3\ \text{s}$$

$$\text{equilibrium time} = 5T = 5 \times 0.3\ \text{s} = 1.5\ \text{s}$$

5-12.3 Differentiators

A differentiator is obtained by interchanging the resistor and the capacitor of an integrator. See Fig. 5-17(a). This circuit performs the mathematical operation of differentiation. The output voltage is the negative of the derivative of the input signal as given in Fig. 5-17(a),

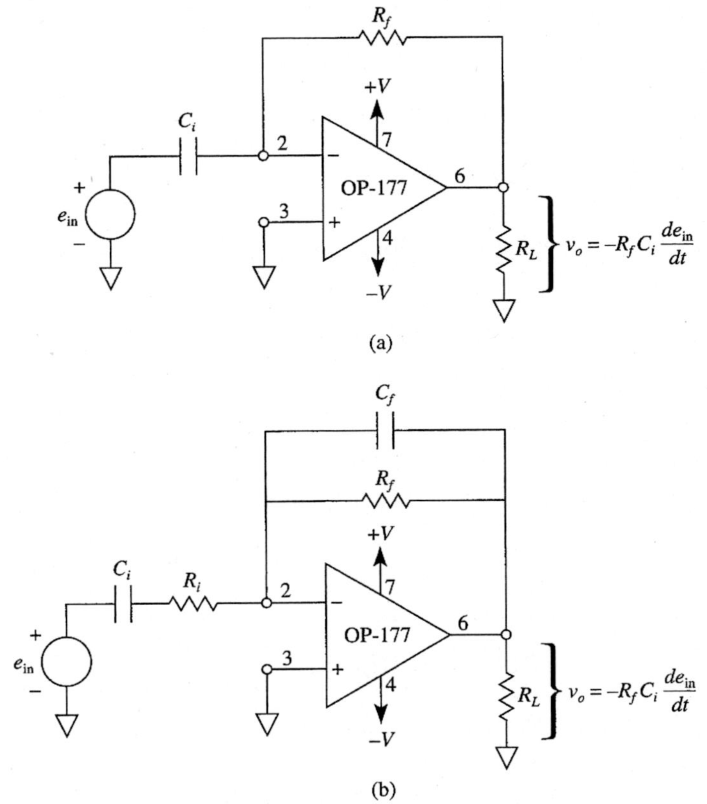

FIGURE 5-17 (a) The basic differentiator; (b) a practical differentiator circuit.

however, this basic differentiator has some inherent problems, primarily, that of instability—the circuit may oscillate. Another problem is that the gain of the circuit increases with an increase in frequency. Thus, high frequency noise can be amplified and become the dominant factor in the output signal. Although this circuit may oscillate, it is this function that we need when designing signal generators. Hence, we will use the circuit in Fig. 5-17(a) with positive feedback (see Chapter 4) to design a multivibrator circuit in the next chapter.

The circuit in Fig. 5-17(b) is a modified differentiator and may be more suited to your application if you wish to design a circuit to differentiate an input signal(s). For this circuit, make the time constants R_iC_i and R_fC_f equal, and the time constant R_fC_i at least 20 times greater than the R_iC_i or R_fC_f time constants. The following procedure may be used to design the differentiator circuit of Fig. 5-17(b).

1. Select $T_1 = R_iC_i = R_fC_f = \dfrac{1}{2\pi f_h}$ (5-16)

   where f_h is the highest frequency to be differentiated.
2. Select $T_2 = R_fC_i \geq 20\ T_1$ (5-17)

Design Example 5-18

(a) Design the circuit in Fig. 5-17(b) to differentiate signals in the range from 500 Hz to 1000 Hz. Choose $C_i = 0.1\ \mu$F. (b) If the input signal is $e_{in} = 0.4 \sin 2\pi(1000)t$ volts, what is the expression of the output voltage?

Solution (a) Rearranging Eq. (5-16)

$$R_i = \frac{1}{2\pi(1 \times 10^3 \text{ Hz})(0.1 \times 10^{-6} \text{ F})} = 1.59 \text{ k}\Omega$$

Let $T_2 = 20T_1$ and solve for R_f from Eq. (5-17).

$$R_f = \frac{20}{2\pi(1 \times 10^3 \text{ Hz})(0.1 \times 10^{-6} \text{ F})} = 31.8 \text{ k}\Omega$$

Use Eq. (5-16) to solve for C_f.

$$C_f = \frac{1}{2\pi(1 \times 10^3 \text{ Hz})(31.8 \text{ k}\Omega)} = 0.005\ \mu\text{F}$$

(b) From Fig. 5-17(a), the output voltage is

$$v_o = -R_f C_i \frac{de_{in}}{dt}$$

Then,

$$\begin{aligned} v_o &= -(31.8 \text{ k}\Omega)(0.1\ \mu\text{F})\frac{d}{dt}[0.4 \sin 2\pi(1000)t] \\ &= -(31.8 \text{ k}\Omega)(0.1\ \mu\text{F})(0.4)(2\pi)(1000)\cos[2\pi(1000)\text{t}] \\ &\cong -8 \cos[2\pi(1000)\text{t}] \text{ V} \end{aligned}$$

Note: With large values of R_f, the output of the op amp may go into saturation because it is multiplied by the input signal's peak value, frequency, and the constant 2π.

5-13 PSPICE SIMULATION

In this section, we will use PSpice and simulate the performance of the differential voltage-to-current converter with a grounded load as shown in Fig. 5-5. Use an **IPROBE** to measure I_L. Place the following parts on the right-hand side of the work area:

Draw => Get New Part

Part	Number	Library
=> uA741	1	**eval.slb**
=> VDC	4	**source.slb**
=> R	5	**analog.slb**
=> GLOBAL	4	**port.slb**
=> AGND	5	**port.slb**
=> IPROBE	1	**special.slb**

Arrange the parts as shown in the schematic of Fig. 5-5 but include the **IPROBE** to measure I_L. Set the four resistors labeled R as $R_1 = R_2 = R_3 = R_4 = 10\ \text{k}\Omega$, $E_1 = 5$ V, $E_2 =$ 3 V, and $R_L = 1\ \text{k}\Omega$. Now select **Analysis => Setup => DC Sweep.** Open **DC Sweep** and select

Swept Var. Type => Voltage Source

and

Sweep Type => Linear

Now set

Name => V1, Start Value => 5V, End Value => 5V, Increment => 1V

Select

Nested Sweep => Swept Var. Type => Voltage Source
=> Sweep Type => Linear

Now set

Name => V2, Start Value => 3V, End Value => 3V, Increment => 1V

In this simulation you do not want to run Probe. Remove Probe execution by

Analysis => Probe Setup =>

Click

Do Not Auto-Run Probe

Save the file with an **.SCH** extension and click **Analysis = > Simulate.** The completed circuit with current value I_L is shown in Fig. 5-18.

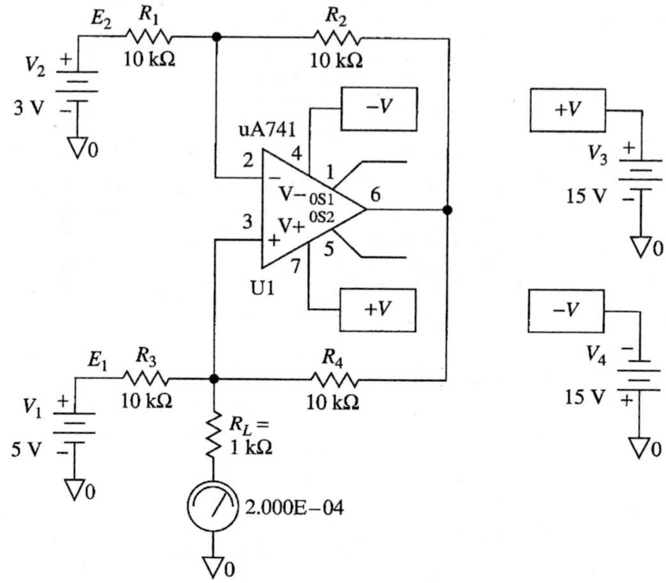

FIGURE 5-18 PSpice model of Fig. 5-5.

PROBLEMS

5-1. Refer to Example 5-1 and Fig. 5-1. Assume that $I_{FS} = 1$ mA and meter winding resistance $R_m = 1$ kΩ. If $E_i = -1.0$ V and $R_i = 1$kΩ, find **(a)** I_m; **(b)** V_o.

5-2. A 1-mA movement, with $R_m = 1$ kΩ, is to be substituted in the circuit in Fig. 5-2. Redesign the R_i resistors for full-scale meter deflection when **(a)** $E_i = \pm 6$ V dc; **(b)** $E_i =$ 6 V rms; **(c)** $E_i = 6$ V p-p; **(d)** $E_i = 6$ V peak.

5-3. In Fig. P5-3 complete the schematic wiring between op amp, diodes, and milliammeter. The current through the meter must be steered from right to left.

5-4. Calculate a value for R_{idc} in Fig. P5-3 so that the meter reads full scale when $E_i = 5$ V and the range switch is on the 5-V position.

5-5. Consider that the range switch is in the 5-V position in Fig. P5-3. Calculate values for the following resistors to give a full-scale meter deflection of 5 V: **(a)** R_{irms} for $E_i = 5$ V rms; **(b)** R_i p-p for $E_i = 5$ V p-p; **(c)** R_i peak for $E_i = 5$ V peak.

5-6. With the circuit conditions shown in Problem 5-4, **(a)** which diodes are conducting? **(b)** Find V_o. Assume that diode drops are 0.6 V.

5-7. For the constant-current source shown in Fig. P5-7, **(a)** draw the current direction, the emitter arrow, and state if the transistor is *npn* or *pnp*; **(b)** find I; **(c)** find V_L.

5-8. If $V_o = 11$ V and $E_i = 5$ V in Fig. 5-3, find V_z.

5-9. I_1 must equal 20 mA in Fig. 5-4 when $E_i = -10$ V. Find R_i.

5-10. Define a floating load.

5-11. In Fig. 5-5, $E_2 = 0$ V, $R = 10$ kΩ, and $R_L = 5$ kΩ. Find I_L, V_L, and V_o for **(a)** $E_i = -2$ V; **(b)** $E_1 = +2$ V.

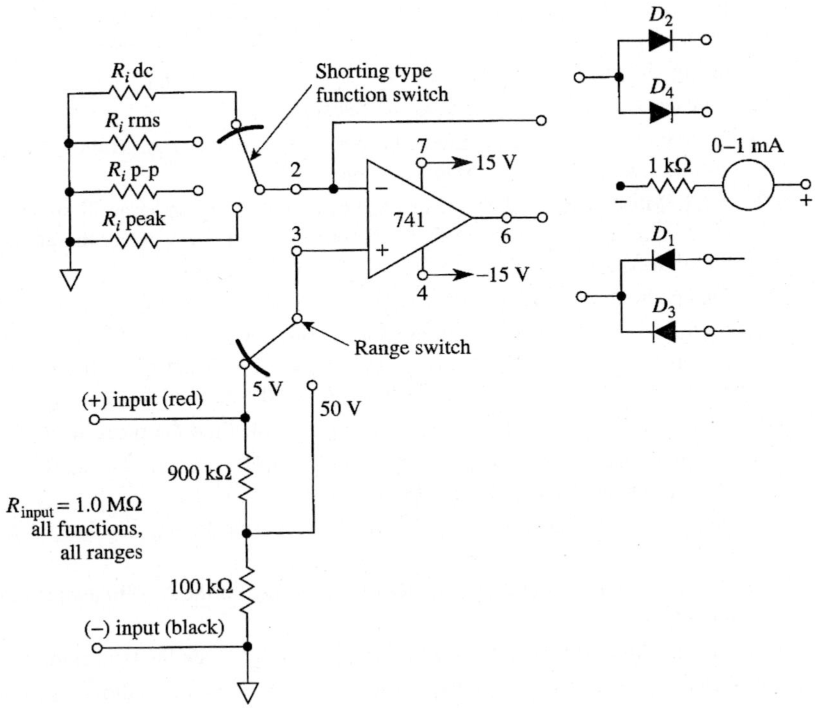

FIGURE P5-3

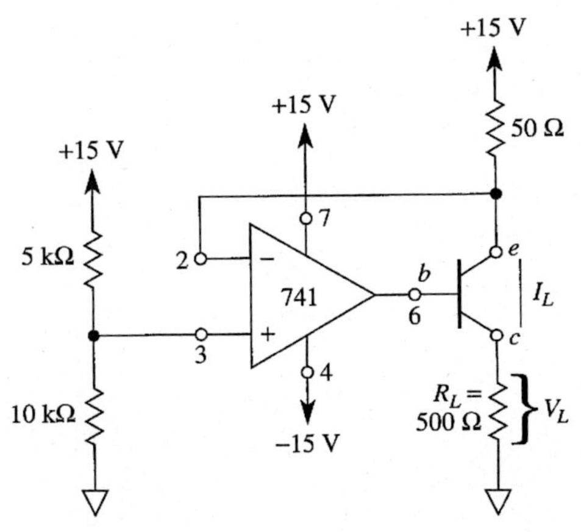

FIGURE P5-7

5-12. In Fig. 5-5, $E_1 = 0$ V, $R = 10$ kΩ, and $R_L = 1$ kΩ. Find I_L, V_L, and V_o for **(a)** $E_2 = -2$ V; **(b)** $E_2 = +2$ V.

5-13. In Fig. 5-5, $E_1 = E_2 = -5$ V, and $R = R_L = 5$ kΩ. Find I_L, V_L, and V_o.

5-14. Replace V_z in Fig. 5-6 with a 900-Ω resistor. Find I_L.

5-15. Sketch an op amp circuit that will draw short-circuit current from a signal source and convert the short-circuit current to a voltage.

5-16. A CL5M9M photocell has a resistance of about 10 kΩ under an illumination of 2 fc. If $E_i = -10$ V in Fig. 5-9, calculate R_f for an output V_o of 0.2 V when the photoconductive cell is illuminated by 2 fc.

5-17. Change multiplier resistor mR in Fig. 5-10 to 49 kΩ. Find I_L.

5-18. A solar cell that has a maximum short-circuit current of 0.1 A = I_{SC} is installed in the circuit in Fig. 5-12. **(a)** Select R_F to give $V_o = 10$ V when $I_{SC} = 0.1$ A. **(b)** A 50-μA meter movement is to indicate full scale when $I_{SC} = 0.1$ A. Find R_{scale} if $R_M = 5$ kΩ.

5-19. Resistor R_i is changed to 10 kΩ in Example 5-14. Find the phase angle θ.

5-20. Design a phase shifter to give a −90° shift at 1 Hz. Choose C_i from 0.001, 0.01, 0.1, or 1.0 μF. R_i must lie between 2 and 100 kΩ.

5-21. Design a −90° phase shift at 1590 Hz. Then for your design, calculate **(a)** θ at 15 Hz; **(b)** θ at 15 kHz.

5-22. Calculate the net current through R_f in Fig. 5-14(a) if the AD590 temperature is 100°C. Then find V_o.

5-23. Calculate the net current through R_f in Fig. 5-14(b) when the temperature is 100°F. Find V_o.

5-24. Calculate the value of R_f in Fig. 5-14(a) to design a signal conditioning circuit that interfaces with a microcontroller's A/D converter. The voltage range of the converter is 0 to 5 V.

5-25. Use a simulation program and design an integrating circuit. The input sugnal is

$$e_{\text{in}} = 1 \sin 2000\pi t \text{ V}$$

5-26. Use a simulation program and design a differentiating circuit. The input signal is **(a)** sine wave of 500 Hz and a peak value of 0.2 V; **(b)** square wave of 500 Hz and a peak value of 0.2 V; **(c)** cosine wave of 500 Hz and a peak value of 0.2 V.

CHAPTER 6

Signal Generators

LEARNING OBJECTIVES

Upon completion of this chapter on signal generators, you will be able to:

- Explain the operation of a multivibrator circuit, sketch its output voltage waveshapes, and calculate its frequency of oscillation.
- Make a one-shot multivibrator and explain the purpose of this circuit.
- Show how two op amps, three resistors, and one capacitor can be connected to form an inexpensive triangle/square-wave generator.
- Predict the frequency of oscillation and amplitude of the voltages in a bipolar or unipolar triangle-wave generator and identify its disadvantages.
- Build a sawtooth wave generator and tell how it can be used as a voltage-to-frequency converter, frequency modulator, or frequency shift key circuit.
- Connect an AD630 balanced modulator/demodulator to operate as a switched gain amplifier.

- Connect the AD630 to an op amp circuit to make a *precision* triangle-wave generator whose output voltage amplitude can be adjusted independently of the oscillating frequency, and vice versa.
- Build, test, measure, and explain the operation of an AD639 universal trigonometric function generator when it is wired to generate sine functions.
- Connect the AD639 to the triangle-wave generator to make a superb *precision* sine-wave generator. Its oscillating frequency can be adjusted over a wide frequency range by a *single* resistor, without changing amplitude.
- Know about the operation of a single IC function generator.

6-0 INTRODUCTION

Up to now our main concern has been to use the op amp in circuits that process signals. In this chapter we concentrate on op amp circuits that *generate* signals. Four of the most common and useful signals are described by their shape when viewed on an oscilloscope. They are the square wave, triangular wave, sawtooth wave, and sine wave. Accordingly, the signal generator is classified by the shape of the wave it generates. Some circuits are so widely used that they have been assigned a special name. For example, the first circuit presented in Section 6-1 is a multivibrator that generates primarily square waves and exponential waves. Some ICs that generate these waveforms from a single IC are available. However, you may need a waveform quickly and not have on hand one of these function generator ICs.

6-1 FREE-RUNNING MULTIVIBRATOR

6-1.1 Multivibrator Action

A *free-running* or *astable multivibrator* is a square-wave generator. The circuit of Fig. 6-1 is a multivibrator circuit and looks something like a comparator with hysteresis (Chapter 4), except that the input voltage is replaced by a capacitor. Resistors R_1 and R_2 form a voltage divider to feed back a fraction of the output to the (+) input. When V_o is at $+V_{sat}$, as shown in Fig. 6-1(a), the feedback voltage is called the upper-threshold voltage V_{UT}. V_{UT} is given in Eq. (4-1) and repeated here for convenience:

$$V_{UT} = \frac{R_2}{R_1 + R_2}(+V_{sat}) \tag{6-1}$$

Resistor R_f provides a feedback path to the (−) input. When V_o is at $+V_{sat}$, current I^+ flows through R_f to charge capacitor C toward V_{UT}. As long as the capacitor voltage V_C is less than V_{UT}, the output voltage remains at $+V_{sat}$.

When V_C charges to a value slightly greater than V_{UT}, the (−) input goes positive with respect to the (+) input. This switches the output from $+V_{sat}$ to $-V_{sat}$. The (+) input is now held negative with respect to ground because the feedback voltage is negative and given by

$$V_{LT} = \frac{R_2}{R_1 + R_2}(-V_{sat}) \tag{6-2}$$

Equation (6-2) is the same as Eq. (4-2). Just after V_o switches to $-V_{sat}$, the capacitor has an initial voltage equal to V_{UT} [see Fig. 6-1(b)]. Now current I^- discharges C to 0 V and recharges C to V_{LT}. When V_C becomes slightly more negative than the feedback voltage V_{LT}, output voltage V_o switches back to $+V_{sat}$. The condition in Fig. 6-1(a) is reestab-

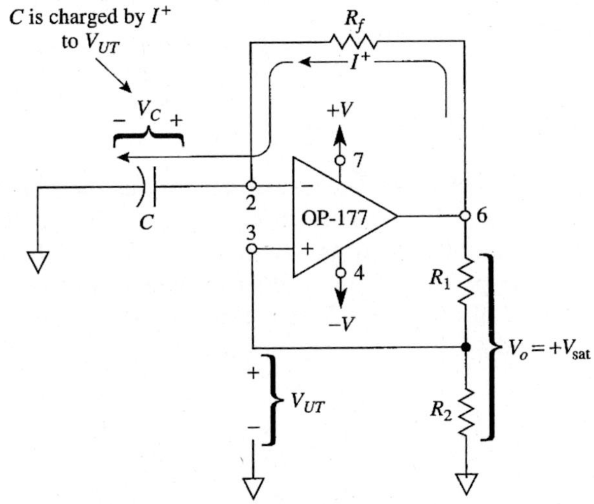

(a) When $V_o = +V_{sat}$, V_C charges toward V_{UT}.

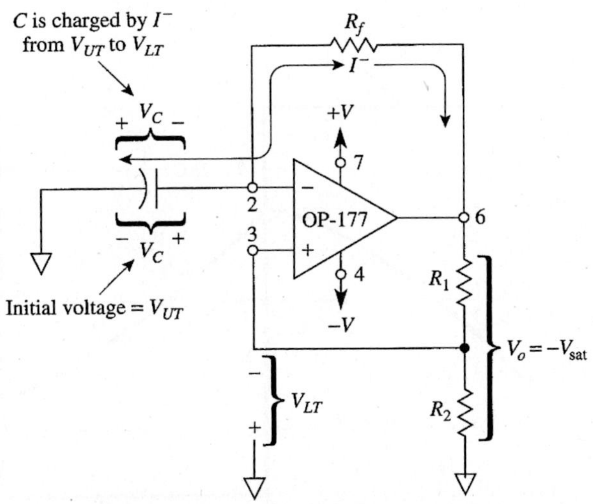

(a) When $V_o = -V_{sat}$, V_C charges toward V_{LT}.

FIGURE 6-1 Free-running multivibrator ($R_1 = 100\,\text{k}\Omega, R_2 = 86\,\text{k}\Omega$). Output-voltage waveforms shown in Fig. 6-2.

lished except that C now has an initial charge equal to V_{LT}. The capacitor will discharge from V_{LT} to 0 V and then recharge to V_{UT}, and the process is repeating. Free-running multivibrator action is summarized as follows:

1. When $V_o = -V_{sat}$, C discharges from V_{UT} to V_{LT} and switches V_o to $+V_{sat}$.
2. When $V_o = +V_{sat}$, C charges from V_{LT} to V_{UT} and switches V_o to $-V_{sat}$.

The time needed for C to charge and discharge determines the frequency of the multivibrator.

6-1.2 Frequency of Oscillation

The capacitor and output-voltage waveforms for the free-running multivibrator are shown in Fig. 6-2. Resistor R_2 is chosen to equal $0.86R_1$ to simplify calculation of capacitor charge time. Time intervals t_1 and t_2 show how V_C and V_o change with time for Figs. 6-1(a) and (b), respectively. Time intervals t_1 and t_2 are equal to the product of R_f and C.

The period of oscillation, T, is the time needed for one complete cycle. Since T is the sum of t_1 and t_2,

$$T = 2R_fC \qquad \text{for } R_2 = 0.86R_1 \tag{6-3a}$$

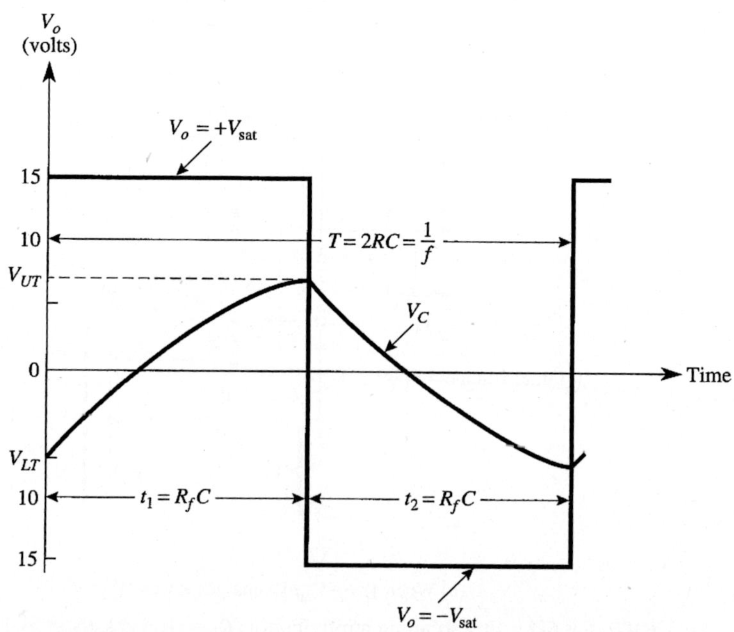

FIGURE 6-2 Voltage waveshapes for the multivibrator of Fig. 6-1.

The frequency of oscillation f is the reciprocal of period T and is expressed by

$$f = \frac{1}{T} = \frac{1}{2R_fC} \tag{6-3b}$$

where T is in seconds, f in hertz, R_f in ohms, and C in farads.

Example 6-1

In Fig. 6-1, if $R_1 = 100$ kΩ, $R_2 = 86$ kΩ, $+V_{sat} = +15$ V, and $-V_{sat} = -15$ V, find (a) V_{UT}; (b) V_{LT}.

Solution (a) By Eq. (6-1),

$$V_{UT} = \frac{86 \text{ k}\Omega}{186 \text{ k}\Omega} \times 15 \text{ V} \cong 7 \text{ V}$$

(b) By Eq. (6-2),

$$V_{LT} = \frac{86 \text{ k}\Omega}{186 \text{ k}\Omega}(-15 \text{ V}) \cong -7 \text{ V}$$

Example 6-2

Find the period of the multivibrator in Example 6-1 if $R_f = 100$ kΩ and $C = 0.1$ μF.

Solution Using Eq. (6-3a), $T = (2)(100 \text{ k}\Omega)(0.1 \text{ }\mu\text{F}) = 0.020 \text{ s} = 20 \text{ ms}$.

Example 6-3

Find the frequency of oscillation for the multivibrator of Example 6-2.

Solution From Eq. (6-3b),

$$f = \frac{1}{20 \times 10^{-3} \text{ s}} = 50 \text{ Hz}$$

Example 6-4

Show why $T = 2\,R_fC$ when $R_2 = 0.86R$, as stated in Eq. (6-3a).

Solution The time required for a capacitor C to charge through a resistor R_f from some *starting* capacitor voltage toward some *aiming* voltage to a *stop* voltage is expressed generally as

$$t = R_fC \ln\left(\frac{\text{aim} - \text{start}}{\text{aim} - \text{stop}}\right)$$

Applying the equation to Fig. 6-2 yields

$$t_1 = R_f C \ln \left(\frac{+V_{sat} - V_{LT}}{+V_{sat} - V_{UT}} \right)$$

If the magnitudes of $+V_{sat}$ and $-V_{sat}$ are equal, the term in parentheses simplifies to

$$\ln \left[\frac{+V_{sat} - \dfrac{R_2}{R_1 + R_2}(-V_{sat})}{+V_{sat} - \dfrac{R_2}{R_1 + R_2}(+V_{sat})} \right] = \ln \left(\frac{R_1 + 2\,R_2}{R_1} \right)$$

Since ln 2.718 = 1, the ln term can be reduced to 1 if

$$\frac{R_1 + 2R_2}{R_1} = 2.718 \qquad \text{or} \qquad R_2 = 0.86R_1$$

Now $t_1 = R_f C$ and $t_2 = R_f C$ if $R_2 = 0.86R_1$. Therefore, $T = t_1 + t_2 = 2R_f C$.

6-2 ONE-SHOT MULTIVIBRATOR

6-2.1 Introduction

A *one-shot multivibrator* generates a single output pulse in response to an input signal. The length of the output pulse depends only on external components (resistors and capacitors) connected to the op amp. As shown in Fig. 6-3, the one-shot generates a single output pulse on the negative-going edge of E_i. The duration of the input pulse can be longer or shorter than the expected output pulse. The duration of the output pulse is represented by τ in Fig. 6-3. Since τ can be changed only by changing resistors or capacitors, the one-shot can be considered a *pulse stretcher.* This is because the width of the pulse can be longer than the input pulse. Moreover, the one-shot introduces an idea of an adjustable delay, that is, the delay between the time when E_i goes negative and the time for V_o to go positive again. Operation of the one-shot will be studied in three parts: (1) the stable state, (2) transition to the timing state, and (3) the timing state.

6-2.2 Stable State

In Fig. 6-4(a), V_o is at $+V_{sat}$. Voltage divider R_1 and R_2 feeds back V_{UT} to the (+) input. V_{UT} is given by Eq. (6-1). The diode D_1 clamps the (−) input at approximately +0.5 V. The (+) input is positive with respect to the (−) input, and the high open-loop gain times the differential input voltage ($E_d = 2.1 - 0.5 = 1.6$ V) holds V_o at $+V_{sat}$.

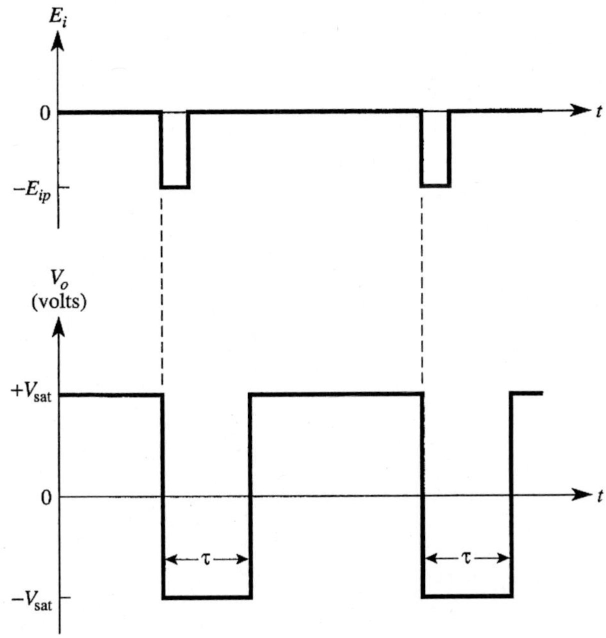

FIGURE 6-3 Input signal E_i and output pulse of a one-shot multivibrator.

6-2.3 Transition to the Timing State

If input signal E_i is at a steady dc potential as in Fig. 6-4(a), the (+) input remains positive with respect to the (−) input and V_o stays at $+V_{sat}$. However, if E_i goes negative by a peak value E_{ip} approximately equal to or greater than V_{UT}, the voltage at the (+) input will be pulled below the voltage at the (−) input. Once the (+) input becomes negative with respect to the (−) input, V_o switches to $-V_{sat}$. With this change, the one-shot is now in its timing state. For best results, the time constant R_iC_i should be 1/10 the time constant R_fC or less.

6-2.4 Timing State

The timing state is an unstable state; that is, the one-shot cannot remain very long in this state for the following reasons. Resistors R_1 and R_2 in Fig. 6-4(b) feed back a negative voltage ($V_{LT} = -2.1$ V) to the (+) input. The diode D_1 is now reverse biased by $-V_{sat}$ and is essentially an open circuit. Capacitor C discharges to 0 and then recharges with a polarity opposite to that in Fig. 6-4(a) [see Fig. 6-4(b)]. As C recharges, the (−) input becomes more and more negative with respect to ground. When the capacitor voltage is slightly more negative than V_{LT}, V_o switches to $+V_{sat}$. The one-shot has now completed its output pulse and is back to the stable state in Fig. 6-4(a). Since the one-shot has only one stable state, it is also called a *monostable multivibrator*.

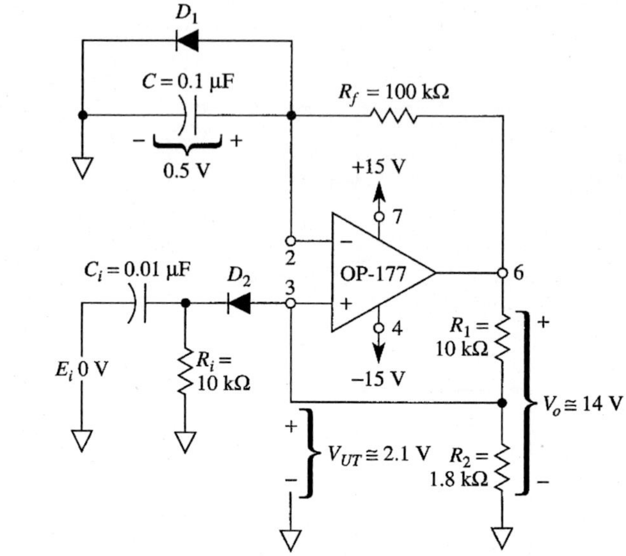

(a) Stable state of a one-shot multivibrator: $R_iC_i \leq (R_fC)/10$.

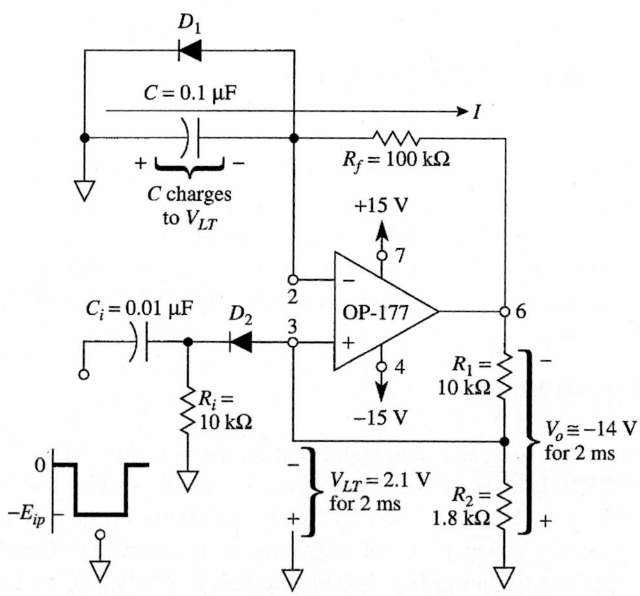

(b) Timing state: When E_i goes negative, V_o goes negative for 2 ms.

FIGURE 6-4 Monostable or one-shot multivibrator.

6-2.5 Duration of Output Pulse

If R_2 is made about one-fifth of R_1, in Fig. 6-4, then the duration of output pulse is given by

$$\tau = \frac{R_f C}{5} \quad \text{if } R_2 = \frac{R_1}{5} \tag{6-4a}$$

The exact equation is

$$\tau = R_f C \ln \left[\frac{-V_{sat} - 0.5\ \text{V}}{-V_{sat} - V_{UT}} \right] \tag{6-4b}$$

Example 6-5

Calculate τ for the one-shot of Fig. 6-4.

Solution By Eq. (6-4a),

$$\tau = \frac{(100\ \text{k}\Omega)(0.1\ \mu\text{F})}{5} = 2\ \text{ms}$$

For test purposes, E_i can be obtained from a square-wave or pulse generator. Diode D_2 prevents the one-shot from coming out of the timing state on positive transitions of E_i. To build a one-shot that has a positive output pulse for a positive input signal, simply reverse the diodes.

6-2.6 Recovery Time

After the timing state is completed, the output returns to $+V_{sat}$. However, the circuit is not ready to be retriggered reliably until C returns to its initial state of 0.5 V because it takes time for C to be discharged from $V_{LT} = -2.1$ V in Fig. 6-4(b) to 0.5 V in Fig. 6-4(a). This time interval is called *recovery time* and is shown in Fig. 6-5(b). Recovery time is approximately τ.

Normally, C is charged back to its initial state by a current through R_f. By adding a discharge resistor R_D in parallel with R_f, as in Fig. 6-5(a), the recovery time is reduced. Typically, if $R_D = 0.1R_f$, recovery time is reduced to one-tenth. Diode D_D prevents R_D from affecting the timing-cycle interval τ.

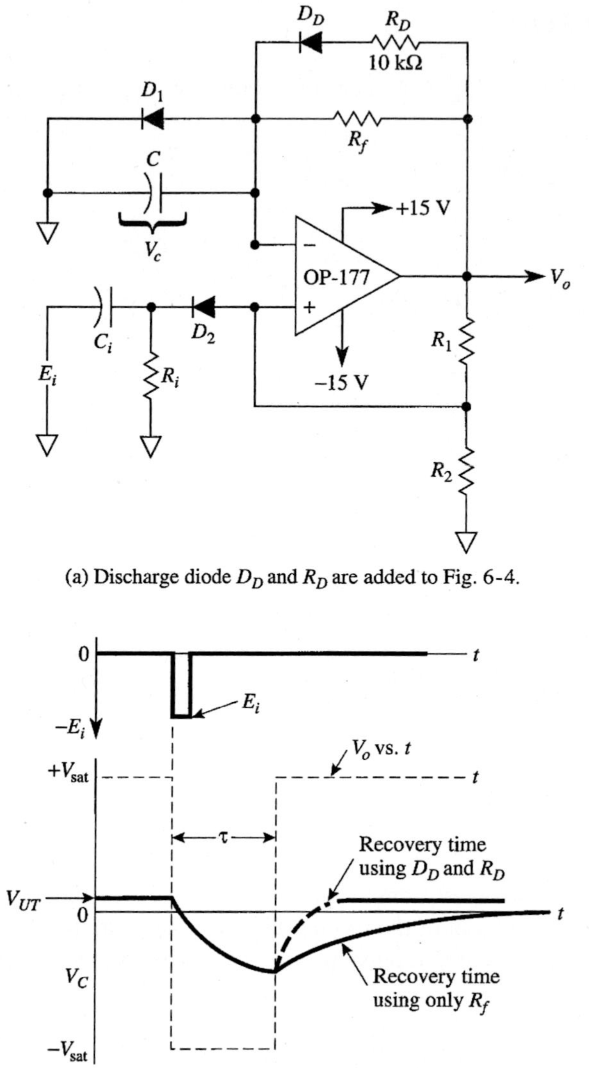

(a) Discharge diode D_D and R_D are added to Fig. 6-4.

(b) Recovery time is reduced by D_D and R_D.

FIGURE 6-5 The recovery time of a one-shot multivibrator is reduced by adding discharge diode D_D and R_D. R_D should be about one-tenth of R_f to reduce recovery time to one-tenth.

6-3 TRIANGLE-WAVE GENERATORS

6-3.1 Theory of Operation

A basic bipolar triangle-wave generator circuit is presented in Fig. 6-6. The triangle wave, V_A, is available at the output of the 741 integrator circuit. An additional square-wave signal, V_B, is available at the output of the 301 comparator.

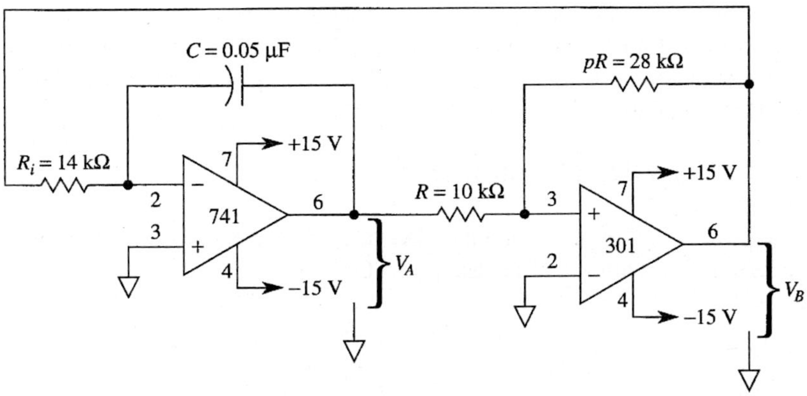

(a) The 741 integrator circuit and 301 comparator circuit are wired to make a triangle-wave generator.

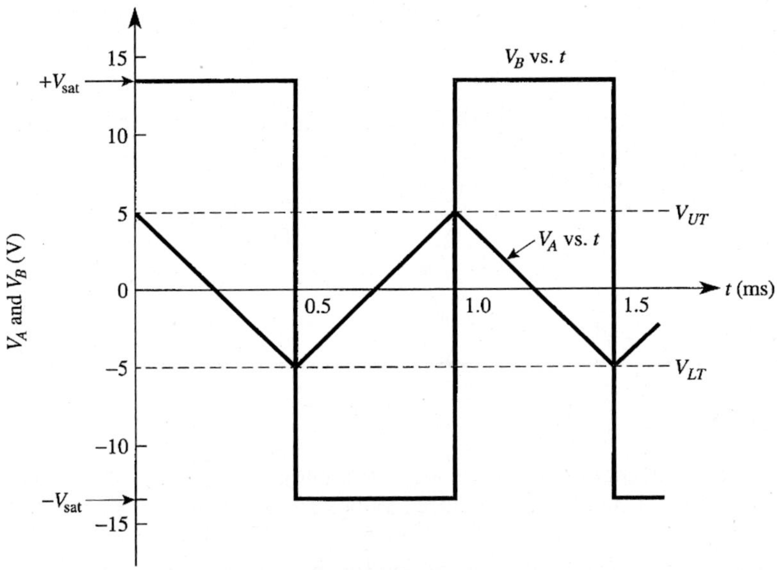

(b) Waveshapes.

FIGURE 6-6 The bipolar triangle-wave generator circuit in (a) generates triangle-wave and square-wave oscillator signals as in (b). (a) Basic bipolar triangle-wave generator oscillator frequency for 1000 Hz; (b) output-voltage waveshapes.

To understand circuit operation, refer to time interval 0 to 1 ms in Fig. 6-6. Assume that V_B is high at $+V_{\text{sat}}$. This forces a constant current (V_{sat}/R_i) through C (left to right) to drive V_A negative from V_{UT} to V_{LT}. When V_A reaches V_{LT}, pin 3 of the 301 goes negative, and V_B snaps to $-V_{\text{sat}}$, and $t = 1$ ms.

When V_B is at $-V_{sat}$, it forces a constant current (right to left) through C to drive V_A positive from V_{LT} toward V_{UT} (see the time interval 1 to 2 ms). When V_A reaches V_{UT} at $t = 2$ ms, pin 3 of the 301 goes positive and V_B snaps to $+V_{sat}$. This initiates the next cycle of oscillation.

6-3.2 Frequency of Operation

The peak values of the triangular wave are established by the *ratio* of resistor pR to R and the saturation voltages. They are given by

$$V_{UT} = -\frac{-V_{sat}}{p} \tag{6-5a}$$

$$V_{LT} = -\frac{+V_{sat}}{p} \tag{6-5b}$$

where

$$p = \frac{pR}{R} \tag{6-5c}$$

If the saturation voltages are reasonably equal, the frequency of oscillation, f, is given by

$$f = \frac{p}{4R_iC} \tag{6-6}$$

Example 6-6

A triangle-wave generator oscillates at a frequency of 1000 Hz with peak values of approximately +5 V. Calculate the required values for pR, R_i, and C in Fig. 6-6.

Solution *First* we work on the calculation for the comparator resistor ratio p that controls peak triangle-wave output voltages, V_{UT} and V_{LT}. $+V_{sat}$ is practically +14.2 V and $-V_{sat}$ is typically −13.8 V for a ±15-V supply. This observation points out one deficiency in our low-cost triangle-wave generator. It does *not* have *precisely* equal positive and negative peak outputs. (We will remedy this problem, at a higher cost, in Section 6-6.) From Eq. (6-5a), solve for p:

$$p = -\frac{-V_{sat}}{V_{UT}} = -\frac{-13.8\text{ V}}{5\text{ V}} = +2.76 \approx 2.8$$

Choose $R = 10\text{ k}\Omega$. Then from Eq. (6-5c) we solve for pR as

$$pR = 2.8(10\text{ k}\Omega) = 28\text{ k}\Omega$$

Next we select R_i and C. Begin by making a *trial* choice for $C = 0.05\ \mu F$. Then calculate a value for R_i to see if R_i is greater than 10 kΩ. From Eq. (6-6),

$$R_i = \frac{p}{4fC} = \frac{2.8}{4(1000\ \text{Hz})(0.05\ \mu\text{F})} = 14\ \text{k}\Omega$$

In practice it would be prudent for R_i to be a 12-kΩ resistor in series with a 0 to 5-kΩ pot. The 5-kΩ pot may then be adjusted for an oscillation frequency of precisely 1.00 kHz.

6-3.3 Unipolar Triangle-Wave Generator

The bipolar triangle-wave generator circuit of Fig. 6-6 can be changed to produce a unipolar triangle wave output. Simply add a diode in series with pR as shown in Fig. 6-7. Circuit operation is studied by reference to the waveshapes in Fig. 6-7(b).

When V_B is at $+V_{\text{sat}}$, the diode stops current flow through pR and sets V_{LT} at 0 V. When V_B is at $-V_{\text{sat}}$, the diode allows current flow through pR and sets V_{UT} at a value of

$$V_{UT} = -\frac{-V_{\text{sat}} + 0.6\ \text{V}}{p} \tag{6-7a}$$

Frequency of oscillation is then given approximately by

$$f \simeq \frac{p}{2R_iC} \tag{6-7b}$$

Example 6-7

Find the approximate peak voltage and frequency for the unipolar triangle-wave generator in Fig. 6-7.

Solution Calculate

$$p = \frac{pR}{R} = \frac{28\ \text{k}\Omega}{10\ \text{k}\Omega} = 2.8$$

Find the peak value of V_A from Eq. (6-7a):

$$V_{UT} = -\left(\frac{-V_{\text{sat}} + 0.6\ \text{V}}{p}\right) = -\left(\frac{-13.8\ \text{V} + 0.6\ \text{V}}{2.8}\right) \simeq 4.7\ \text{V}$$

From Eq. (6-7b),

$$f = \frac{p}{2R_iC} = \frac{2.8}{2(28\ \text{k}\Omega)(0.05\ \mu\text{F})} = 1000\ \text{Hz}$$

Note the change in R_i value from Fig. 6-6(a) to Fig. 6-7(a).

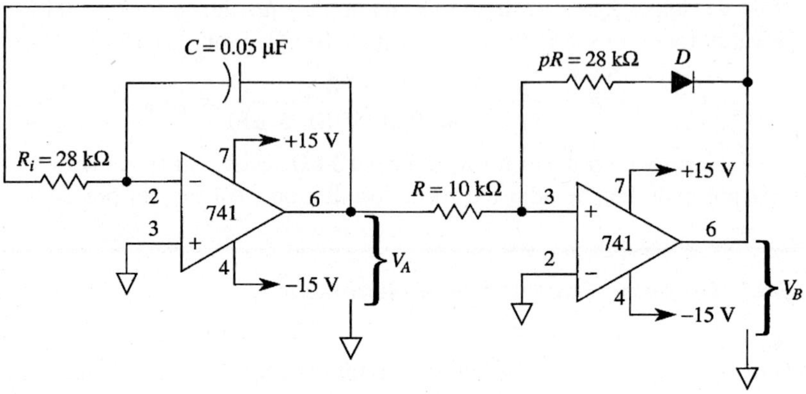

(a) Unipolar triangle-wave generator.

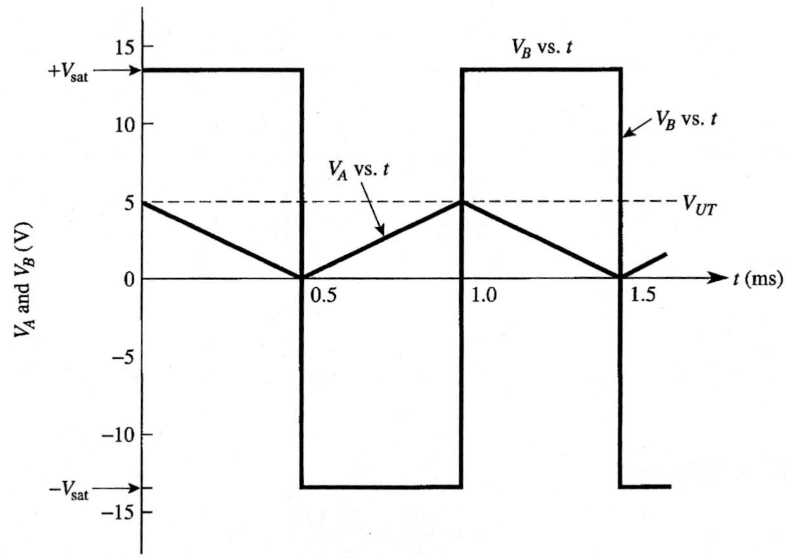

(b) Waveshapes.

FIGURE 6-7 Diode D in (a) converts the bipolar triangle-wave generator into a unipolar triangle-wave generator. Waveshapes are shown in (b). (a) Basic unipolar triangle-wave generator; oscillating frequency is 1000 Hz. (b) Output-voltage waveshapes.

6-4 SAWTOOTH-WAVE GENERATOR

6-4.1 Circuit Operation

A low-parts-count sawtooth-wave generator circuit is shown in Fig. 6-8(a). Op amp A is a ramp generator. Since E_i is negative, $V_{o\,\mathrm{ramp}}$ can only ramp up. The rate of rise of the ramp voltage is constant at

$$\frac{V_{o\,\mathrm{ramp}}}{t} = \frac{E_i}{R_i C} \tag{6-8}$$

The ramp voltage is monitored by the (+) input of comparator 301B. If $V_{o\,\mathrm{ramp}}$ is below V_{ref}, the comparator's output is negative. Diodes protect the transistors against excessive reverse bias.

When $V_{o\,\mathrm{ramp}}$ rises to just exceed V_{ref}, the output $V_{o\,\mathrm{comp}}$ goes to positive saturation. This forward biases "dump" transistor Q_D into saturation. The saturated transistor acts as a short circuit across the integrating capacitor C. C discharges quickly through Q_D to essentially 0 V. When $V_{o\,\mathrm{comp}}$ goes positive, it turns on Q_1 to short-circuit the 10-kΩ potentiometer. This drops V_{ref} to almost zero volts.

As C discharges toward 0 V, it drives $V_{o\,\mathrm{ramp}}$ rapidly toward 0 V. $V_{o\,\mathrm{ramp}}$ drops below V_{ref}, causing $V_{o\,\mathrm{comp}}$ to go negative and turn off Q_D. C begins charging linearly, and generation of a new sawtooth wave begins.

6-4.2 Sawtooth Waveshape Analysis

The ramp voltage rises at a rate of 1 V per millisecond in Fig. 6-8(b). Meanwhile, $V_{o\,\mathrm{comp}}$ is shown to be negative. When the ramp crosses V_{ref}, $V_{o\,\mathrm{comp}}$ snaps positive to drive the ramp voltage quickly toward 0 V. As $V_{o\,\mathrm{ramp}}$ snaps to 0 V, the comparator's output is reset to negative saturation. Ramp operation is summarized in Fig. 6-8(c).

6-4.3 Design Procedure

The time for one sawtooth-wave period can be derived most efficiently by analogy with a familiar experience.

$$\text{time (of rise)} = \frac{\text{distance (of rise)}}{\text{speed (of rise)}} \tag{6-9a}$$

$$\text{period } T = \frac{V_{\mathrm{ref}}}{E_i/R_i C} \tag{6-9b}$$

Since frequency is the reciprocal of the period

$$f = \left(\frac{1}{R_i C}\right)\frac{E_i}{V_{\mathrm{ref}}} \tag{6-9c}$$

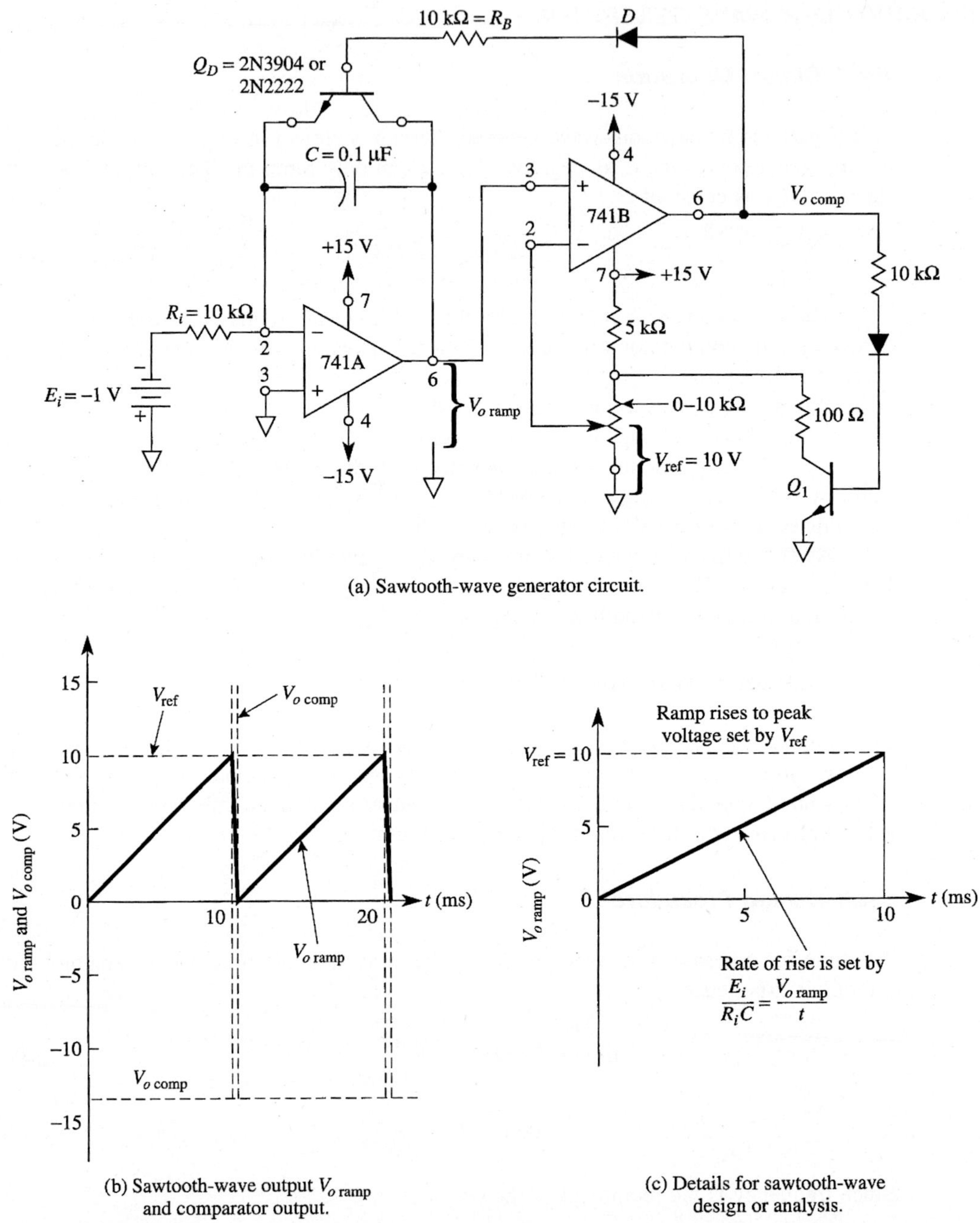

FIGURE 6-8 The sawtooth-wave generator circuit in (a) has the waveshapes shown in (b) and (c). Oscillating frequency is 100 Hz or $f = (1/R_iC)(E_i/V_{ref})$.

Design Example 6-8

Design a sawtooth-wave generator to have a 10-V peak output and a frequency of 100 Hz. Let $E_i = 1$ V.

Design Procedure

1. Design a voltage divider to give a reference voltage $V_{ref} = +10$ V for op amp B in Fig. 6-8(a).
2. Let's select a ramp rate rise of 1 V/ms. Pick any R_iC combination to give 1.0 ms. Therefore, let's select $R_i = 10$ kΩ and $C = 0.1$ μF.
3. The resulting circuit is shown in Fig. 6-8(a).
4. E_i may be made from a voltage divider and voltage follower to make an ideal voltage source (see Section 3.7).
5. Alternatively, you could pick a trial value for R_iC and solve for E_i in Eq. (6-9b).
6. Check the design values in Eq. 6-9c.

$$f = \frac{1}{(10\ \text{k}\Omega)(0.1\ \mu\text{F})}\left(\frac{1\ \text{V}}{10\ \text{V}}\right) = 100\ \text{Hz}$$

6-4.4 Voltage-to-Frequency Converter

There are two ways to change or modulate the oscillating frequency of Fig. 6-8. We see from Eq. (6-9c) that the frequency is directly proportional to E_i and inversely proportional to V_{ref}. The advantages and disadvantages of each method are examined with an example.

This type of frequency modulation by V_{ref} has two disadvantages with respect to control of frequency by E_i. First, the relationship between input voltage V_{ref} and output frequency is *not* linear. Second, the sawtooth's peak output voltage is not constant, since it varies directly with V_{ref}.

6-4.5 Frequency Modulation and Frequency Shift Keying

Examples 6-9 and 6-10 indicate one way of achieving *frequency modulation* (FM). Thus, if the amplitude of E_i varies, the frequency of the sawtooth oscillator will be changed or modulated. If E_i is keyed between two voltage levels, the sawtooth oscillator changes frequencies. This type of application is called *frequency shift keying* (FSK) and is used for data transmission. These two preset frequencies correspond to "0" and "1" states (commonly called *space* and *mark*) in binary.

Example 6-9

If E_i is doubled to -2 V in Fig. 6-8, find the new frequency of oscillation.

Solution In Eq. (6–9c) use $|E_i|$

$$f = \left(\frac{1}{R_iC}\right)\frac{E_i}{V_{\text{ref}}} = \frac{1}{(10 \times 10^3\ \Omega)(0.1 \times 10^{-6}\ \text{F})}\frac{E_i}{10\ \text{V}}$$

$$= \frac{1}{1.0 \times 10^{-3}\ \text{s}} = \frac{E_i}{10\ \text{V}} = \left(\frac{100\ \text{Hz}}{\text{V}}\right)E_i$$

For $E_i = -2$ V, $f = (2\ \text{V})(100\ \text{Hz/V}) = 200$ Hz. Thus as E_i changes from 0 V to -10 V, frequency changes from 0 Hz to 1 kHz. The peak amplitude of the sawtooth wave remains equal to V_{ref} (10 V) for all frequencies.

Example 6-10

Keep E_i, R_i, and C at their value shown in Fig. 6-8(a). Reduce V_{ref} by one-half to 5 V. Is the frequency doubled or halved?

Solution From Example 6-8 and Eq. (6-9c),

$$f = \frac{1\ \text{V}}{(\text{ms})V_{\text{ref}}} = \frac{(1000\ \text{Hz})/\text{V}}{V_{\text{ref}}}$$

For $V_{\text{ref}} = 10$ V, $f = 100$ Hz. For $V_{\text{ref}} = 5$ V the frequency is doubled to 200 Hz. As V_{ref} is *reduced* from 10 V to 0 V, the frequency is increased from 100 Hz to a very high value.

6-4.6 Disadvantages

The triangle-wave generators of Section 6-3 are inexpensive and reliable. However, they have two disadvantages. The rates of rise and fall of the triangle wave are unequal. Also, the peak values of both triangle-wave and square-wave outputs are unequal, because the magnitudes of $+V_{\text{sat}}$ and $-V_{\text{sat}}$ are unequal.

In the next section we substitute an AD630 for the comparator. This will give the equivalent of precisely equal square-wave ± voltages that will *also* be equal to the ± peak values of triangle-wave voltage. Once we have made a precision triangle-wave generator, we will use it to drive a new state-of-the-art trigonometric function generator to make a precision sine-wave generator.

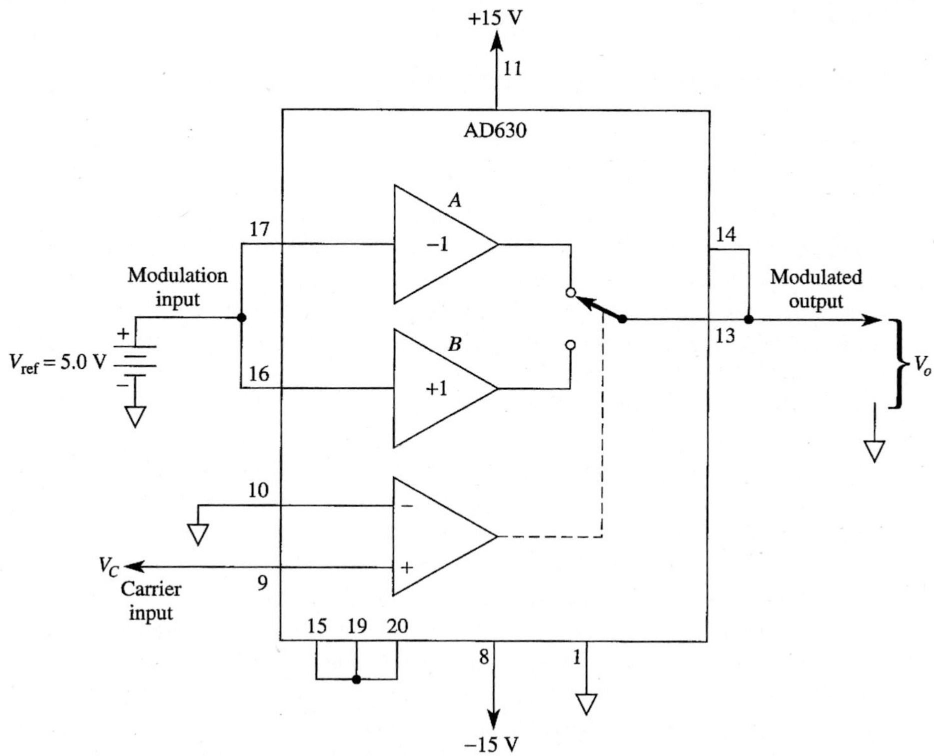

(a) The AD630 is wired as a switch gain amplifier.

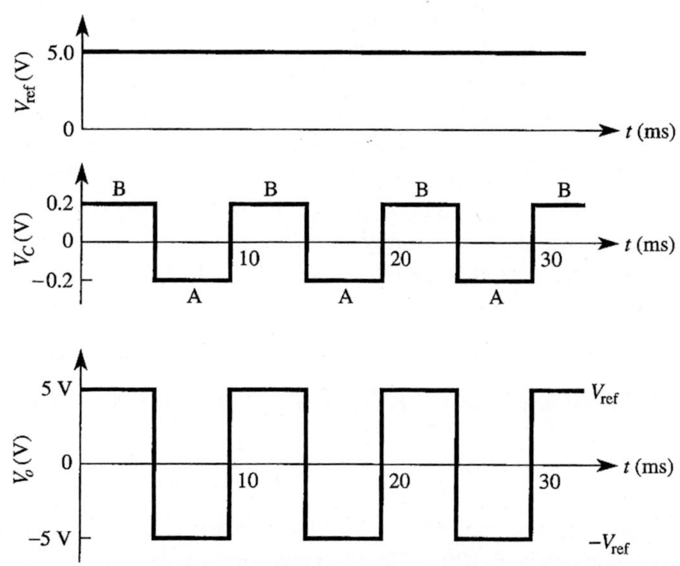

(b) Output and input waveshapes.

FIGURE 6-9 Operation of the AD630 balanced modulator/demodulator as a switched gain amplifier. (a) Wiring for switched gains of +1 or −1; (b) carrier V_C selects gains of +1 or −1 for input V_{ref}. Output V_o is equal to precisely V_{ref} or $-V_{ref}$.

6-5 BALANCED MODULATOR/DEMODULATOR, THE AD630

6-5.1 Introduction

The AD630 is an advanced integrated circuit. It has 20 pins, which allows this versatile switched voltage gain IC to act as a modulator, demodulator, phase detector, and multiplexer, as well as perform other signal conditioning tasks. We connect the AD630, as in Fig. 6-9(a), as a controlled switched gain ($+1$ or -1) amplifier. This particular application will be examined by discussing the role performed by the dominant terminals.

6-5.2 Input and Output Terminals

The input signal V_{ref} is connected to *modulation* pins 16 and 17 in Fig. 6-9, and thus to the inputs of two amplifiers, A and B. The gain of A is programmed for -1 and B for $+1$ by shorting terminals (1) 13 to 14, (2) 15 to 19 to 20, (3) 16 to 17, and (4) grounding pin 1.

The carrier input terminal, pin 9 (in this application), determines which amplifier, A or B, is connected to the output terminal. If pin 9 is *above* the voltage at pin 10 (ground), amplifier B is selected. Voltage at output pin 13 then equals V_{ref} times $(+1)$.

If pin 9 voltage is *below* ground (negative), amplifier A is selected and output pin 13 equals V_{ref} times (-1). (Note that in communication circuits, V_{ref} is called the *analog data* or *signal voltage,* V_C is called a chopper or *carrier voltage,* and V_o is the *modulated* output. That is, the *amplitude* of the low-frequency signal voltage is impressed upon the higher-frequency carrier wave—hence the names selected for the AD630's input and output terminals.)

6-5.3 Input–Output Waveforms

V_{ref} is a dc voltage of 5.0 V in Fig. 6-9(b). V_C is a 100-Hz square wave with peak amplitudes that must exceed ± 1 mV. Output voltage V_o is shown in Fig. 6-9(c) to switch synchronously with V_C from $+V_{ref}$ to $-V_{ref,}$ and vice versa. We are going to replace the unpredictable $\pm V_{sat}$ of the 301 comparator in Fig. 6-6 with precisely + or $-V_{ref}$. Moreover, V_{ref} can be adjusted easily to any required value. As shown in the next section, V_{ref} will set the positive and negative peak values of both triangle-wave and square-wave generators.

6-6 PRECISION TRIANGLE/SQUARE-WAVE GENERATOR

6-6.1 Circuit Operation

Only six parts plus a voltage source, V_{ref}, make up the versatile precision triangle- and square-wave generator in Fig. 6-10(a). Circuit operation is explained by referencing the waveshapes in Fig. 6-10(b). We begin at time zero. Square-wave output V_{os} begins at $-V_{ref}$ or -5 V. This forces the triangle wave V_{ot} to go positive from a starting point of $-V_{ref} = -5$ V. During this time, pin 9 is below ground to select an AD630 gain of -1 and holds V_{os} at -5 V.

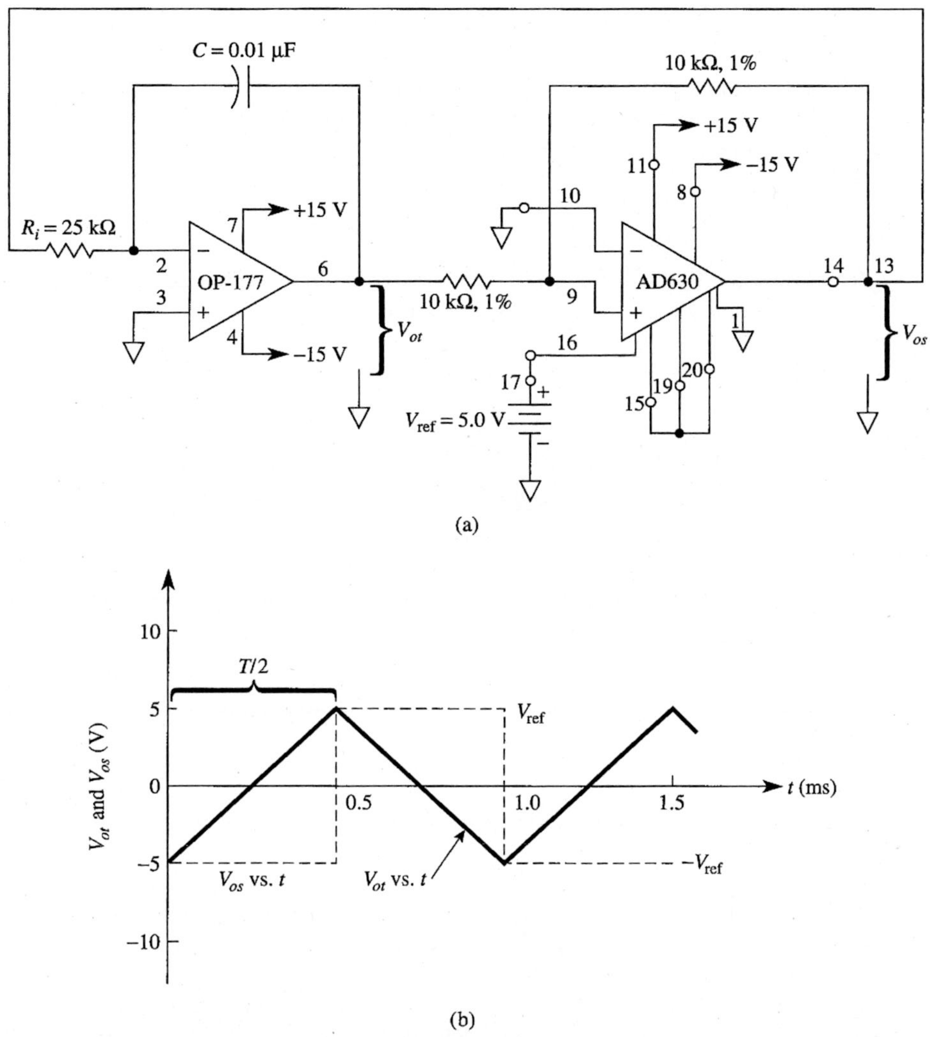

FIGURE 6-10 The precision triangle/square-wave oscillator in (a) has the output waveshapes in (b). V_{ref} should be buffered for a low-impedance source voltage. (a) Precision triangle/square-wave oscillator (compare with Fig. 6-6). V_{ref} must be a low-impedance source. V_{ref} sets the ± peak values and R_i adjusts the frequency. (b) Square- and triangular-wave output waveshapes.

At time $T/2 = 0.5$ ms, V_{ot} reaches +5 V, where pin 9 is driven slightly positive to select an AD630 gain of +1. This snaps V_{os} to $V_{ref} = +5$ V. V_{os} then drives V_{ot} negative. When V_{ot} reaches −5 V, pin 9 goes negative at $T = 1.0$ ms and snaps V_{os} negative to −5 V. This completes one cycle of oscillation and begins another.

6-6.2 Frequency of Oscillation

The easiest way to find the frequency of oscillation is to begin.with the rate of rise of the triangle wave, V_{ot}/t, in volts per second. The rate of rise of the triangle wave, from 0 to 0.5 ms in Fig. 6-10(b), is found from

$$\frac{V_{ot}}{t} = \frac{V_{\text{ref}}}{R_i C} \tag{6-10}$$

The time t for a half-cycle is $T/2$, and during this time, V_{ot} changes by $2V_{\text{ref}}$. Substituting these for t and V_{ot} into Eq. (6-10), we obtain

$$\frac{2V_{\text{ref}}}{T/2} = \frac{V_{\text{ref}}}{R_i C} \tag{6-11}$$

and solve for both period T and frequency of oscillation f:

$$T = 4R_i C \qquad \text{and} \qquad f = \frac{1}{T} = \frac{1}{4R_i C} \tag{6-12}$$

Note that V_{ref} cancels out in Eqs. (6-11) and (6-12). This is a very important advantage. The peak output voltages of both square- and triangle-wave signals are set by $+V_{\text{ref}}$. As V_{ref} is adjusted, the frequency of oscillation does *not* change.

Example 6-11

Make a triangle/square-wave generator that has peak voltages of $\pm$ 5 V and oscillates at a frequency of 1.0 kHz.

Solution Choose $V_{\text{ref}} = 5.0$ V. For low impedance, V_{ref} should be the output of an op amp. Arbitrarily choose $C = 0.01\ \mu$F. From Eq. (6-12),

$$R_i = \frac{1}{4\,fC} = \frac{1}{4(1000)(0.01\ \mu F)} = 25.0\ k\Omega$$

For a fine adjustment of the output frequency, make R_i from a 22-kΩ resistor in series with a 5- or 10-kΩ variable resistor.

6-7 SINE-WAVE GENERATION SURVEY

Commercial function generators produce triangular, square, and sinusoidal signals whose frequency and amplitude can be changed by the user. To obtain a sine-wave output, the triangle wave is passed through a shaping network made of carefully selected resistors and diodes (see Fig. 7-19). The sine waves thus produced are reasonably good. However, there is inevitably some distortion, particularly at the peaks of the sine wave. Another solution is to use an IC function generator chip such as MAXIM's MAX038, which is capable of producing sine, square, and triangle waveforms from 1 Hz to 20 MHz.

When an application requires a single-frequency sine wave, conventional oscillators use phase-shifting techniques that usually employ (1) two *RC* tuning networks, and (2) complex amplitude limiting circuitry. To minimize distortion, the limit circuit must usually be custom-adjusted for each oscillator. The frequency of this oscillator is difficult to vary because two *RC* networks must be varied and their values must track within $\pm 1\%$.

Waveforms may also be generated by using the AD630 with a universal trigonometric function generator, the AD639. The AD630 has already been used to generate a precision triangle wave whose frequency and amplitudes are precise and easy to adjust. We will connect the triangle-wave output V_{ot} of Fig. 6-10(a) to an AD639 universal trigonometric function generator. The resulting circuit will have the best qualities of a *precision sine-wave generator whose frequency will be easily adjustable.*

6-8 UNIVERSAL TRIGONOMETRIC FUNCTION GENERATOR, THE AD639

6-8.1 Introduction

The AD639 is a state-of-the-art trigonometric function generator. It will perform all trigonometric functions *in real time,* including sin, cos, tan, cosec, sec, and cotan. When a calculator performs a trig function, the operator punches in a number corresponding to the number of angular degrees and punches SIN. The calculator pauses, then displays a number indicating the sine of the angle. That is, a number for angle θ is entered and the calculator produces a number for sin θ.

The AD639 accepts an input voltage that represents the angle. It is called the *angle voltage,* V_{ang}. For the AD639, the angle voltage is found from

$$V_{\text{ang}} = \left(\frac{20\text{ mV}}{1^\circ}\right)\theta = \left(\frac{1\text{ V}}{50^\circ}\right)\theta \tag{6-13}$$

Four input terminals are available. However, we shall look at only the single active input that generates sin functions. The output voltage will equal sin θ or 10 sin θ, depending how the internal gain control is pin programmed.

6-8.2 Sine Function Operation

The AD639 is wired to output $V_o = 1 \sin\theta$ in Fig. 6-11. There are four input terminals: 1, 2, 7, and 8. Wired as shown, the chip performs a sine function. Pins 3, 4, and 10 control gain. Normally, 3 and 4 are grounded so that pin 10 can activate the internal gain control. A gain of 1 results when pin 10 is wired to $-V_s$ or pin 9. Wire pin 10 to $+V_s$ or pin

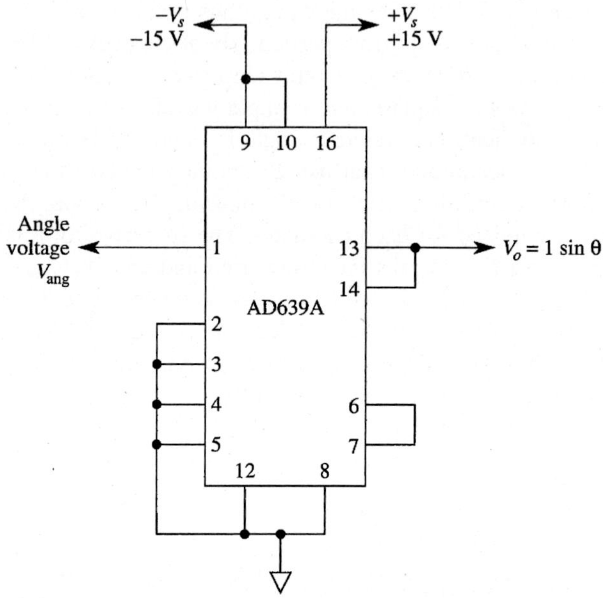

(a)

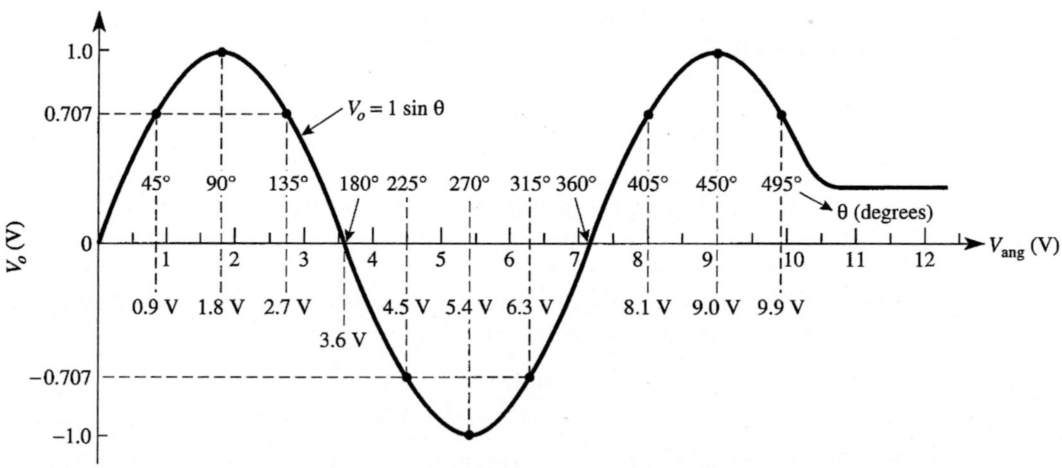

(b)

FIGURE 6-11 The AD639 is pin-programmed in (a) to act as a sine function generator. Each ± 20 mV of input angle voltage corresponds to an input angle of $\theta = \pm 1°$. Output V_o equals $1 \times \sin \theta$. (a) The AD639A is pin-programmed to output the sine of the angle voltage; (b) output voltage V_o equals the sine of θ if θ is represented by an angle voltage of 20 mV per angular degree.

16 and obtain a gain of +10. Then $V_o = 10 \sin \theta$. Pin 6 is a precision 1.80-V reference voltage that corresponds to an angle voltage of 90° (see Eq. (6-13)). We analyze sine function operation by an example.

Example 6-12

Calculate the required input angle voltage and resultant output voltage for angles of (a) ±45°; (b) ±90°; (c) ±225°; (d) ±405°.

Solution From Eq. (6-13) and Fig. 6-11,

(a) $V_{\text{ang}} = \left(\frac{20 \text{ mV}}{1°}\right)(\pm 45°) = \pm 0.90 \text{ V}$, $V_o = 1 \sin(\pm 45°) = \pm 0.707 \text{ V}$.

(b) $V_{\text{ang}} = \left(\frac{20 \text{ mV}}{1°}\right)(\pm 90°) = \pm 1.80 \text{ V}$, $V_o = 1 \sin(\pm 90°) = \pm 1.0 \text{ V}$.

(c) $V_{\text{ang}} = \left(\frac{20 \text{ mV}}{1°}\right)(\pm 225°) = \pm 4.50 \text{ V}$, $V_o = 1 \sin(\pm 225°) = \pm 0.707 \text{ V}$.

(d) $V_{\text{ang}} = \left(\frac{20 \text{ mV}}{1°}\right)(\pm 405°) = \pm 8.10 \text{ V}$, $V_o = 1 \sin(\pm 405°) = \pm 0.707 \text{ V}$.

Example 6-12 clearly illustrates that the AD639, remarkable as it is, cannot output the sine of, for example, 36,000°. This would require an angle voltage of 720 V. The normal ±15-V supply limits the guaranteed usable input angle to ±500°, or ±10.000 V. We extend the results of Example 6-12 to summarize briefly the performance of the sine function generator in Table 6-1 and Fig. 6-11(b).

In Fig. 6-11(b), V_o is plotted against both V_{ang} and θ. A study of this figure shows that if V_{ang} *could be varied linearly* by a triangle wave, V_o *would vary sinusoidally*. Further, if the frequency of the triangle wave could be varied easily, the sine-wave frequency could *easily* be tuned, adjusted, or varied. We pursue this observation in the next section.

6-9 PRECISION SINE-WAVE GENERATOR

6-9.1 Circuit Operation

Connect the precision triangle-wave oscillator in Fig. 6-10 to the sine function generator in Fig. 6-11 to construct the precision sine-wave generator in Fig. 6-12. As a bonus, we also have precision triangle-wave and square-wave outputs. The 1.80-V reference voltage of the AD639 is connected to modulation inputs 16 and 17 of the AD630 modulator (Fig. 6-9). Circuit operation is now examined by reference to Fig. 6-12.

TABLE 6-1 AD639 Sine Functions[a]

Input		Output (V)	
θ (angular degrees)	Angle voltage, V_{ang} (V)	$V_o = 1 \sin \theta$ (wire pin 10 to 9)	$V_o = 10 \sin \theta$ (wire pin 10 to 16)
0	0.00	0.000	0.000
±45	±0.90	±0.707	±7.07
±90	±1.80	±1.000	±10.07
±135	±2.70	±0.707	±7.07
±180	±3.60	0.000	0.000
±225	±4.50	±0.707	±7.07
±270	±5.40	±1.000	±10.00
±315	±6.30	±0.707	±7.07
±360	±7.20	0.000	0.00
±405	±8.10	±0.707	±7.07
±450	±9.00	±1.000	±10.00
±495	±9.90	±0.707	±7.07
±500	±10.00	±0.643	±6.43

[a] Connect terminal 10 to 9 to pin program $V_o = \sin \theta$; or connect pin 10 to 16 to pin program $V_o = 10 \sin \theta$. Input angle voltage $V_{ang} = (20\text{ mV}/1°\text{C})\ \theta$.

Triangle-wave rise time, 0 to T/2 in Fig. 6-12(b)

1. *AD630*
   a. Pin 13 is at $-V_{ref} = -1.8$ V, causing
   b. Pin 9 to select gain $= -1$ to hold 13 at -1.8 V and
   c. Op amp output voltage to ramp up.
2. *Op amp*
   a. Pin 6 ramps from $-V_{ref} = -1.8$ V *toward* $+V_{ref} = 1.8$ V to
   b. Hold pin 9 of the AD630 negative and
   c. Drive input 1 of the AD639 with an angle voltage linearly from -1.8 to 1.8 V.
3. *AD639*
   a. Pin 1's input angle voltage corresponds to an input angle that varies linearly from $-90°$ to $+90°$.
   b. Pin 13 outputs $V_o = 10 \sin \theta$ from -10 to $+10$ V.

When op amp pin 6 reaches $+1.8$ V, pin 9 of the AD630 goes positive to select a gain of $+1$. Its output, in 13, snaps to $+1.8$ V. This begins the *fall time*.

Triangle-wave fall time, T/2 to T in Fig. 6-12(b)

1. *AD630:* Causes the triangle wave to ramp down from $+1.8$ V to -1.8 V. At -1.8 V, gain is switched to -1 and a new cycle begins.

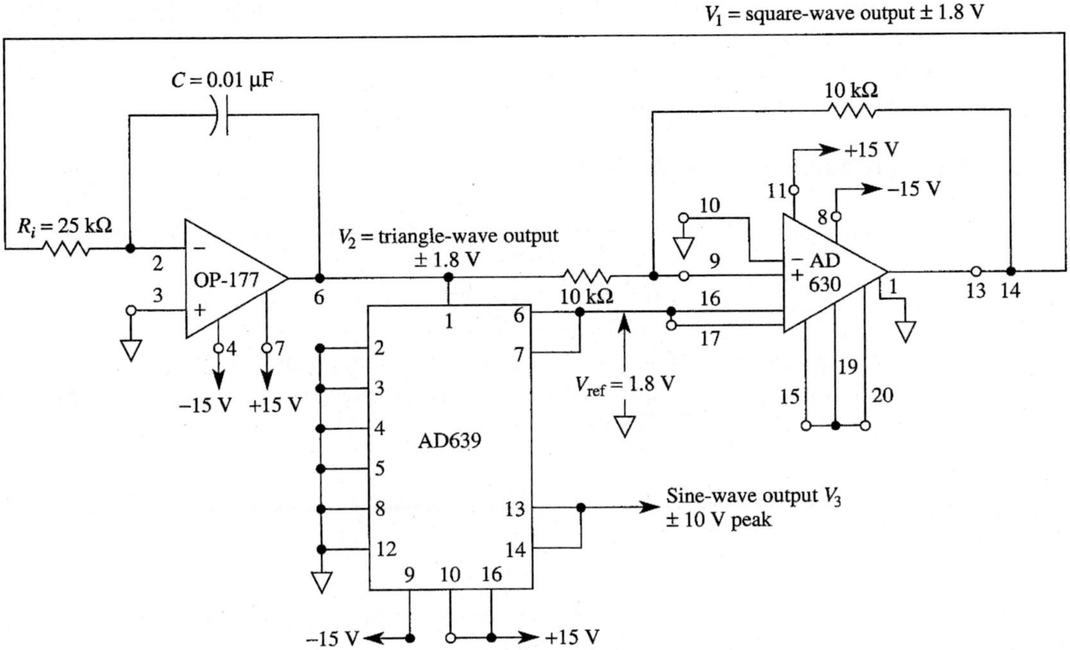

(a) Precision sine-wave generator circuit.

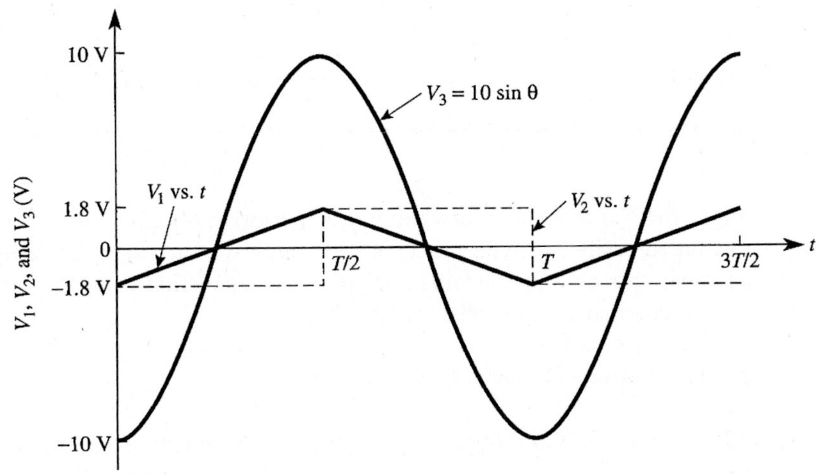

(b) Waveshapes.

FIGURE 6-12 Frequency of the precision sine-square-triangle-wave generator in (a) can be easily changed by adjusting R_i. Output waveshapes are shown in (b). Their amplitudes are independent of frequency.

2. *Op amp:* Applies an angle voltage to input pin 1 of the AD639 that varies linearly from +1.8 to −1.8 V.
3. *AD639:* Its input angle voltage corresponds to an input angle of $\theta = +90°$ to $-90°$. Pin 13 outputs a sine wave that varies from +10 to −10 V.

6-9.2 Frequency of Oscillation

The frequency of oscillation, f, is determined by R_i, C, and the op amp in Fig. 6-12(a) from

$$f = \frac{1}{4R_iC} \tag{6-14}$$

Peak amplitudes of the triangle wave and square wave are precisely equal to ±1.8 V. The sine wave has peak amplitude of ±10 V and is synchronized to the triangle wave (for the ±1-V peak, change the AD639 pin 10 connection to $-V_s$).

Example 6-13

Let $C = 0.025$ μF in Fig. 6-12(a) (two 0.05-μF capacitors in series). How does frequency change as R_i is changed from 10 kΩ to 100 kΩ?

Solution From Eq. (6-14),

$$f = \frac{1}{4(10\text{ k}\Omega)(0.025\ \mu\text{F})} = 1\text{ kHz} \qquad f = \frac{1}{4(100\text{ k}\Omega)(0.025\ \mu\text{F})} = 100\text{ Hz}$$

Example 6-13 shows the overwhelming superiority of this multiwave generator. Frequency is tuned easily *and* with precision. Although we have used the AD639 to generate a sine wave, this chip is a universal trigonometric function generator and could be rewired to produce other trigonometric waveforms.

6-9.3 High-Frequency Waveform Generator

Maxim's MAC038 is a 20-pin, high-frequency, precision function generator whose frequency can be controlled over a wide range from 0.1 Hz to 20 MHz. It can produce sine, triangle, or square waves at outputs that are selected by an appropriate code at two (transistor-transistor logic) TTL-compatible select inputs. It can also produce sawtooth or pulse outputs as well as a synchronizing output.

Applications for this versatile chip include function or FSK generation, VCOs (voltage controlled oscillator), frequency modulators, and synthesizers, as well as pulse-width modulation.

6-10 PSPICE SIMULATION OF SIGNAL GENERATOR CIRCUITS

In this section, we will use PSpice to model and simulate the performance of four signal generator circuits studied in this chapter: the free-running multivibrator, the one-shot multivibrator, the bipolar triangle-wave generator, and the unipolar triangle-wave generator.

6-10.1 Free-Running Multivibrator

Refer to Fig. 6-1 and create the PSpice model of the circuit. Set the resistor and capacitor values as given in Examples 6-1 and 6-2. Use the 741 op amp if you are using the evaluation software package of PSpice. Obtain a plot of V_C and V_o versus time. For the circuit to begin oscillating in a simulation it is necessary to provide a sudden impulse at the beginning of the simulation. This stimulus can be generated by using two pulse sources instead of two dc sources to power the op amp. The pulse width will be set to a much longer time interval than the period of oscillation and will have a fast rise time to simulate suddenly applying power to the circuit. To begin, place the following parts in the work area.

Draw => Get New Part

Part	Number	Library
=> **uA741**	1	eval.slb
=> **VPULSE**	2	source.slb
=> **R**	3	analog.slb
=> **C**	1	analog.slb
=> **GLOBAL**	4	port.slb
=> **AGND**	4	port.slb

(*Note:* We are using **VPULSE** instead of **VDC** for the op amp supplies.) Arrange the parts and wire the circuit as shown in Fig. 6-1. Change the attributes of the parts as given in Examples 6-1 and 6-2. Set up each of the **VPULSE** attributes by double-clicking on the symbol. In the pop-up box, set the values for **V1** (minimum input voltage), **V2** (maximum input voltage), **TD** (time delay), **TR** (rise time), **TF** (fall time), **PW** (pulse width), and **PER** (period):

V1 => 0V => Save Attr
V2 => 15V => Save Attr => Change Display => Both name and value
TD => 0 => Save Attr
TR => 1ns => Save Attr
TF => 1ns => Save Attr
PW => 50s => Save Attr
PER => 51s => Save Attr

Each pulse attribute pop-up box is set up the same because the $-V$ supply has been rotated 180 degrees, as we have done with the dc supplies connected to pin 4 of other op amps. Double-click on the lead from the capacitor and label it **Vc.** Double-click on the lead from the output terminal of the op amp and label it **Vo** (see Fig. 6-13).

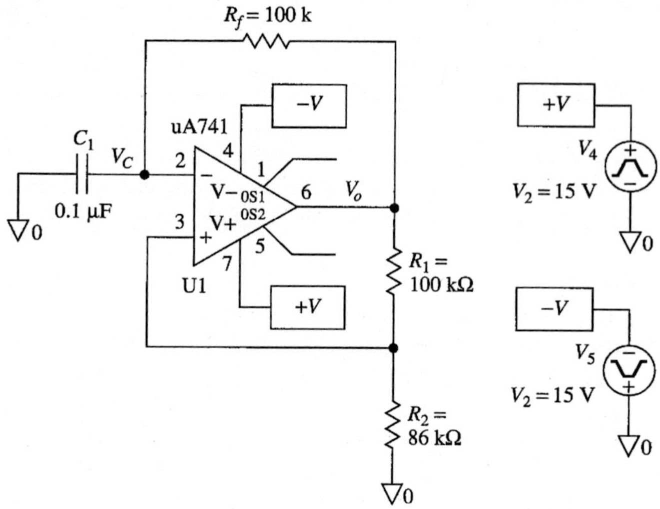

FIGURE 6-13 PSpice model of the free-running multivibrator.

In order to obtain a plot of V_C and V_o versus time, we must initialize the transient menu.

Analysis => **Setup** = > Select **Transient**
Click **Transient** => **Print Step:** => **0.2ms**
=> **Final Time:** => **60ms**

Save the circuit as a file with the **.SCH** extension. Run the simulation

Analysis => **Simulate**

In the Probe window, select

Trace => **Add** => **V[Vc]**
=> **V[Vo]**

Label the plots and obtain a printout as shown in Fig. 6-14.

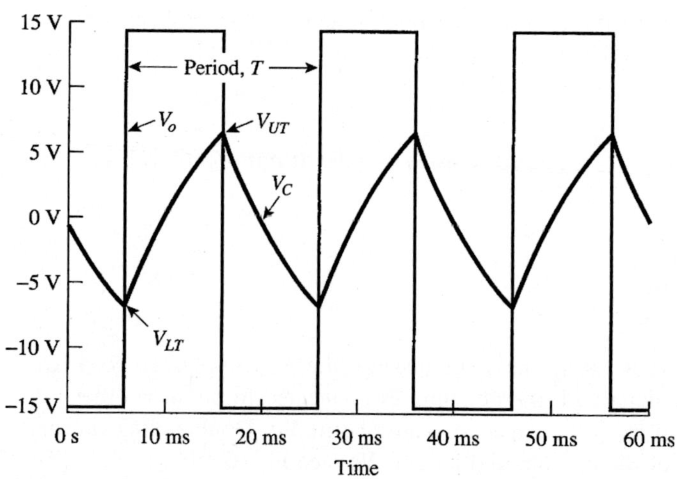

FIGURE 6-14 Capacitor and output voltage waveforms generated from the PSpice model of Fig. 6-13.

6-10.2 One-Shot Multivibrator

Create the PSpice model of the one-shot multivibrator shown in Fig. 6-4. Use the 741 op amp if you are using the evaluation software package of PSpice. Obtain a plot of E_i and V_o versus time. Place the following parts in the work area.

Draw => Get New Part

Part	Number	Library
=> **uA741**	1	eval.slb
=> **D1N4002**	2	eval.slb
=> **VPULSE**	1	source.slb
=> **VDC**	2	source.slb
=> **R**	4	analog.slb
=> **C**	2	analog.slb
=> **GLOBAL**	4	port.slb
=> **AGND**	6	port.slb

Arrange the parts and wire the circuit as shown in Fig. 6-4. Change the attributes as given in Fig. 6-4, but set $R_2 = 2\ \text{k}\Omega$ [so that results closely approximate Eq. (6-4a)] and set **VPULSE** for a 10-V negative pulse with a pulse width of 0.1 ms. The pulse attributes are shown in Fig. 6-15. Double-click the **VPULSE** symbol and set the values accordingly. Double-click on the lead from **VPULSE** and label it **Ei**. Label the output of the op amp as **Vo**. See Fig. 6-15.

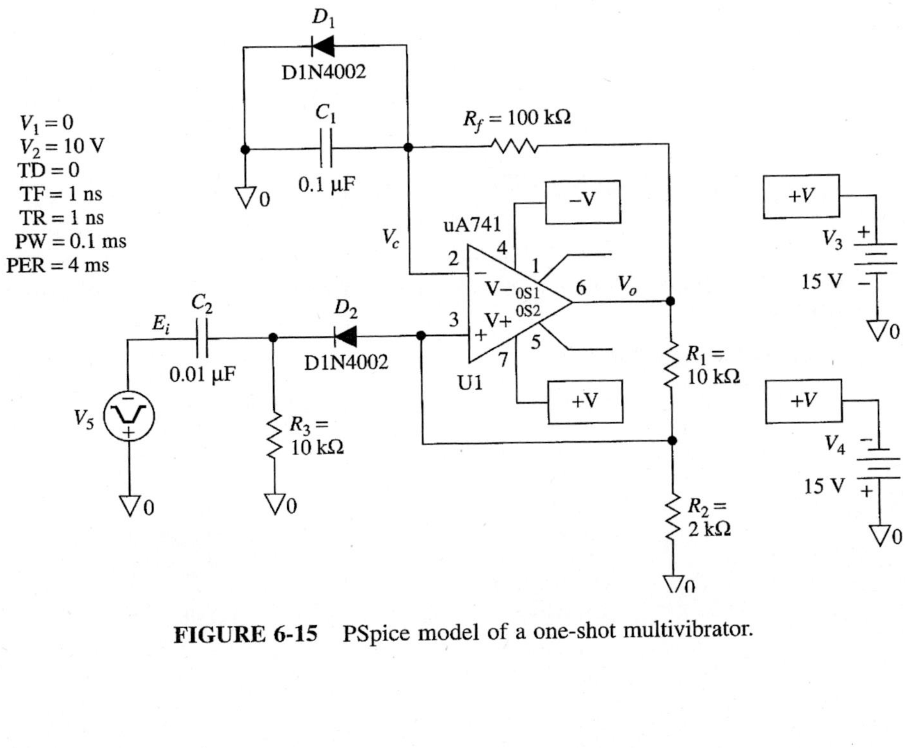

FIGURE 6-15 PSpice model of a one-shot multivibrator.

In order to obtain a plot of E_i and V_o versus time, we must initialize the transient menu.

Analysis => **Setup** => Select **Transient**
Click **Transient** => **Print Step:** => **100us**
=> **Final Time:** => **10ms**

Save the circuit as a file with the **.SCH** extension. Run the simulation

Analysis => **Simulate**

In the Probe window, select

Trace => **Add** => **V[Ei]**
=> **V[Vo]**

Label the plots and obtain a printout as shown in Fig. 6-16.

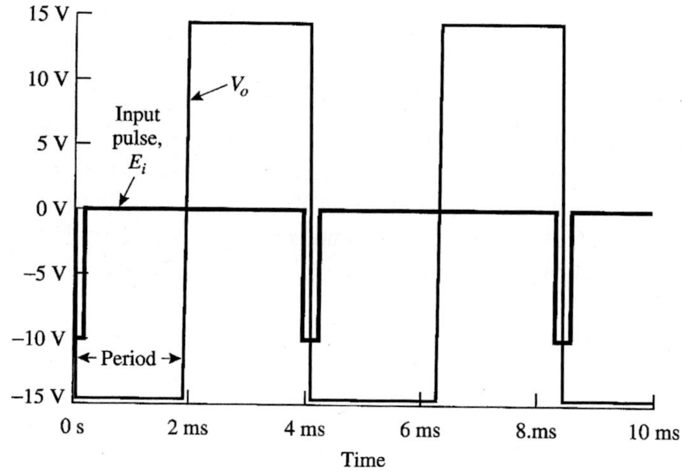

FIGURE 6-16 Waveforms for the one-shot multivibrator modeled in PSpice in Fig. 6-15.

6-10.3 Bipolar Triangle-Wave Generator

Create the PSpice model of the bipolar triangle-wave generator shown in Fig. 6-6. Use two 741 op amps. Obtain a plot of V_A and V_o versus time. Place the following parts in the work area.

Draw => **Get New Part**

Part	Number	Library
=> **uA741**	2	eval.slb
=> **VPULSE**	2	source.slb
=> **R**	4	analog.slb
=> **C**	1	analog.slb
=> **GLOBAL**	6	port.slb
=> **AGND**	5	port.slb

Arrange the parts, change the attributes, and wire the circuit as shown in Fig. 6-6. The **VPULSE** attributes are set the same as they are in Section 6-10.1 (see Fig. 6-17).

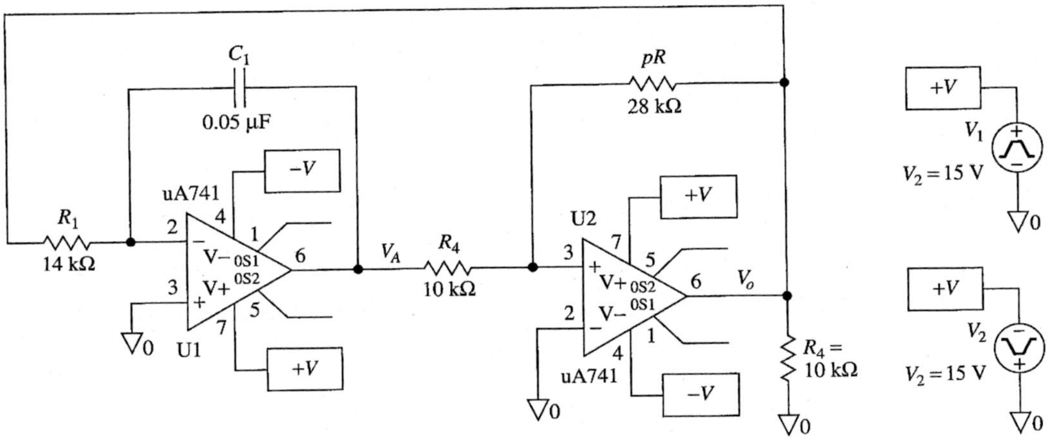

FIGURE 6-17 Bipolar triangle-wave generator modeled in PSpice.

In order to obtain a plot of V_A and V_o versus time, set the transient menu.

Analysis => **Setup** => Select **Transient**
Click **Transient** => **Print Step:** => **0.01ms**
=> **Final Time:** => **3ms**

Save the circuit as a file with the **.SCH** extension. Run the simulation

Analysis = > **Simulate**

In the Probe window, select

Trace => **Add** => **V[VA]**
=> **V[Vo]**

Label the plots and obtain a printout as shown in Fig. 6-18.

6-10.4 Unipolar Triangle-Wave Generator

Modify the PSpice model of Fig. 6-17 to create the unipolar triangle-wave generator shown in Fig. 6-7. From the PSpice parts list, obtain a 1N4002 diode and place it in series with pR (see Fig. 6-19). Save the circuit with the **.SCH** extension.

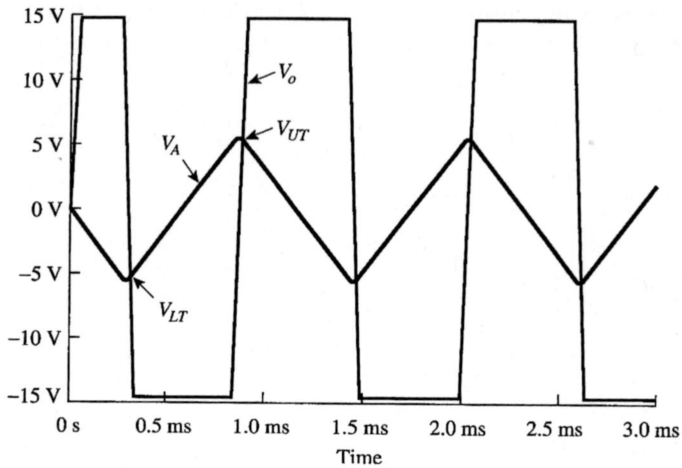

FIGURE 6-18 Waveforms generated from the bipolar triangle-wave generator shown in Fig. 6-17.

Follow the procedure given in Section 6-10.3 to obtain a plot of V_A and V_o (see Fig. 6-20).

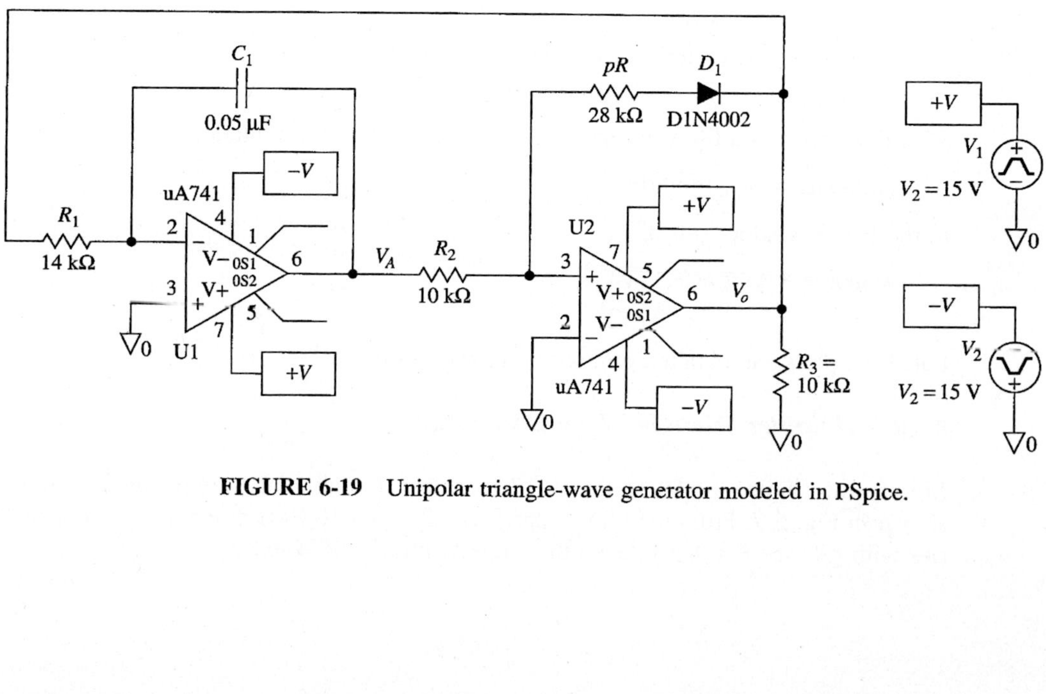

FIGURE 6-19 Unipolar triangle-wave generator modeled in PSpice.

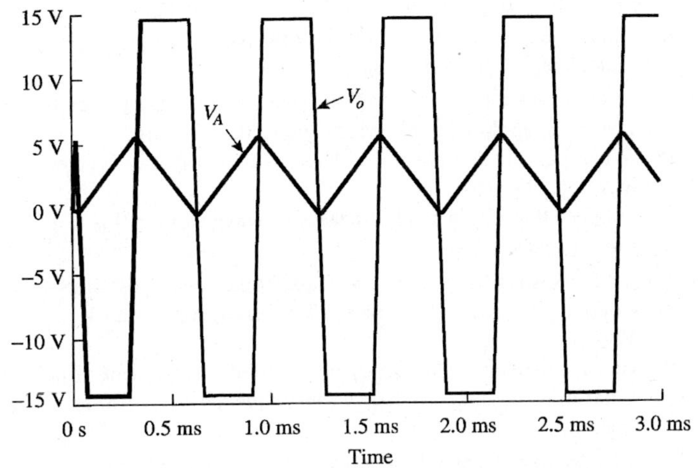

FIGURE 6-20 Waveforms V_A and V_o for the unipolar triangle-wave generator of Fig. 6-19.

PROBLEMS

6-1. Make two drawings of a multivibrator circuit with $R_1 = 100\ \text{k}\Omega$, $R_2 = 86\ \text{k}\Omega$, $R_f = 10\ \text{k}\Omega$, and $C = 0.01\ \mu\text{F}$. Show the direction of current through C and calculate both V_{UT} and V_{LT} for **(a)** $V_o = +V_{\text{sat}} = 15$ V; **(b)** $V_o = -V_{\text{sat}} = -15$ V.

6-2. Calculate the frequency of oscillation for the multivibrator circuit in Problem 6-1.

6-3. In Problem 6-1, if C is changed to 0.1 μF, do you expect the output frequency to oscillate at 500 Hz? (See Example 6-3.) What could you do to R_f to increase frequency to 1000 Hz?

6-4. The monostable multivibrator of Figs. 6-4 and 6-5 generates a negative output pulse in response to a negative-going input signal. How would you change these circuits to get a positive output pulse for a positive-going input edge?

6-5. Explain what is meant by *monostable recovery time.*

6-6. Sketch a one-shot multivibrator circuit whose output will deliver a negative pulse lasting 1 ms with a recovery time of about 0.1 ms.

6-7. Assume for simplicity that saturation voltages in the triangle-wave oscillator of Fig. 6-6 are ± 15 V, $R_i = R = 10\ \text{k}\Omega$, $C = 0.1\ \mu\text{F}$, and $pR = 50\ \text{k}\Omega$. Find the peak triangle-wave voltages and oscillating frequency.

6-8. Refer to the triangular-wave oscillator circuit of Fig. 6-6. What happens to peak output voltages and oscillating frequency if you **(a)** double pR only; **(b)** double R_i only; **(c)** double capacitor C only?

6-9. Change pR to 14 kΩ and C to 0.1 μF in the unipolar triangle-wave generator of Fig. 6-7. Find the resulting peak output voltage and frequency of oscillation. (See Example 6-7.)

6-10. In the sawtooth-wave generator of Fig. 6-8(a), let $V_{ref} = 1$ V, $R_i = 10$ kΩ, and $C = 0.1$ μF.
(a) Find an expression for frequency f in terms of E_i.
(b) Calculate f for $E_i = 1$ V and $E_i = 2$ V.

6-11. These questions refer to the AD630 balanced modulator circuit in Fig. 6-9.
(a) Name the application for which the AD630 is wired.
(b) When pin 9 is at a positive voltage, which amplifier is selected, and what is the value of V_o?
(c) Suppose that V_{ref} is a ±1-V-peak sine wave and pin 9 is at 1 V; what happens at V_o when pin 9 is changed to −1 V?

6-12. Figure 6-10 shows a precision triangle/square-wave oscillator. Three components control peak output voltages and oscillating frequency, R_i, C, and V_{ref}.
(a) Which does what?
(b) Can the oscillating frequency be adjusted independent of peak outputs, and vice versa?
(c) What must be done to change the frequency from 100 to 500 Hz and the peak voltages from ±5 V to ±1 V?

6-13. $V_o = 0.866$ V in the sine function generator circuit of Fig. 6-11.
(a) What angle does this represent?
(b) What is the value of the input angle voltage?

6-14. Calculate V_o in Fig. 6-11 when the input angle is 30° and pin 10 is wired to **(a)** pin 9; **(b)** pin 16.

6-15. Design a sine-wave oscillator whose frequency can be varied from 0.5 Hz to 50 Hz with just a single variable resistor.

CHAPTER 7

Op Amps with Diodes

LEARNING OBJECTIVES

Upon completing this chapter on op amps with diodes, you will be able to:

- Draw the circuit for a precision (or linear) half-wave rectifier.
- Show current flow and circuit voltages in a precision half-wave rectifier for either positive or negative inputs.
- Do the same for precision full-wave rectifiers.
- Sketch two types of precision full-wave rectifier circuits.
- Explain the operation of a peak detector circuit.
- Add one capacitor to a precision half-wave rectifier to make an ac-to-dc converter (mean-average-value) circuit.
- Explain the operation of dead-zone circuits.
- Draw the circuit for and explain the operation of precision clipper circuits.

- Name at least five application areas for precision rectifiers.
- Simulate op amp circuits with diodes using PSpice.

7-0 INTRODUCTION TO PRECISION RECTIFIERS

The major limitation of ordinary silicon diodes is that they cannot rectify voltages below 0.6 V. For example, Fig. 7-1(a) shows that V_o does not respond to positive inputs below 0.6 V in a half-wave rectifier built with an ordinary silicon diode. Figure 7-1(b) shows the waveforms for a half-wave rectifier built with an ideal diode. An output voltage occurs for all positive input voltages, even those below 0.6 V. A circuit that acts like an ideal

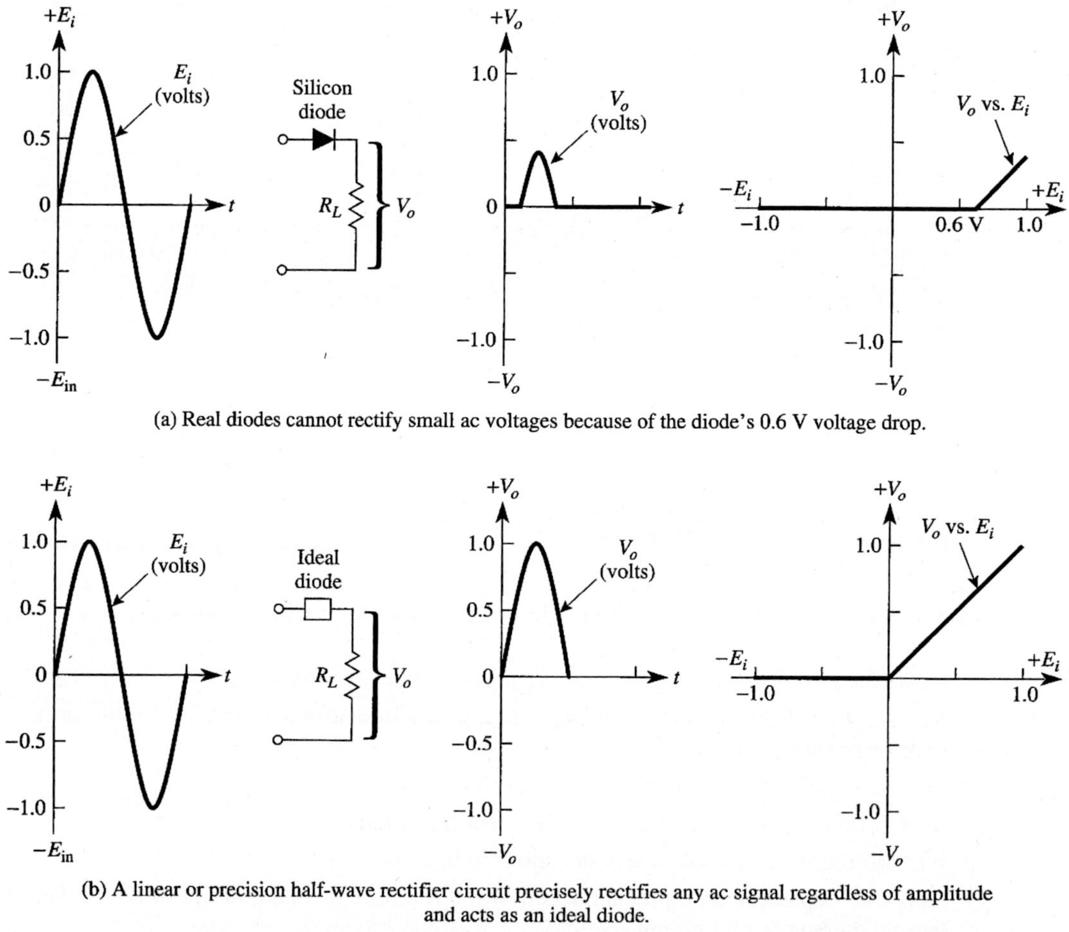

FIGURE 7-1 The ordinary silicon diode requires about 0.6 V of forward bias in order to conduct. Therefore, it cannot rectify small ac voltages. A precision half-wave rectifier circuit overcomes this limitation.

diode can be designed using an op amp and two ordinary diodes. The result is a powerful circuit capable of rectifying input signals of only a few millivolts.

The low cost of this equivalent ideal diode circuit allows it to be used routinely for many applications. They can be grouped loosely into the following classifications: linear half-wave rectifiers and precision full-wave rectifiers.

1. *Linear half-wave rectifiers.* The linear half-wave rectifier circuit delivers an output that depends on the magnitude *and polarity* of the input voltage. It will be shown that the linear half-wave rectifier circuit can be modified to perform a variety of signal-processing applications. The linear half-wave rectifier is also called a *precision half-wave rectifier* and acts as an ideal diode.
2. *Precision full-wave rectifiers.* The precision full-wave rectifier circuit delivers an output proportional to the magnitude but *not* the polarity of the input. For example, the output can be positive at 2 V for inputs of either +2 V or −2 V. Since the absolute value of +2 V and −2 V is equal to +2 V, the precision full-wave rectifier is also called an *absolute-value circuit.*

Applications for both linear half-wave and precision full-wave rectifiers includes:

1. Detection of amplitude-modulated signals
2. Dead-zone circuits
3. Precision bound circuits or *clippers*
4. Current switches
5. Waveshapers
6. Peak-value indicators
7. Sample-and-hold circuits
8. Absolute-value circuits
9. Averaging circuits
10. Signal polarity detectors
11. Ac-to-dc converters

The functions listed are often necessary to condition signals before they are applied to an input of a microcontroller.

7-1 LINEAR HALF-WAVE RECTIFIERS

7-1.1 Introduction

Linear half-wave rectifier circuits transmit only one-half cycle of a signal and eliminate the other by *bounding* the output to 0 V. The input half-cycle that is transmitted can be either inverted or noninverted. It can also experience gain or attenuation, or remain unchanged in magnitude, depending on the choice of resistors and placement of diodes in the op amp circuit.

7-1.2 Inverting Linear Half-Wave Rectifier, Positive Output

The inverting amplifier is converted into an ideal (linear precision) half-wave rectifier by adding two diodes as shown in Fig. 7-2. When E_i is positive in Fig. 7-2(a), diode D_1 conducts, causing the op amp's output voltage, V_{OA}, to go negative by one diode drop ($\simeq$0.6 V). This forces diode D_2 to be reverse biased. The circuit's output voltage V_o equals zero because input current I flows through D_I. For all practical purposes, no current flows through R_f and therefore $V_o = 0$.

Note the load is modeled by a resistor R_L and must always be resistive. If the load is a capacitor, inductor, voltage, or current source, then V_o will *not* equal zero.

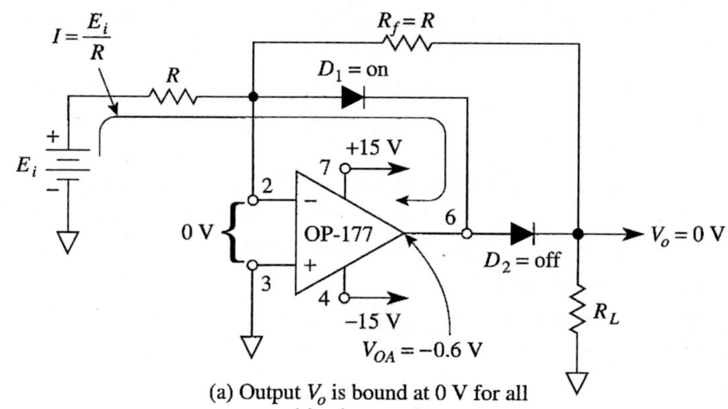

(a) Output V_o is bound at 0 V for all positive input voltages.

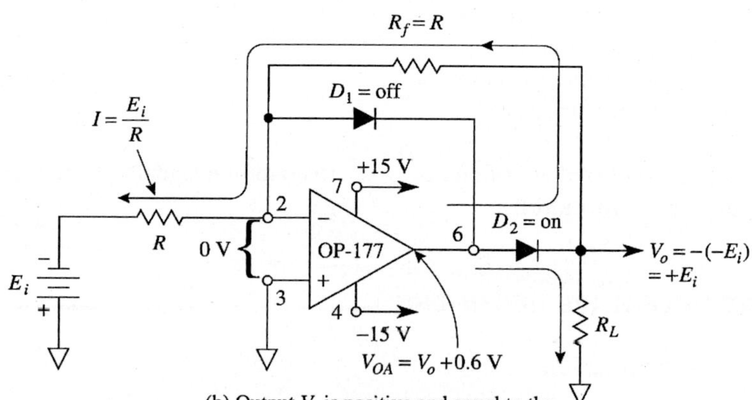

(b) Output V_o is positive and equal to the magnitude of E_i for all negative inputs.

FIGURE 7-2 Two diodes convert an inverting amplifier into a positive-output, inverting, linear (ideal) half-wave rectifier. Output V_o is positive and equal to the magnitude of E_i for negative inputs, and V_o equals 0 V for all positive inputs. Diodes are 1N914 or 1N4154.

In Fig. 7-2(b), negative input E_i forces the op amp output V_{OA} to go positive. This causes D_2 to conduct. The circuit then acts like an inverter, since $R_f = R_i$ and $V_o = -(-E_i) = +E_i$. Since the (−) input is at ground potential, diode D_1 is reverse biased. Input current is set by E_i/R_i and gain by $-R_f/R_i$. Remember that this gain equation applies only for negative inputs, and V_o can only be positive or zero.

Circuit operation is summarized by the waveshapes in Fig. 7-3. V_o can only go positive in a linear response to negative inputs. The most important property of this linear half-wave rectifier will now be examined. An ordinary silicon diode or even a hot-carrier diode requires a few tenths of volts to become forward biased. Any signal voltage below this threshold voltage cannot be rectified. However, by connecting the diode in the feedback loop of an op amp, the threshold voltage of the diode is essentially eliminated. For example, in Fig. 7-2(b) let E_i be a low voltage of −0.1 V. E_i and R_i convert this low voltage to a current that is conducted through D_2. V_{OA} goes to whatever voltage is required to supply the necessary diode drop plus the voltage drop across R_f. Thus millivolts of input voltage can be rectified, since the diode's forward bias is supplied automatically by the negative feedback action of the op amp.

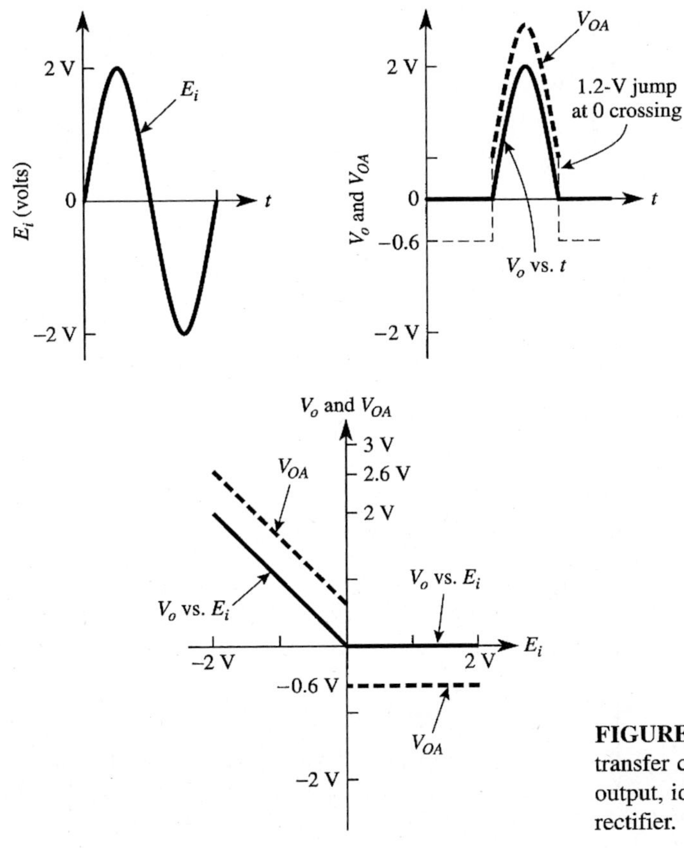

FIGURE 7-3 Input, output, and transfer characteristics of a positive-output, ideal, inverting half-wave rectifier.

Finally, observe the waveshape of op amp output V_{OA} in Fig. 7-3. When E_i crosses 0 V (going negative), V_{OA} jumps quickly from −0.6 V to +0.6 V as it switches from supplying the drop for D_2 to supplying the drop for D_1. This jump can be monitored by a differentiator to indicate the zero crossing. During the jump time the op amp operates open loop.

7-1.3 Inverting Linear Half-Wave Rectifier, Negative Output

The diodes in Fig. 7-2 can be reversed as shown in Fig. 7-4. Now only positive input signals are transmitted and inverted. The output voltage V_o equals 0 V for all negative inputs. Circuit operation is summarized by the plot of V_o and V_{OA} versus E_i in Fig. 7-4(b).

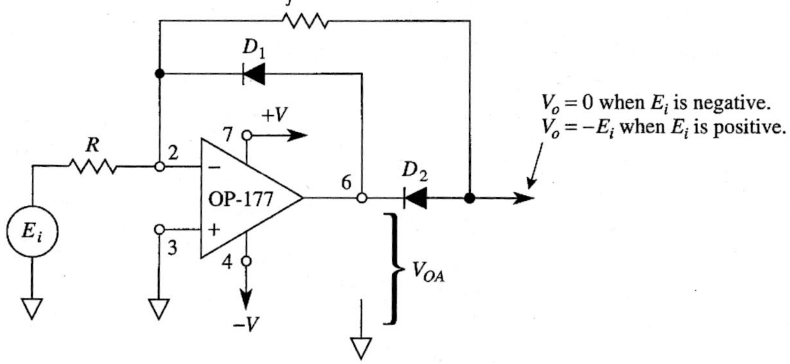

(a) Inverting linear half-wave rectifier: negative output.

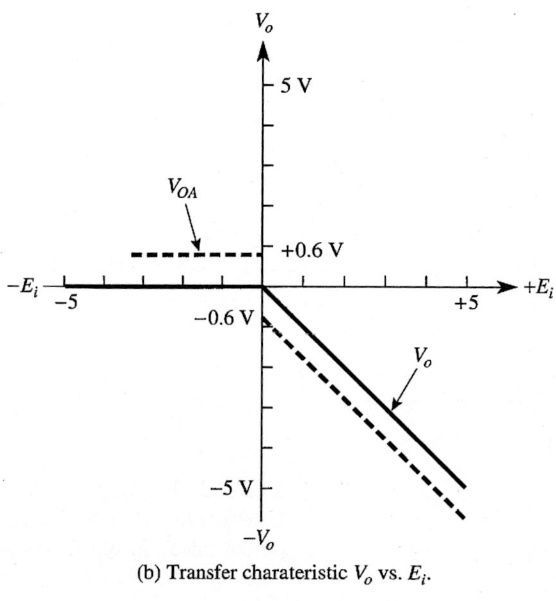

(b) Transfer charateristic V_o vs. E_i.

FIGURE 7-4 Reversing the diodes in Fig. 7-2 gives an inverting linear half-wave rectifier. This circuit transmits and inverts only positive input signals.

7-1.4 Signal Polarity Separator

The circuit of Fig. 7-5 is an expansion of the circuits in Figs. 7-2 and 7-4. When E_i is positive in Fig. 7-5(a), diode D_1 conducts and an output is obtained only on output V_{o_1}. V_{o_2} is bound at 0 V. When E_i is negative, D_2 conducts, $V_{o_2} = -(-E_i) = +E_i$, and V_{o_1} is bound at 0 V. This circuit's operation is summarized by the waveshapes in Fig. 7-6.

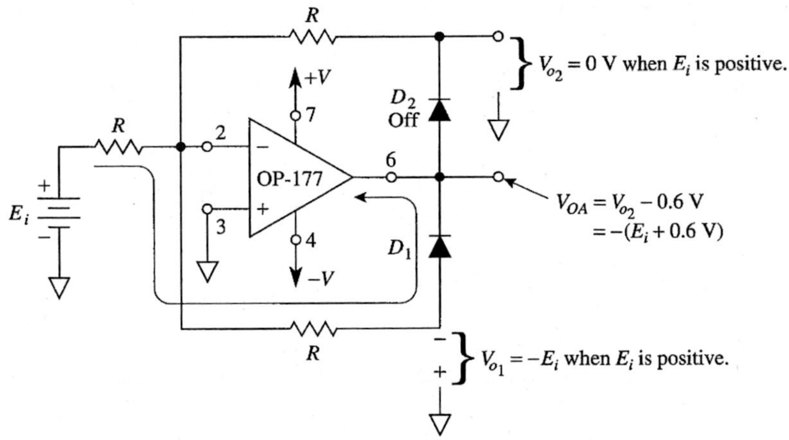

(a) When E_i is positive, V_{o1} is negative and V_{o2} is bound at 0 V.

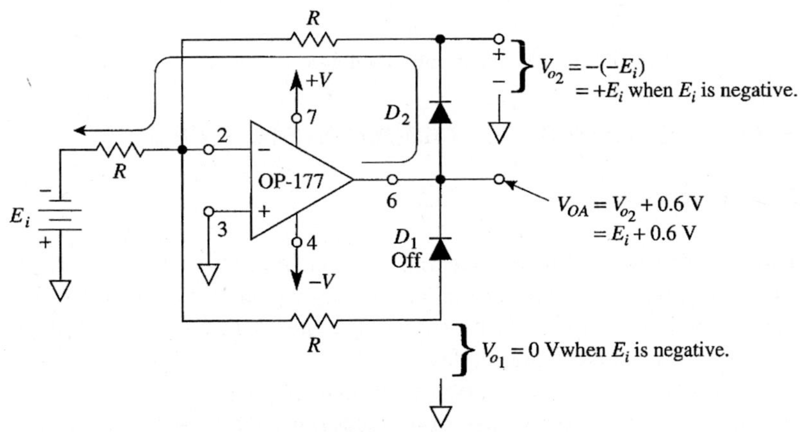

(b) When E_i is negative, $V_{o1} = 0$ V and V_{o2} goes positive.

FIGURE 7-5 This circuit inverts and separates the polarities of input signal E_i. A positive output at V_{o_2} indicates that E_i is negative, and a negative output at V_{o_1} indicates that E_i is positive. These outputs should be buffered.

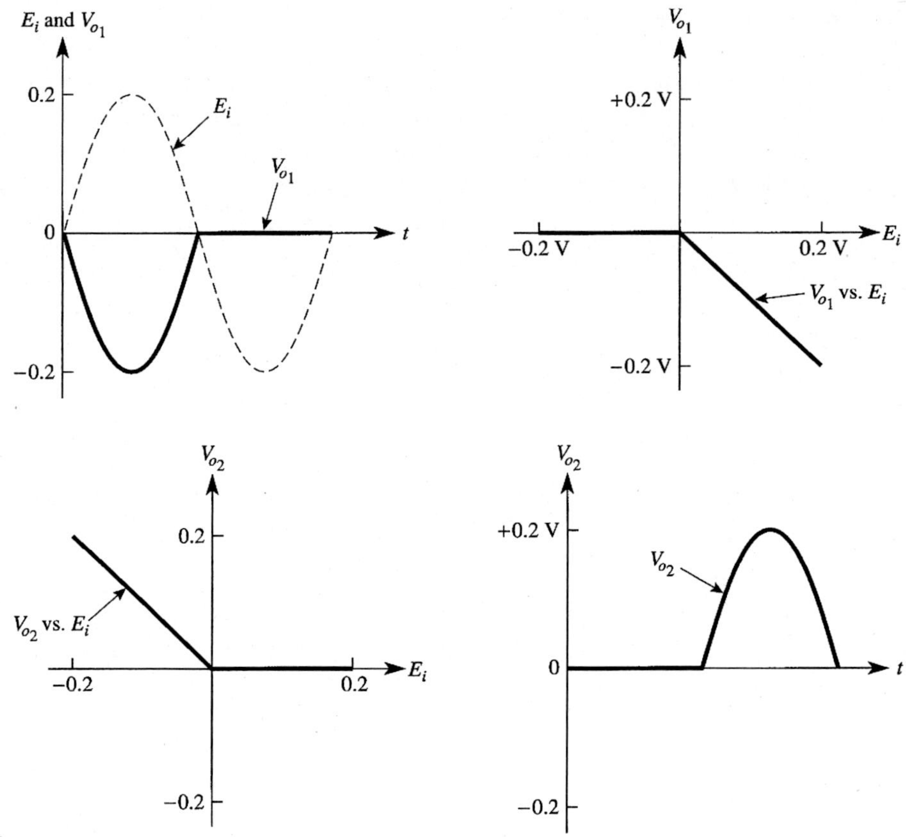

FIGURE 7-6 Input and output voltages for the polarity separator of Fig. 7-5.

7-2 PRECISION RECTIFIERS: THE ABSOLUTE-VALUE CIRCUIT

7-2.1 Introduction

The precision full-wave rectifier transmits one polarity of the input signal and inverts the other. Thus both half-cycles of an alternating voltage are transmitted but are converted to a single polarity of the circuit's output. The precision full-wave rectifier can rectify input voltages with millivolt amplitudes.

This type of circuit is useful to prepare signals for multiplication, averaging, or demodulation. The characteristics of an ideal precision rectifier are shown in Fig. 7-7.

The precision rectifier is also called an *absolute-value* circuit. The absolute value of a number (or voltage) is equal to its magnitude regardless of sign. For example, the absolute values of $|+2|$ and $|-2|$ are $+2$. The symbol $|\,|$ means "absolute value of." Figure 7-7 shows that the output equals the absolute value of the input. In a precision rectifier circuit the output is either negative or positive, depending on how the diodes are installed.

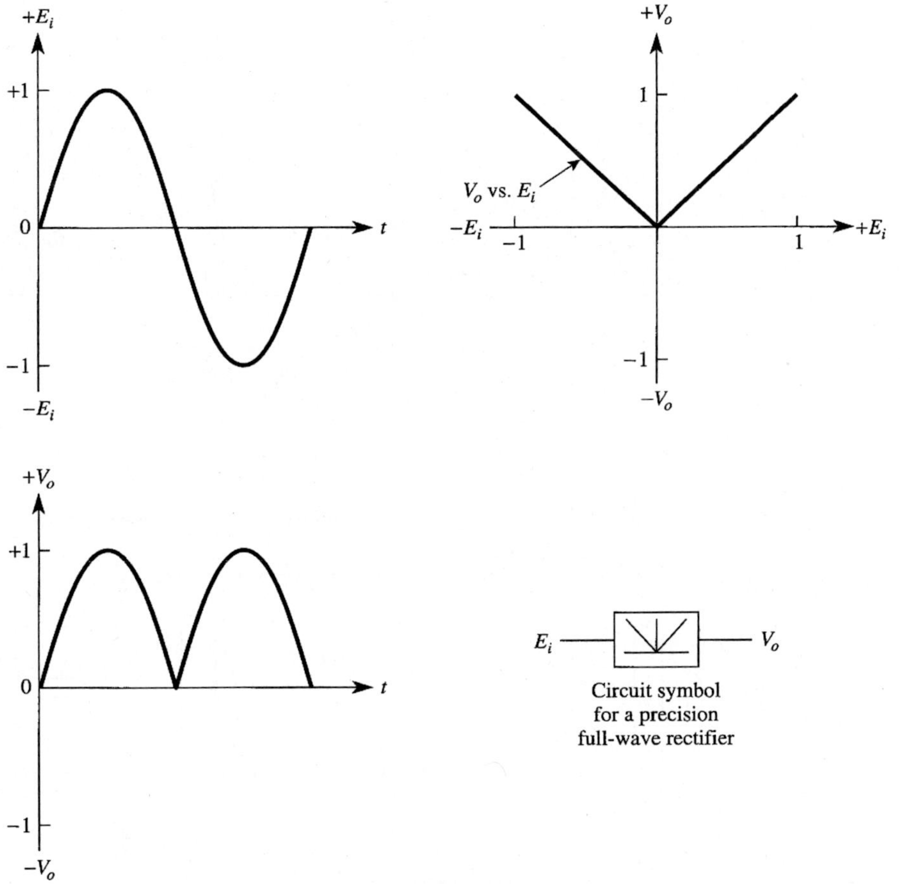

FIGURE 7-7 The precision full-wave rectifier fully rectifies input voltages, including those with values less than a diode threshold voltage.

7-2.2 Types of Precision Full-Wave Rectifiers

Three types of precision rectifiers will be presented. The first is inexpensive because it uses two op amps, two diodes, and five *equal* resistors. Unfortunately, it does not have high input resistance, so a second type is given that does have high input resistance but requires resistors that are precisely proportioned but *not* all equal. Neither type has a summing node at virtual ground potential, so a third type will be presented in Section 7-4.2 to allow averaging.

Full-wave precision rectifier with equal resistors. The first type of precision full-wave rectifier or absolute-value circuit is shown in Fig. 7-8. This circuit uses equal resistors and has an input resistance equal to R. Figure 7-8(a) shows current direc-

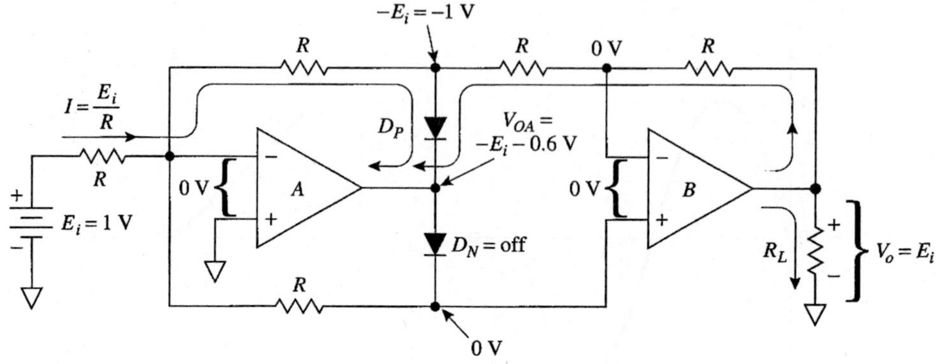

(a) For positive inputs D_P conducts; op amps A and B act as inverting amplifiers.

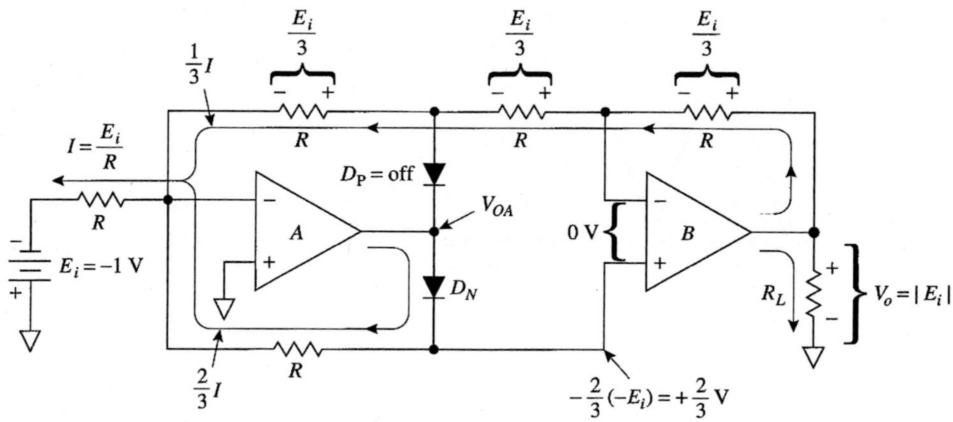

(b) For negative inputs, D_N conducts.

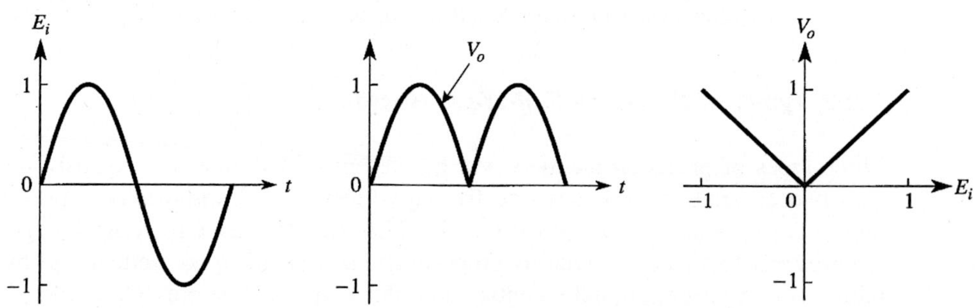

(c) Waveshapes.

FIGURE 7-8 Absolute-value circuit or precision full-wave rectifier, $V_o = |E_i|$.

tions and voltage polarities for positive input signals. Diode D_P conducts so that both op amps, A and B, act as inverters, and $V_o = +E_i$.

Figure 7-8(b) shows that for negative input voltages, diode D_N conducts. Input current I divides as shown, so that op amp B acts as an inverter. Thus output voltage V_o is positive for either polarity of input E_i and V_o is equal to the absolute value of E_i.

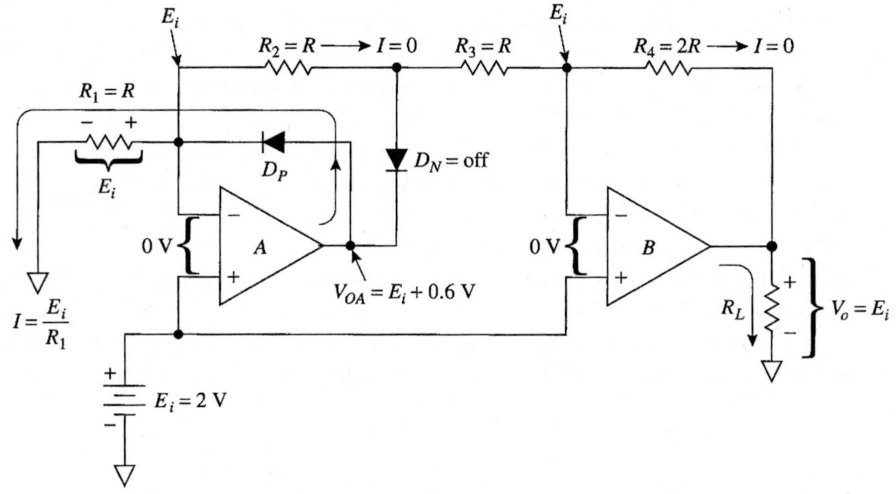

(a) Voltage levels for positive inputs: $V_o = +E_i$ for all positive E_i.

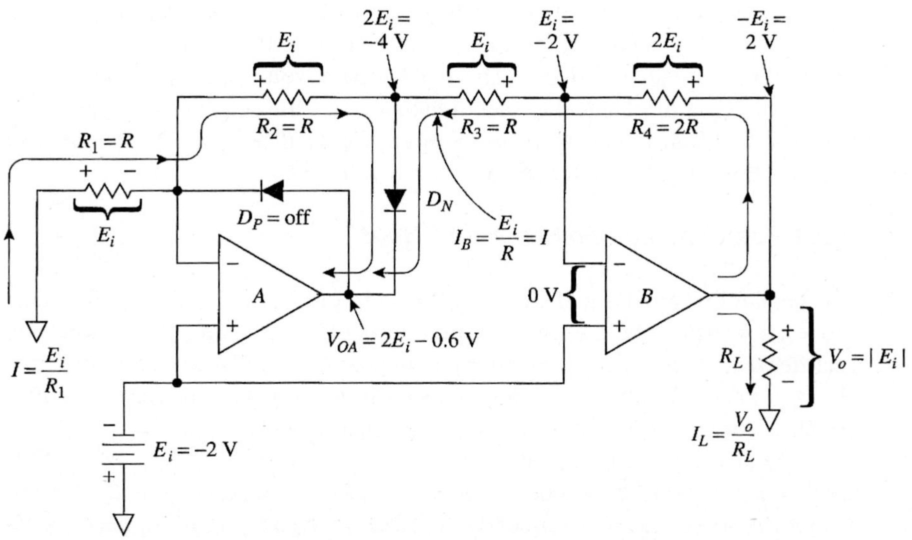

(b) Voltage levels for negative inputs: $V_o = -(-E_i) = |E_i|$.

FIGURE 7-9 Precision full-wave rectifier with high input impedance. $R = 10\text{ k}\Omega$, $2R = 20\text{ k}\Omega$.

The waveshapes in Fig. 7-8(c) show that V_o is always of positive polarity and equal to the absolute value of the input voltage. To obtain negative outputs for either polarity of E_i, simply reverse the diodes.

High-impedance precision full-wave rectifier. The second type of precision rectifier is shown in Fig. 7-9. The input signal is connected to the noninverting op amp inputs to obtain high input impedance. Figure 7-9(a) shows what happens for positive inputs. E_i and R_i set the current through diode D_P. The (−) inputs of both op amps are at a potential equal to E_i so that no current flows through R_2, R_3, and R_4. Therefore, $V_o = E_i$ for all positive input voltages.

When E_i goes negative in Fig. 7-9(b), E_i and R_1 set the current through both R_1 and R_2 to turn on diode D_N. Since $R_1 = R_2 = R$, the anode of D_N goes to $2E_i$ or $2(-E_i) = -4$ V. The (−) input of op amp B is at $-E_i$. The voltage drop across R_3 is $2E_i - E_i$ or $(-4\text{ V}) - (-2) = -2$ V. This voltage drop and R_3 establishes a current I_3 through both R_3 and R_4 equal to the input current I. Consequently, V_o is positive when E_i is negative. Thus V_o is always positive despite the polarity of E_i, so $V_o = |E_i|$.

The waveshapes for this circuit are the same as in Fig. 7-8(c). Note that the maximum value of E_i is limited by the negative saturation voltage of the op amps.

7-3 PEAK DETECTORS

In addition to rectifying a signal precisely, diodes and op amps can be interconnected to build a peak detector circuit. This circuit follows the voltage peaks of a signal and stores the highest value (almost indefinitely) on a capacitor. If a higher peak signal value comes along, this new value is stored. The highest peak voltage is stored until the capacitor is discharged by a mechanical or electronic switch. This peak detector circuit is also called a *follow-and-hold* circuit or *peak follower.* We shall also see that reversing two diodes changes this circuit from a peak to a *valley follower.*

7-3.1 Positive Peak Follower and Hold

The peak follower-and-hold circuit is shown in Fig. 7-10. It consists of two op amps, two diodes, a resistor, a hold capacitor, and a reset switch. Op amp A is a precision half-wave rectifier that charges C only when input voltage E_i exceeds capacitor voltage V_C. Op amp B is a voltage follower whose output signal is equal to V_C. The follower's high input impedance does not allow the capacitor to discharge appreciably.

To analyze circuit operation, let us begin with Fig. 7-10(a). When E_i exceeds V_C, diode D_P is forward biased to charge hold capacitor C. As long as E_i is greater than V_C, C charges toward E_i. Consequently, V_C follows E_i as long as E_i exceeds V_C. When E_i drops below V_C, diode D_N turns on as shown in Fig. 7-10(b). Diode D_P turns off and disconnects C from the output of op amp A. Diode D_P must be a low-leakage-type diode or the capacitor voltage will discharge (droop). To minimize droop, op amp B should require

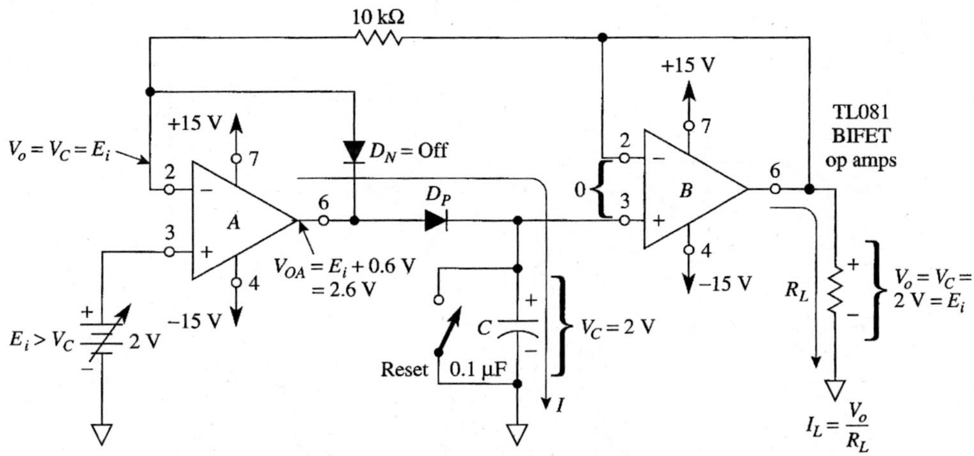

(a) When E_i exceeds V_C, C is charged toward E_i via D_P.

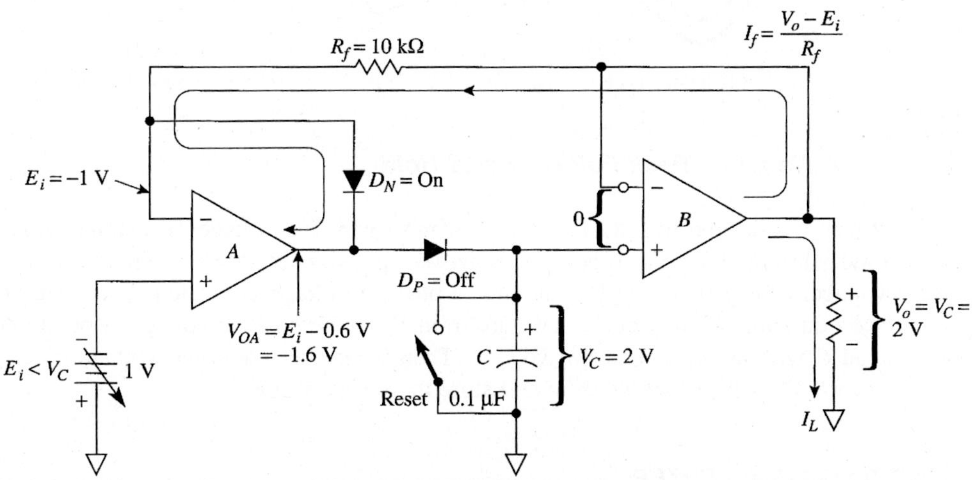

(b) When E_i is less than V_C, C holds its voltage at the highest previous value of E_i.

FIGURE 7-10 Positive peak follower-and-hold or peak detector circuit. Op amps are BiFETs (or the OP-177 can be used for most applications).

small input bias currents (see Chapter 9). For that reason op amp B should be a metal-oxide-semiconductor (MOS) or bipolar-field-effect (BiFET) op amp.

Figure 7-11 shows an example of voltage waveshapes for a positive voltage follower-and-hold circuit. To reset the hold capacitor voltage to zero, connect a discharge path across it with a 2-kΩ resistor.

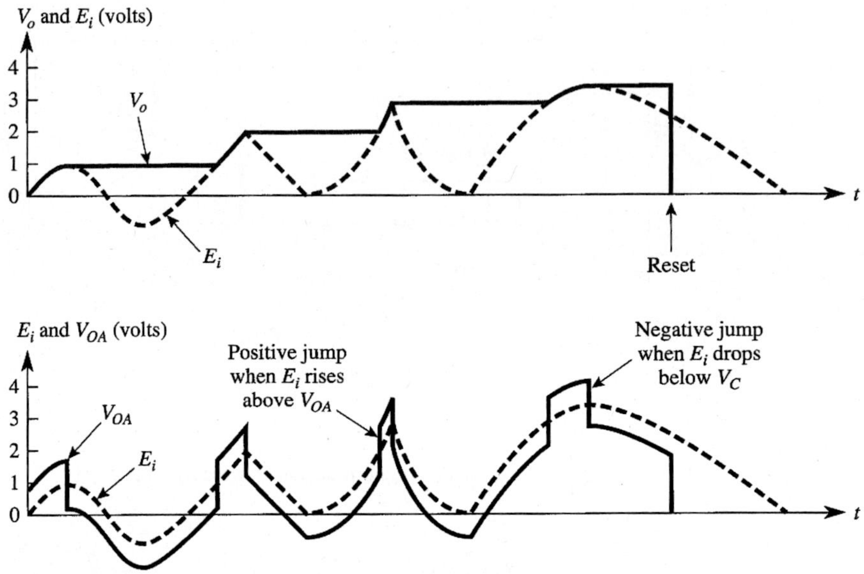

FIGURE 7-11 Waveshapes for the positive detector of Fig. 7-10(a).

7-3.2 Negative Peak Follower and Hold

When it is desired to hold the lowest or most negative voltage of a signal, reverse both diodes in Fig. 7-10. For bipolar or negative input signals, V_o will store the most negative voltage. It may be desired to monitor a positive voltage and catch any negative dips of short duration. Simply connect a wire from V_C to the positive voltage to be monitored to load C with an equal positive voltage. Then when the monitored voltage drops and recovers, V_C will follow the drop and store the lowest value.

7-4 AC-TO-DC CONVERTER

7-4.1 AC-to-DC Conversion or MAV Circuit

In this section we show how to design and build an op amp circuit that computes the average value of a rectified ac voltage. This type of circuit is called an ac-to-dc converter. Since a full-wave rectifier circuit is also known as an absolute-value circuit and since an average value is also called a mean value, the ac-to-dc converter is also referred to as a *mean-absolute-value* (MAV) circuit.

To see how the MAV circuit is useful, refer to Fig. 7-12. Sine, triangle, and square waves are shown with equal maximum (peak) values. Therefore, a peak detector could not distinguish between them. The positive and negative half-cycles are equal for each particular wave. Therefore, the average value of each signal is zero, so you could not dis-

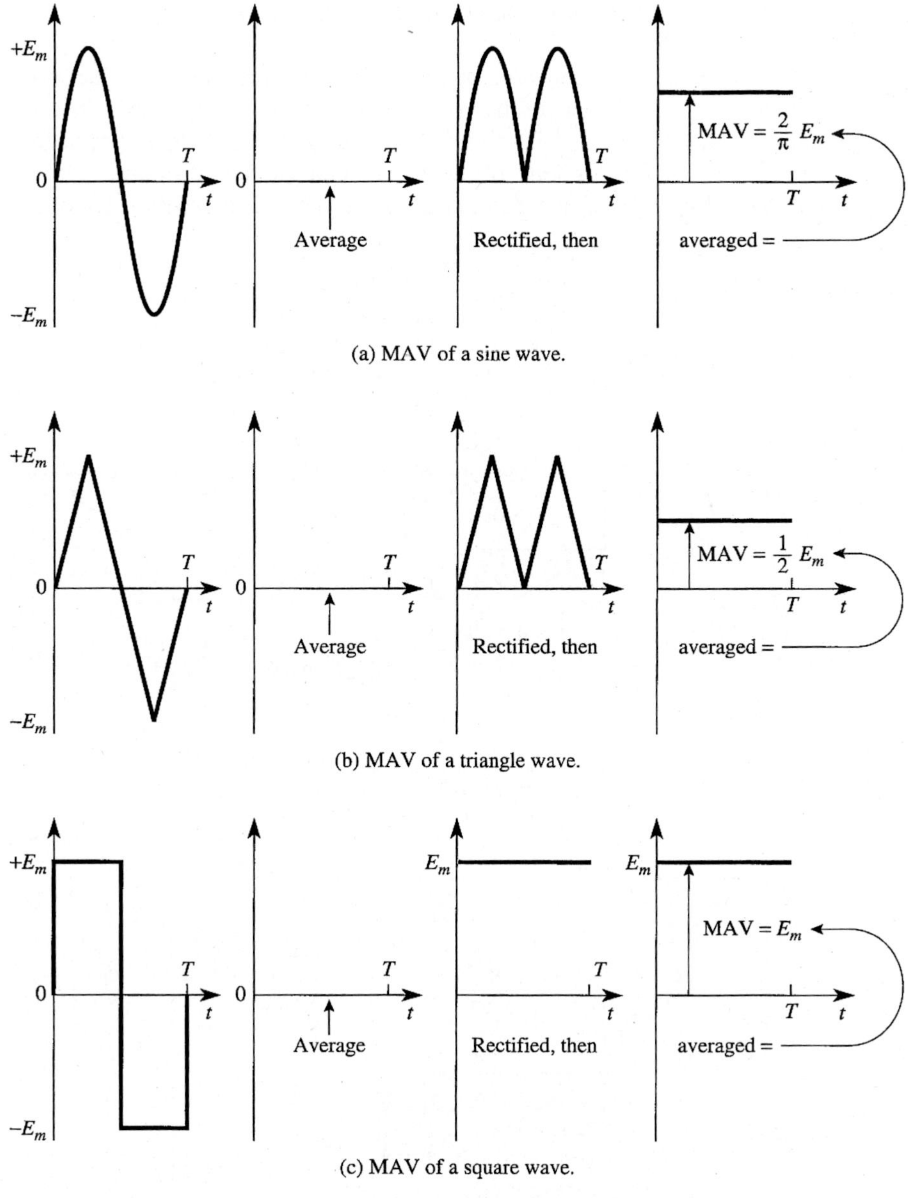

FIGURE 7-12 Mean absolute value of alternating sine, triangular, and square waves.

tinguish one from another with an averaging circuit or device such as a dc voltmeter. However, the MAV of each voltage is different (see Fig. 7-12).

The MAV of a voltage wave is approximately equal to its rms value. Thus an inexpensive MAV circuit can be used as a substitute for a more expensive rms calculating circuit.

7-4.2 Precision Rectifier with Grounded Summing Inputs

To construct an ac-to-dc converter, we begin with the precision rectifier or absolute-value amplifier of Fig. 7-13. For positive inputs in Fig. 7-13(a), op amp A inverts E_i. Op amp B sums the output of A and E_i to give a circuit output of $V_o = E_i$. For negative inputs as shown in Fig. 7-13(b), op amp B inverts $-E_i$ and the circuit output V_o is $+E_i$. Thus the circuit output V_o is positive and equal to the rectified or absolute value of the input.

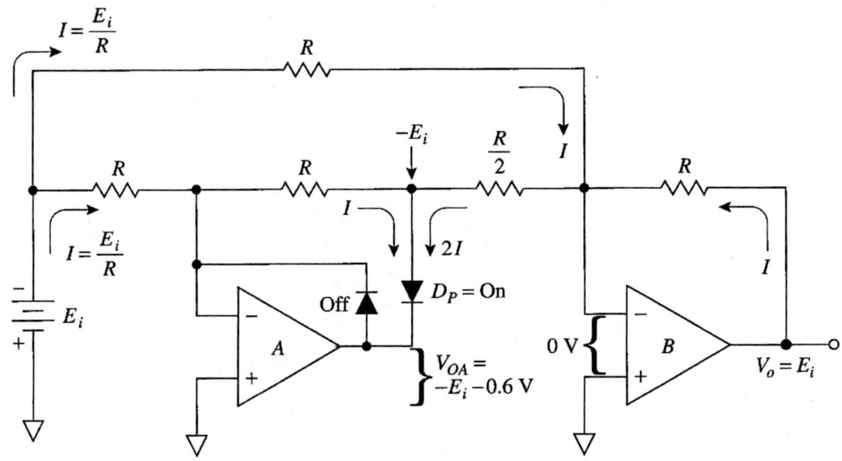

(a) For positive inputs, op amp A inverts E_i; op amp B is an inverting adder, so $V_o = E_i$.

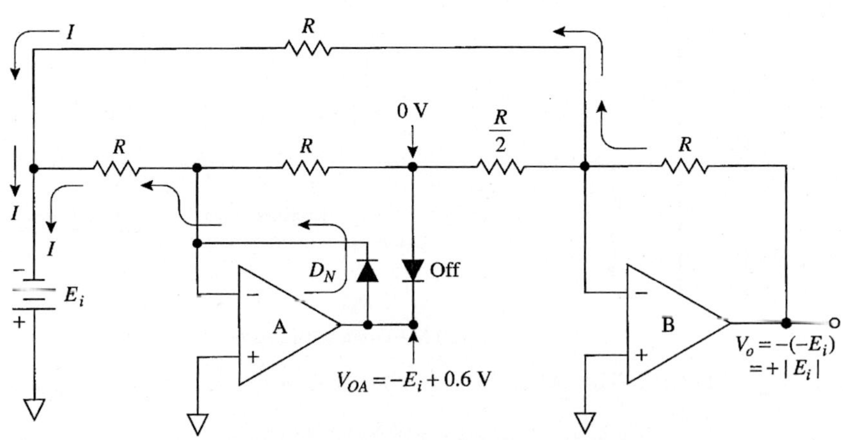

(b) For negative inputs, the output of A is rectified to 0; op amp B inverts E_i, so $V_o = +E_i$.

FIGURE 7-13 This absolute-value amplifier has both summing nodes at ground potential during either polarity of input voltage. $R = 20\ \text{k}\Omega$.

7-4.3 AC-to-DC Converter

A large-value low-leakage capacitor (10-μF tantalum) is added to the absolute-value circuit of Fig. 7-13. The resultant circuit is the MAV amplifier or ac-to-dc converter shown in Fig. 7-14. Capacitor C does the averaging of the rectified output of op amp B. It takes about 50 to 500 cycles of input voltage before the capacitor voltage settles down to its final reading. If the waveshapes of Fig. 7-12 are applied to the ac-to-dc converter, its output will be the MAV of each input signal.

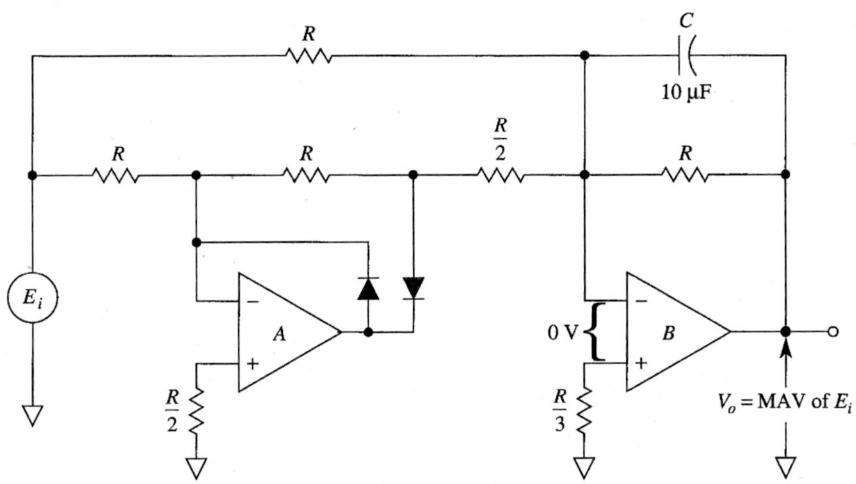

FIGURE 7-14 Add one capacitor to the absolute-value amplifier of Fig. 7-13 to get this ac-to-dc converter or mean-absolute-value amplifier.

7-5 DEAD-ZONE CIRCUITS

7-5.1 Introduction

Comparator circuits tell *if* a signal is below or above a particular reference voltage. In contrast with the comparator, a dead-zone circuit tells *by how much* a signal is below or above a reference voltage.

7-5.2 Dead-Zone Circuit with Negative Output

Analysis of a dead-zone circuit begins with the circuit of Fig. 7-15. A convenient regulated supply voltage $+V$ and resistor mR establish a reference voltage V_{ref}. V_{ref} is found from the equation $V_{ref} = +V/m$. As will be shown, the *negative* of V_{ref}, $-V_{ref}$, will establish the dead zone. In Fig. 7-15(a), current I is determined by $+V$ and resistor mR at $I = +V/mR$.

Diode D_N will conduct for all positive values of E_i, clamping V_{OA} and V_{OB} to 0 V. Therefore, all positive inputs are eliminated from affecting the output. In order to get any

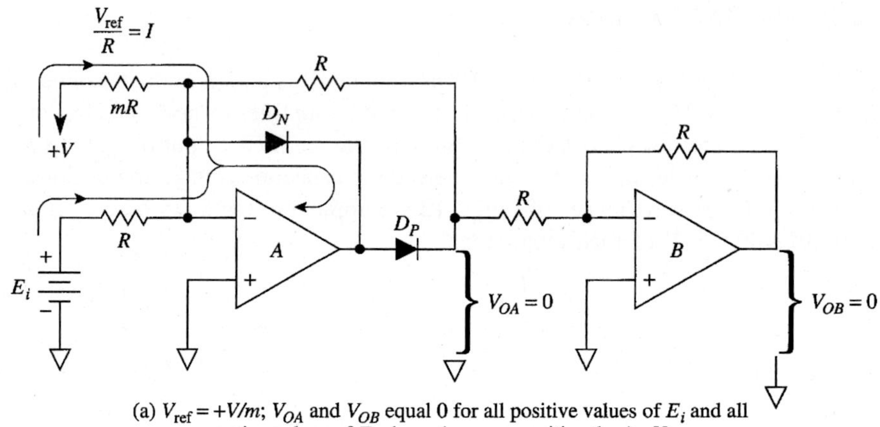

(a) $V_{ref} = +V/m$; V_{OA} and V_{OB} equal 0 for all positive values of E_i and all negative values of E_i above (or more positive than) $-V_{ref}$.

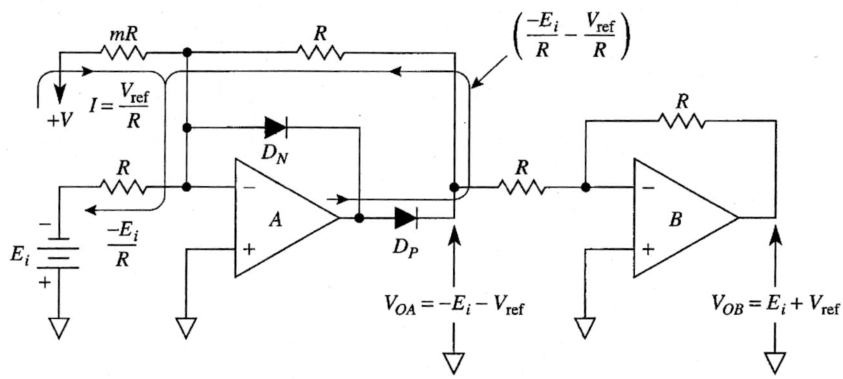

(b) When E_i is negative and below $-V_{ref}$, V_{OA} goes positive to a value of $-(E_i + V_{ref})$ and V_{OB} goes negative to $E_i + V_{ref}$.

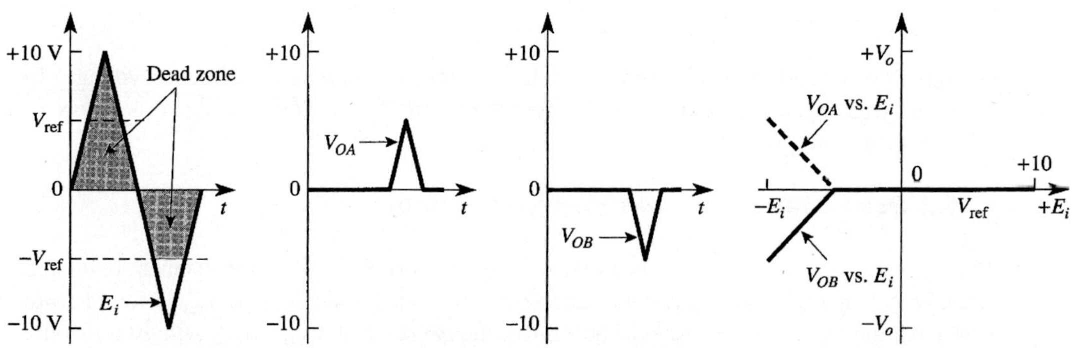

(c) Waveshapes for (a) and (b).

FIGURE 7-15 The dead-zone circuit output V_{OB} eliminates all portions of the signal above $-V_{ref}$ where $V_{ref} = +V/m$.

output at V_{OA}, E_i must go negative, as shown in Fig. 7-15(b). Diode D_P will conduct when the loop current E_i/R through E_i exceeds the loop current V/mR through resistor mR.

The value of E_i necessary to turn on D_P in Fig. 7-15(b) is equal to $-V_{ref}$. This conclusion is found by equating the currents

$$-\frac{E_i}{R} = \frac{+V}{mR}$$

and solving for E_i:

$$E_i = -\frac{+V}{m} = -V_{ref} \tag{7-1a}$$

where

$$V_{ref} = \frac{+V}{m} \tag{7-1b}$$

Thus all values of E_i above $-V_{ref}$ will lie in a dead zone where they will not be transmitted [see Fig. 7-15(c)]. Outputs V_{OA} and V_{OB} will be zero.

When E_i is below V_{ref}, E_i and V_{ref} are added and their sum is inverted at output V_{OA}. V_{OA} is reinverted by op amp B. Thus V_{OB} only has an output when E_i goes below V_{ref}. V_{OB} tells you by how many volts E_i lies below V_{ref}.

Circuit operation is summarized by the waveshapes of Fig. 7-15(c) and illustrated by an example.

Example 7-1

In the circuit of Fig. 7-15, $+V = +15$ V, $mR = 30$ kΩ, and $R = 10$ kΩ, so that $m = 3$. Find (a) V_{ref}; (b) V_{OA} when $E_i = -10$ V; (c) V_{OB} when $E_i = -10$ V.

Solution (a) From Eq. (7-1b), $V_{ref} = +15\text{ V}/3 = 5$ V. (b) V_{OA} and V_{OB} will equal zero for all values of E_i above $-V_{ref} = -5$ V, from Eq. (7-1a). Therefore, $V_{OA} = -E_i - V_{ref} = -(-10\text{ V}) - 5\text{ V} = +5$ V. (c) Op amp B inverts the output of V_{OA} so that $V_{OB} = -5$ V. Thus, V_{OB} indicates how much E_i goes below $-V_{ref}$. All input signals *above* $-V_{ref}$ fall in a dead zone and are eliminated from the output.

7-5.3 Dead-Zone Circuit with Positive Output

If the diodes in Fig. 7-15 are reversed, the result is a positive-output dead-zone circuit as shown in Fig. 7-16. Reference voltage V_{ref} is found from Eq. (7-1b): $V_{ref} = -15\text{ V}/3 = -5$ V. Whenever E_i goes above $-V_{ref} = -(-5\text{ V}) = +5$ V, the output V_{OB} tells by how much E_i exceeds $-V_{ref}$. The dead zone exists for all values of E_i *below* $-V_{ref}$.

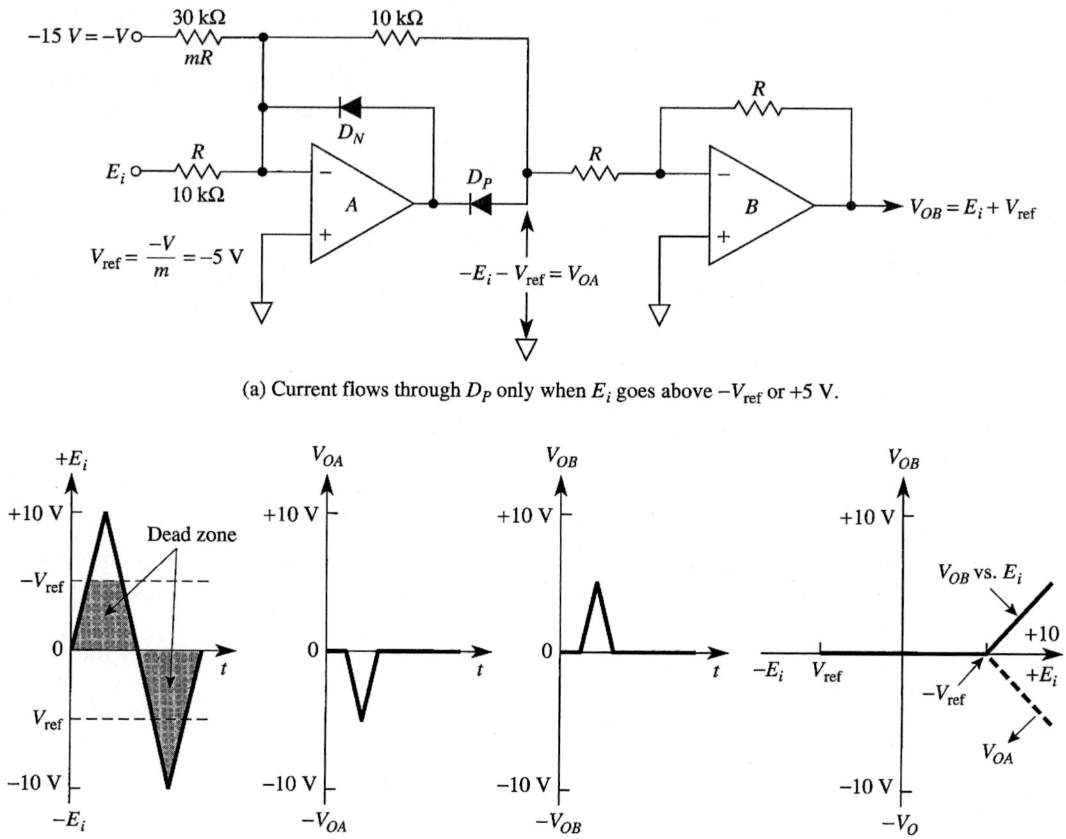

(a) Current flows through D_P only when E_i goes above $-V_{ref}$ or +5 V.

(b) Waveshapes for the positive-output dead-zone circuit.

FIGURE 7-16 Positive-output dead-zone circuit.

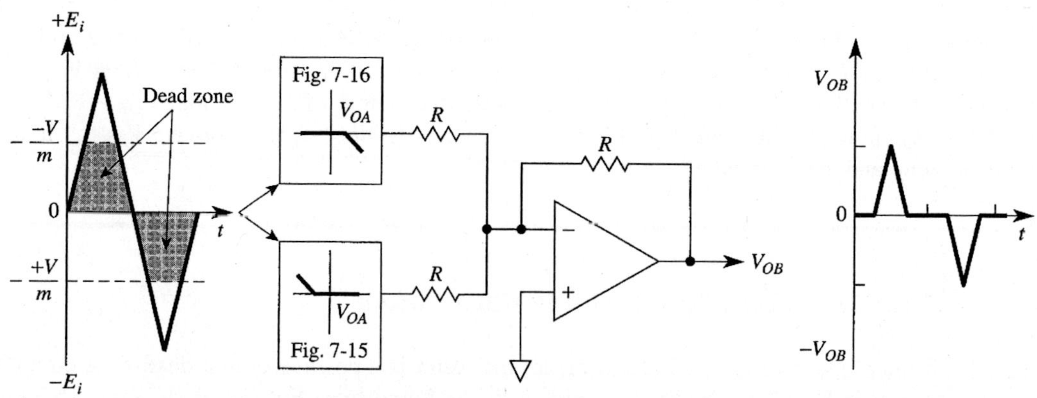

FIGURE 7-17 The V_{OA} outputs of Figs. 7-15 and 7-16 are combined by an inverting adder to give a bipolar output dead-zone circuit.

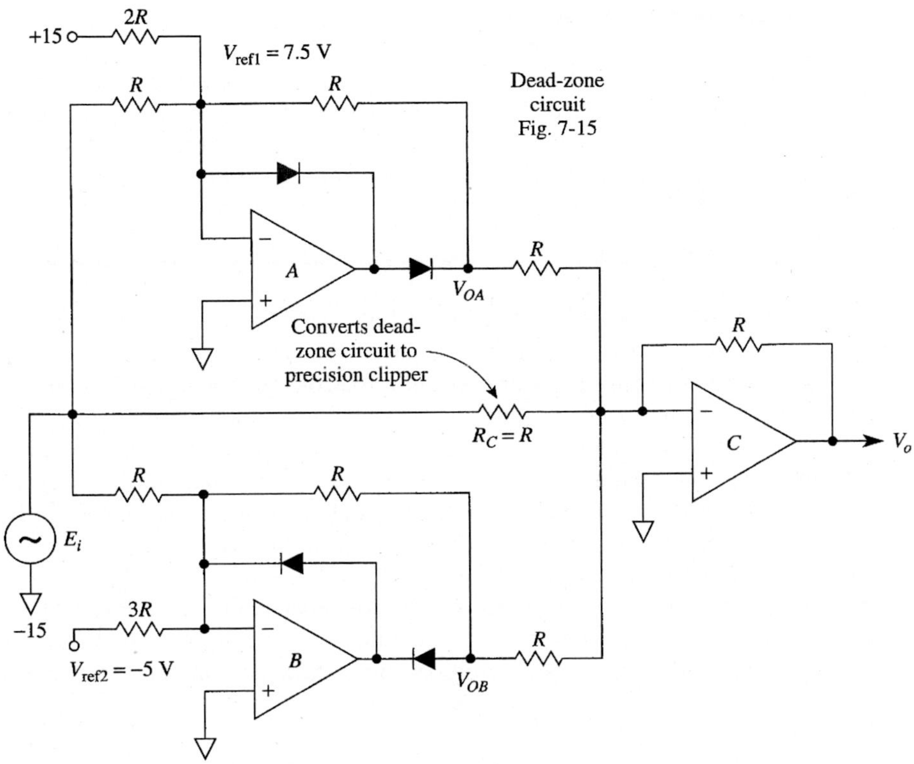

(a) Adding a resistor R_C to the dead-zone circuit of Fig. 7-17 gives a precision clipper.

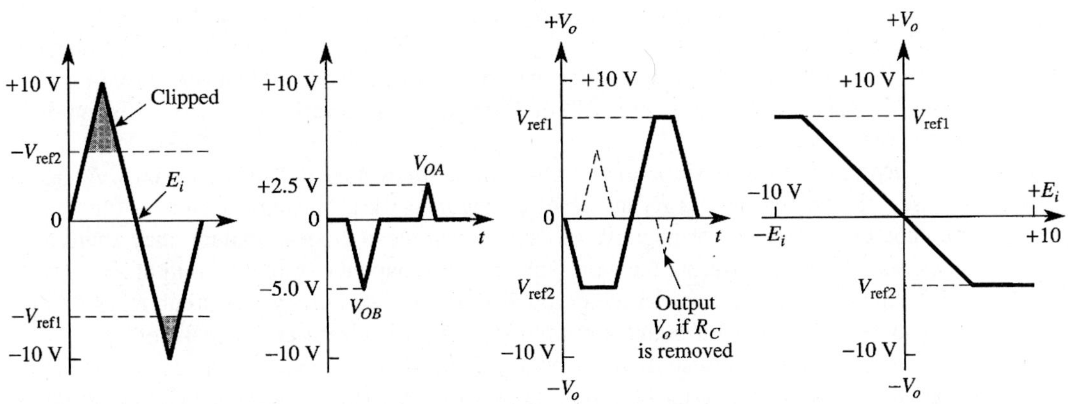

(b) Waveshapes for precision clipper.

FIGURE 7-18 A precision clipper is made from a bipolar dead-zone circuit plus an added resistor R_C.

7-5.4 Bipolar-Output Dead-Zone Circuit

The positive and negative output dead-zone circuits can be combined as shown in Fig. 7-17 and discussed in Fig. 7-18. The V_{OA} outputs from Figs. 7-15 and 7-16 are connected to an inverting adder. The adder output V_{OB} tells how much E_i goes above one positive reference voltage and also how much E_i goes below a different negative reference voltage.

7-6 PRECISION CLIPPER

A *clipper* or *amplitude limiter* circuit clips off all signals above a positive reference voltage and all signals below a negative reference voltage. The reference voltages can be made symmetrical or nonsymmetrical around zero. Construction of a precision clipper circuit is accomplished by adding a single resistor, R_C, to a bipolar output dead-zone circuit as shown in Fig. 7-18. The outputs of op amps A and B are each connected to the input of the inverting adder. Input signal E_i is connected to a third input of the inverting adder, via resistor R_C. If R_C is removed, the circuit would act as a dead-zone circuit. However, when R_C is present, input voltage E_i is subtracted from the dead-zone circuit's output and the result is an inverting precision clipper.

Circuit operation is summarized by the waveshapes in Fig. 7-18(b). Outputs V_{OA} and V_{OB} are inverted and added to $-E_i$. The plot of V_o versus time shows by solid lines how the clipped output appears. The dashed lines show how the circuit acts as a dead-zone circuit if R_C is removed.

7-7 TRIANGULAR-TO-SINE-WAVE CONVERTER

Variable-frequency sine-wave oscillators are much harder to build than variable-frequency triangular-wave generators. The circuit of Fig. 7-19 converts the output of a triangular-wave generator into a sine wave that can be adjusted for less than 5% distortion. The triangle-to-sine-wave converter is an amplifier whose gain varies inversely with amplitude of the output voltage.

R_1 and R_3 set the slope of V_o at low amplitudes near the zero crossings. As V_o increases, the voltage across R_3 increases to begin forward biasing D_1 and D_3 for positive outputs, or D_2 and D_4 for negative outputs. When these diodes conduct, they shunt feedback resistance R_3, lowering the gain. This tends to shape the triangular output above about 0.4 V into a sine wave. In order to get rounded tops for the sine-wave output, R_2 and diodes D_5 and D_6 are adjusted to make amplifier gain approach zero at the peaks of V_o.

The circuit is adjusted by comparing a 1-kHz sine wave and the output of the triangle/sine-wave converter on a dual-trace CRO. R_1, R_2, R_3, and the peak amplitude of E_i are adjusted in sequence for best sinusoidal shape. The adjustments interact, so they should be repeated as necessary. (*Note:* Although the circuit of Fig. 7-19 will shape a triangular wave to a sine wave, the parts count is high, but you may need to generate such a waveform with readily available parts. A better solution is to purchase an IC chip that generates triangle, square, and sine waves in a single package.)

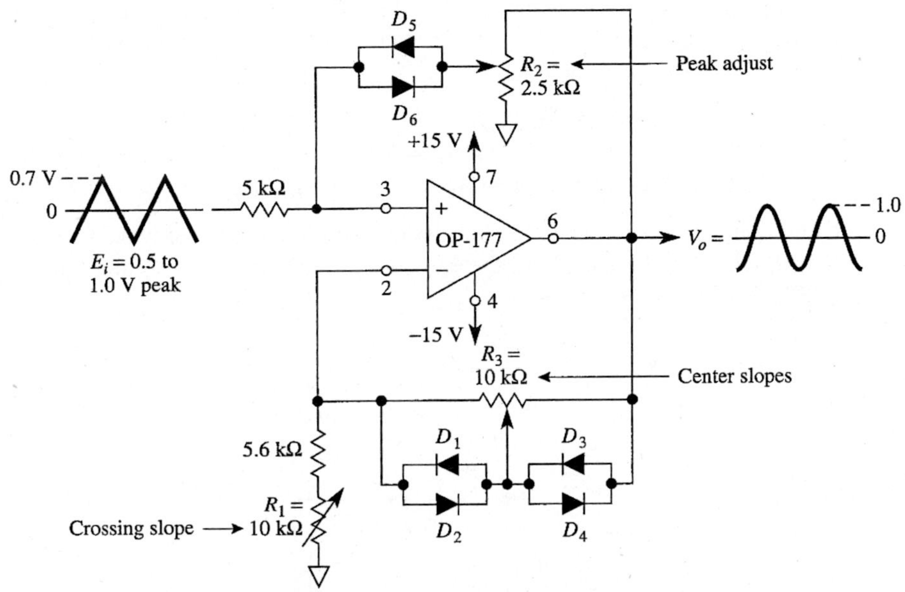

FIGURE 7-19 Triangle-to-sine-wave shaper.

7-8 PSPICE SIMULATION OF OP AMPS WITH DIODES

In this section, we will use PSpice and simulate the performance of three circuits studied in this chapter: the linear half-wave rectifier, precision full-wave rectifier, and mean-absolute-value amplifier.

7-8.1 Linear Half-Wave Rectifier

Refer to Fig. 7-2 and create the PSpice model of the circuit. Set the input voltage to a sine wave with a peak value of 2 V and a frequency of 1 kHz. Obtain a plot of E_i and V_o versus time. To begin, place the following parts in the work area.

Draw => Get New Part

Part	Number	Library
=> **uA741**	1	**eval.slb**
=> **D1N4002**	2	**eval.slb**
=> **VSIN**	1	**source.slb**
=> **VDC**	2	**source.slb**
=> **R**	3	**analog.slb**
=> **GLOBAL**	4	**port.slb**
=> **AGND**	5	**port.slb**

Arrange the parts as shown and change the attributes of the parts as given in Fig. 7-2. Set up the sine-wave attributes by double-clicking the symbol. In the pop-up window change **VOFF, VAMPL,** and **FREQ**.

> **VOFF => 0 => Save Attr**
> **VAMPL => 2V => Save Attr => Change Display => Both name and value**
> **FREQ => 1kHz => Save Attr => Change Display => Both name and value**

Close the pop-up box. Double-click on the lead from the sine wave to R_1 and label it E_i. Double-click on the lead from the cathode terminal of diode D_2 and label it V_o (see Fig. 7-20).

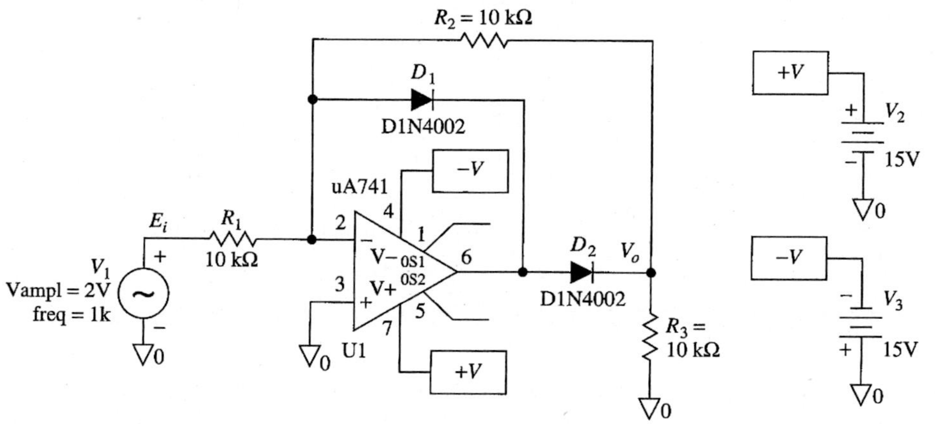

FIGURE 7-20 PSpice model of Fig. 7-2.

In order to obtain a plot of E_i and V_o versus time, we must initialize the Transient menu.

> **Analysis => Setup =>** Select **Transient**
> Click **Transient => Print Step: => 1μs**
> **=> Final Time: => 2ms**

Save the circuit as a file with the **.SCH** extension. Run the simulation

> **Analysis => Simulate**

In the Probe window, select

> **Trace => Add => V[Ei]**
> **=> V[Vo]**

Label the plots and obtain a printout as shown in Fig. 7-21.

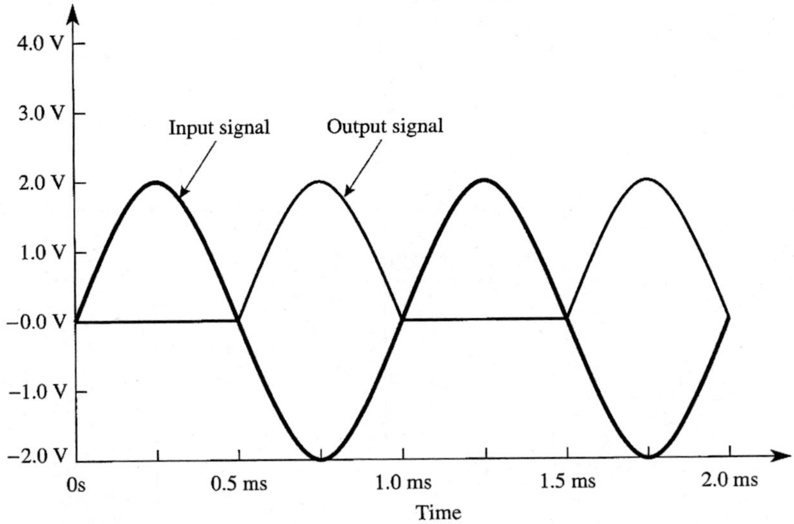

FIGURE 7-21 Input and output waveforms, for a linear half-wave rectifier, using PSpice.

7-8.2 Precision Full-Wave Rectifier

We will use PSpice and model the precision full-wave rectifier circuit of Fig. 7-9. The input voltage will be set to a sine wave with a peak value of 2 V and a frequency of 1 kHz. To compare input and output waveforms, obtain a plot of E_i and V_o versus time. To begin, place the following parts in the work area.

Draw => Get New Part

Part	Number	Library
=> **uA741**	2	**eval.slb**
=> **D1N4002**	2	**eval.slb**
=> **VSIN**	1	**source.slb**
=> **VDC**	2	**source.slb**
=> **R**	5	**analog.slb**
=> **GLOBAL**	6	**port.slb**
=> **AGND**	5	**port.slb**

Arrange the parts as shown in Fig. 7-2 and set the attributes of the parts as $R_1 = R_2 = R_3 = 10\ \text{k}\Omega$, $R_4 = 20\ \text{k}\Omega$, and $R_L = 10\ \text{k}\Omega$. Set the sine-wave attributes by double-clicking the symbol. In the pop-up window change **VOFF**, **VAMPL**, and **FREQ**.

VOFF => 0 => Save Attr
VAMPL = > 2V => Save Attr => Change Display => Both name and value
FREQ => 1kHz => Save Attr => Change Display => Both name and value

Close the pop-up box. Double-click on the lead from the positive side of the sine wave generator and label it **Ei**. Double-click on the output lead of op amp B and label it **Vo** (see Fig. 7-22).

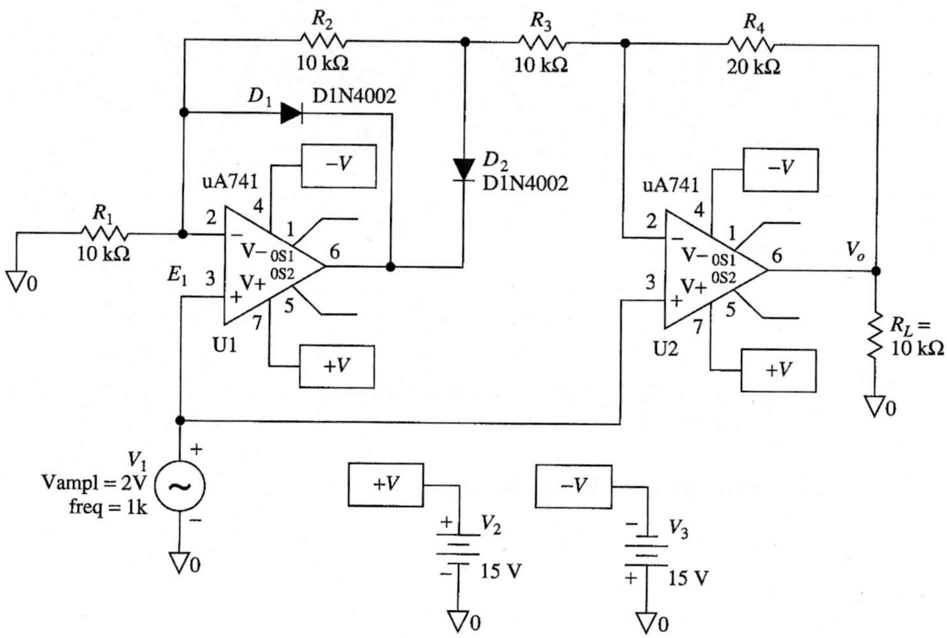

FIGURE 7-22 PSpice model of Fig. 7-9.

In order to obtain a plot of E_i and V_o versus time, we must initialize the Transient menu.

Analysis => **Setup** => Select **Transient**
Click **Transient** => **Print Step:** => **1μs**
=> **Final Time:** => **2ms**

Save the circuit as a file with the **.SCH** extension. Run the simulation

Analysis => **Simulate**

In the Probe window, select

Trace => **Add** => **V[Ei]**
=> **V[Vo]**

Label the plots and obtain a printout as shown in Fig. 7-23.

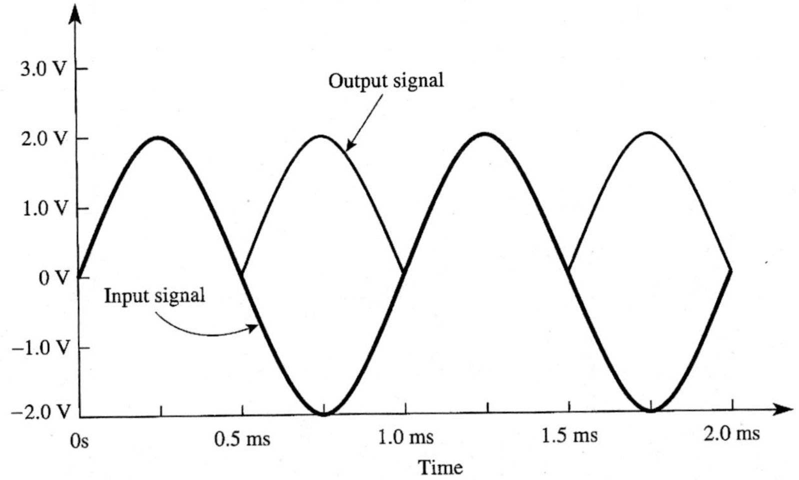

FIGURE 7-23 Input and output waveforms, for a precision full-wave rectifier, using PSpice.

7-8.3 Mean-Absolute-Value Amplifier

We will simulate the output of the MAV amplifier given in Fig. 7-14 using PSpice. The input voltage will be set to a sine wave with a peak value of 2 V and a frequency of 1 kHz. We wish to obtain a plot of V_o versus time. To begin, place the following parts in the work area.

Draw => Get New Part

Part	Number	Library
=> **uA741**	2	**eval.slb**
=> **D1N4002**	2	**eval.slb**
=> **VSIN**	1	**source.slb**
=> **VDC**	2	**source.slb**
=> **R**	8	**analog.slb**
=> **C**	1	**analog.slb**
=> **GLOBAL**	6	**port.slb**
=> **AGND**	5	**port.slb**

Arrange the parts as shown in Fig. 7-14 and set the attributes of the parts as $R_1 = R_2 = R_3 = R_6 = 10\ \text{k}\Omega$, $R_7 = 5\ \text{k}\Omega$, $R_8 = 3.3\ \text{k}\Omega$, and $R_L = 10\ \text{k}\Omega$. Set $C = 10\mu\text{F}$. Set the sine-wave attributes by double-clicking the symbol. In the pop-up window change **VOFF**, **VAMPL**, and **FREQ**.

VOFF => 0 => Save Attr
VAMPL => 2V => Save Attr => Change Display => Both name and value
FREQ => 1kHz => Save Attr => Change Display => Both name and value

Close the pop-up box. Double-click on the lead from the positive side of the sine wave generator and label it **Ei**. Double-click on the output lead of op amp *B* and label it **Vo** (see Fig. 7-24).

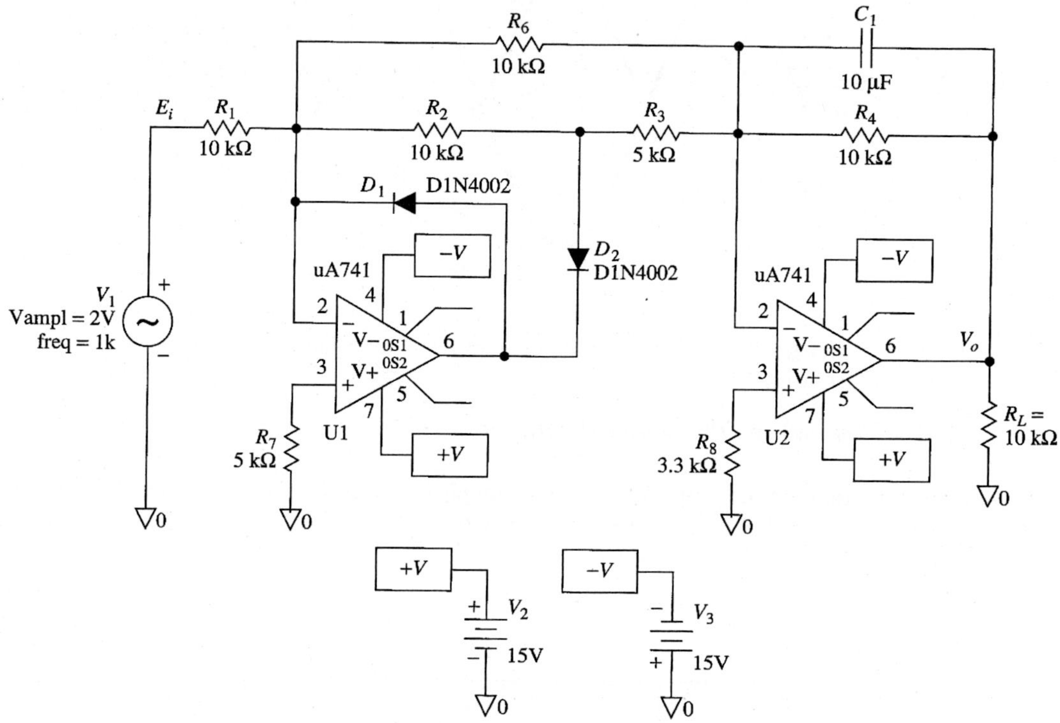

FIGURE 7-24 PSpice model of Fig. 7-14.

In order to obtain a plot of E_i and V_o versus time, we must initialize the Transient menu.

Analysis => **Setup** => Select **Transient**
Click **Transient** => **Print Step:** => **100μs**
=> **Final Time:** => **500ms**

Save the circuit as a file with the **.SCH** extension. Run the simulation

Analysis = > **Simulate**

In the Probe window, select

Trace => **Add** => **V[Vo]**

Label the plot and obtain a printout as shown in Fig. 7-25.

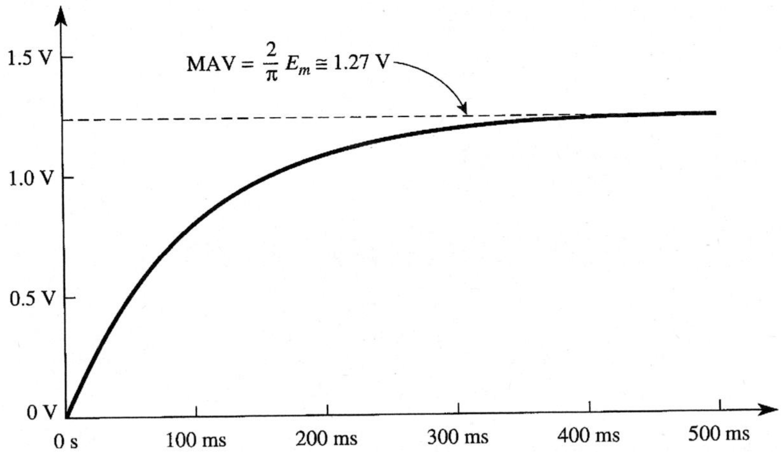

FIGURE 7-25 Plot of V_o versus time for the mean-absolute-value amplifier in Fig. 7-24.

PROBLEMS

7-1. What is the absolute value of +3 V and −3 V?

7-2. If the peak value of $E_i = 0.5$ V in Fig. 7-1, sketch the waveshapes of V_o vs. t and V_o vs. E_i for both a silicon and an ideal diode.

7-3. If E_i is a sine wave with a peak value of 1 V in Figs. 7-2 and 7-3, sketch the waveshapes of V_o vs. t and V_o vs. E_i.

7-4. If diodes D_1 and D_2 are reversed in Fig. 7-2, sketch V_o vs. E_i and V_o vs. t.

7-5. Sketch the circuit for a signal polarity separator.

7-6. Let both diodes be reversed in Fig. 7-8. What is the value of V_o if $E_i = +1$ V or $E_i = -1$ V?

7-7. What is the name of a circuit that follows the voltage peaks of a signal and stores the highest value?

7-8. How do you reset the hold capacitor's voltage to zero in a peak follower-and-hold circuit?

7-9. How do you convert the absolute-value amplifier of Fig. 7-13 to an ac-to-dc converter?

7-10. If resistor mR is changed to 50 kΩ in Example 7-1, find **(a)** V_{ref}; **(b)** V_{OA} when $E_i = 10$ V; **(c)** V_{OB} when $E_i = 10$ V.

7-11. If resistor R_C is removed in Fig. 7-18, sketch V_o vs. E_i.

CHAPTER 8

Differential, Instrumentation, and Bridge Amplifiers

LEARNING OBJECTIVES

When you complete this chapter on differential, instrumentation, and bridge amplifiers, you will be able to:

- Draw the circuit for a basic differential amplifier, state its output–input equation, and explain why it is superior to a single-input amplifier.
- Define common-mode and differential input voltage.
- Draw the circuit for a differential input to differential output voltage amplifier and add a differential amplifier to make a three-op-amp instrumentation amplifier (IA).
- Calculate the output voltage of a three-op-amp instrumentation amplifier if you are given the input voltages and resistance values.
- Use commercially available instrumentation amplifiers.
- Explain how the sense and reference terminals of an IA allow you to (1) eliminate the effects of connecting-wire resistance on load voltage, (2) obtain load current boost, or (3) make a differential voltage-to-current converter (ac current source).

- Explain how a strain gage converts tension or compression forces into a change in resistance.
- Connect strain gages into a passive bridge resistance network to convert gage resistance change into an output voltage.
- Amplify the strain gage bridge's differential output with an instrumentation amplifier.
- Measure pressure, force, or weight.
- Draw the circuit for a bridge amplifier and show how it converts a change in transducer resistance to an output voltage.
- Use the bridge amplifier to make a temperature-to-voltage converter.

8-0 INTRODUCTION

The most useful amplifier for measurement, instrumentation, or control is the *instrumentation amplifier.* It is designed with several op amps and precision resistors, which make the circuit extremely stable and useful where accuracy is important. There are now many integrated circuits available in single packages. Although these packages are more expensive than a single op amp, when performance and precision are required, the instrumentation amplifier is well worth the price, because its performance cannot be matched by the average op amp.

A first cousin and basic block within the instrumentation amplifier is the *differential amplifier,* also referred to as a *subtractor circuit.* This chapter begins with the differential amplifier, showing the applications in which it is superior to the ordinary inverting or noninverting amplifier. The differential amplifier, with some additions, leads into the instrumentation amplifier, which is discussed in the second part of this chapter. The final sections consider *bridge amplifiers,* which involve both instrumentation and basic differential amplifiers.

8-1 BASIC DIFFERENTIAL AMPLIFIER

8-1.1 Introduction

The differential amplifier can measure as well as amplify small signals that are buried in much larger signals. How the differential amplifier accomplishes this task will be studied in Section 8-2, but first, let us build and analyze the circuit performance of the basic differential amplifier.

Four precision (1%) resistors and an op amp make up a differential amplifier, as shown in Fig. 8-1. There are two input terminals, labeled (−) input and (+) input, corresponding to the closest op amp terminal. If E_1 is replaced by a short circuit, E_2 sees an inverting amplifier with a gain of $-m$. Therefore, the output voltage due to E_2 is $-mE_2$. Now let E_2 be short-circuited; E_1 divides between R and mR to apply a voltage of $E_1m/(1 + m)$ at the op amp's (+) input. This divided voltage sees a noninverting amplifier with a gain of $(m + 1)$. The output voltage due to E_1 is the divided voltage, $E_1m/(1 + m)$, times the noninverting amplifier gain, $(1 + m)$, which yields mE_1. Therefore, E_1 is amplified at

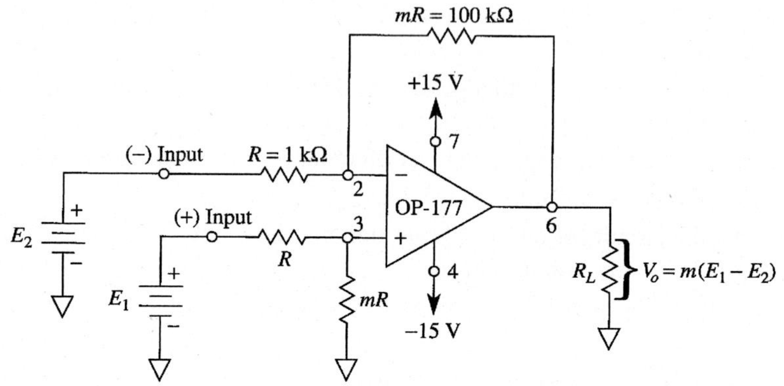

FIGURE 8-1 Basic differential amplifier (subtractor circuit).

the output by the multiplier m to mE_1. When both E_1 and E_2 are present at the (+) and (−) inputs, respectively, V_o is $mE_1 - mE_2$, or

$$V_o = mE_1 - mE_2 = m(E_1 - E_2) \tag{8-1}$$

Equation (8-1) shows that the output voltage of the differential amplifier, V_o, is proportional to the *difference* in voltage applied to the (+) and (−) inputs. Multiplier m is called the *differential gain* and is set by the resistor ratios.

Example 8-1

In Fig. 8-1, the differential gain is found from

$$m = \frac{mR}{R} = \frac{100\ \text{k}\Omega}{1\ \text{k}\Omega} = 100$$

Find V_o for $E_1 = 10$ mV and (a) $E_2 = 10$ mV, (b) $E_2 = 0$ mV, and (c) $E_2 = -20$ mV.

Solution By Eq. (8-1), (a) $V_o = 100(10 - 10)$ mV $= 0$; (b) $V_o = 100(10 - 0)$ mV $=$ 1.0 V; (c) $V_o = 100[10 - (-20)]$ mV $= 100(30$ mV$) = 3$ V.

As expected from Eq. (8-1) and shown from part (a) of Example 8-1, when $E_1 = E_2$ the output voltage is 0. To put it another way, when a common (same) voltage is applied to the input terminals, $V_o = 0$. Section 8-1.2 examines this idea of a common voltage in more detail.

8-1.2 Common-Mode Voltage

The output of the differential amplifier should be 0 when $E_1 = E_2$. The simplest way to apply equal voltages is to wire the inputs together and connect them to the voltage source (see Fig. 8-2a). For such a connection, the input voltage is called the *common-mode input voltage*, E_{CM}. Now V_o will be 0 if the resistor ratios are equal (mR to R for the inverting amplifier gain equals mR to R of the voltage-divider network). Practically, the resistor ratios are equalized by installing a potentiometer in series with one resistor, as shown in Fig. 8-2a. The potentiometer is trimmed until V_o is reduced to a negligible value. This causes the *common-mode voltage gain*, V_o/E_{CM}, to approach 0. It is this characteristic of a differential amplifier that allows a small signal voltage to be picked out of a

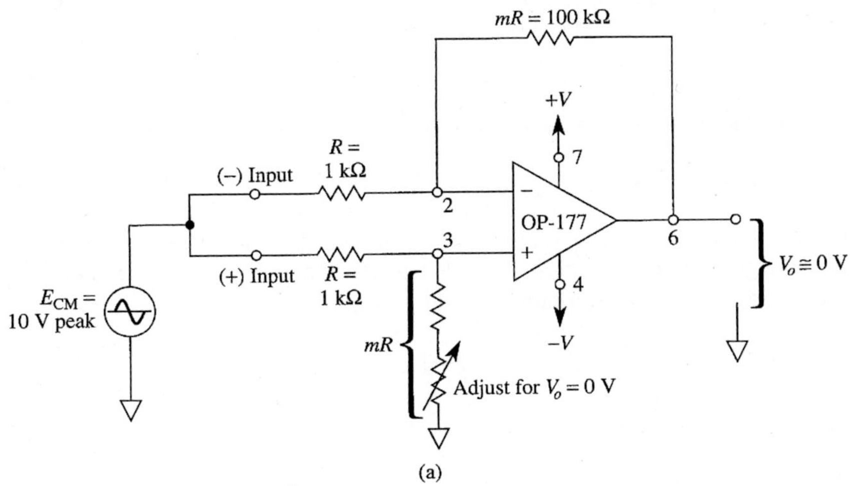

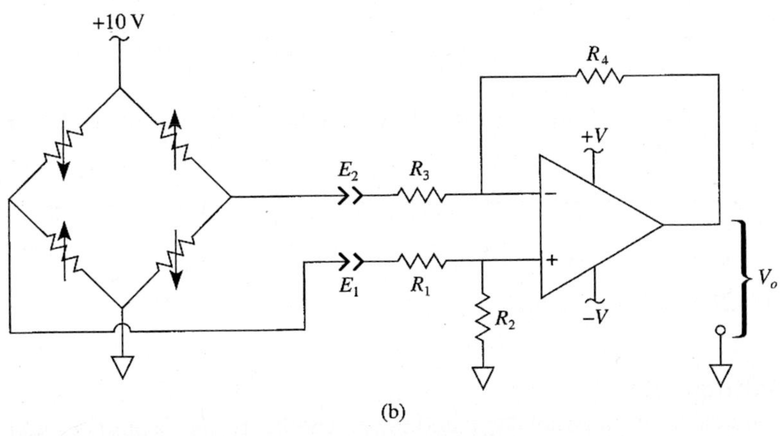

FIGURE 8-2 (a) The common-mode voltage gain should be zero. (b) Basic differential amplifier connected to a Wheatstone bridge.

larger noise voltage. It may be possible to arrange the circuit so that the larger undesired signal is the common-mode input voltage and the small signal is the differential input voltage. Then the differential amplifier's output voltage will contain only an amplified version of the differential input voltage.

Operational amplifier circuits, as well as op amps, have common mode error, and the common mode voltage is different for the circuit and the op amp. In this section, we will introduce common mode voltage and common mode rejection for a basic differential amplifier circuit, which is the fundamental building block for the instrumentation amplifier. In Chapter 10, common mode rejection is revisited, but for the op amp only.

Refer to the basic differential amplifier circuit in Fig. 8-2(b). The difference voltage, $E_1 - E_2$, may be defined as input difference signal voltage, E_{ds}, and the average of the two inputs, $(E_1 + E_2)/2$, is defined as the common mode voltage, E_{cm}. In this circuit, the differential amplifier is being used to measure the voltage difference between the two output nodes of a Wheatstone bridge.

Example 8-2

For the circuit of Fig. 8-2(b), the bridge is slightly unbalanced and the input voltages with respect to ground are $E_1 = 5\text{ V} + 2\text{ mV}$ and $E_2 = 5\text{ V} - 2\text{ mV}$. Determine (a) the input signal voltage and (b) the common mode voltage.

Solution (a) The difference input signal voltage is

$$E_{ds} = E_1 - E_2 = (5\text{ V} + 2\text{ mV}) - (5\text{ V} - 2\text{ mV}) = 4\text{ mV}$$

(b) The common mode voltage is

$$E_{cm} = \frac{E_1 + E_2}{2} = \frac{(5\text{ V} + 2\text{ mV}) + (5\text{ V} - 2\text{ mV})}{2} = 5\text{ V}$$

Ideally, the circuit design is to reject the common mode voltage and amplify the input signal voltage.

8-1.3 Common Mode Rejection

Introduction

Circuits or op amps are compared to one another by their common-mode rejection (CMR) or common mode rejection ratio (CMRR). CMR is a measure of the change in output voltage when both inputs are changed by an equal amount. CMRR is a ratio expression, while common-mode rejection is a logarithm of that ratio. For example, if CMRR is 10,000, the CMR is calculated by

$$\text{CMR} = 20 \log_{10} \text{CMRR} \tag{8-2}$$

For this example,

$$\text{CMR} = 20 \log_{10} (10{,}000) = 80 \text{ dB}$$

CMR of a Differential Amplifier Circuit

Referring to the circuit in Fig. 8-2(b) and using the superposition principle discussed in Section 8-1, the general equation for output voltage is

$$V_o = E_1\left(\frac{R_2}{R_1 + R_2}\right)\left(\frac{R_3 + R_4}{R_3}\right) - E_2\frac{R_4}{R_3} \tag{8-3}$$

For this circuit to work properly all resistors have to be carefully ratio-matched to maintain excellent common-mode rejection. For example, if all the resistors are equal, then the output voltage is the difference of the input signals, $V_o = E_1 - E_2$, and if the input voltages are equal, then V_o is 0. This results in an infinite common-mode rejection. However, let's consider the situation where one resistor is mismatched and the circuit values are $E_1 = E_2 = E$ and $R_1 = R_3 = R_4 = R$. The mismatched resistor is R_2 where $R_2 = R + 0.1\%R$. Applying Eq. (8.3) yields

$$V_o = \left[\left(\frac{1.001R}{2.001R}\right)\left(\frac{2R}{R}\right) - \left(\frac{R}{R}\right)\right] \times E = 0.0005 \times E$$

Although this expression shows the input voltage is reduced by 2000 (1/0.0005), the common-mode rejection of the circuit has been degraded to

$$\text{CMR} = 20 \log_{10} (2000) = 66 \text{ dB}$$

In this application, we have discussed only the common-mode rejection of the circuit and considered the op amp to be ideal. A disadvantage with the basic differential amplifier circuit in Fig. 8.2(b) is that a slight mismatch of the resistors causes a degradation of the CMR. In this example, the mismatch was R_2, but a more common mismatch is the input resistors R_1 or R_3 because of loading effect(s). A solution to this problem will be covered in Section 8-3.

8-2 DIFFERENTIAL VERSUS SINGLE-INPUT AMPLIFIERS

8-2.1 Measurement with a Single-Input Amplifier

A simplified wiring diagram of an inverting amplifier is shown in Fig. 8-3. The power common terminal is shown connected to earth ground. Earth ground comes from a connection to a water pipe on the street side of the water meter. Ground is extended via conduit or a bare Romex wire to the third (green) wire of the instrument line cord and finally to the chassis of the amplifier. This equipment or chassis ground is made to ensure the

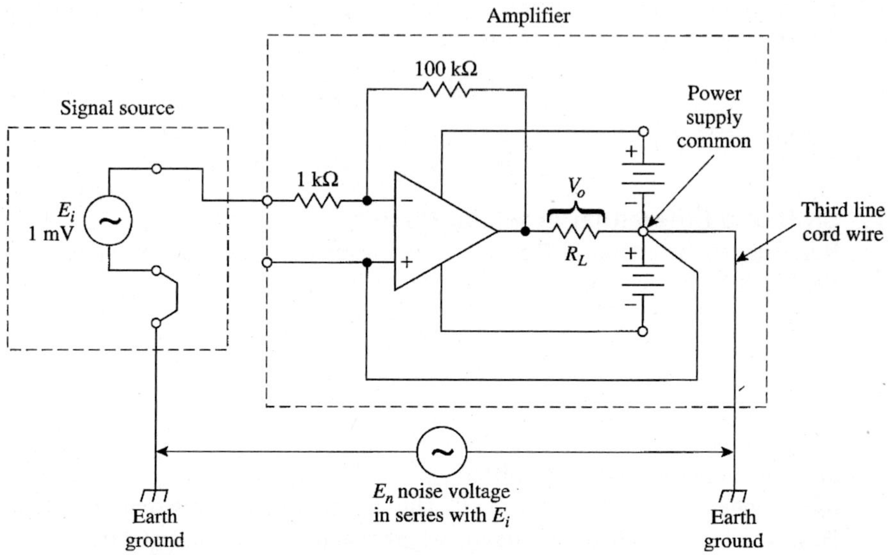

FIGURE 8-3 Noise voltages act as if they are in series with the input signal E_i. Consequently, both are amplified equally. This arrangement is unworkable if E_n is equal or greater than E_i.

safety of human operators. It also helps to drain off static charges or any capacitive coupled noise currents to earth.

The signal source is also shown in Fig. 8-3 to be connected to earth ground. Even if it were not grounded, there would be a leakage resistance or capacitance coupling to earth, to complete a ground loop.

Inevitably, noise currents and noise voltages abound from a variety of sources that are often not easily identifiable. The net effect of all this noise is modeled by noise voltage source E_n in Fig. 8-3. Using the power supply common as the reference, it is evident that E_n is in series with signal voltage E_i, so that both are amplified by a factor of -100 due to the inverting amplifier. E_n may be much larger than E_i. For example, the skin signal voltage due to heart beats is less than 1 mV, whereas the body's noise voltage may be tenths of volts or more; it would be impossible to make an EKG measurement with a single-input amplifier. What is needed is an amplifier that can distinguish between E_i and E_n and amplify only E_i. Such a circuit is the differential amplifier.

8-2.2 Measurement with a Differential Amplifier

A differential amplifier is used to measure only the signal voltage (see Fig. 8-4). The signal voltage E_i is connected across the (+) and (−) inputs of the differential amplifier. Therefore, E_i is amplified by a gain of -100. Noise voltage E_n becomes the common-mode voltage input voltage to the differential amplifier as shown in Fig. 8-2. (*Note*: Apply superposition.) Therefore, the noise voltage is *not* amplified and has been effectively eliminated from having any significant effect on the output V_o, as long as the resistors are matched as shown.

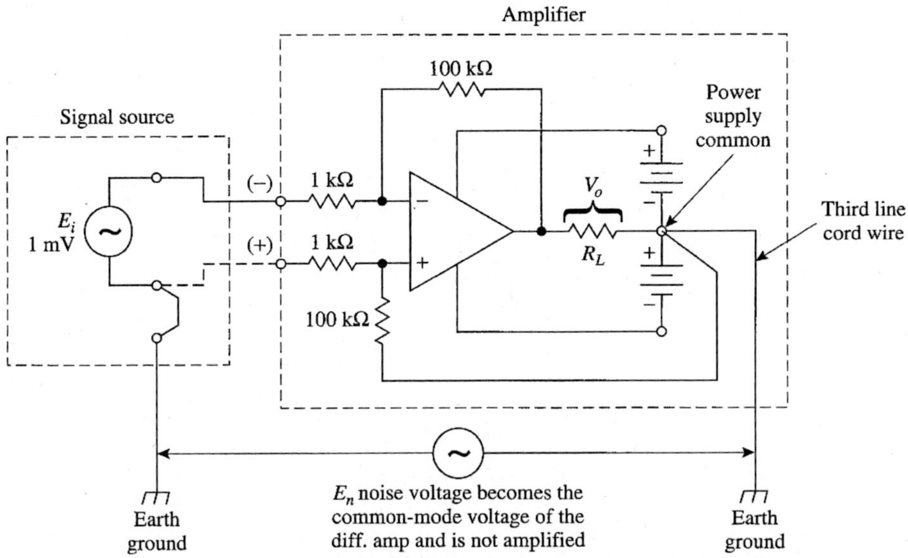

FIGURE 8-4 The differential amplifier is connected so that noise voltage becomes the common-mode voltage and is not amplified. Only the signal voltage E_i is amplified because it has been connected as the differential input voltage.

8-3 IMPROVING THE BASIC DIFFERENTIAL AMPLIFIER

8-3.1 Increasing Input Resistance

There are two disadvantages to the basic differential amplifier studied thus far: It has low input resistance, and changing gain is difficult because the resistor ratios must be closely matched. The first disadvantage is eliminated by *buffering* or isolating the inputs with voltage followers. This is accomplished with two op amps connected as voltage followers in Fig. 8-5(a). The output of op amp A_1 with respect to ground is E_1, and the output of op amp A_2 with respect to ground is E_2. The differential output voltage V_o is developed across the load resistor R_L. V_o equals the difference between E_1 and E_2 ($V_o = E_1 - E_2$). Note that the output of the basic differential amplifier of Fig. 8-1 is a single-ended output; that is, one side of R_L is connected to ground, and V_o is measured from the output pin of the op amp to ground. The buffered differential amplifier of Fig. 8-5(a) is a differential output; that is, neither side of R_L is connected to ground, and V_o is measured only across R_L.

8-3.2 Adjustable Gain

The second disadvantage of the basic differential amplifier is the lack of adjustable gain. This problem is eliminated by adding three more resistors to the buffered amplifier. The resulting buffered, differential-input to differential-output amplifier with adjustable gain is shown in Fig. 8-5(b). The high input resistance is preserved by the voltage followers.

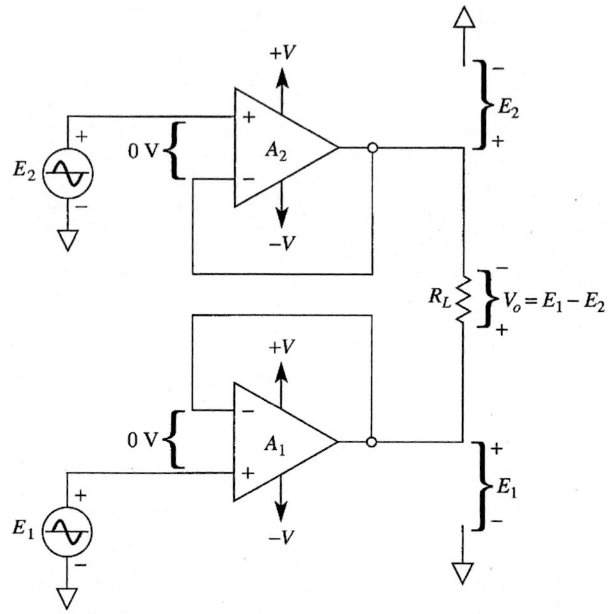

(a) Buffered differential-input to differential-output amplifier.

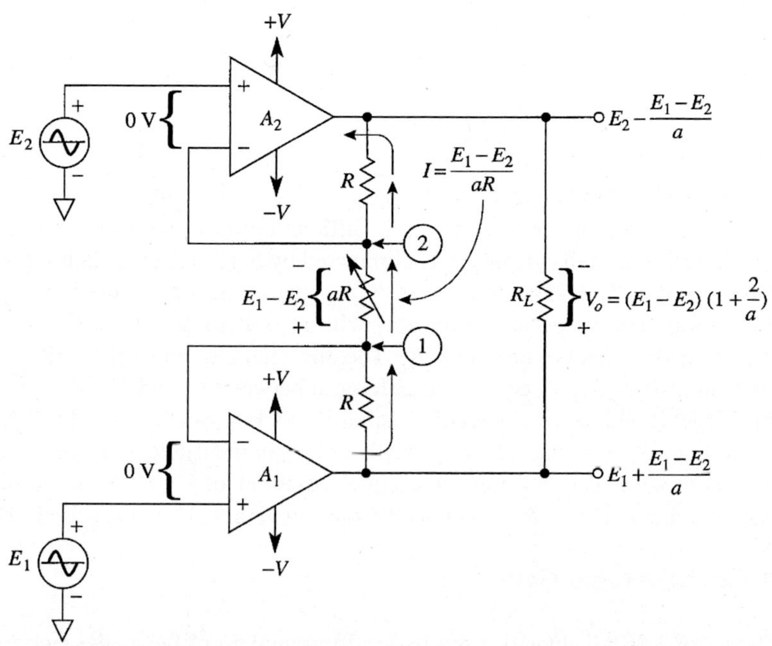

(b) Buffered differential-input to differential-output amplifier with adjustable gain.

FIGURE 8-5 Improving the basic differential amplifier.

Since the differential input voltage of each op amp is 0 V, the voltages at points 1 and 2 (with respect to ground) are equal to E_1 and E_2, respectively. Therefore, the voltage across resistor aR is $E_1 - E_2$. Resistor aR may be a fixed resistor or a potentiometer that is used to adjust the gain. The current through aR is

$$I = \frac{E_1 - E_2}{aR} \tag{8-4}$$

When E_1 is above (more positive than) E_2, the direction of I is as shown in Fig. 8-5(b). I flows through both resistors labeled R, and the voltage across all three resistors establishes the value of V_o. In equation form,

$$V_o = (E_1 - E_2)\left(1 + \frac{2}{a}\right) \tag{8-5}$$

where

$$a = \frac{aR}{R}$$

Example 8-2

In Fig. 8-5(b), $E_1 = 10$ mV and $E_2 = 5$ mV. If $aR = 2$ kΩ and $R = 9$ kΩ, find V_o.

Solution Since $aR = 2$ kΩ and $R = 9$ kΩ,

$$\frac{aR}{R} = \frac{2\text{ k}\Omega}{9\text{ k}\Omega} = \frac{2}{9} = a$$

From Eq. (8-5),

$$1 + \frac{2}{a} = 1 + \frac{2}{2/9} = 10$$

Finally,

$$V_o = (10\text{ mV} - 5\text{ mV})(10) = 50\text{ mV}$$

Conclusion. To change the amplifier gain, *only a single resistor aR* now has to be adjusted. However, the buffered differential amplifier has one disadvantage: It can only drive floating loads. *Floating loads* are loads that have neither terminal connected to ground. To drive grounded loads, a circuit must be added that converts a differential input voltage to a single-ended output voltage. Such a circuit is the basic differential amplifier. The resulting circuit configuration, to be studied in Section 8-4, is called an *instrumentation amplifier.*

8-4 INSTRUMENTATION AMPLIFIER

8-4.1 Circuit Operation

The instrumentation amplifier (IA) is one of the most useful, precise, and versatile amplifiers available today. You will find at least one in every data acquisition unit. The basic IA is made from three op amps and seven resistors, as shown in Fig. 8-6. To simplify circuit analysis, note that the instrumentation amplifier is actually made by connecting a buffered amplifier [Fig. 8-5(b)] to a basic differential amplifier (Fig. 8-1). Op amp A_3 and its four equal resistors, R, form a differential amplifier with a gain of 1. Only the A_3 resistors have to be matched. The primed resistor, R', can be made variable to balance out any common-mode voltage, as shown in Fig. 8-2. Only *one resistor, aR,* is used to set the gain according to Eq. (8-6), repeated here for convenience:

$$\frac{V_o}{E_1 - E_2} = 1 + \frac{2}{a} \tag{8-6}$$

where $a = aR/R$.

E_1 is applied to the (+) input and E_2 to the (−) input. V_o is proportional to the difference between input voltages. Characteristics of the instrumentation amplifier are summarized as follows:

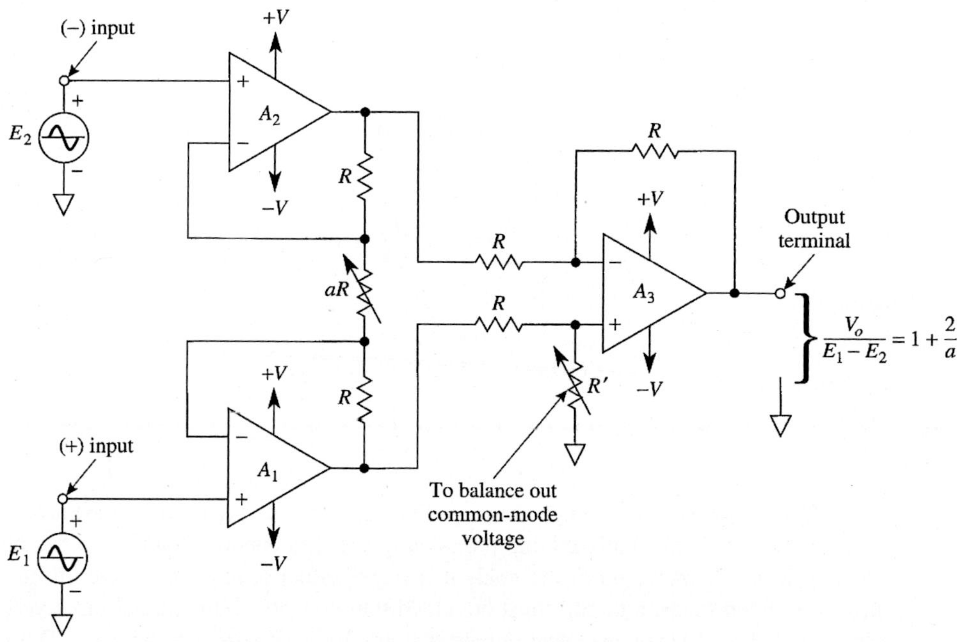

FIGURE 8-6 Basic instrumentation amplifier model.

1. The voltage gain, from differential input ($E_1 - E_2$) to single-ended output, is set by *one* resistor.
2. The input resistance of both inputs is very high and does not change as the gain is varied.
3. V_o does *not* depend on the voltage common to both E_1 and E_2 (common-mode voltage), only on their difference.

Example 8-3

In Fig. 8-6, $R = 25\text{ k}\Omega$ and $aR = 50\ \Omega$. Calculate the voltage gain.

Solution From Eq. (8-6),

$$\frac{aR}{R} = \frac{50}{25{,}000} = \frac{1}{500} = a$$

$$\frac{V_o}{E_1 - E_2} = 1 + \frac{2}{a} = 1 + \frac{2}{1/500} = 1 + (2 \times 500) = 1001$$

Example 8-4

If aR is removed in Fig. 8-6 so that $aR = \infty$, what is the voltage gain?

Solution $a = \infty$, so

$$\frac{V_o}{E_1 - E_2} = 1 + \frac{2}{\infty} = 1$$

Example 8-5

In Fig. 8-6, the following voltages are applied to the inputs. Each voltage polarity is given with respect to ground. Assuming the gain of 1001 from Example 8-3, find V_o for (a) $E_1 = 5.001$ V and $E_2 = 5.002$ V; (b) $E_1 = 5.001$ V and $E_2 = 5.000$ V; (c) $E_1 = -1.001$ V, $E_2 = -1.002$ V.

Solution (a)

$$V_o = 1001(E_1 - E_2) = 1001(5.001 - 5.002)\text{ V}$$

$$= 1001(-0.001)\text{ V} = -1.001\text{ V}$$

(b) $V_o = 1001(5.001 - 5.000)\text{ V} = 1001(0.001)\text{ V} = 1.001\text{ V}$
(c) $V_o = 1001[-1.001 - (-1.002)]\text{ V} = 1001(0.001)\text{ V} = 1.001\text{ V}$

8-4.2 Referencing Output Voltage

In some signal conditioning applications, it is desirable to offset the output voltage to a reference level other than 0 V. (For example, it would be convenient to position a pen on a chart recorder or an oscilloscope trace at a level other than zero.) This can be done quite easily by adding a reference voltage in series with one resistor of the basic differential amplifier. Assume that E_1 and E_2 are set equal to 0 V in Fig. 8-6(a). The outputs of A_1 and A_2 will equal 0 V. Thus, we can show the inputs of the differential amplifier, A_3 as 0 V in Fig. 8-7(a).

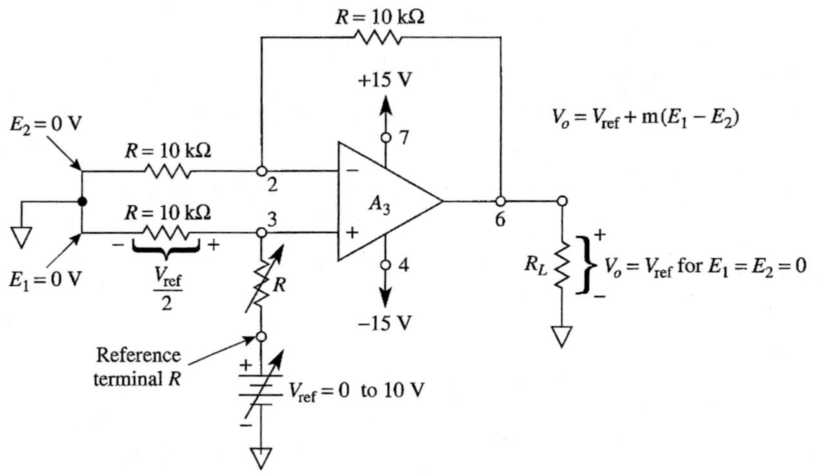

(a) Op amp 3 of the IA in Fig. 8-6 has its "normally grounded" terminal brought out; the new terminal is called "reference terminal," *R*.

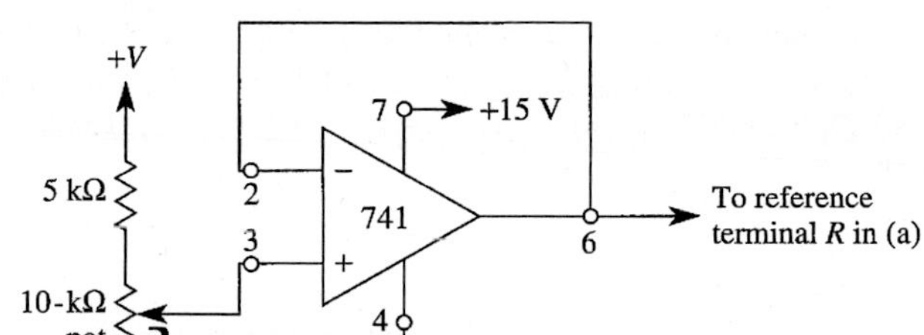

(b) Practically, the reference voltage in (a) must have a very low output impedance; a buffering op amp solves the problem.

FIGURE 8-7 The output voltage of an instrumentation amplifier (*IA*) may be offset by connecting the desired offset voltage (+ or −) to the reference terminal.

An offset voltage or reference voltage V_{ref} is inserted in series with *reference* terminal R. V_{ref} is divided by 2 and applied to the A_3 op amp's (+) input. Then the noninverting amplifier gives a gain of 2 so that V_o equals V_{ref}. Now V_o can be set to any desired offset value by adjusting V_{ref}. In practice V_{ref} is the output of a voltage-follower circuit as shown in Fig. 8-7(b).

8-5 SENSING AND MEASURING WITH THE INSTRUMENTATION AMPLIFIER

8-5.1 Sense Terminal

The versatility and performance of the instrumentation amplifier in Fig. 8-6 can be improved by breaking the negative feedback loop around op amp A_3 and bringing out three terminals. As shown in Fig. 8-8, these terminals are *output* terminal 0, *sense* terminal S, and *reference* terminal R. If long wires or a current-boost transistor are required between the instrumentation amplifier and load, there will be voltage drops across the connecting wires. To eliminate these voltage drops, the sense terminal and reference terminal are wired *directly to the load.* Now, wire resistance is added equally to resistors in series with the sense and reference terminals to minimize any mismatch. Even more important, by sensing voltage at the load terminals and *not* at the amplifier's output terminal, feedback acts to hold load voltage constant. If only the basic differential amplifier is used, the out-

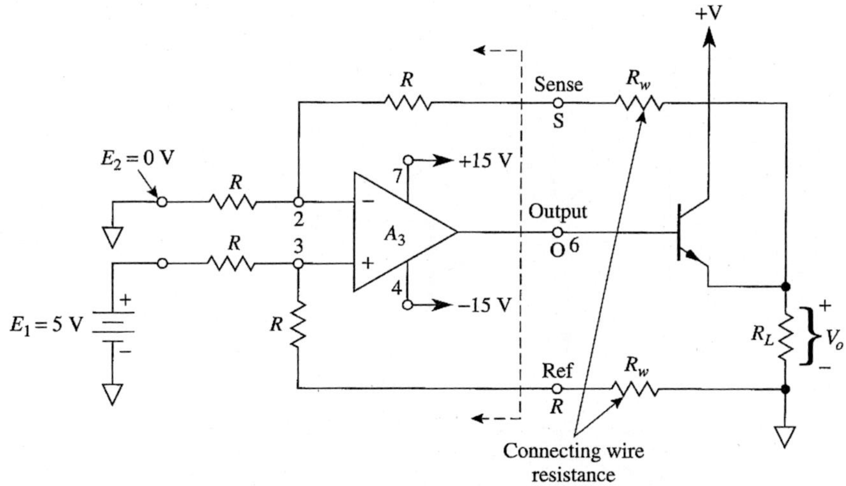

FIGURE 8-8 Extending the sense and reference terminals to the load terminals makes V_o depend on the amplifier gain and the input voltages, not on the load current or load resistance.

put voltage is found from Eq. (8-1) with $m = 1$. If the instrumentation amplifier is used, the output voltage is determined from Eq. (8-6).

This technique is also called *remote voltage sensing;* that is, you sense and control the voltage at the remote load and *not* at the amplifier's output terminals.

8-5.2 Differential Voltage Measurements

The schematic drawing of a low-cost instrumentation amplifier without a sense input terminal is presented in Fig. 8-9. The AD620 requires only one external resistor to set gains from 1 to 1000. The device can be purchased either in an 8-pin SOIC or DIP as shown in Fig. 8-9(a). The manufacturer gives the relationship between gain and R_G as

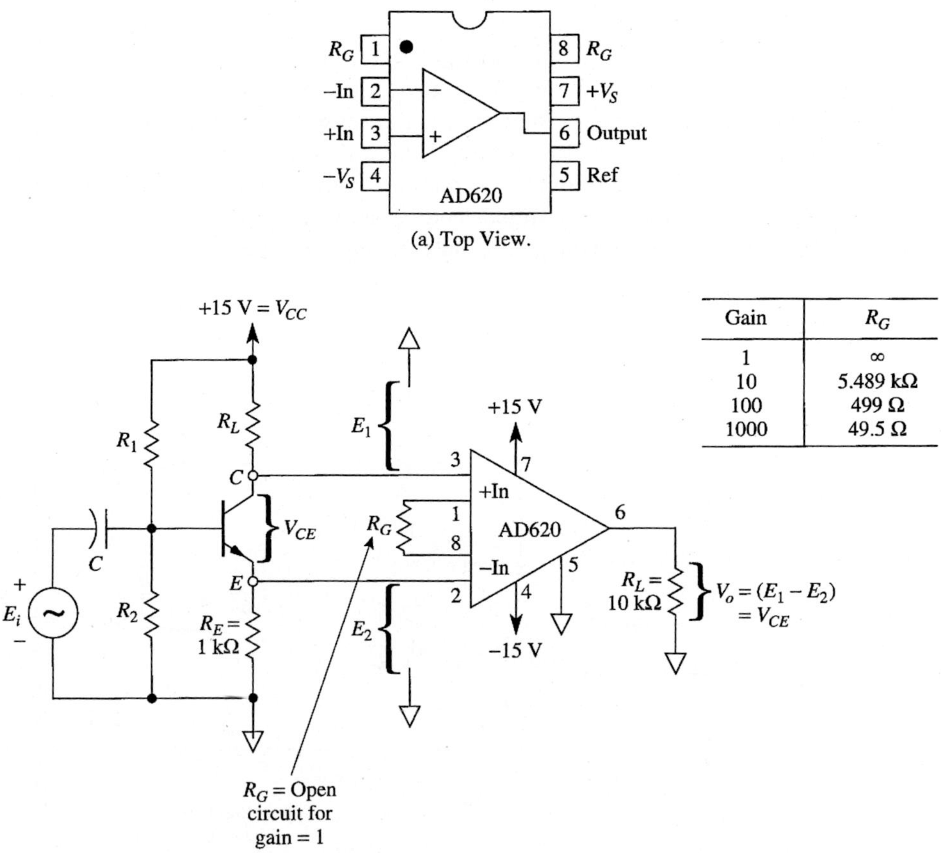

Gain	R_G
1	∞
10	5.489 kΩ
100	499 Ω
1000	49.5 Ω

FIGURE 8-9 Package style in (a) and an IA are used to measure a floating differential voltage in (b).

$$\text{Gain} = 1 + \left(\frac{49{,}400}{R_G}\right) \tag{8-7}$$

For gain values of 1, 10, 100, and 1000, a table listing the R_G values is given in Fig. 8-9(b).

The usual way to measure V_{CE} of a working common-emitter amplifier circuit is to (1) measure collector voltage (with respect to ground), (2) measure emitter voltage (with respect to ground), and (3) calculate the difference. The IA allows you to make the measurement in one step, as shown in Fig. 8-9(b). Since $E_1 = V_{\text{collector}}$ and $E_2 = V_{\text{emitter}}$,

$$V_o = (1)(E_1 - E_2) = (1)(V_{\text{collector}} - V_{\text{emitter}}) = V_{CE} \tag{8-8}$$

Example 8-6

Given $V_o = 5$ V in Fig. 8-9(b), find V_{CE}.

Solution From Eq. (8-8),

$$5\text{ V} = (E_1 - E_2) = V_{CE}$$

Example 8-7

Extend Example 8-6 as follows. Connect + In to the emitter and − In to ground, assume V_o measures 1.2 V, and calculate (a) emitter current I_E; (b) the voltage across R_L or V_{RL}.

Solution (a) Since $V_o = 1.2$ V, $E_i - E_2 = 1.2$ V, and therefore $V_{RE} = 1.2$ V. Use Ohm's law to find I_E.

$$I_E = \frac{V_{RE}}{R_E} = \frac{1.2\text{ V}}{1\text{ k}\Omega} = 1.2\text{ mA}$$

(b)

$$V_{\text{collector}} = V_{CE} + V_{RE} = 5\text{ V} + 1.2\text{ V} = 6.2\text{ V}$$

$$V_{RL} = V_{CC} - V_{\text{collector}} = 15\text{ V} - 6.2\text{ V} = 8.8\text{ V}$$

Part (a) of this example shows how to measure current in a working circuit by measuring the voltage drop across a known resistor.

8-5.3 Differential Voltage-to-Current Converter

The AD620 instrumentation amplifier does not have a sense terminal. Therefore, if your application requires this terminal, choose another IA such as the AD524 or AD624. Figure 8-10 shows how to make an excellent current source that can sink or source dc current into a grounded load. It can also be an ac current source.

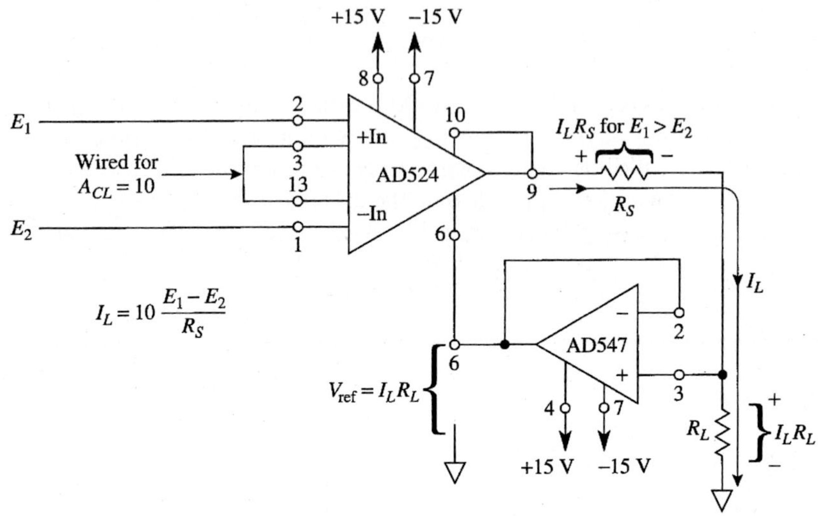

FIGURE 8-10 A differential voltage-to-current converter is made from an IA, op amp, and resistor.

To understand how this circuit operates, one must understand that the IA's output voltage at pin 9 depends on load current, I_L, load resistor, R_L, and *current set resistor,* R_s. In equation form

$$V_9 = I_L R_s + I_L R_L \tag{8-9a}$$

The output voltage of an IA can also be expressed generally by

$$V_9 = V_{\text{ref}} + \text{gain}(E_1 - E_2) \tag{8-9b}$$

The AD547 voltage follower forces the reference voltage to equal load voltage or $V_{\text{ref}} = I_L R_L$. Since the IA's gain is set for 10 in Fig. 8-10, we can rewrite Eq. (8-9b) as

$$V_9 = I_L R_L + 10(E_1 - E_2) \tag{8-9c}$$

Equate Eqs. (8-9a) and (8-9c) to solve for I_L, which yields

$$I_L = 10\left(\frac{E_1 - E_2}{R_s}\right) \tag{8-9d}$$

Equation (8-9d) indicates that load resistor, R_L, does *not* control load current; this is true as long as neither amplifier is forced to saturation. I_L is controlled by R_s and the difference between E_1 and E_2.

Example 8-8

In the circuit of Fig. 8-10, $R_s = 1\text{ k}\Omega$, $E_1 = 100\text{ mV}$, $E_2 = 0\text{ V}$, and $R_L = 5\text{ k}\Omega$. Find (a) I_L; (b) V_{R_s}; (c) V_{ref}; (d) V_9.

Solution (a) From Eq. (8-9d),

$$I_L = 10\left(\frac{0.1\text{ V} - 0\text{ V}}{1000\ \Omega}\right) = 1\text{ mA}$$

(b) $V_{R_s} = I_L R_s = (1\text{ mA})(1\text{ k}\Omega) = 1\text{ V}$
(c) $V_{\text{ref}} = I_L R_L = (1\text{ mA})(5\text{ k}\Omega) = 5\text{ V}$
(d) From Eq. (8-9a) or (8-9c),

$$V_9 = I_L R_s + I_L R_L = 1\text{ V} + 5\text{ V} = 6\text{ V}$$

or

$$V_9 = V_{\text{ref}} + \text{gain}(E_1 - E_2) = 5\text{ V} + 10(0.1\text{ V}) = 6\text{ V}$$

8-6 THE INSTRUMENTATION AMPLIFIER AS A SIGNAL CONDITIONING CIRCUIT

In Chapter 3, we mentioned that some sensor circuits are designed with a differential output. However, the input voltage to a microcontroller is with respect to ground and hence is single ended. Therefore we need a signal conditioning circuit (SCC) that has a differential input and a single-ended output—the instrumentation amplifier. We begin by introducing a sensor circuit with a differential output.

8-6.1 Introduction to the Strain Gage

A strain gage is a conducting wire whose resistance changes by a small amount when it is lengthened or shortened. The change in length is small, a few millionths of an inch. The strain gage is bonded to a structure so that the percent change in length of the strain gage and structure are identical.

A foil-type gage is shown in Fig. 8-11(a). The active length of the gage lies along the transverse axis. The strain gage must be mounted so that its transverse axis lies in the same direction as the structure motion that is to be measured [see Figs. 8-11(b) and (c)]. Lengthening the bar by tension lengthens the strain gage conductor and increases its resistance. Compression reduces the gage's resistance because the normal length of the strain gage is reduced.

8-6.2 Strain-Gage Material

Strain gages are made from metal alloy such as constantan, Nichrome V, Dynaloy, Stabiloy, or platinum alloy. For high-temperature work they are made of wire. For moderate temperature, strain gages are made by forming the metal alloy into very thin sheets by a photoetching process. The resultant product is called a foil-type strain gage and a typical example is shown in Fig. 8-11(a).

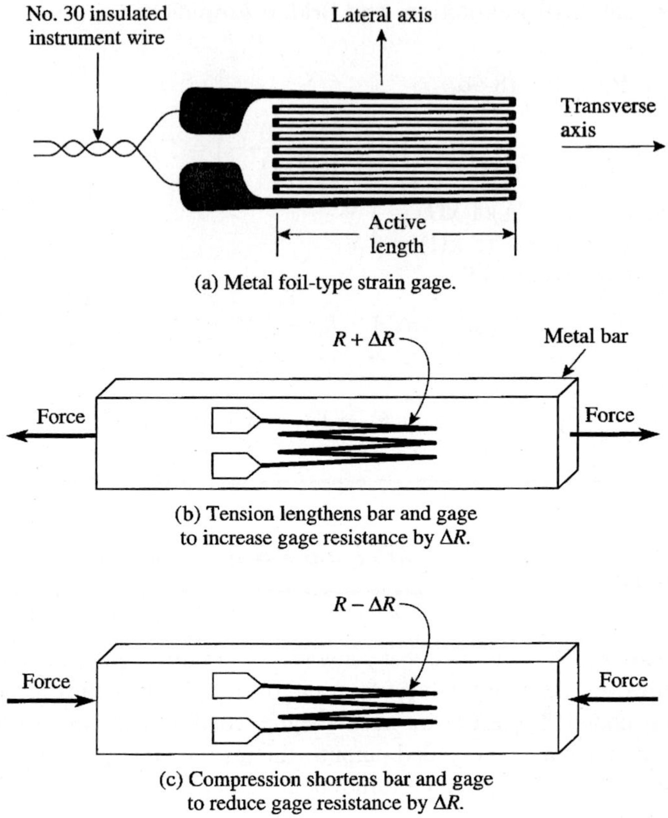

FIGURE 8-11 Using a strain gage to measure the change in length of a structure.

8-6.3 Using Strain-Gage Data

In the next section, we show that our instrumentation measures only the gage's *change* in resistance ΔR. The manufacturer specifies the unstrained gage's resistance R. Once ΔR has been measured, the ratio $\Delta R/R$ can be calculated. The manufacturer also furnishes a specified *gage factor* (*GF*) for each gage. The gage factor is the ratio of the percent change in resistance of a gage to its percent change in length. These percent changes may also be expressed as decimals. If the ratio $\Delta R/R$ is divided by gage factor G, the result is the ratio of the *change* in length of the gage ΔL to its original length L. Of course the structure where the gage is mounted has the same $\Delta L/L$. An example will show how gage factor is used.

Example 8-9

A 120-Ω strain gage with a gage factor of 2 is affixed to a metal bar. The bar is stretched and causes a ΔR of 0.001 Ω. Find $\Delta L/L$.

Solution

$$\frac{\Delta L}{L} = \frac{\Delta R/R}{GF} = \frac{0.001\ \Omega/120\ \Omega}{2}$$

$$\simeq 4.1 \text{ microinches per inch}$$

The ratio $\Delta L/L$ has a name. It is called *unit strain.* It is the unit strain data (we have developed from a measurement of ΔR) that mechanical engineers need. They can use this unit strain data together with known characteristics of the structural material (modulus of elasticity) to find the *stress* on the beam. *Stress is the amount of force acting on a unit area.* The unit for stress is pounds per square inch (psi). If the bar in Example 8-9 were made of mild steel, its stress would be about 125 psi. *Strain is the deformation of a material resulting from stress,* or $\Delta L/L$.

8-6.4 Strain-Gage Mounting

Before mounting a strain gage the surface of the mounting beam must be cleaned, sanded, and rinsed with alcohol, Freon, or methyl ethyl ketone (MEK). The gage is then fastened permanently to the cleaned surface by Eastman 910, epoxy, polymide adhesive, or ceramic cement. The manufacturer's procedures should be followed carefully.

8-6.5 Strain-Gage Resistance Changes

It is the *change* of resistance in a strain gage ΔR that must be measured and this change is *small.* ΔR has values of a few milliohms. The technique employed to measure small resistance change is discussed next.

8-7 MEASUREMENT OF SMALL RESISTANCE CHANGES

8-7.1 Need for a Resistance Bridge

To measure resistance, we must first find a technique to convert the *resistance change* to a current or voltage for display on an ammeter or voltmeter. If we must measure a small *change* of resistance, we will obtain a very small voltage *change.* For example, if we passed 5 mA of current through a 120-Ω strain gage, the voltage across the gage would be 0.600 V. If the resistance *changed* by 1 mΩ, the voltage *change* would be 5 μV. To display the 5-μV change, we would need to amplify it by a factor of, for example, 1000 to 5 mV. However, we would also amplify the 0.6 V by 1000 to obtain 600 V plus 5 mV. It is difficult to detect a 5-mV difference in a 600-V signal. Therefore, we need a circuit that allows us to amplify only the *difference* in voltage across the strain gage caused by a *change* in resistance. The solution is found in the Wheatstone bridge circuit.

8-7.2 Basic Resistance Bridge

The strain gage is placed in one arm of a resistance bridge, as shown in Fig. 8-12. Assume that the gage is unstrained, so that its resistance = R. Also assume that R_1, R_2, and R_3 are all precisely equal to R. (This unlikely assumption is dealt with in Section 8-8.) Under these conditions $E_1 = E_2 = E/2$ and $E_1 - E_2 = 0$. The bridge is said to be *balanced.* If the strain gage is *compressed,* R would decrease by ΔR and the differential voltage $E_1 - E_2$ would be given by

$$E_1 - E_2 = E\frac{\Delta R}{4R} \qquad \text{(8-10)}$$

This approximation is valid because $2\ \Delta R << 4R$ for strain gages.

Equation (8-10) shows that E should be made large to maximize the bridge differential output voltage, $E_1 - E_2$.

Example 8-10

If $\Delta R = 0.001\ \Omega$, $R = 120\ \Omega$, and $E = 1.0$ V in Fig. 8-12, find the output of the bridge, $E_1 - E_2$.

Solution From Eq. (8-10),

$$E_1 - E_2 = 1.0\text{ V} \times \frac{0.001\ \Omega}{(4)(120)\ \Omega} = 2.2\ \mu\text{V}$$

If E is increased to 10 V, then $E_1 - E_2$ will be increased to 22 μV.

An instrumentation amplifier can then be used to amplify the differential voltage $E_1 - E_2$ by 1000 to give an output of about 22 mV per milliohm of ΔR.

We conclude that a voltage E and bridge circuit plus an instrumentation amplifier can convert a change in resistance of 1 mΩ to an output voltage change of 22 mV.

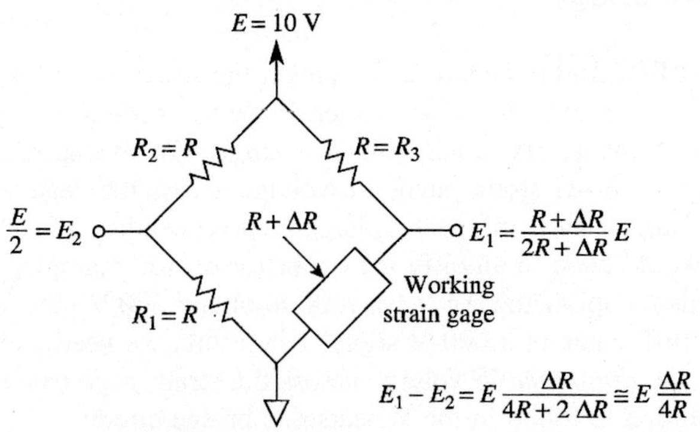

FIGURE 8-12 The resistor bridge arrangement and supply voltage E convert a resistance change in the strain gage ΔR to a differential output voltage $E_1 - E_2$. If $R = 120\ \Omega$, $E = 10$ V, and $R = 1$ mΩ, $E_1 - E_2 = 22\ \mu$V.

8-7.3 Thermal Effects on Bridge Balance

Even if you succeed in balancing the bridge circuit of Fig. 8-12, it will not stay in balance because slight temperature changes in the strain gage cause resistance change equal to or greater than those caused by strain. This problem is solved by mounting another identical strain gage immediately adjacent to the working strain gage so that both share the same thermal environment. Therefore, as temperature changes, the added gage's resistance changes exactly as the resistance of the working gage. The added gage provides automatic temperature compensation, and is appropriately called the *temperature-compensation* or *dummy gage.*

The temperature-compensation gage is mounted with its transverse axis perpendicular to the transverse axis of the working gage, as shown in Fig. 8-13. This type of standard gage arrangement is available from manufacturers. The new gage is connected in place of resistor R_1 in the bridge circuit of Fig. 8-12. Once the bridge has been balanced, R of the temperature-compensation gage and working gage track one another to hold the bridge in balance. Any unbalance is caused strictly by ΔR of the working gage due to strain.

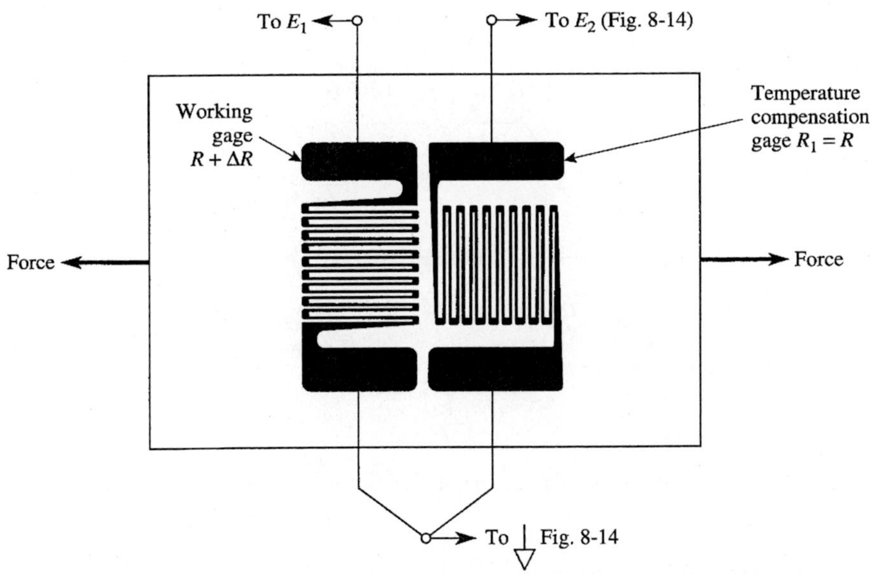

FIGURE 8-13 The temperature compensation gage has the same resistance changes as the working gage with changes in temperature. Only the working gage changes resistance with strain. By connecting in the bridge circuit of Fig. 8-12 as shown, resistance changes due to temperature changes are automatically balanced out.

8-8 BALANCING A STRAIN-GAGE BRIDGE

8-8.1 The Obvious Technique

Suppose that you had a working gage and temperature-compensation gage in Fig. 8-14 that are equal to within 1 mΩ. To complete the bridge, you install two 1%, 120-Ω resistors. One is high by 1% at 121.200 Ω and one is low by 1% at 118.800 Ω. They must be equalized to balance the bridge. To do so, a 5-Ω, 20-turn balancing pot is installed, as shown in Fig. 8-14. Theoretically, the pot should be set as shown to equalize resistances in the top branches of the bridge at 122.500 Ω.

Further assume that an instrumentation amplifier with a gain of 1000 is connected to the bridge of Fig. 8-14. From Example 8-10, the *output* of the instrumentation amplifier (IA) will be about 22 mV per milliohm of unbalance. This means that the 5-Ω pot must be adjusted to within 1 mΩ of the values shown, so that $E_1 - E_2$ and consequently V_o of the IA will equal 0 V ± 22 mV.

Unfortunately, it is very difficult in practice to adjust for balance. This is because each turn of the pot is worth 5 Ω/20 turns = 250 mΩ. When you adjust the pot it is normal to expect a backlash of $\pm\frac{1}{50}$ of a turn. Therefore, your best efforts result in an unbalance at the pot of about ±5 mΩ. You observe this unbalance at the IA's output, where V_o changes by ±0.1 V on either side of zero as you fine-tune the 20-turn pot. It turns out there is a better technique that uses an ordinary linear potentiometer ($\frac{3}{4}$ turn) and a single resistor.

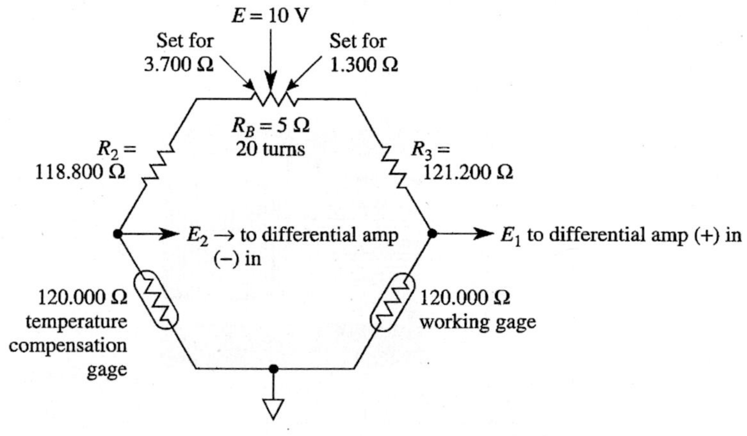

FIGURE 8-14 Balance pot R_B is adjusted in an attempt to make $E_1 - E_2 = 0$ V.

8-8.2 The Better Technique

To analyze operation of the balance network in Fig. 8-15, assume that the R_2 and R_3 bridge resistors are reasonably equal, to within ±1%. The strain gage's resistance should have equal resistances within several milliohms if the working gage is not under strain.

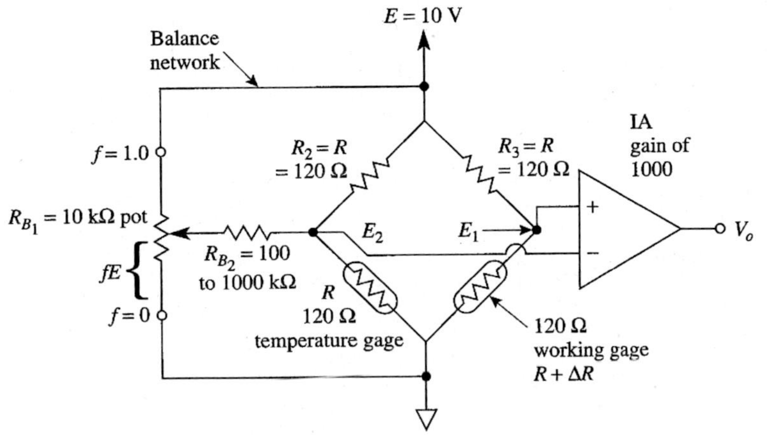

FIGURE 8-15 Improved balance network. R_{B_1} and R_{B_2} allow easy adjustment of V_o to 0 V.

Resistor R_{B_1} is an ordinary 3/4-turn linear pot. Its resistance should be about 1/10 or less than resistor R_{B_2} so that the voltage fE depends only on E and the decimal fraction f. Values of f vary from 0 to 1.0 as the pot is adjusted from one limit to the other. R_{B_1} should be 10 or more times the gage resistance.

Resistor R_{B_2} is chosen to be greater than 10 or more times R_{B_1}. Under these conditions R_{B_2} does not load down the voltage-divider action of R_{B_1}. Also, the size of R_{B_2} determines the maximum balancing current that can be injected into, or extracted from, the E_2 node. The pot setting f determines how much of that maximum current is injected or extracted.

Balancing action is summarized by observing that if $f > 0.5$, a small current is injected into the E_2 node and flows through the temperature gage to ground, this makes E_2 more positive. If $f < 0.5$ current is extracted from the E_2 node, this increases current through R_2 to make E_2 less positive.

In a real bridge setup, begin with $R_{B_2} = 100$ kΩ and $R_{B_1} = 10$ kΩ. Monitor V_o of the IA and check the balancing action. If the variation in V_o is larger than you want, increase R_{B_2} to 1000 kΩ and recheck the balance action. The final value of R_{B_2} is selected by experiment and depends on the magnitude of unbalance between R_2 and R_3.

8-9 INCREASING STRAIN-GAGE BRIDGE OUTPUT

A single working gage and a temperature-compensation gage were shown to give a differential bridge output in Fig. 8-12 of

$$E_1 - E_2 = E\frac{\Delta R}{4R} \tag{8-11}$$

This bridge circuit and placement of the gages is shown again in Fig. 8-16(a).

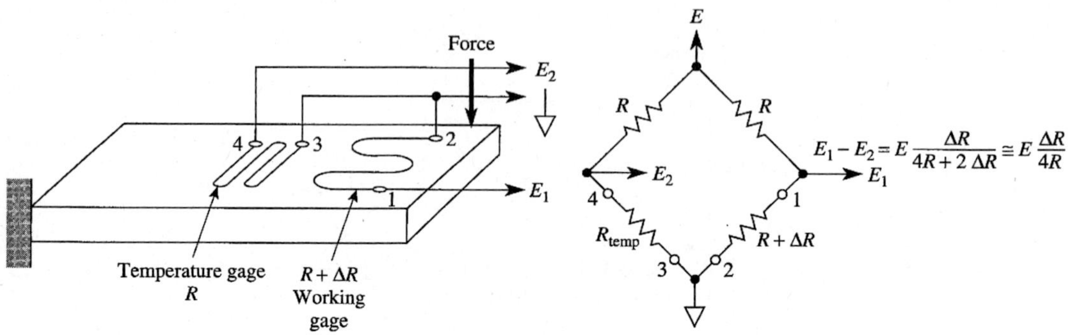

(a) Single working gage gives $E_1 - E_2 = E \frac{\Delta R}{4R}$.

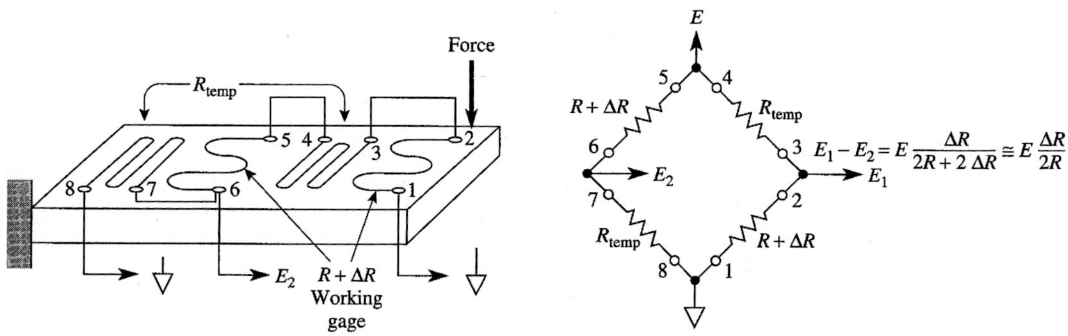

(b) Two working gages double the $E_1 - E_2$ output over that of (a).

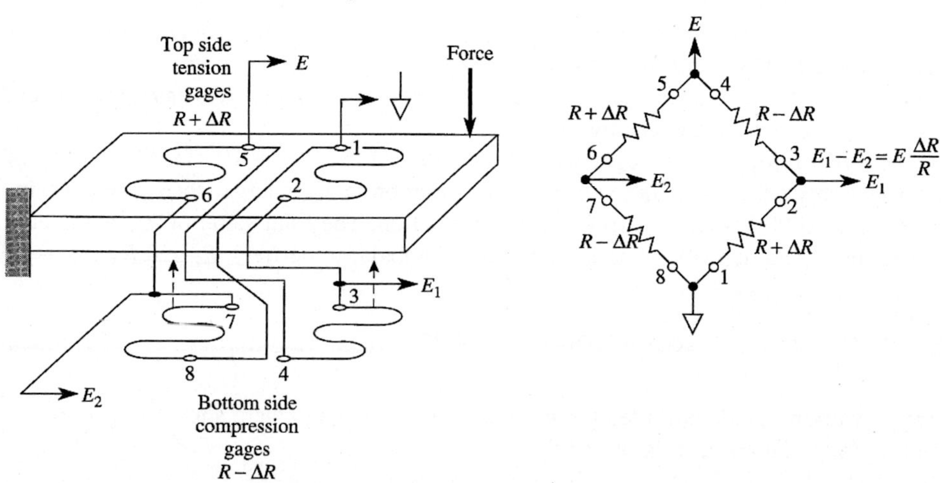

(c) Four working gages quadruple the $E_1 - E_2$ over that of (a).

FIGURE 8-16 Comparison of sensitivity for three strain-gage bridge arrangements. (ΔR is small with respect to R for foil strain gages.)

The bridge output voltage $E_1 - E_2$ can be doubled by doubling the number of working gages, as in Fig. 8-16(b). Gages 1–2 and 5–6 are the working gages and will increase resistance (tension) if force is applied as shown. By arranging the working gages in opposite arms of the bridge and the temperature gages in the other arms, the bridge output is

$$E_1 - E_2 = E\frac{\Delta R}{2R + \Delta R} \simeq E\frac{\Delta R}{2R} \tag{8-12}$$

If the structural member experiences bending as shown in Fig. 8-16(c), even greater bridge sensitivity can be obtained. The upper side of the bar will lengthen (tension) to increase the resistance of the working strain gages by $+\Delta R$. The lower side of the bar will shorten (compression) to decrease the working strain gages by $-\Delta R$.

The tension gages 1–2 and 5–6 are connected in opposite arms of the bridge. Compression gages 3–4 and 7–8 are connected in the remaining opposite arms of the bridge. The gages also temperature-compensate one another. The output of the four-strain-gage arrangement in Fig. 8-16(c) is quadrupled over the single-gage bridge to

$$E_1 - E_2 = E\frac{\Delta R}{R} \tag{8-13}$$

Of course, each bridge arrangement in Fig. 8-16 should be connected to a balance network (which, for clarity, was not shown) (see Fig. 8-15 and Section 8-8). Note that the output of the four-gage bridge is linear.

8-10 A PRACTICAL STRAIN-GAGE APPLICATION

As shown in Fig. 8-17, an AD620 (Analog Devices) instrumentation amplifier (IA) is connected to a bridge arrangement of four strain gages. The gages are 120-Ω, SR4, foil-type strain gages. They are mounted on a steel bar in accordance with Fig. 8-16(c). Also the balance network of Fig. 8-15 is connected to the strain-gage bridge. R_{B_2} was selected, after experiment, as 100 kΩ. Strain gages were mounted in strict accordance with the manufacturer's instructions (BLH Electronics, Inc.). From the table in Fig. 8-9(b) R_G is selected for a gain of 1000.

Example 8-11

The SCC of Fig. 8-17 is used to measure the strain resulting from deflection of a steel bar. V_o is measured to be 100 mV. Calculate (a) ΔR; (b) $\Delta R/R$; (c) $\Delta L/L$. The gain is 1000. The gage factor is 2.0.

Solution (a) Find $E_1 - E_2$ from

$$E_1 - E_2 = \frac{V_o}{\text{gain}} = \frac{100 \text{ mV}}{1000} = 0.1 \text{ mV}$$

From Eq. (8-13):

$$\Delta R = \frac{R(E_1 - E_2)}{E} = \frac{120\ \Omega(0.1 \times 10^{-3}\ \text{V})}{5.0\ \text{V}}$$

$$= 0.0024\ \Omega = 2.4\ \text{m}\Omega$$

(b)

$$\frac{\Delta R}{R} = \frac{0.0024\ \Omega}{120\ \Omega} = 0.000020 = 20 \times 10^{-6}\ \mu\Omega/\Omega$$

(c) From gage factor = $(\Delta R/R)/(\Delta L/L)$, we obtain

$$\frac{\Delta L}{L} = \frac{20 \times 10^{-6}}{2} = 10 \times 10^{-6} = 10\ \mu\text{in/in.}$$

Note: Resistor R_{B_3} is selected to restrict gage current *below* 25 mA to limit self-heating.

Since we now know the value of strain from $\Delta L/L$, we can look up the *modulus of elasticity* for steel, $E = 30 \times 10^6$. The *stress* can be calculated from

$$\text{stress} = E \times \text{strain} = (30 \times 10^6)(10 \times 10^{-6}) = 300\ \text{psi}$$

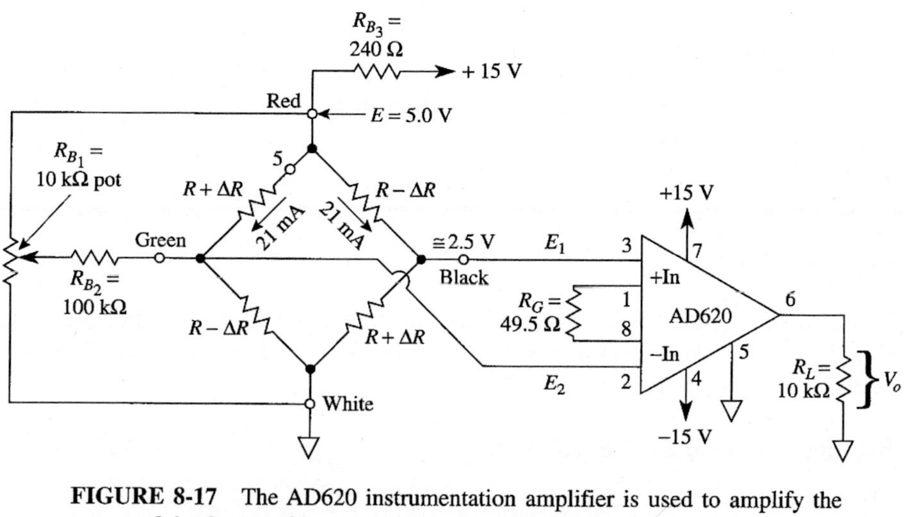

FIGURE 8-17 The AD620 instrumentation amplifier is used to amplify the output of the four working strain gages [see Fig. 8-16(c)].

8-11 MEASUREMENT OF PRESSURE, FORCE, AND WEIGHT

Example 8-11 illustrated how pressure could be measured by a strain-gage system. The mechanical engineers can be given $\Delta L/L$ by electrical personnel, who can measure $\Delta R/R$ and look up the gage factor. From the value of $\Delta L/L$, the mechanical engineers and technicians can calculate pressure on a structure. Since pressure is force per unit area, they can calculate force by measuring the structure's area.

Furthermore, the weight of an object exerts a force on any supporting structure. By installing a strain gage on the supporting structure, you can weigh very heavy objects such as a gravel-filled truck or a 747 aircraft.

8-12 BASIC BRIDGE AMPLIFIER

8-12.1 Introduction

Some transducers can be connected into a lower-parts-count signal conditioning circuit called a bridge amplifier. An op amp, four resistors, and a transducer form the basic bridge amplifier in Fig. 8-18(b). The transducer in this case is any device that converts an environmental change to a resistance change. For example, a *thermistor* is a transducer whose resistance decreases as temperature increases. A *photoconductive* cell is a transducer whose resistance decreases as light intensity increases. For circuit analysis, the transducer is represented by a resistor R plus a *change* in resistance ΔR. R is the resistance value at the desired reference, and ΔR is the amount of change in R. For example, a UUA 41J1 thermistor has a resistance of 10,000 Ω at a reference of 25°C. A temperature change of +1° to 26°C results in a thermistor resistance of 9573 Ω. ΔR is found to be negative from

$$R_{\text{transducer}} = R_{\text{reference}} + \Delta R$$

$$9573\ \Omega = 10{,}000\ \Omega + \Delta R$$

$$\Delta R = -427\ \Omega$$

Since we have defined 25°C to be the reference temperature, we define the reference resistance to be $R_{\text{ref}} = 10{,}000\ \Omega$. Our definitions force ΔR to have a negative sign if the transducer's resistance is less than R_{ref}.

To operate the bridge, we need a stable bridge voltage E, which may be either ac or dc. E should have an internal resistance that is small with respect to R. The simplest way to generate E is to use a voltage divider across the stable supply voltages as shown in Fig. 8-18(a). Then connect a simple voltage follower to the divider. For the resistor values shown, E can be adjusted between +10 and −10 V.

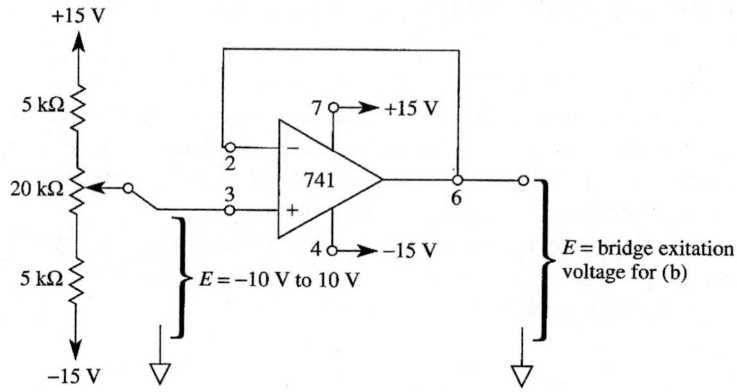

(a) A bridge exitation voltage must have the low output resistance of a voltage follower and be derived from a regulated, stable voltage source.

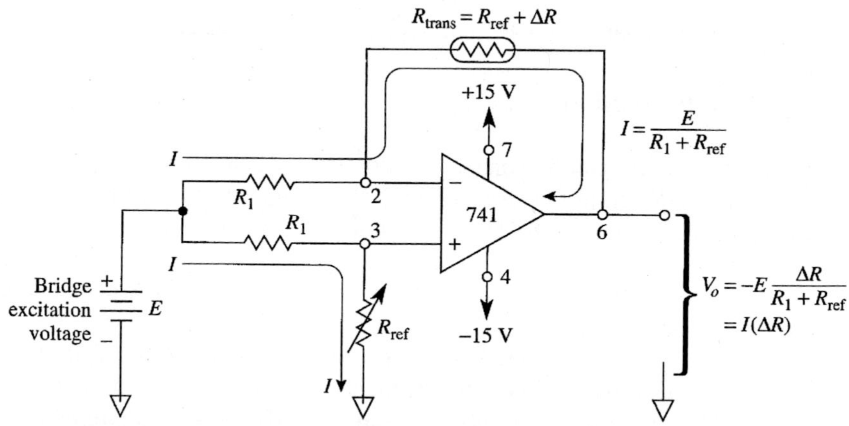

(b) Practical bridge amplifier.

FIGURE 8-18 This bridge amplifier outputs a voltage that is directly proportional to the *change* in transducer resistance.

8-12.2 Basic Bridge Circuit Operation

A basic low-parts-count bridge circuit is presented in Fig. 8-18(b). Resistors R_1 are 1% low Tempco (metal-film) resistors. Current I is constant and set by R_1, R_{ref}, and E. That is, $I = E/(R_1 + R_{ref})$. Note that transducer current is *constant* and equal to I because the voltage drops across both R_1 resistors are equal ($E_d = 0$ V).

The resistor from (+) input to ground is always chosen to equal the reference resistance of the transducer. We want V_o to be zero volts when $R_{trans} = R_{ref}$. This will allow us to calibrate or check operation of the bridge. For Fig. 8-18(b), ΔR is the input and V_o is the output. The output–input relation is given by

$$V_o = -E\left(\frac{\Delta R}{R_{\text{ref}} + R}\right) = -I\,\Delta R \tag{8-14}$$

where $I = E/(R_{\text{ref}} + R)$, $\Delta R = R_{\text{trans}} - R_{\text{ref}}$.

Zeroing procedure

1. Place the transducer in the reference environment: for example, 25°C.
2. Adjust R_{ref} until $V_o = 0$ V.

Normally, it is too costly to control an environment for test or calibration of a single circuit. Therefore, (1) replace the transducer, R_{trans}, with a resistor equal to R_{ref}; (2) now, ΔR equals zero; (3) from Eq. (8-14), V_o should also be equal to zero.

Suppose that V_o is close to, but not equal to, zero. You want to adjust V_o to precisely zero volts.

1. Check that the R_1 resistors are equal to within 1%.
2. Check that the replacement transducer R_{ref} equals the value of R_{ref} [from (+) input to ground in Fig. 8-18(b)] within 1%.
3. Use an op amp with a low dc offset such as the OP-177.

8-12.3 Temperature Measurement with a Bridge Circuit

In this section we show how to design a low-parts-count temperature-measuring system to illustrate a design procedure.

Design Example 8-12

Design a temperature-to-voltage converter that will measure temperatures between 25° and 50°C.

Design Procedure

1. *Select any thermistor* on a trial basis. The thermistor converts a temperature change to a resistance change. Select the Fenwal UUA41J1 and list its corresponding temperature versus resistance as in Table 8-1. (Note the nonlinearity between temperature and resistance.)
2. *Select the reference temperature.* At the reference temperature, V_o must equal zero. Select either the low limit of 25°C or the high limit of 50°C. We shall select the low limit of 25°C for this example. *We have just defined R_{ref}.* R_{ref} is equal to the transducer's resistance at the reference temperature. Specifically, $T_{\text{ref}} = 25$°C; therefore, $R_{\text{ref}} = 10{,}000\ \Omega$. Now calculate ΔR for each temperature from

$$R_{trans} = R_{ref} + \Delta R$$

At 50°C,

$$3603\ \Omega = 10{,}000\ \Omega + \Delta R$$

$$\Delta R = -6397\ \Omega$$

Note the negative sign for ΔR.

TABLE 8-1 Resistance versus Temperature of a UUA41J1 Thermistor

Temp. (°C)	R_{trans}		
25	10,000	1983	Ohmic change for 5°C change
30	8,057		
35	6,530		
			nonlinear
40	5,327		
45	4,370	767	
50	3,603		Ohmic change for 5°C change

3. *Predict the voltage–temperature characteristics.* We shall select the bridge circuit of Fig. 8-18 because it converts a resistance change ΔR into an output voltage [see Eq. (8-14)].
   a. Select resistors R_1 to equal 10 kΩ, 1%.
   b. Make a trial choice for $E = 1.0$ V.

   If you are wondering why we should make these particular choices, the answer is that (1) 10-kΩ resistor sizes are readily available, and (2) a 1-V selection will give us an idea of the size of V_o. If later you want to double or triple V_o, simply double or triple E.

   c. Calculate I from Eq. (8-14).

$$I = \frac{E}{R_{ref} + R_1} = \frac{1\text{ V}}{10\text{ k}\Omega + 10\text{ k}\Omega} = 0.050\text{ mA}$$

   d. Calculate V_o for each value of R and tabulate the results (see Table 8-2). From Eq. (8-14),

$$V_o = -I\,\Delta R$$

   For 50°C,

$$V_o = -(0.050\text{ mA})(-6397\ \Omega) = 310\text{ mV}$$

4. *Document performance.* V_o is plotted against temperature in Fig. 8-19, where the design circuit is also drawn.

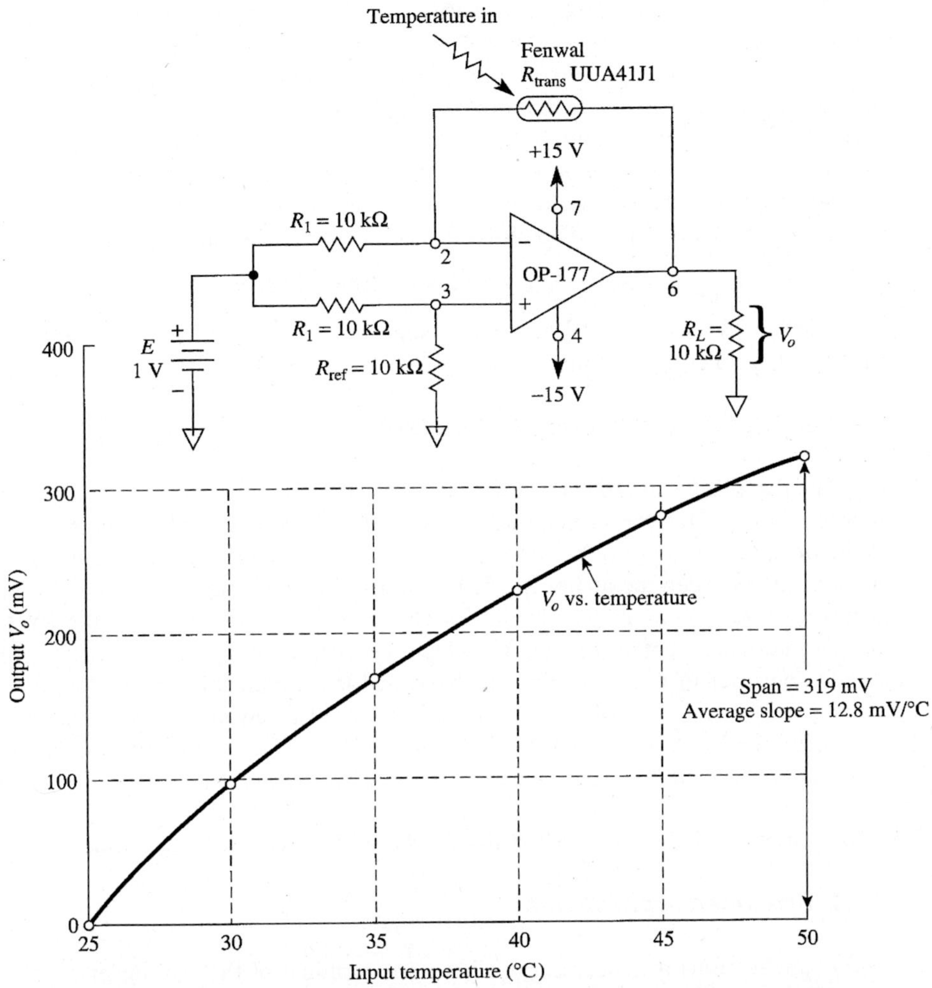

FIGURE 8-19 Solution to Design Example 8-12. An input temperature change of 25° to 50°C gives an output voltage change of 0 to 319 mV. The circuit is a temperature-to-voltage converter.

TABLE 8-2 Calculations for Temperature-to-Voltage Converter

Temp. (°C)	R_{trans} (Ω)	ΔR (Ω)	V_o (mV)
25	10000	0	0
30	8057	−1943	97
35	6530	−3470	173
40	5327	−4673	233
45	4370	−5630	281
50	3603	−6397	319

Summary review and comments. Example 8-12 shows how a bridge circuit converts the resistance change of a transducer into a voltage change. The circuit output voltage is linear with respect to ΔR [see Eq. (8-14)]. However, ΔR is *not* linear with respect to temperature (see Table 8-1 and Fig. 8-19). Therefore, V_o is *not* linear with respect to temperature. The bridge simply transmits the nonlinearity of the thermistor.

The sensitivity of the temperature-to-voltage converter can be increased easily by increasing E. The maximum value of E is set by the maximum thermistor current to avoid self-heating, typically 1 mA. Therefore, E has a maximum value of

$$E = I(R_{\text{ref}} + R_1) = (1 \text{ mA})(10 + 10) \text{ k}\Omega = 20 \text{ V}$$

If we want to increase the 319-mV output span ($E = 1$ V) to a 5.0-V span for the input to a microcontroller, simply increase E by 15.67 to 15.67 V [(5.0 V/0.319 V = 15.67)].

8-12.4 Bridge Amplifiers and Computers

Thus far, the bridge amplifier shown converts temperature changes to a voltage. In Chapter 2 you studied the principles of how a voltage change could be converted into a change of high time by a pulse-width modulator. Those principles can be used again to communicate between the analog world of continuous temperature variation and the digital world of the computer. The bridge amplifier provides measurement; the PWM provides an interface. Another method is to have the output voltage of Fig. 8-19 applied to the A/D converter of a microcontroller. Note that the nonlinearity of the thermistor can be compensated for by a software program known as piecewise linearization. (See *Data Acquisition and Process Control with the HC11 Microcontroller* 2nd Edition by F. Driscoll, R. Coughlin, and R. Villanucci, published by Prentice Hall.)

8-13 ADDING VERSATILITY TO THE BRIDGE AMPLIFIER

8-13.1 Grounded Transducers

In some applications it is necessary to have one terminal of the transducer connected to ground. The standard technique is shown in Fig. 8-20(a). Note that current I depends on transducer resistance (in Figs. 8-18 and 8-19, the current was constant). Note also that V_o is not linear with ΔR because ΔR appears in the denominator of the equation for V_o versus ΔR. Finally, in contrast with Design Example 8-10, if E_i is positive and T_{ref} is at the low end of the scale, V_o goes negative for negative values of ΔR. That is, if R_{trans} is a thermistor, V_o goes more negative as temperature increases.

8-13.2 High-Current Transducers

If the current required by the transducer is higher than the current capability of the op amp (5 mA), use the circuit of Fig. 8-20(b). Transducer current is furnished from E. Resistors mR are chosen to hold their currents to about 1 to 4 mA. Transducer current and output voltage may be found from the equation in Fig. 8-20(b). If the transducer current

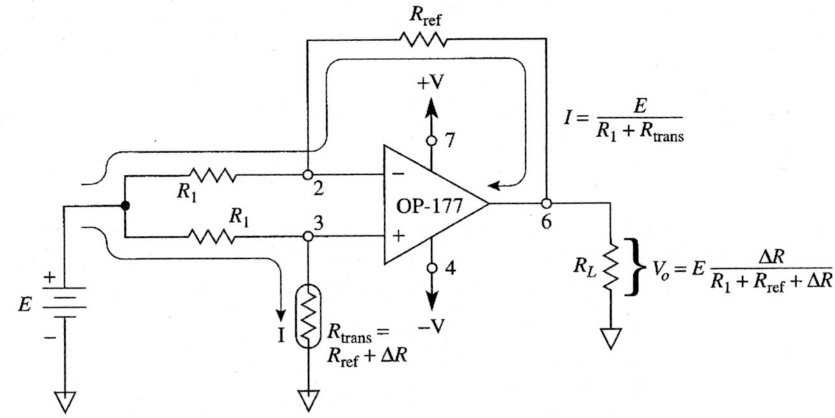

(a) Bridge amplifier with grounded transducer.

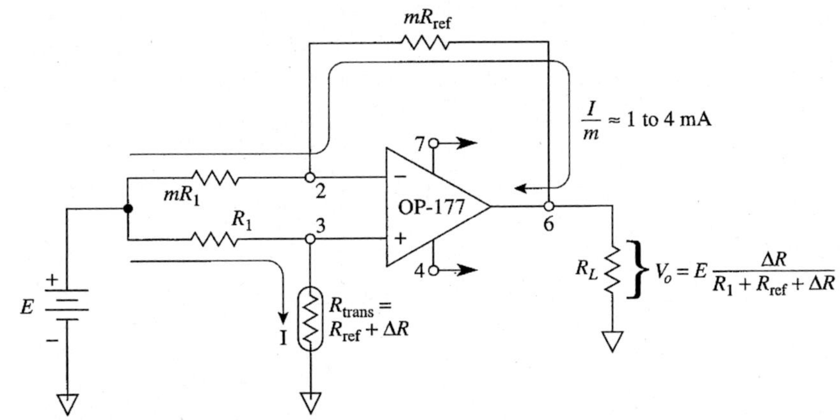

(b) Some transducers require currents larger than the op amp can supply. The transducer's current is scaled down by the multiplying factor m so that feedback current supplied by the op amp is I/m.

FIGURE 8-20 The bridge amplifier is used with a grounded transducer in (a) and with a high current transducer in (b).

is very small (high-resistance transducers), the same circuit can be used except that the mR resistors will be smaller than R to hold output current of the op amp at about 1 mA. BiFET op amps such as the AD548 have small bias currents (see Chapter 9).

PROBLEMS

8-1. In Fig. 8-1, $m = 20$, $E_1 = 0.2$ V, and $E_2 = 0.25$ V. Find V_o.

8-2. If $V_o = 10$ V in Fig. 8-1, $E_1 = 7.5$ V, and $E_2 = 7.4$ V, find m.

8-3. If $E_{cm} = 5.0$ V in Fig. 8-2a, find V_o.

8-4. For the differential amplifier circuit of Fig. 8-2(b), all the resistors are equal to 10 kΩ except R_2, which is mismatched by 5 Ω (R_2 = 10 kΩ + 5 Ω). What is the common mode rejection of the circuit?

8-5. Design the differential amplifier circuit of Fig. 8-2(b) for a gain of 4. The input resistors R_1 and R_3 are equal to 25 kΩ.

8-6. A circuit has a common-mode rejection ratio of 100,000. What is the common-mode rejection?

8-7. A circuit has a CMR of 75 dB. What is the common-mode rejection ratio?

8-8. In Fig. 8-3, E_i = 2 mV and E_n = 50 mV. What is the output voltage due to **(a)** E_i; **(b)** E_n?

8-9. In Fig. 8-4, E_i = 2 mV and E_n = 50 mV. What is the output voltage due to **(a)** E_i; **(b)** E_n?

8-10. What is the main advantage of a differential amplifier over an inverting amplifier with respect to an input noise signal voltage?

8-11. Find V_o in Fig. 8-5(b) if $E_1 = -5$ V and $E_2 = -3$ V.

8-12. In Fig. 8-5(b), R = 10 kΩ and aR = 2 kΩ. If E_1 = 1.5 V and E_2 = 0.5 V, find V_o.

8-13. In Fig. 8-6 the overall gain is 21 and V_o = 3 V. Determine **(a)** $E_1 - E_2$; **(b)** a.

8-14. In Fig. 8-6, R = 25 kΩ, aR = 100 Ω, E_1 = 1.01 V, and E_2 = 1.02 V. Find V_o.

8-15. If V_{ref} = 5.0 V in Fig. 8-7, find **(a)** V_o; **(b)** the voltage at the (+) input with respect to ground.

8-16. Refer to the circuit of Fig. P8-16. Complete the table below for each input condition.

	E_1 (V)	E_2 (V)	E_3 (V)	V_o	V at (+) input
(a)	−2	−2	0		
(b)	−2	−2	2		
(c)	2	−2	−2		
(d)	2	0	2		

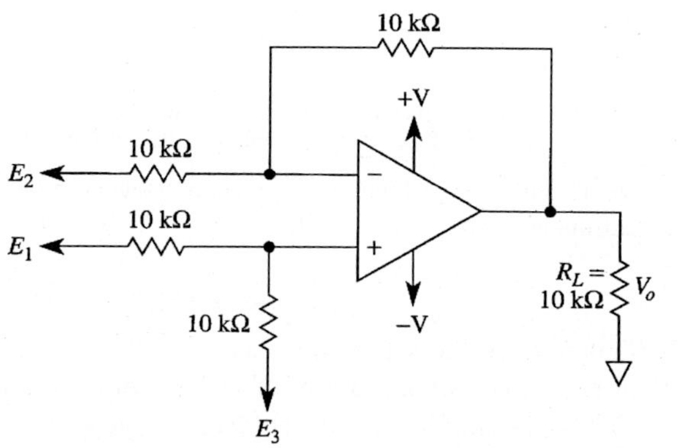

FIGURE P8-16

8-17. Refer to the voltage-to-current converter of Fig. 8-10. Assume that the AD524 is wired for a gain of 1 [no wires on pins 13 and 3]. The load current is now $I_L = (E_1 - E_2)/R_s$. Let R_s = 1 kΩ, E_2 = 0 V or ground, and E_1 = 1 V. **(a)** Will the direction of I_L be up or down in Fig. 8-10? **(b)** Find I_L. **(c)** Find the voltage across R_L if R_L = 100 Ω. **(d)** Find the output voltage of the IA (V_9) if R_L = 3 kΩ.

8-18. Repeat Problem 8-17 except change E_2 = ground and $E_1 = 1$ V. (Note that Section 8-5.3 tells you how to make an ac voltage-controlled current source for a grounded load.)

8-19. Change E_1 to -1 V in Problem 8-17. **(a)** Would V_o be positive or negative with respect to ground? **(b)** Would V_o decrease or increase in *magnitude* as temperature increased?

8-20. In Fig. 8-16, the value for $R = 120.00\ \Omega$, $\Delta R = 1.2\ \text{m}\Omega$, and $E = 10.0$ V. Find $(E_1 - E_2)$ for the strain-gage arrangement of **(a)** Fig. 8-16(a); **(b)** Fig. 8-16(b); **(c)** Fig. 8-16(c).

8-21. Assume that an IA with a gain of 1000 is wired to the bridges of Problem 8-20. Find V_o for each of the three bridge arrangements.

8-22. Consider a gage factor of 2 in Problems 8-20 and 8-21 and calculate $\Delta L/L$ for each bridge arrangement.

8-23. Refer to Section 8-12.3. To gain experience with this type of bridge circuit, repeat Design Example 8-13 (except change only your reference temperature to 50°C. R_1 remains at 10 kΩ and $E = 1$ V). Present your solution in the same format as shown in Table 8-2 and Fig. 8-19. Redraw the new design schematic like that of Fig. 8-19. [Remember that R_{ref} will now be 3603 Ω so that $I = 1\ \text{V}/(10{,}000 + 3603)\ \Omega = 73.51\ \mu\text{A}$.]

8-24. You want a circuit that has an increasing magnitude of output voltage as temperature of a thermistor increases. You put the thermistor in the feedback loop [see Fig. 8-18(b)]. Would you choose R_{ref} at the low or the high end of the temperature scale? (*Hint:* Compare V_o vs. V_T of Design Example 8-13 with the solution of Problem 8-23.) This problem forces you to face briefly the issue of "human engineering." People want to see an increasing voltage as temperature increases.

CHAPTER 9

DC Performance: Bias, Offsets, and Drift

LEARNING OBJECTIVES

Upon completion of this chapter on dc performance, you will be able to:

- Name the op amp characteristics that add dc error components to the output voltage.
- Show how an op amp requires a small bias current at both (−) and (+) inputs to activate its internal transistors.
- Give the definition for input offset voltage and show how it is modeled in an op amp circuit.
- Write the equation for input offset current in terms of the bias currents.
- Calculate the effect of input offset voltage on the output voltage of either an inverting or noninverting amplifier.
- Calculate the effects of bias currents on the output voltage of an inverting or noninverting amplifier.
- Calculate the value of and install a compensating resistor to minimize the errors in output voltage caused by bias currents.

- Connect a nulling circuit to null out any errors due to bias currents and input offset voltage.
- Measure offset voltage and bias currents.
- Understand the dc characteristics of present-day op amps versus earlier versions.
- Understand the difference between CMRR of an op amp and an op amp circuit.
- Calculate power supply rejection ratio.

9-0 INTRODUCTION

The op amp is widely used in amplifier circuits to amplify dc or ac signals or combinations of them. In dc amplifier applications, certain electrical characteristics of the op amp can cause large errors in the output voltage. The ideal output voltage should be equal to the product of the dc input signal and the amplifier's closed-loop voltage gain. However, the output voltage may have an added error component. This error is due to differences between an ideal op amp and a real op amp. If the ideal value of output voltage is large with respect to the error component, then we can usually ignore the op amp characteristic that causes it, but if the error component is comparable to or even larger than the ideal value, we must try to minimize the error. Op amp characteristics that add error components to the dc output voltage are

1. Input bias currents
2. Input offset current
3. Input offset voltage
4. Drift

When the op amp is used in an ac amplifier, coupling capacitors eliminate dc output-voltage error. Therefore, characteristics 1 to 4 are often unimportant in ac applications. However, there are new problems for ac amplifiers:

5. Frequency response
6. Slew rate

Frequency response refers to how voltage gain varies as frequency changes. The most convenient way to display such data is by a plot of voltage gain versus frequency. Op amp manufacturers give such a plot for open-loop gain versus frequency. A glance at the plot quickly shows how much gain is obtainable at a particular frequency.

If the op amp has sufficient gain at a particular frequency, there is still a possibility of an error being introduced in V_o. This is because there is a fundamental limit imposed by the op amp (and certain circuit capacitors) on *how fast the output voltage can change.* If the input signal "tells" the op amp output to change faster than it can, distortion is introduced in the output voltage. The op amp characteristic responsible for this type of error is its internal capacitance. This type of error is called *slew-rate limiting.*

Op amp characteristics and the circuit applications that each type of error *may* affect are summarized in Table 9-1. The first four characteristics can limit dc performance; the last two can limit ac performance.

Op amp characteristics that cause errors primarily in dc performance are studied in this chapter. Those that cause errors in ac performance are studied in Chapter 10. We begin with input bias currents and ways in which they cause errors in the dc output voltage of an op amp circuit.

TABLE 9-1 OP AMP Applications and Characteristics That Affect Operation

	Op amp application			
	Dc amplifier		Ac amplifier	
Op amp characteristic that may affect performance	Small output	Large output	Small output	Large output
1. Input bias current	Yes	Maybe	No	No
2. Offset current	Yes	Maybe	No	No
3. Input offset voltage	Yes	Maybe	No	No
4. Drift	Yes	No	No	No
5. Frequency response	No	No	Yes	Yes
6. Slew rate	No	Yes	No	Yes

In addition to these characteristics the user may need to consider common-mode and power supply rejection ratios. These values may be specified under static test but vary dynamically.

9-1 INPUT BIAS CURRENTS

Transistors within the op amp must be *biased* correctly before any signal voltage is applied. Biasing correctly means that the transistor has the right value of base and collector current as well as collector-to-emitter voltage. Until now, we have considered that the input terminals of the op amp conduct no signal or bias current. This is the ideal condition. Practically, however, the input terminals do conduct a small value of dc current to bias the op amps' transistors (see Appendix 1). A simplified diagram of the op amp is shown in Fig. 9-1(a). To discuss the effect of input bias currents, it is convenient to model them as current sources in series with each input terminal, as shown in Fig. 9-1(b).

The (−) input's bias current, I_{B-}, will usually not be exactly equal to the (+) input's bias current, I_{B+}. Manufacturers specify an *average* input bias current I_B, which is found by adding the *magnitudes* of I_{B+} and I_{B-} and dividing this sum by 2. In equation form,

$$I_B = \frac{|I_{B+}| + |I_{B-}|}{2} \tag{9-1}$$

where $|I_{B+}|$ is the magnitude of I_{B+} and $|I_{B-}|$ is the magnitude of I_{B-}. The range of I_B is from 1 μA or more for general-purpose op amps to 1 pA or less for op amps that

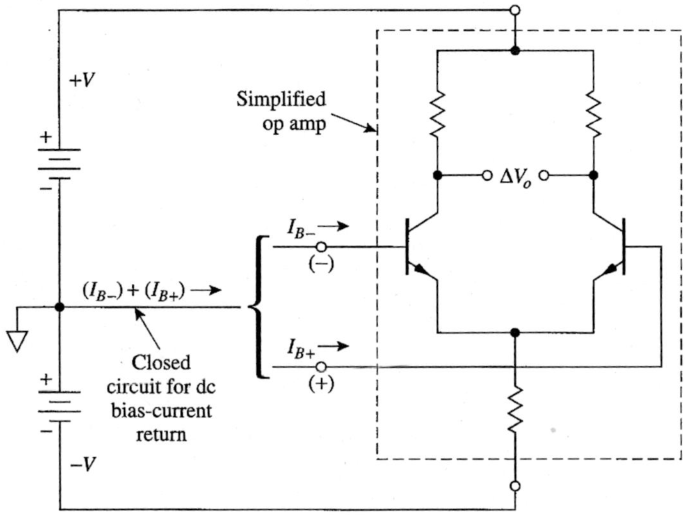

(a) Simplified op amp input circuit.

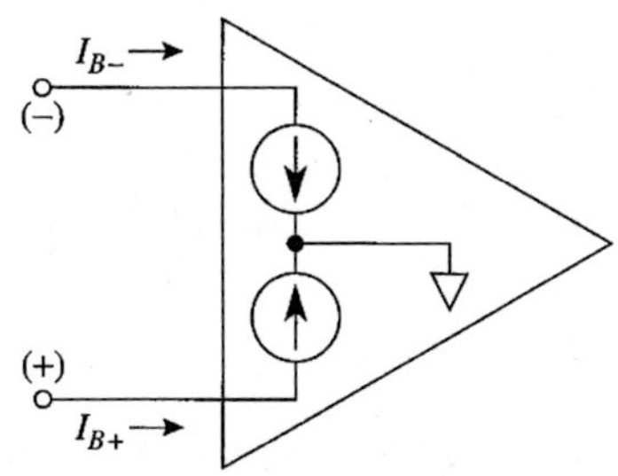

(b) Model for bias currents.

FIGURE 9-1 Origin and model of dc input bias currents.

have field-effect transistors at the input. There are op amps referred to as electrometer operational amplifiers that have ultralow input bias currents less than 60 fA (fA—femto amperes—10^{-15}A). These devices are used in special signal conditioning circuits such as those required for pH measurements. They require careful attention to circuit board design and input signal connections.

9-2 INPUT OFFSET CURRENT

The difference in magnitudes between I_{B+} and I_{B-} is called the *input offset current* I_{os}:

$$I_{os} = |I_{B+}| - |I_{B-}| \tag{9-2}$$

Manufacturers specify I_{os} for a circuit condition where the output is at 0 V and the temperature is 25°C. The typical I_{os} is less than 25% of I_B for the average input bias current (see Appendix 1).

Example 9-1

Consider a 741 op amp that has $I_{B+} = 0.4\ \mu\text{A}$ and $I_{B-} = 0.3\ \mu\text{A}$. Find (a) the average bias current I_B; (b) the offset current I_{os}.

Solution (a) By Eq. (9-1),

$$I_B = \frac{(0.4 + 0.3)\ \mu\text{A}}{2} = 0.35\ \mu\text{A}$$

(b) By Eq. (9-2),

$$I_{os} = (0.4 - 0.3)\ \mu\text{A} = 0.1\ \mu\text{A}$$

A general-purpose op amp such as the OP-177 has typical values of $I_B = 2.4$ nA and $I_{os} = 0.5$ nA. Newer versions of the 741 op amp have typical values lower than those given in Example 9-1.

9-3 EFFECT OF BIAS CURRENTS ON OUTPUT VOLTAGE

9-3.1 Simplification

In this section it is assumed that bias currents are the only op amp characteristic that will cause an undesired component in the output voltage. The effects of other op amp characteristics on V_o will be dealt with individually.

9-3.2 Effect of (−) Input Bias Current

Output voltage should ideally equal 0 V in each circuit of Fig. 9-2, because input voltage E_i is 0 V. The fact that a voltage component will be measured is due strictly to I_{B-}. (Assume for simplicity that V_{io}, input offset voltage, is zero. V_{io} is discussed in Section 9-5.) In Fig. 9-2(a), the bias current is furnished from the output terminal. Since negative feedback forces the differential input voltage to 0 V, V_o must rise to supply the voltage drop across R_f. Thus, the output voltage error due to I_{B-} is found from $V_o = R_f I_{B-}$. I_{B+} flows through 0 Ω, so it causes no voltage error. Signal source E_i must contain a dc path to ground.

The circuit of Fig. 9-2(b) has the same output-voltage error expression, $V_o = R_f I_{B-}$. No current flows through R_i, because there is 0 V on each side of R_i. Thus all of I_{B-} flows through R_f. [Recall that an ideal amplifier with negative feedback has 0 voltage between the (+) and (−) inputs.]

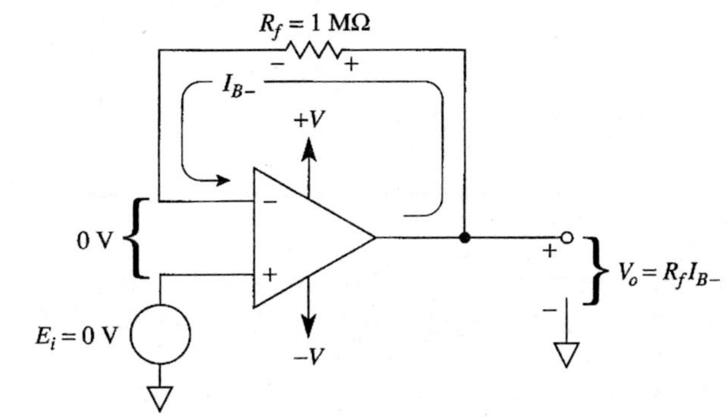

(a) Basic follower circuit.

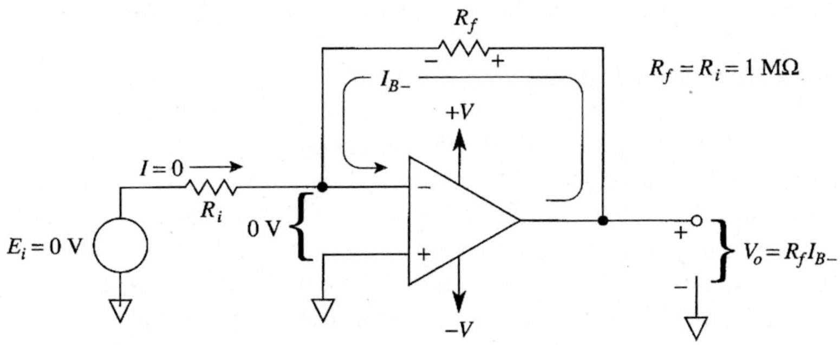

(b) Basic inverting circuit.

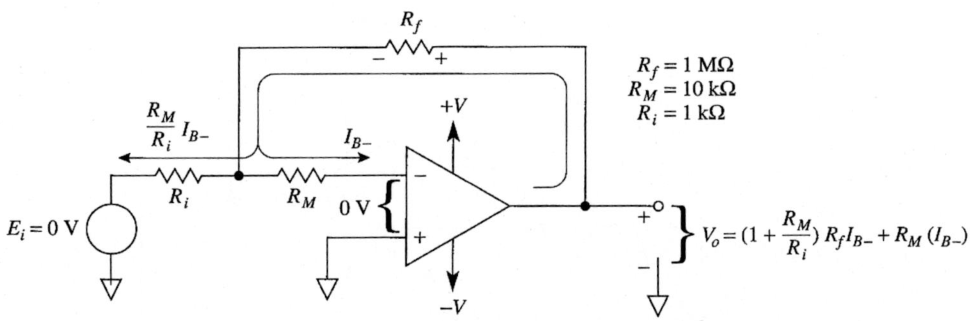

(c) The multiplier resistor R_M increases the effect of I_{B-} on V_o.

FIGURE 9-2 Effects of (−) input bias current on output voltages.

Example 9-2

In Fig. 9-2(a), $V_o = 0.4$ V. Find I_{B-}.

Solution

$$I_{B-} = \frac{V_o}{R_f} = \frac{0.4\text{ V}}{1\text{ M}\Omega} = 0.4\ \mu\text{A}$$

Placing a multiplying resistor R_M in series with the (−) input in Fig. 9-2(c) multiplies the effect of I_{B-} on V_o. I_{B-} sets up a voltage drop across R_M that establishes an equal drop across R_i.

Now both the R_i current and I_{B-} must be furnished through R_f. Thus the error in V_o will be much larger. R_M would be undesirable in a normal circuit; however, if we want to measure low values of the bias current, Fig. 9-2(c) shows a way of doing it. For the resistor values shown, $V_o \simeq 11R_fI_{B-}$; I_{B-} acts to drive the output positive.

9-3.3 Effect of (+) Input Bias Current

Since $E_i = 0$ V in Fig. 9-3, V_o should ideally equal 0 V. However, the positive input bias current I_{B+} flows through the internal resistance of the signal generator. Internal generator resistance is modeled by resistor R_G in Fig. 9-3. I_{B+} sets up a voltage drop of R_GI_{B+} across R_G and applies it to the (+) input. The differential input voltage is 0 V, so the (−) input is also at R_GI_{B+} in Fig. 9-3. Since there is no resistance in the feedback loop, V_o equals R_GI_{B+}. (The return path for I_{B+} is through the $-V$ supply and back to ground.) I_{B+} acts to drive the output negative.

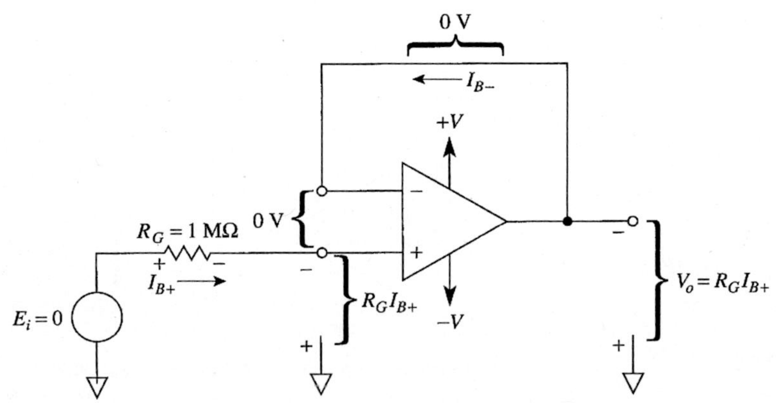

FIGURE 9-3 Effect of (+) input bias current on output voltages.

Example 9-3

In Fig. 9-3, $V_o = -0.3$ V. Find I_{B+}.

Solution

$$I_{B+} = -\frac{V_o}{R_G} = -\frac{-0.3\text{ V}}{1\text{ M}\Omega} = 0.3\ \mu\text{A}$$

9-4 EFFECT OF OFFSET CURRENT ON OUTPUT VOLTAGE

9-4.1 Current Compensating the Voltage Follower

If I_{B+} and I_{B-} were always equal, it would be possible to compensate for their effects on V_o. For example, in the voltage follower of Fig. 9-4(a), I_{B+} flows through the signal generator resistance R_G. If we insert $R_f = R_G$ in the feedback loop, I_{B-} will develop a volt-

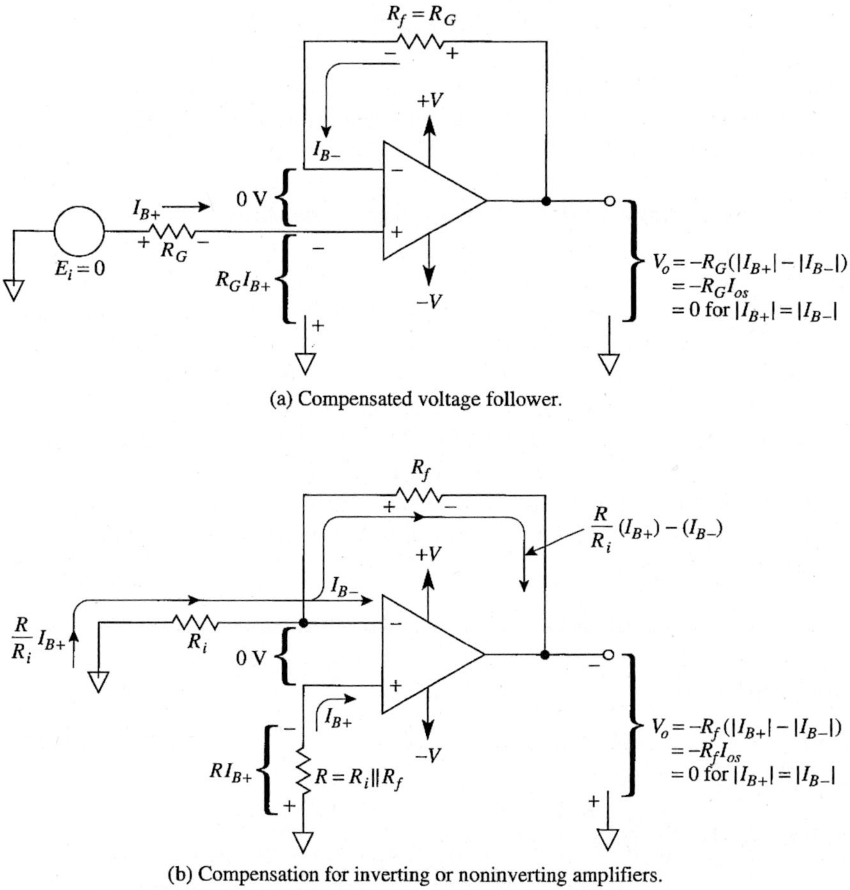

FIGURE 9-4 Balancing-out effects of bias current in V_o.

age drop across R_f of $R_f I_{B-}$. If $R_f = R_G$ and $I_{B+} = I_{B-}$, their voltage drops will cancel each other and V_o will equal 0 V when $E_i = 0$ V. Unfortunately, I_{B+} is seldom equal to I_{B-}. V_o will then be equal to R_G times the difference between I_{B+} and I_{B-} ($I_{B+} - I_{B-} = I_{os}$). Therefore, by making $R_f = R_G$, we have reduced the error in V_o from $R_G I_{B+}$ in Fig. 9-3 to $-R_G I_{os}$ in Fig. 9-4(a). Recall that I_{os} is typically 25% of I_B. If the value of I_{os} is too large, an op amp with a smaller value of I_{os} is needed.

9-4.2 Current Compensating Other Amplifiers

To minimize errors in V_o due to bias currents for either inverting or noninverting amplifiers, resistor R as shown in Fig. 9-4(b) must be added to the circuit. With no input signal applied, V_o depends on R_f times I_{os} [where I_{os} is given by Eq. (9-2)]. Resistor R is called the *current-compensating resistor* and is equal to the parallel combination of R_i and R_f, or

$$R = R_i \| R_f = \frac{R_i R_f}{R_i + R_f} \qquad (9\text{-}3)$$

R_i and R should include any signal generator resistance. By inserting resistor R, the error voltage in V_o will be reduced more than 25%, from $R_f I_{B-}$ in Fig. 9-2(b) to $-R_f I_{os}$ in Fig. 9-4(b). In the event that $I_{B-} = I_{B+}$, then $I_{os} = 0$ and $V_o = 0$.

9-4.3 Summary of Bias-Current Compensation

Always add a bias-current compensating resistor R in series with the (+) *input terminal* (except for FET input op amps). The value of R should equal the parallel combination of all resistance branches connected to the (−) terminal. Any internal resistance in the signal source should also be included in the calculations.

In circuits where more than a single resistor is connected to the (+) input, bias-current compensation is accomplished by observing the following principle. *The dc resistance seen from the* (+) *input to ground should equal the dc resistance seen from the* (−) *input to ground.* In applying this principle, signal sources are replaced by their internal dc resistance and the op amp output terminal is considered to be at ground potential.

Example 9-4

(a) In Fig. 9-4(b), $R_f = 100\ \text{k}\Omega$ and $R_i = 10\ \text{k}\Omega$. Find R. (b) If $R_f = 100\ \text{k}\Omega$ and $R_i = 100\ \text{k}\Omega$, find R.

Solution (a) By Eq. (9-3),

$$R = \frac{(100\ \text{k}\Omega)(10\ \text{k}\Omega)}{100\ \text{k}\Omega + 10\ \text{k}\Omega} = 9.1\ \text{k}\Omega$$

(b) By Eq. (9-3),

$$R = \frac{(100\ \text{k}\Omega)(100\ \text{k}\Omega)}{100\ \text{k}\Omega + 100\ \text{k}\Omega} = 50\ \text{k}\Omega$$

9-5 INPUT OFFSET VOLTAGE

9-5.1 Definition and Model

In Fig. 9-5(a), the output voltage V_o should equal 0 V. However, there will be a small error-voltage component present in V_o. Its value can range from microvolts to millivolts and is caused by very small but unavoidable imbalances inside the op amp. The easiest way to study the *net effect* of all these internal imbalances is to visualize a small dc voltage in *series* with one of the input terminals. This dc voltage is modeled by a battery in Fig. 9-5(b) and is called *input offset voltage,* V_{io} (see Appendix 1 for typical values). Note that

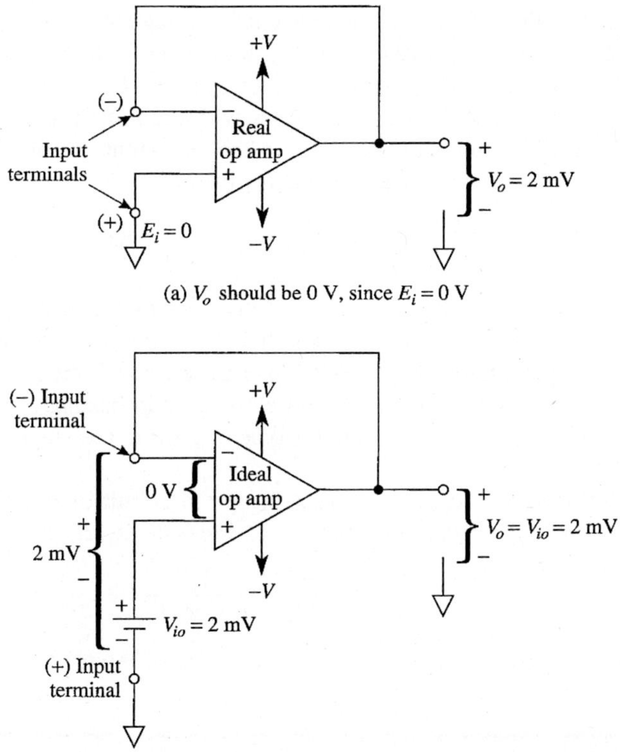

(a) V_o should be 0 V, since $E_i = 0$ V

(b) Error in V_o is modeled by dc voltage V_{io} in series with (+) input.

FIGURE 9-5 Effect of input offset voltage in the real op amp of (a) is modeled by an ideal op amp plus battery V_{io} in (b).

V_{io} is shown in series with the (+) input terminal of the op amp. It makes no difference whether V_{io} is modeled in series with the (−) input or the (+) input. But it is easier to determine the polarity of V_{io} if it is placed in series with the (+) input. For example, if the output terminal is positive (with respect to ground) in Fig. 9-5(b), V_{io} should be drawn with its (+) battery terminal connected to the ideal op amp's (+) input.

9-5.2 Effect of Input Offset Voltage on Output Voltage

Fig. 9-6(a) shows that V_{io} and the large value of the open-loop gain of the op amp act to drive V_o to negative saturation. Contrast the polarity of V_{io} in Figs. 9-5(b) and 9-6(a). If you buy several op amps and plug them into the test circuit of Fig. 9-6(a), some will drive V_o to $+V_{sat}$ and the remainder will drive V_o to $-V_{sat}$. Therefore, the magnitude and polarity of V_{io} varies from op amp to op amp. To learn how V_{io} affects amplifiers with negative feedback, we study how to measure V_{io}.

9-5.3 Measurement of Input Offset Voltage

For simplicity, the effects of bias currents are neglected in the following discussion. Figure 9-6(b) shows how to measure V_{io}. It also shows how to predict the magnitude of error that V_{io} will cause in the output voltage. Since $E_i = 0$ V, V_o should equal 0 V, but V_{io} acts exactly as would a signal in series with the noninverting input. Therefore, V_{io} is amplified exactly as any signal applied to the (+) input of a noninverting amplifier (see Section 3-5). The error in V_o due to V_{io} is given by

$$V_o = \text{error voltage due to } V_{io} = V_{io}\left(1 + \frac{R_f}{R_i}\right) \qquad (9\text{-}4)$$

The output error voltage in Fig. 9-6(b) is given by Eq. (9-4) whether the circuit is used as an inverting or as a noninverting amplifier. That is, E_i could be inserted in series with R_i (inverting amplifier) for a gain of $-(R_f/R_i)$ or in series with the (+) input (noninverting amplifier) for a gain of $1 + (R_f/R_i)$. A bias-current compensating resistor (a resistor in series with the (+) input) has no effect on this type of error in the output voltage due to V_{io}.

Conclusion. To measure V_{io}, set up the circuit of Fig. 9-6(b). The capacitor is installed across R_f to minimize noise in V_o. Measure V_o, R_f, and R_i. Calculate V_{io} from

$$V_{io} = \frac{V_o}{1 + R_f/R_i} \qquad (9\text{-}5)$$

Note that R_f is made small to minimize the effect of input bias current.

Example 9-5

V_{io} is specified to be 1 mV for a 741-type op amp. Predict the value of V_o that would be measured in Fig. 9-6(b).

Solution From Eq. (9-5),

$$V_o = \left(1 + \frac{1000}{10}\right)(1 \text{ mV}) = 101 \text{ mV}$$

Note: Today you can purchase general-purpose op amps, such as the OP-177, with a typical input offset voltage of 10 μV.

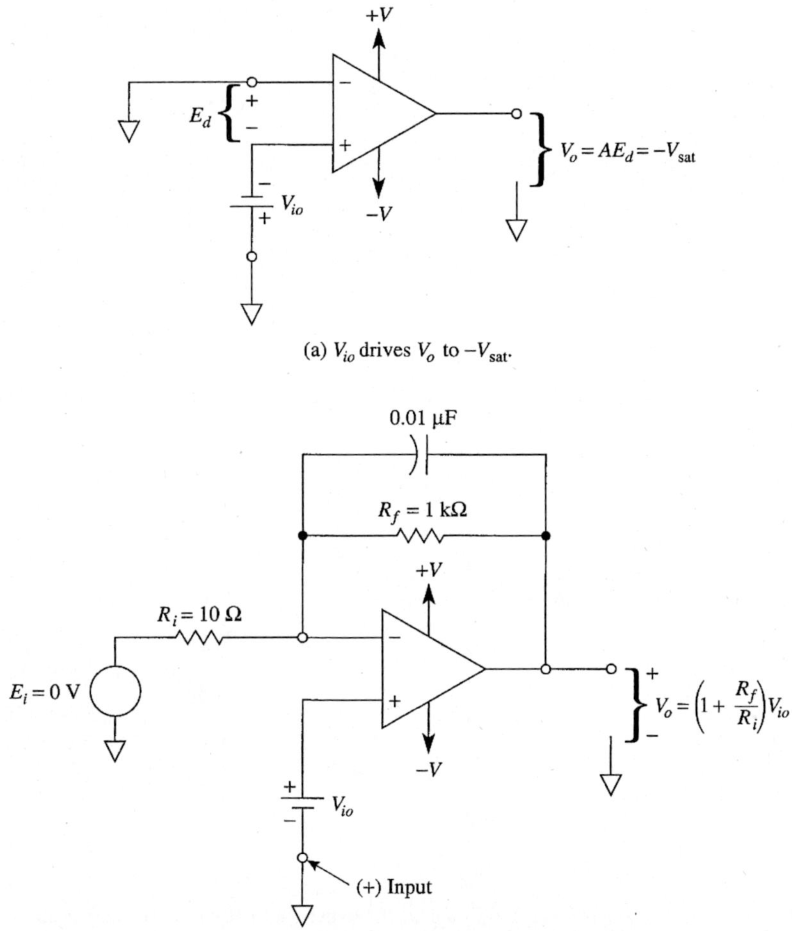

(a) V_{io} drives V_o to $-V_{sat}$.

(b) V_{io} is amplified, causing a large error in V_o.

FIGURE 9-6 V_o should be 0 V in (a) and (b) but contains a dc error voltage due to V_{io}. (The error component due to bias current is neglected.)

9-6 INPUT OFFSET VOLTAGE FOR THE ADDER CIRCUIT

9-6.1 Comparison of Signal Gain and Offset Voltage Gain

In both inverting and noninverting amplifier applications, the input offset voltage V_{io} is multiplied by $(1 + R_f/R_i)$. The input signal in either circuit is multiplied by a different gain. R_f/R_i is the gain for the inverter and $(1 + R_f/R_i)$ for the noninverter. In the inverting adder circuit of Fig. 9-7(a) (neglecting bias currents), V_{io} is multiplied by a larger number than the signal at each input.

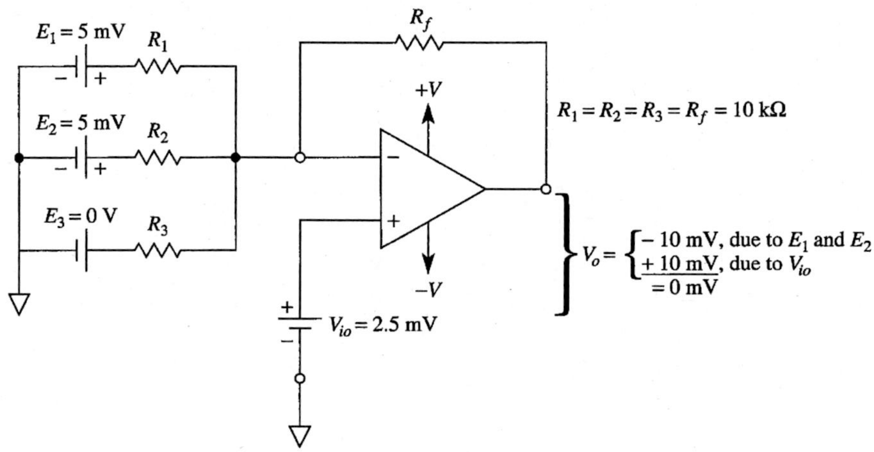

(a) V_o has a −10-mV component due to E_1 and E_2 plus a 10-mV error component due to V_{io}.

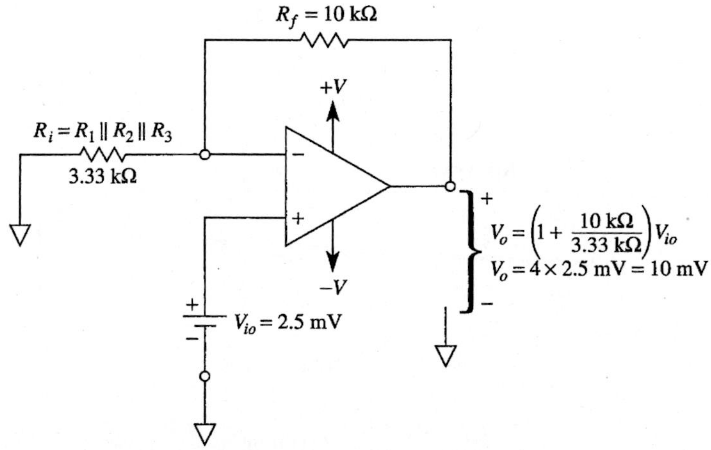

(b) V_{io} is multiplied by a gain of 4 to generate a 10-mV error component in V_o.

FIGURE 9-7 Each input voltage of the inverting adder in (a) is multiplied by a gain of −1. V_{io} is multiplied by a gain of +4.

For example, in Fig. 9-7(a) signals E_1 and E_2 are each larger than V_{io} but E_1 is multiplied by $-R_f/R_1 = -1$ and develops a component of -5 mV in V_o. E_2 is likewise multiplied by -1 and adds a -5-mV component to V_o. Thus the correct value of V_o should be -10 mV. Since E_3 is 0 its contribution to V_o is 0 (see Section 3-2).

If we temporarily let E_1 and $E_2 = 0$ V in Fig. 9-7(a), the $(-)$ input sees three equal resistors forming parallel paths to ground. The single equivalent series resistance, R_i, is shown in Fig. 9-7(b). For three equal 10-kΩ resistors in parallel, the equivalent resistance R_i is found by 10 kΩ/3 = 3.33 kΩ. V_{io} is amplified just as in Fig. 9-6(b) to give an output error of $+10$ mV. Therefore, the total output voltage in Fig. 9-7(a) is 0 instead of -10 mV.

Conclusions. In an adder circuit, the input offset voltage has a gain of *1 plus the number of inputs.* The more inputs, the greater the error component in the output voltage. Since the gain for the inputs is -1, *the offset voltage gain always exceeds the signal voltage gain.*

9-6.2 How Not to Eliminate the Effects of Offset Voltage

One might be tempted to add an adder input such as E_3 in Fig. 9-7(a) to balance out the effect of V_{io}. For example, if E_3 is made equal to 10 mV, then E_3, R_3, and R_f will add a -10 mV component to V_o and balance out the $+10$ mV due to V_{io}. There are two disadvantages to this approach. First, such a small value of E_3 would have to be obtained from a resistor-divider network between the power supply terminals of $+V$ and $-V$. The second disadvantage is that any resistance added between the $(-)$ input and ground raises the *noise gain.* This situation is treated in Sections 10-4.3 and 10-4.4. In Section 9-7 we show how to minimize the output voltage errors caused by both bias currents and input offset voltage.

9-7 NULLING-OUT EFFECT OF OFFSET VOLTAGE AND BIAS CURRENTS

9-7.1 Design or Analysis Sequence

To minimize dc error voltages in the output voltage, follow this sequence:

1. Select a bias-current compensating resistor in accordance with the principles set forth in Section 9-4.3.
2. Get a circuit for minimizing effects of the input offset voltage from the op amp manufacturer's data sheet. This principle is treated in more detail in Section 9-7.2 and in Appendix 1.
3. Go through the output-voltage nulling procedure given in Section 9-7.3.

Data sheets published by op amp manufacturers usually show an offset nulling circuit. In most of your applications this circuit is not necessary because you can purchase an op amp with an input offset voltage small enough to cause negligible effects. However, let us study the operation of these nulling circuits.

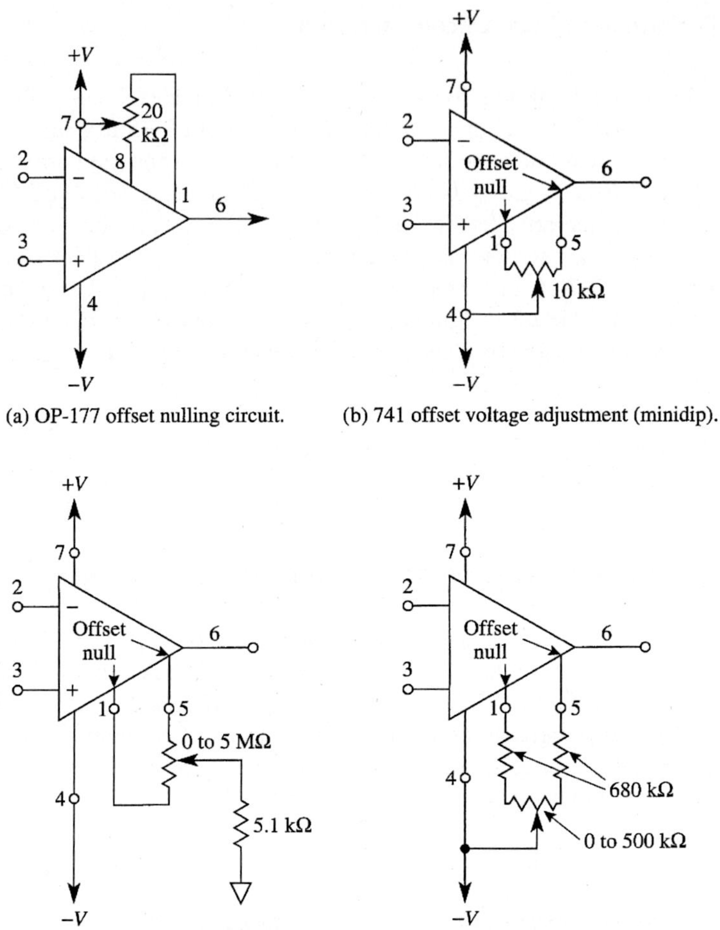

(a) OP-177 offset nulling circuit.

(b) 741 offset voltage adjustment (minidip).

(c) 301 or 748 offset voltage adjustment (TO-99 case).

(d) 537 offset voltage adjustment.

FIGURE 9-8 Typical circuits to minimize errors in output voltage due to input offset voltage and offset current (see also Appendix 1).

9-7.2 Null Circuits for Offset Voltage

It is possible to imagine a fairly complex resistor-divider network that would inject a small variable voltage into the (+) or (−) input terminal. This would compensate for the effects of *both* input offset voltage and offset current. However, the extra components are more costly and bulky than necessary. It is far better to go to the op amp manufacturer for guidance. The data sheet for your op amp will have a *voltage offset null circuit* recommended by the manufacturer. The op amp will have *null* terminals brought out for connection to the null circuit. Experts have designed the null circuit to minimize offset errors at the lowest cost to the user (see Appendix 1).

Some typical output-voltage null circuits are shown in Fig. 9-8. In Fig. 9-8(a), one variable resistor is connected between the $+V$ supply and a *trim* terminal. For an expensive op amp, the manufacturer may furnish a metal-film resistor selected especially for that op amp. In Fig. 9-8(b), a 10-kΩ pot is connected between terminals called *offset null.* More complicated null circuits are shown in Figs. 9-8(c) and (d). Note that only the offset-voltage compensating resistors are shown by the manufacturer. They assume that a current-compensating resistor will be installed in series with the (+) input.

9-7.3 Nulling Procedure for Output Voltage (If Necessary)

1. Build the circuit. Include (a) the current-compensating resistor (see Section 9-4.3) and (b) the voltage offset null circuit (see Section 9-7.2).
2. Reduce all generator signals to 0. If their output cannot be set to 0, replace them with resistors equal to their internal resistance. This step is unnecessary if their internal resistance is negligible with respect to (more than about 1% of) any series resistor R_i connected to the generator.
3. Connect the load to the output terminal.
4. Turn on the power and wait a few minutes for things to settle down.
5. Connect a dc voltmeter or an oscilloscope (dc coupled) across the load to measure V_o. (The voltage sensitivity should be capable of reading down to less than a few millivolts.)
6. Vary the offset voltage adjustment resistor until V_o reads 0 V. Note that output voltage errors due to *both* input offset voltage and input offset current are now minimized.
7. Install the signal sources and do *not* touch the offset-voltage adjustment resistor again.

9-8 DRIFT

It has been shown in this chapter that dc error components in V_o can be minimized by installing a current-compensating resistor in series with the (+) input and by trimming the offset-voltage adjustment resistor. It must also be emphasized that the zeroing procedure holds only at one temperature and at one time.

The offset current and offset voltage change with time because of aging of components. The offsets will also be changed by temperature changes in the op amp. In addition, if the supply voltage changes, bias currents, and consequently the offset current, change. By use of a well-regulated power supply, the output changes that depend on supply voltage can be eliminated. However, the offset changes with temperature can only be minimized by (1) holding the temperature surrounding the circuit constant, or (2) selecting op amps with offset current and offset voltage ratings that change very little with temperature changes.

The changes in offset current and offset voltage due to temperature are described by the term *drift.* Drift is specified for offset current in nA/°C (nanoamperes per degree Celsius). For offset voltage, drift is specified in μV/°C (microvolts per degree Celsius). Drift rates may differ at different temperatures and may even reverse; that is, at low temperatures V_{io} may drift by +20 μV/°C (increase), and at high temperatures V_{io} may change by −10 μV/°C (decrease). For this reason, manufacturers may specify either an average or maximum drift between two temperature limits. Even better is to have a plot of drift vs. temperature. An example is shown to calculate the effects of drift.

Example 9-6

A 741 op amp in the circuit of Fig. 9-9 has the following drift specifications. As temperature changes from 25°C to 75°C, I_{os} changes by a *maximum* of 0.1 nA/°C and V_{io} changes by a *maximum* of 20 μV/°C. Assume that V_o has been zeroed at 25°C and then the surrounding temperature is raised to 75°C. Find the maximum error in output voltage due to drift in (a) V_{io}; (b) I_{os}.

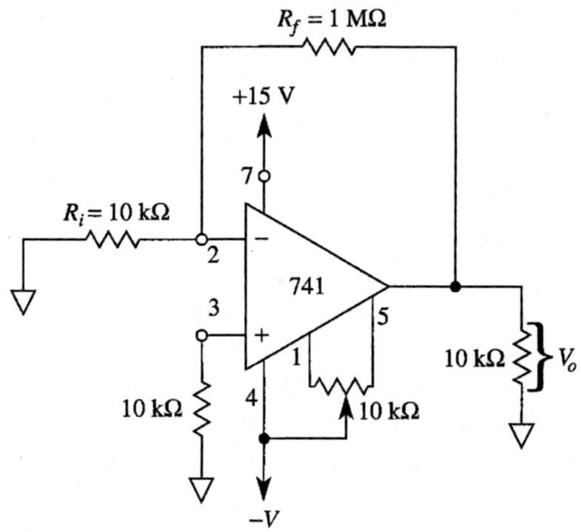

FIGURE 9-9 Circuit for Example 9-6.

Solution (a) V_{io} will change by

$$\pm\frac{20\ \mu\text{V}}{^\circ\text{C}} \times (75 - 25)^\circ\text{C} = \pm 1.0\ \text{mV}$$

From Eq. (9-4), the change in V_o due to the change in V_{io} is

$$1.0\ \text{mV}\left(1 + \frac{R_f}{R_i}\right) = 1.0\ \text{mV}(101) \simeq \pm\ 101\ \text{mV}$$

(b) I_{os} will change by

$$\pm\frac{0.1\ \text{nA}}{^\circ\text{C}} \times 50^\circ\text{C} = \pm 5\ \text{nA}$$

From Section 9-4, the change in V_o due to the change in I_{os} is ± 5 nA $\times R_f = \pm 5$ nA (1 MΩ) $= \pm 5$ mV.

The changes in V_o due to both V_{io} and I_{os} can either add or subtract from one another. Therefore, the worst possible change in V_o is either +106 mV or −106 mV from the 0 value at 25°C.

As a contrast to the 741, the OP-177 op amp has typical drift values of 1.5 pA/°C for I_{os} and 0.1 μV/°C for V_{io}.

9-9 MEASUREMENT OF OFFSET VOLTAGE AND BIAS CURRENTS

The effects of offset voltage and bias currents have been discussed separately to simplify the problem of understanding how error voltage components appear in the dc output voltage of an op amp. However, their effects are *always* present simultaneously.

In order to measure V_{io}, I_{B+}, and I_{B-} of general-purpose op amps as inexpensively as possible, the following procedure is recommended.

1. As shown in Fig. 9-10(a), measure V_o with a digital voltmeter and calculate input offset voltage V_{io}:

$$V_{io} = \frac{V_o}{(R_f + R_i)/R_i} = \frac{V_o}{101} \tag{9-6}$$

Note that R_i and R_f are small. Therefore, by adding the 50-Ω current-compensation resistor, we force the output voltage error component due to I_{os} to be negligible.

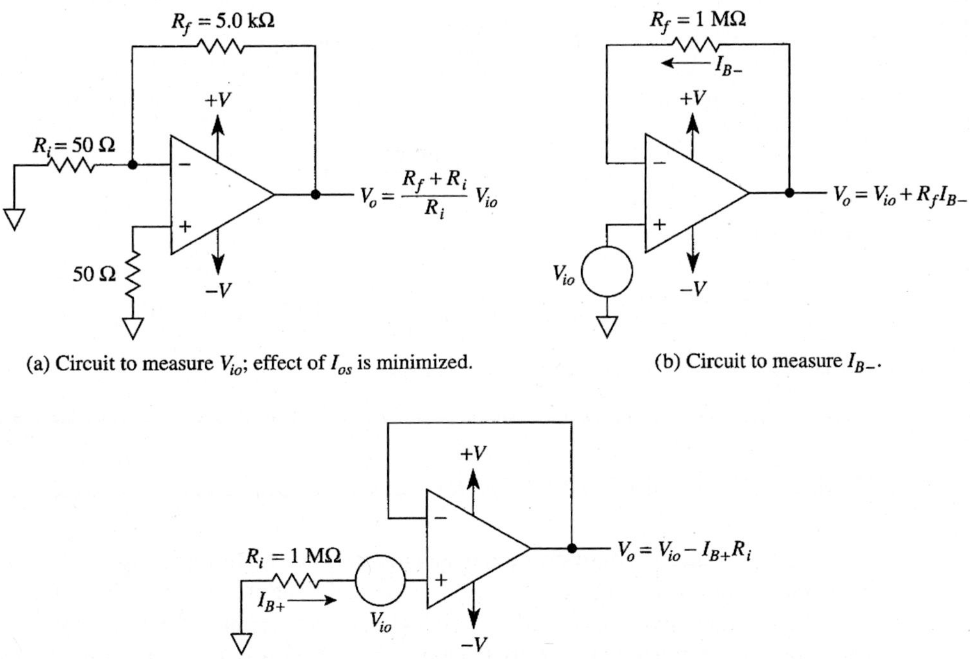

(a) Circuit to measure V_{io}; effect of I_{os} is minimized.

(b) Circuit to measure I_{B-}.

(c) Circuit to measure I_{B+}.

FIGURE 9-10 Procedure to measure offset voltage, then bias currents for a general-purpose op amp.

2. To measure I_{B-}, set up the circuit of Fig. 9-10(b). Measure V_o. Using the value of V_{io} found in step 1, calculate I_{B-} from

$$I_{B-} = \frac{V_o - V_{io}}{R_f} \tag{9-7}$$

3. To measure I_{B+}, measure V_o in Fig. 9-10(c) and calculate I_{B+} from

$$I_{B+} = -\left(\frac{V_o - V_{io}}{R_f}\right) \tag{9-8}$$

Example 9-7

The circuits of Fig. 9-10 are used with the resistance values shown for a 741 op amp. Results are $V_o = +0.421$ V for Fig. 9-10(a), $V_o = 0.097$ V for Fig. 9-10(b), and $V_o = -0.082$ V for Fig. 9-10(c). Find (a) V_{io}; (b) I_{B-}; (c) I_{B+}.

Solution (a) From Eq. (9-6),

$$V_{io} = \frac{0.421 \text{ V}}{101} = 4.1 \text{ mV}$$

(b) From Eq. (9-7),

$$I_{B-} = \frac{(97 - 4.1) \text{ mV}}{1 \text{ M}\Omega} = 93 \text{ nA}$$

(c) From Eq. (9-8),

$$I_{B+} = -\frac{(-82 - 4.1) \text{ mV}}{1 \text{ M}\Omega} = 86 \text{ nA}$$

Note that I_{os} is found to be $I_{B+} - I_{B-} = -7$ nA.

9-10 COMMON-MODE REJECTION RATIO

Chapter 8 introduced common-mode rejection ratio (CMRR) as it applied to an op amp subtractor circuit (also referred to as the basic differential amplifier circuit). For that circuit, the op amp was considered to be ideal and we saw the effects of a mismatch of a single resistor. A mismatch of only 0.1% in one of four resistors caused the CMRR of the circuit to be degraded to 66 dB. However, the common-mode rejection ratio of the operational amplifier is different than that of the op amp circuit. For example, the CMRR of an operational amplifier is usually greater than 100 dB and for the OP-177 a typical value is 130 dB. In this section, we shall consider only an operational amplifier's common-mode rejection ratio specification and not the circuit's CMRR.

First remember that an ideal op amp is a pure differential amplifier and is insensitive to the absolute voltages on the inputs with respect to ground. The feedback connection for inverting and noninverting amplifiers maintains the differential voltage, E_d, near 0 V. In some applications, such as comparators or in amplifier circuits where the input voltages exceed the level required to saturate the output, the differential voltage may exceed a maximum value. In order not to cause permanent damage to the device, this differential voltage must be within the range bounded by the supply voltages.

For differential amplifiers, the voltage at both inputs can be greater than or less than ground potential but must be within the supply voltage limits. Common-mode voltage, E_{cm}, is defined as the same voltage being applied to both inputs simultaneously. An ideal operational amplifier responds only to the difference between the inputs ($E^+ - E^-$) and produces zero output for a common-mode voltage—both inputs at the same potential. However, due to slight differences between the input transistors that make up the input differential pair of an operational amplifier, the common-mode voltage is not subtracted entirely. Hence a common-mode output voltage exists.

The common-mode rejection ratio (CMRR) is defined as the ratio of the change in common-mode voltage to the resulting change in input offset voltage. It is often convenient to specify this parameter logarithmically in dB:

$$\text{CMR} = 20 \log_{10} (\text{CMRR}) \tag{9-9}$$

For example, consider a common-mode voltage of 10 V results in a 10 μV change in input offset voltage. For this application, the CMRR is 10 V/10 μV = 10^6:1 or 120 dB. This expression means that a 10 V common-mode voltage passes through the device as if it were a 10 μV differential input signal.

CMRR is specified at dc but it is frequency dependent. Therefore, some manufacturers will supply curves of CMRR vs. frequency and you will see that it is a nonlinear curve.

Note: Common-mode rejection is sometimes defined as the ratio of the open loop gain to the common-mode gain, where common-mode gain is the ratio of the output voltage to the common-mode voltage. In equation form,

$$\text{CMR} = 20 \log_{10} \frac{A_{OL}}{A_{cm}} \tag{9-10}$$

where A_{OL} is the op amp's open loop gain and A_{cm} is the common-mode gain and equals the ratio of the output voltage due to the input common-mode voltage.

9-11 POWER SUPPLY REJECTION RATIO

Power supplies connected to the op amp should have low noise and be well regulated because all of the op amp's internal circuitry is sensitive to changes in supply voltages. The maximum offset versus supply value is a measure of this sensitivity. The reciprocal of the sensitivity is the power supply rejection ratio (PSRR). This parameter is conveniently expressed logarithmically as

$$\text{PSR} = 20 \log_{10} (\text{PSRR}) \tag{9-11}$$

Consider that an op amp has an offset change of 10 μV per volt. This results in a PSRR of 100,000 (1 V/10 μV) and a power supply rejection of 100 dB. As a comparison, the typical value of power supply rejection for the OP-177 op amp is 125 dB, and for the 741 the sensitivity is only 30 μV/V or a PSR of 90.5 dB.

PROBLEMS

9-1. Which op amp characteristics normally have the most effect on **(a)** dc amplifier performance; **(b)** ac amplifier performance?

9-2. If $I_{B+} = 0.2\ \mu A$ and $I_{B-} = 0.1\ \mu A$, find **(a)** the average bias current I_B; **(b)** the offset current I_{os}.

9-3. In Example 9-2, $V_o = 0.2$ V. Find I_{B-}.

9-4. In Example 9-3, $V_o = -0.2$ V. Find I_{B+}.

9-5. I_{B-} is 0.2 μA in Fig. 9-2(c). Find V_o.

9-6. In Fig. 9-4(a), $R_f = R_G = 100\ k\Omega$. $I_{B+} = 0.3\ \mu A$ and $I_{B-} = 0.2\ \mu A$. Find V_o.

9-7. In Fig. 9-4(b), $R_f = R_i = 25\ k\Omega$ and $R = 12.5\ k\Omega$. If $I_{os} = 0.1\ \mu A$, find V_o.

9-8. In Fig. 9-4(b), $R_i = R_f = 25\ k\Omega$ and $R = 12.5\ k\Omega$. If $I_{os} = -0.1\ \mu A$, find V_o.

9-9. In Fig. 9-6(b), $V_o = 200$ mV. Find V_{io}.

9-10. Resistors R_1, R_2, R_3, and R_f all equal 20 kΩ in Fig. 9-7(a). $E_1 = E_2 = E_3 = V_{io} = 2$ mV. Find **(a)** the actual value of V_o; **(b)** V_o assuming that $V_{io} = 0$.

9-11. What value of current-compensating resistor should be added in Problem 9-10?

9-12. What is the general procedure to null the dc output voltage of an op amp to 0 V?

9-13. In Fig. 9-9, V_{io} changes by ± 0.5 mV when the temperature changes by 50°C. What is the change in V_o due to the change in V_{io}?

9-14. I_{os} changes by ± 2 nA in Fig. 9-9 for a temperature change of 50°C. What is the resulting change in V_o?

9-15. $V_o = 101$ mV in the circuit in Fig. 9-10(a), $V_o = 201$ mV in Fig. 9-10(b), and $V_o = -99$ mV in Fig. 9-10(c). Find **(a)** V_{io}; **(b)** I_{B-}; **(c)** I_{B+}.

9-16. Refer to Fig. P9-16. $V_{io} = 3$ mV, $I_{B-} = 0.4\ \mu A$, and $I_{B+} = 0.1\ \mu A$. **(a)** What is the best value for resistor R? Calculate the individual error in the output voltage, V_o, due to **(b)** V_{io} only; **(c)** I_{B+} only; **(d)** I_{B-} only; **(e)** I_{os} only. The ideal value of V_o should be 1.00 V because of E_i. **(f)** What is the actual value of V_o when both input offset voltage and current are present along with E_i?

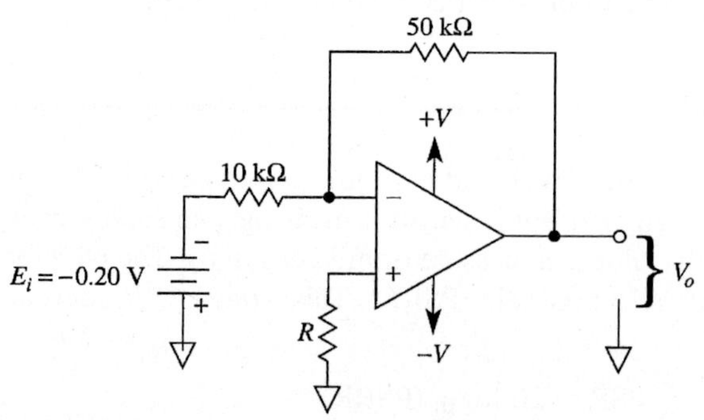

FIGURE P9-16

9-17. An op amp's typical common-mode rejection is 110 dB for the change in common-mode voltage of 12 V. What is the change of input offset voltage?

9-18. Refer to Maxim Integrated Products' Web site and compare their MAX4281 op amp CMRR and PSRR with the OP-177 device.

CHAPTER 10

AC Performance: Bandwidth, Slew Rate, and Noise

LEARNING OBJECTIVES

Upon completion of this chapter on ac performance of an op amp, you will be able to:

- Recognize an op amp's frequency response graph in the manufacturer's data sheet, determine (1) the dc open-loop gain A_{OL}, (2) the small-signal unity-gain bandwidth B from it, and (3) read the magnitude of A_{OL} at any frequency.
- Calculate the unity-gain bandwidth if rise time is given, and vice versa.
- Predict the open-loop gain of an op amp at any frequency if you know the unity-gain bandwidth.
- Measure the rise time.
- Show how closed-loop gain, A_{CL}, of either an inverting or noninverting amplifier depends on open-loop gain, A_{OL}.
- Measure the frequency response of an inverting or noninverting amplifier.

- Predict the bandwidth or upper cutoff frequency for an inverting amplifier if you know the external resistor values and the op amp's small-signal unity.
- Calculate the maximum sinusoidal frequency that can be obtained from an op amp at a given peak output voltage if you know its slew rate.
- Calculate the maximum peak output voltage at any given sine frequency if the op amp's slew rate is known.
- Calculate noise gain.
- Understand the difference between open loop, closed loop, and loop gain.

10-0 INTRODUCTION

When the op amp is used in a circuit that amplifies only ac signals, we must consider whether ac output voltages will be small signals (below about 1 V peak) or large signals (above 1 V peak). If only small ac *output* signals are present, the important op amp characteristics that limit performance are *noise* and *frequency response.* If large ac output signals are expected, then an op amp characteristic called *slew-rate limiting* determines whether distortion will be introduced by the op amp, and may further limit frequency response.

Bias currents and offset voltages affect dc performance and usually do not have to be considered with respect to ac performance. This is especially true if a coupling capacitor is in the circuit to pass ac signals and block dc currents and voltages. We begin with an introduction to the frequency response of an op amp.

10-1 FREQUENCY RESPONSE OF THE OP AMP

10-1.1 Internal Frequency Compensation

General-purpose op amps and specialized op amps are *internally compensated;* that is, the manufacturer has installed within such op amps a small capacitor, usually 30 pF. This *internal frequency compensation capacitor* prevents the op amp from oscillating at high frequencies. Oscillations are prevented by decreasing the op amp's gain as frequency increases. Otherwise, there would be sufficient gain and phase shift at some high frequency where enough output signal could be fed back to the input and cause oscillations (see Appendix 1).

From basic circuit theory it is known that the reactance of a capacitor goes down as frequency goes up: $X_C = 1/(2\pi fC)$. For example, if the frequency is increased by 10, the capacitor reactance decreases by 10. Thus, it is no accident that the voltage gain of an op amp goes down by 10 as the frequency of the input signal is increased by 10. A change in frequency of 10 is called a *decade.* Manufacturers show how the open-loop gain of the op amp is related to the frequency of the differential input signal by a curve called *open-loop voltage gain versus frequency.* The curve may also be called *small-signal response.*

10-1.2 Frequency–Response Curve

A typical curve is shown in Fig. 10-1 for internally compensated op amps such as the 741. At low frequencies (below 0.1 Hz), the open-loop voltage gain is very high. A typical value is 200,000 (106 dB), and it is this value that is specified on data sheets where a curve is not given. See "Large-Signal Voltage Gain" equals 200,000 in Appendix 1. In comparison, the open-loop gain for the OP-177 is greater than 140 dB or 10,000,000.

Point *A* in Fig. 10-1 locates the *break frequency* at which the open-loop voltage gain of the op amp is 0.707 times its value at very low frequencies. Therefore, the voltage gain at point *A* (where the frequency of E_d is 5 Hz) is about 140,000, or 0.707 × 200,000.

Points *C* and *D* show how gain drops by a factor of 10 as frequency rises by a factor of 10. Changing frequency or gain by a factor of 10 is expressed more efficiently by the term *per decade* ("decade" signifies 10). The right-hand vertical axis of Fig. 10-1 is a plot of voltage gain in decibels (dB). The voltage gain decreases by 20 dB for an increase in frequency of 1 decade. This explains why the frequency-response curve from *A*

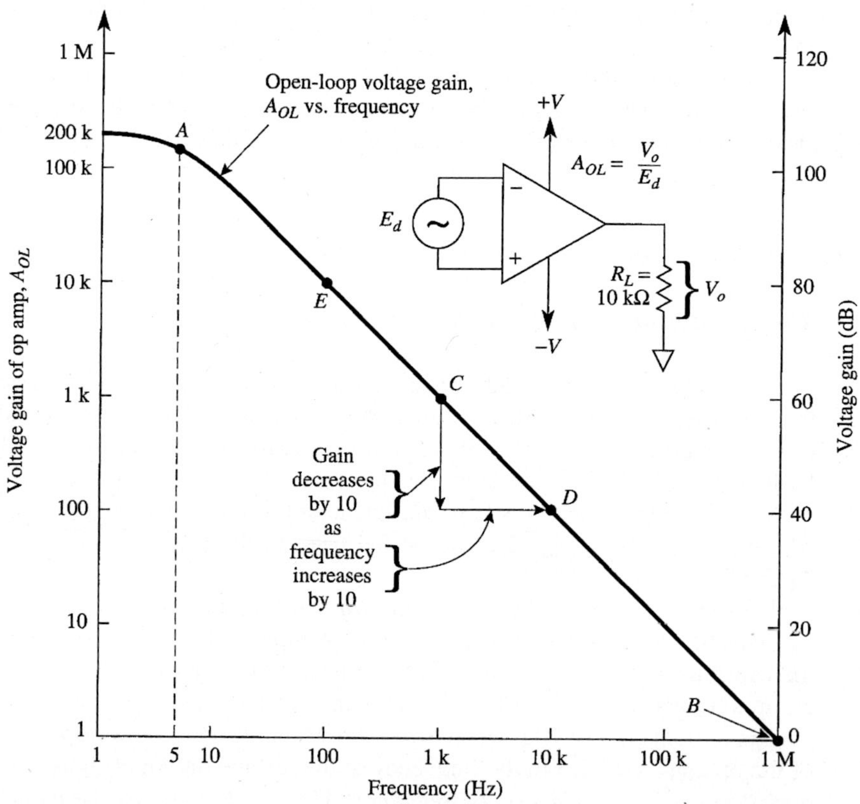

FIGURE 10-1 Open-loop voltage gain of a 741 op amp versus frequency.

to B is described as *rolling off at* 20 *dB/decade*. An alternative description is 6 *dB/octave roll-off* ("octave" signifies a frequency change of 2). Therefore, each time the frequency doubles, the voltage gain decreases by 6 dB.

10-1.3 Unity-Gain Bandwidth

When an amplifier is made from an op amp and a few resistors, the frequency response of the amplifier depends on the frequency response of the op amp. The key op amp characteristic is defined as that frequency where the op amp's gain equals unity. We will use the symbol B for this op amp characteristic, which is called *small-signal unity-gain bandwidth*. Later in this chapter we will need a value for B of the op amp to predict the high-frequency response of an amplifier constructed with this op amp.

Three ways to obtain B from a manufacturer's data sheet are presented in this section. First, if you have the manufacturer's plot of A_{OL} versus frequency, look for that frequency where $A_{OL} = 1$ (see point B in Fig. 10-1, $B = 1$ MHz). Second, some data sheets may not give a specification called unity-gain bandwidth or a curve like Fig. 10-1. Instead, they give a specification called *transient response rise time* (*unity gain*). For a 741 op amp it is typically 0.25 μs and 0.8 μs at maximum. The bandwidth B is calculated from the rise-time specification by

$$B = \frac{0.35}{\text{rise time}} \tag{10-1}$$

where B is in hertz and rise time is in seconds. Rise time is defined in Section 10-1.4. (See the "Electrical Characteristics" tables in Appendix 1, "Transient Response (Unity Gain)" = 0.3 μs typical.)

Example 10-1

A 741 op amp has a rise time of 0.35 μs. Find the small-signal or unity-gain bandwidth.

Solution From Eq. (10-1),

$$B = \frac{0.35}{0.35\ \mu\text{s}} = 1\text{ MHz}$$

Example 10-2

What is the open-loop voltage gain for the op amp of Example 10-1 at 1 MHz?

Solution From the definition of B, the voltage gain is 1.

Example 10-3

What is the open-loop voltage gain at 100 kHz for the op amp in Examples 10-1 and 10-2?

Solution By inspection of Fig. 10-1, if the frequency goes down by 10, the gain goes up by 10. Therefore, since the frequency goes down by a decade (from 1 MHz to 100 kHz), the gain must go up by a decade from 1 at 1 MHz to 10 at 100 kHz.

Example 10-3 leads to the conclusion that if you divide the frequency of the signal, *f*, into the unity-gain bandwidth, *B*, the result is the op amp's gain at the signal frequency. Expressed mathematically,

$$\text{open-loop gain at } f = \frac{\text{bandwidth at unity gain}}{\text{input signal frequency, } f} \tag{10-2}$$

Example 10-4

What is the open-loop gain of an op amp that has a unity-gain bandwidth of 1.5 MHz for a signal of 1 kHz?

Solution From Eq. (10-2), the open-loop gain at 1 kHz is

$$\frac{1.5 \text{ MHz}}{1 \text{ kHz}} = 1500$$

Equation (10-2) gives a third way to find *B*. If you know the op amp's open-loop gain at one frequency (in the roll-off region), simply multiply the two values to obtain *B*. Let's consider Example 10-4 again. If $A_{OL} = 1000$ at a frequency of 1500 Hz, then $B = 1500 \times 1000 = 1.5$ MHz.

The data shown in Fig. 10-1 are useful for learning but probably do not apply to your op amp. For example, while 200,000 is a specified typical open-loop gain for the 741, the manufacturer guarantees only a minimum gain of 20,000 for general-purpose op amps. Still, 20,000 may be enough to do the job. Section 10-2 deals with this question.

10-1.4 Rise Time

Assume that the input voltage E_i of a unity-gain amplifier is changed very rapidly by a square wave or pulse signal. Ideally, E_i should be changed from 0 V + 20 mV in 0 time; practically, a few nanoseconds are required to make this change (see Appendix 1, "Transient Response" in the "Typical Performance Curves"). At unity gain, the output should change from 0 to +20 mV in the same few nanoseconds. However, it takes time for the signal to propagate through all the transistors in the op amp. It also takes time for the output voltage to rise to its final value. *Rise time* is defined as the time required for the output voltage to rise from 10% of its final value to 90% of its final value. From Section 10-1.3, the rise time of a 741 is 0.35 μs. Therefore, it would take 0.35 μs for the output voltage to change from 2 mV to 18 mV.

10-2 AMPLIFIER GAIN AND FREQUENCY RESPONSE

10-2.1 Effect of Open-Loop Gain on Closed-Loop Gain of an Amplifier, DC Operation

It is necessary to learn how open-loop gain A_{OL} affects the actual closed-loop gain of an amplifier with dc signal voltages (zero frequency). First, we must define *ideal* closed-loop gain of an amplifier as that gain which should be determined only by external resistors. However, the *actual* dc closed loop of an amplifier is determined by *both* the external resistors and open-loop gain of an op amp.

The actual dc closed-loop gain of a *noninverting* amplifier is

$$\text{actual } A_{CL} = \frac{(R_f + R_i)/R_i}{1 + \dfrac{1}{A_{OL}}\left(\dfrac{R_f + R_i}{R_i}\right)} \tag{10-3a}$$

where

$$\frac{R_f + R_i}{R_i} = \text{ideal } A_{CL} \text{ for } \textit{noninverting} \text{ amplifiers} \tag{10-3b}$$

If A_{OL} is very large, the denominator of Eq. (10-3a) approaches unity. Then the amplifier gain will *not* depend on the open-loop gain of the op amp but rather only on the external resistors and can be calculated from Eq. (10-3b).

The actual dc gain of an *inverting amplifier* depends on A_{OL} according to

$$\text{actual } A_{CL} = \frac{-R_f/R_i}{1 + \dfrac{1}{A_{OL}}\left(\dfrac{R_f + R_i}{R_i}\right)} \tag{10-3c}$$

where

$$\frac{-R_f}{R_i} = \text{ideal } A_{CL} \text{ for } \textit{inverting} \text{ amplifiers} \tag{10-3d}$$

Equation (10-3d) is valid if A_{OL} is large with respect to $(R_f + R_i)/R_i$.

Example 10-5

Find the actual gain for a dc noninverting amplifier if ideal $A_{CL} = 100$ and A_{OL} is (a) 10,000; (b) 1000; (c) 100; (d) 10; (e) 1. Repeat for a dc inverting amplifier with an ideal gain of -100.

Solution (a) For the noninverting amplifier: $(R_f + R_i)/R_i = 100$ = ideal gain. From Eq. (10-3a),

$$\text{actual } A_{CL} = \frac{100}{1 + \left(\frac{1}{10{,}000}\right)100} = 99.0099$$

For the inverting amplifier, $R_f/R_i = |\text{ideal gain}| = 100$. Therefore, $(R_f + R_i)/R_i = 101$. From Eq. (10-3c),

$$\text{actual } A_{CL} = \frac{-100}{1 + \left(\frac{1}{10{,}000}\right)101} = -99.0000$$

If these steps are repeated for parts (b) through (e), the results may be tabulated as follows:

	A_{OL}					
	1	10	100	10^3	10^4	10^5
Actual A_{CL}, noninverting	0.99	9.09	50	90.9	99.0	99.9
Actual A_{CL}, inverting	− 0.98	− 9.01	− 49.7	− 90.8	− 99.0	− 99.9

The results of Example 10-5 are shown by plot A_{CL} versus A_{OL} in Fig. 10-2. There are two important lessons to be learned from Example 10-5 and Fig. 10-2. First

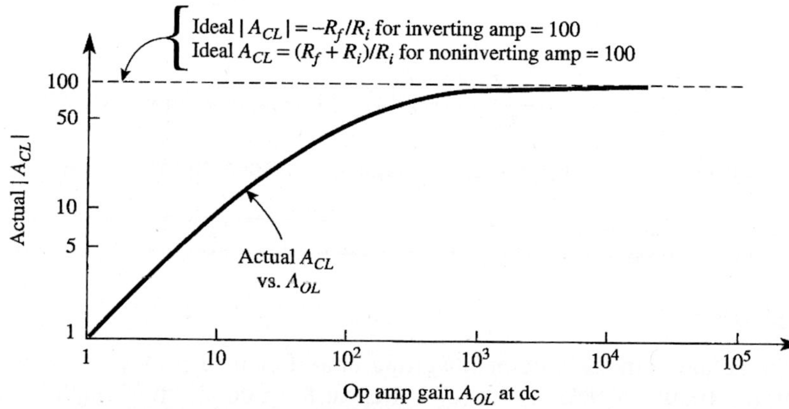

FIGURE 10-2 The actual closed-loop gain of an inverting or noninverting amplifier depends on both the ideal gain that is set by resistor ratios, and the open-loop gain of the op amp at dc (see Example 10-5).

the actual gains of both noninverting and inverting amplifiers are of approximately the same magnitudes for the same value of open-loop gain. Second, we would like the actual closed-loop gain to be equal to the ideal closed-loop gain. An examination of Eqs. (10-3a) and (10-3c) shows that this will be true *if* the open-loop gain of the op amp A_{OL} is large with respect to the ideal closed-loop gain of the op amp. Practically, we would like A_{OL} to be 100 or more times the ideal A_{CL}, so that the external precision resistors and *not* the op amp's A_{OL} determine the actual gain, to within 1%.

We already learned in Section 10-1.2 that A_{OL} depends on frequency. Since A_{OL} of the op amp also determines A_{CL} of an amplifier, then A_{CL} of the amplifier will also depend on frequency. But before we look at the amplifier's frequency response we must define it and also define bandwidth.

10-2.2 Small-Signal Bandwidth, Low- and High-Frequency Limits

The useful frequency range of any amplifier (closed- or open-loop) is defined by a high-frequency limit f_H and a low-frequency limit f_L. At f_L and f_H, the voltage gain is down to 0.707 times its maximum value in the middle of the useful frequency range. In terms of decibels, the voltage gain is down 3 dB at both f_L and f_H. These statements are summarized on the general frequency response curve in Fig. 10-3 and in Appendix 1.

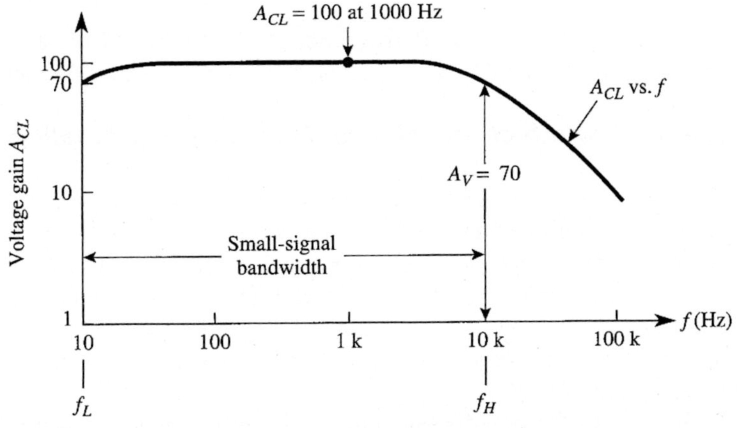

FIGURE 10-3 Small-signal bandwidth.

Small-signal bandwidth is the difference between f_H and f_L. Often f_L is very small with respect to f_H, or f_L is 0 for a dc amplifier. Therefore, the small-signal bandwidth approximately equals the high-frequency limit f_H. From point A of Fig. 10-1, we see that the small-signal bandwidth of an op amp is 5 Hz.

10-2.3 Measuring Frequency Response

You can learn a lot about frequency response by learning how to measure frequency response at a test bench.

Laboratory procedure. The frequency–response curve of Fig. 10-3 would have been obtained in the following manner:

1. Adjust the input voltage E_i of an op amp to some convenient value, let's say 30 mV rms.
2. Set the sinusoidal frequency of E_i to some convenient midband value, let's say 1000 Hz.
3. Measure the midband output voltage; assume that it equals 3.0 V.
4. Calculate the midband voltage gain $A_{CL} = 3\text{ V}/0.030\text{ V} = 100$.
5. Calculate the expected value of V_o at f_L and f_H, $V_o = (0.707)$ (V_o midband). Thus $V_o = (0.707)\ 3\text{ V} = 2.1$ V rms where $A_{CL} = 70.7$.
6. Hold E_i *constant* in magnitude at 30 mV. Reduce the oscillator frequency until $V_o = 2.1$ V. Read the oscillator dial frequency to obtain the *lower cutoff* frequency f_L.
7. Hold E_i constant in amplitude at 30 mV. Increase the oscillator frequency (beyond 1 kHz) until V_o again drops to 2.1 V. Read f_H from the oscillator dial.
8. Calculate bandwidth B from $B = f_H - f_L$.

Note: For dc amplifiers, $f = 0$; therefore, $B = f_H$.

The low and high cutoff frequencies are also called the *corner* frequencies, the 3-dB frequencies, the 0.707 frequencies, or simply the *cutoff* frequencies.

10-2.4 Bandwidth of Inverting and Noninverting Amplifiers

In this section let's stipulate that all amplifiers are direct coupled. Next, observe that both inverting and noninverting amplifiers are made from exactly the same structure. They have an op amp, a feedback resistor R_f, and an input resistor R_i. An amplifier only assumes an identity when you choose which input will experience the input signal. If you connect E_i via R_i to the (−) input and ground the (+) input, you define the amplifier to be inverting. If E_i is wired to (+) input and ground to R_i, the same structure becomes a noninverting amplifier.

In view of the observation above, it is perhaps not surprising that the *upper cutoff frequency f_H for both inverting and noninverting amplifiers is given by*

$$f_H = \frac{B}{(R_f + R_i)/R_i} \tag{10-4}$$

where B = op amp small-signal bandwidth
R_f = feedback resistance
R_i = input resistor

Example 10-6

Given that $R_f = R_i = 10\ \text{k}\Omega$ for an inverting amplifier and also for a noninverting amplifier, find the gain and bandwidth of (a) the inverting amplifier; (b) the noninverting amplifier. (c) What are the gain and bandwidth of a voltage follower? The op amp is a 741 with a small-signal gain–bandwidth product of $B = 1$ MHz.

Solution (a) From Eq. (3-2b) or (10-3d), $A_{CL} = -R_f/R_i = -1$. From Eq. (10-4),

$$f_H = \frac{1 \times 10^6\ \text{Hz}}{(10\ \text{k}\Omega + 10\ \text{k}\Omega)/10\ \text{k}\Omega} = 500\ \text{kHz}$$

(b) From Eq. (3-11b) or (10-3b), $A_{CL} = (R_f + R_i)/R_i = 2$. f_H is the same as in part (a). The noninverting amplifier has a higher gain–bandwidth product than the inverting amplifier. (c) The voltage follower has a gain of 1 [see Eq. (3-9b)]. In Eq. (3-11b), $R_f = 0$ and R_i is an open circuit approaching an infinite resistance for a voltage follower. Therefore, $(R_f + R_i)/R_i = 1$. Hence the upper cutoff frequency f_H is calculated from Eq. (10-4) as

$$f_H = \frac{1 \times 10^6\ \text{Hz}}{(R_f + R_i)/R_i} = \frac{10^6\ \text{Hz}}{1} = 1\ \text{MHz}$$

10-2.5 Finding Bandwidth by a Graphical Method

There is a graphical technique for obtaining the frequency response of a noninverting amplifier. An example is shown in Fig. 10-4. Let the amplifier gain equal 1000 at low and middle frequencies. From Eq. (10-4), $f_H = 999\ \text{Hz} \simeq 1\ \text{kHz}$. At f_H, the amplifier gain is approximately 700 ($0.707 \times 1000 \simeq 700$). For all frequencies above f_H, the frequency response of the amplifier and op amp coincide. For another example, use Fig. 10-4 and draw a horizontal line starting at $A_{CL} = 100$. The ending point where it intercepts the curve of A_{OL} versus f shows the amplifier's bandwidth. For this case, $f_H \simeq 10$ kHz. The conclusion is that the gain–bandwidth product of a noninverting amplifier is equal to B of the op amp. There is a direct trade-off. If you want more closed-loop gain, you must sacrifice bandwidth.

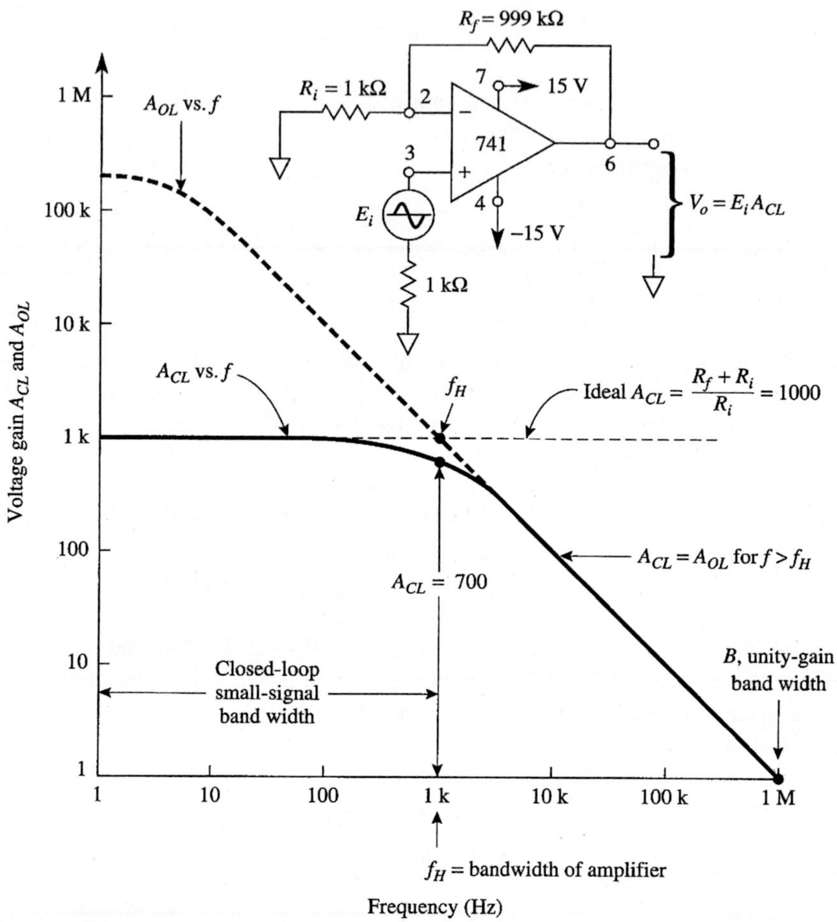

FIGURE 10-4 Op amp small-signal bandwidth and amplified closed-loop bandwidth.

10-3 SLEW RATE AND OUTPUT VOLTAGE

10-3.1 Definition of Slew Rate

The slew rate of an op amp tells how fast its output voltage can change. For a general-purpose op amp such as the 741, the maximum slew rate is 0.5 V/μs. This means that the output voltage can change a maximum of $\frac{1}{2}$ V in 1 μs. Slew rate depends on many factors: the amplifier gain, compensating capacitors, and even whether the output voltage is going positive or negative. The worst case, or slowest slew rate, occurs at unity gain. Therefore, slew rate is usually specified at unity gain (see Appendix 1).

10-3.2 Cause of Slew-Rate Limiting

Within general-purpose op amps at least one capacitor is required to prevent oscillation (see Section 10-1.1). Connected to this capacitor is a portion of the op amp's internal circuitry that can furnish a maximum current that is limited by op amp design. The ratio of this maximum current I to the compensating capacitor C is the *slew rate*. For example, a 741 can furnish a maximum of 15 μA to its internal 30-pF compensating capacitor (see Appendix 1). Therefore,

$$\text{slew rate} = \frac{\text{output voltage change}}{\text{time}} = \frac{I}{C} = \frac{15\ \mu\text{A}}{30\ \text{pF}} = 0.5\ \frac{\text{V}}{\mu\text{s}} \tag{10-5}$$

From Eq. (10-5), a faster slew rate requires the op amp to have either a higher maximum current or a smaller compensating capacitor. Op amps with slew rates greater than 100 V/μs are referred to as high-speed operational amplifiers. For special applications such as video systems, op amps with slew rates greater than 1000 V/μs are available.

Example 10-7

An instantaneous input change of 10 V is applied to a unity-gain inverting amplifier. If the op amp is a 741, how long will it take for the output voltage to change by 10 V?

Solution By Eq. (10-5),

$$\text{slew rate} = \frac{\text{output voltage change}}{\text{time}}$$

$$\frac{0.5\ \text{V}}{\mu\text{s}} = \frac{10\ \text{V}}{\text{time}}, \qquad \text{time} = \frac{10\ \text{V} \times \mu\text{s}}{0.5\ \text{V}} = 20\ \mu\text{s}$$

10-3.3 Slew-Rate Limiting of Sine Waves

In the voltage follower of Fig. 10-5, E_i is a sine wave with peak amplitude E_p. The maximum rate of change of E_i, depends on both its frequency f and the peak amplitude. It is given by $2\pi f E_p$. If this rate of change is larger than the op amp's slew rate, the output V_o will be distorted. That is, output V_o tries to follow E_i but cannot do so because of slew-rate limiting. The result is distortion, as shown by the triangular shape of V_o in Fig. 10-5. The maximum frequency f_{max} at which we can obtain an undistorted output voltage with a peak value of V_{op} is determined by the slew rate in accordance

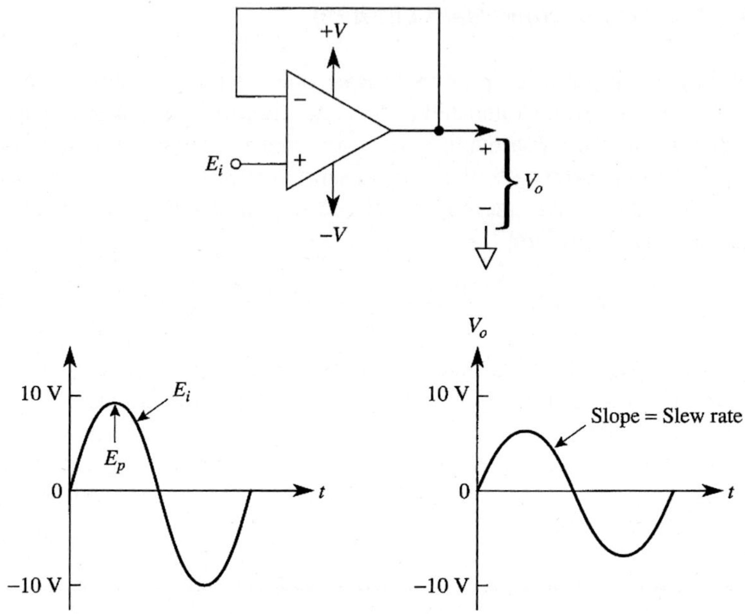

FIGURE 10-5 Example of slew-rate limiting of output voltage V_o.

with

$$f_{\text{max}} = \frac{\text{slew rate}}{6.28 \times V_{op}} \tag{10-6a}$$

where f_{max} is the maximum frequency in Hz, V_{op} is the maximum undistorted output voltage in volts, and the slew rate is in volts per microsecond.

The maximum peak sinusoidal output voltage $V_{op\ \text{max}}$ that can be obtained at a given frequency f is found from

$$V_{op\ \text{max}} = \frac{\text{slew rate}}{6.28 \times f} \tag{10-6b}$$

Example 10-8

The slew rate for a 741 is 0.5 V/μs. At what maximum frequency can you get an undistorted sine-wave output voltage of (a) 10-V peak; (b) 1-V peak?

Solution (a) From Eq. (10-6a),

$$f_{\text{max}} = \frac{1}{6.28 \times 10\ \text{V}} \times \frac{0.5\ \text{V}}{\mu\text{s}} = 8\ \text{kHz}$$

(b) From Eq. (10-6a),

$$f_{max} = 80 \text{ kHz}$$

In the next example, we learn that the slew rate *and* bandwidth must *both* be considered before we can predict the highest frequency at which we can obtain an undistorted output voltage.

Example 10-9

In Example 10-6, the small-signal bandwidth was 500 kHz for *both* an inverting amplifier with a gain of -1 and a noninverting amplifier with a gain of 2. Find (a) the maximum peak and undistorted sine wave output voltage at $f_H = 500$ kHz; (b) the maximum frequency at which you can obtain a peak output voltage of 10 V.

Solution Since the op amp is a 741, its maximum slew rate is 0.5 V/μs. (a) From Eq. (10-6b),

$$V_{op\ max} = \frac{0.5 \text{ V}/\mu\text{s}}{6.28(500 \times 10^3) \text{ Hz}} = 160 \text{ mV}$$

(b) From Eq. (10-6a),

$$f_{max} = \frac{0.5 \text{ V}/\mu\text{s}}{(6.28)(10 \text{ V})} \simeq 8 \text{ kHz}$$

f_{max} is defined as *full-power output frequency* at *full-power output.* The meaning of these new terms will become clear after a brief introduction.

A prudent amplifier design would restrict V_o to limits of ± 10 V. Then you have a safety margin of $\pm 20\%$ if the amplifier is overdriven at ± 12 V (almost into $\pm V_{sat}$). Manufacturers of op amps specify the ± 10 V output voltage level as *full-power output.* Note that a *full-power output frequency* specification is often supplied by the manufacturer (see Appendix 1, "Output Voltage Swing as a Function of Frequency" in the "Typical Performance Curves"). Examples 10-8 and 10-9 showed that the op amp's slew rate limits the upper frequency of large-amplitude output voltages. As the peak output voltage required from the op amp is reduced, the upper-frequency limitation imposed by the slew rate increases.

Recall that the upper-frequency limitation imposed by small-signal response increases as the closed-loop gain decreases. For each amplifier application, the upper-frequency limit imposed by slew-rate limiting (Section 10-3.3) and small-signal bandwidth

(Section 10-2.3) must be calculated. The *smaller* value determines the actual upper-frequency limit. In general, the slew rate is a large-signal frequency limitation and small-signal frequency response is a small-signal frequency limitation.

10-3.4 Slew Rate Made Easy

Figure 10-6 simplifies the problem of finding f_{max} at any peak output voltage for slew rates between 0.5 and 5 V/μs. For example, to do part (b) of Example 10-8, locate where the horizontal line V_{op} = 10 V intersects the slew-rate line 0.5 V/μs. Below the intersection, read f_{max} = 8 kHz.

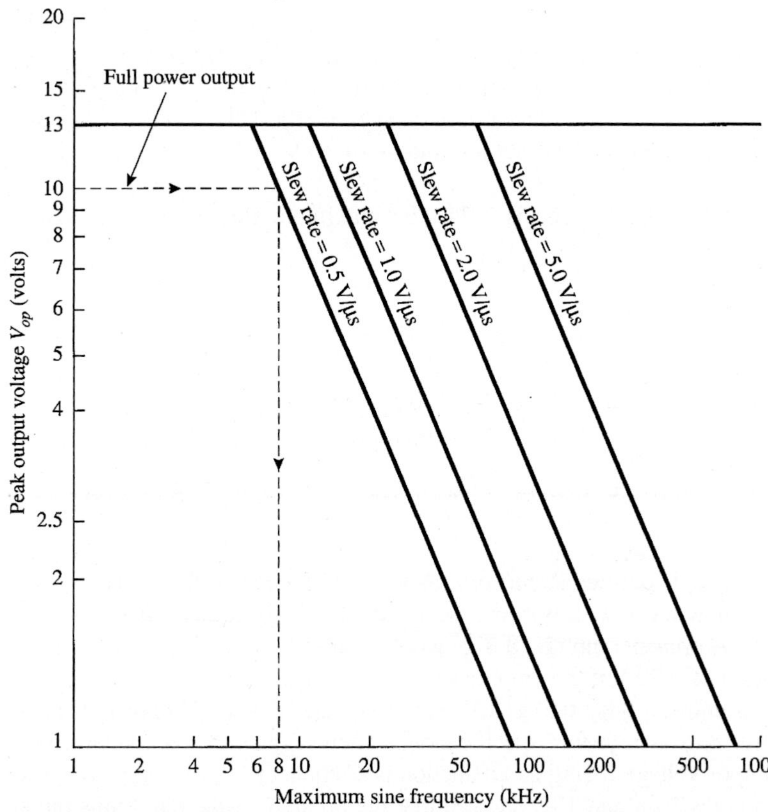

FIGURE 10-6 Slew rate made easy. Any point on a slew-rate line shows the maximum sinusoidal frequency allowed for the corresponding peak output voltage.

10-4 NOISE IN THE OUTPUT VOLTAGE

10-4.1 Introduction

Undesired electrical signals present in the output voltage are classified as *noise*. Drift (see Chapter 9) and offsets can be considered as very-low-frequency noise. If you view the output voltage of an op amp amplifier with a sensitive oscilloscope setting (1 mV/cm), you will see a random display of noise voltages called *hash*. The frequencies of these noise voltages range from 0.01 Hz to megahertz.

Noise is generated in any material that is above absolute zero (−273°C). Noise is also generated by all electrical devices and their controls. For example, in an automobile the spark plugs, voltage regulator, fan motor, air conditioner, and generator all generate noise. Even when headlights are switched on (or off), there is a sudden change in current that generates noise. This type of noise is external to the op amp. Effects of external noise can be minimized by proper construction techniques and circuit selection (see Sections 10-4.3 to 10-4.5).

10-4.2 Noise in Op Amp Circuits

Even if there was no external noise, there would still be noise in the output voltage caused by the op amp. This internal op amp noise is modeled most simply by a noise voltage source E_n. As shown in Fig. 10-7, E_n is placed in series with the (+) input. On data sheets, noise voltage is specified in microvolts or nanovolts (rms) for different values of source resistance over a particular frequency range. For example, the 741 op amp has 2 μV of *total* noise over a frequency of 10 Hz to 10 kHz. This noise voltage is valid for source

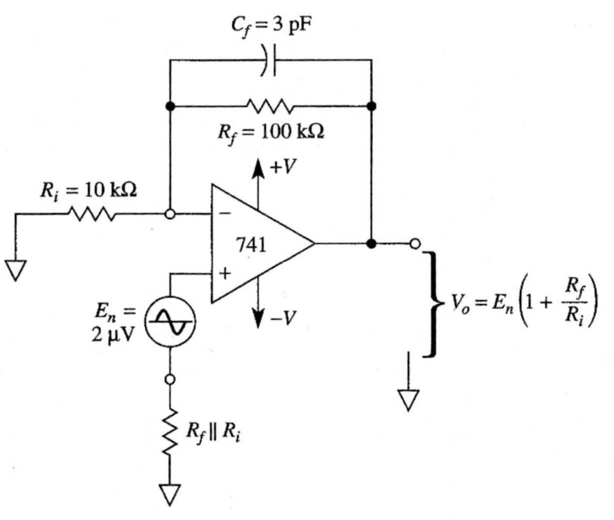

FIGURE 10-7 Op amp noise is modeled by a noise voltage in series with the (+) input.

resistors (R_i) between 100 Ω and 20 kΩ. The noise voltage goes up directly with R_i once R_i exceeds 20 kΩ. Thus R_i should be kept below 20 kΩ to minimize noise in the output (see Appendix 1). Data sheets may also specify an input current value over the same frequency range.

10-4.3 Noise Gain

Noise voltage is amplified just as offset voltage is. That is, *noise voltage gain* is the same as the gain of a noninverting amplifier:

$$\text{noise gain} = 1 + \frac{R_f}{R_i} \tag{10-7}$$

What can you do about minimizing output voltage errors due to noise? First, avoid, if possible, large values of R_i and R_f. Install a small capacitor (3-pF) across R_f to shunt it at high noise frequencies. Then the higher noise frequencies will not be amplified as much. Next, do not shunt R_i with a capacitor; otherwise, the R_iC combination will have a smaller impedance at higher noise frequencies than R_i alone, and gain will increase with frequency and aggravate the situation. Finally, try to keep R_i at about 10 kΩ or below.

Noise currents, like bias currents, are also present at each op amp input terminal. If a bias-current compensation resistor is installed (see Chapter 9), the effect of noise currents on output voltage will be reduced. As with offset current, the effects of noise currents also depend on the feedback resistor. So if possible, reduce the size of R_f to minimize the effects of noise currents.

10-4.4 Noise in the Inverting Adder

In the inverting adder (see Section 3-2), each signal input voltage has a gain of 1. However, the noise gain will be 1 plus the number of inputs; for example, a four-input adder would have a noise gain of 5. Thus noise voltage has five times as much gain as each input signal. Therefore, low-amplitude signals should be preamplified before connecting them to an adder.

10-4.5 Summary

To reduce the effects of op amp noise:

1. *Never* connect a capacitor across the input resistor or from (−) input to ground. There will always be a few picofarads of stray capacitance from (−) input to ground due to wiring, so
2. *Always* connect a small capacitor (3 pF) across the feedback resistor. This reduces the noise gain at high frequencies.
3. If possible, avoid large resistor values.

10-5 LOOP GAIN

Loop gain ($A\beta$) is defined as the ratio of open-loop gain to closed-loop gain.

$$A\beta = \frac{A_{OL}}{A_{CL}} \tag{10-8}$$

See Fig. 10-8. Loop gain determines circuit performance. For 1% gain accuracy, the loop gain must be 100 times the closed-loop gain. This is the same conclusion stated in Section 10-2. Hence the greater the loop gain at a particular frequency, the closer the closed-loop gain depends only on external circuit values.

β is that fraction of the output fed back to the input

$$\text{Noise gain} \equiv \frac{1}{\beta} = \frac{R_i + R_f}{R_i} \tag{10-9}$$

Figure 10-9 shows that the noise gain may peak in the vicinity of the frequency at which $A\beta = 1$ even if the signal gain (the normal closed-loop gain) rolls off at a much lower frequency. This is because the loop gain may have more than 90° phase shift and the amplifier is lightly damped at this frequency.

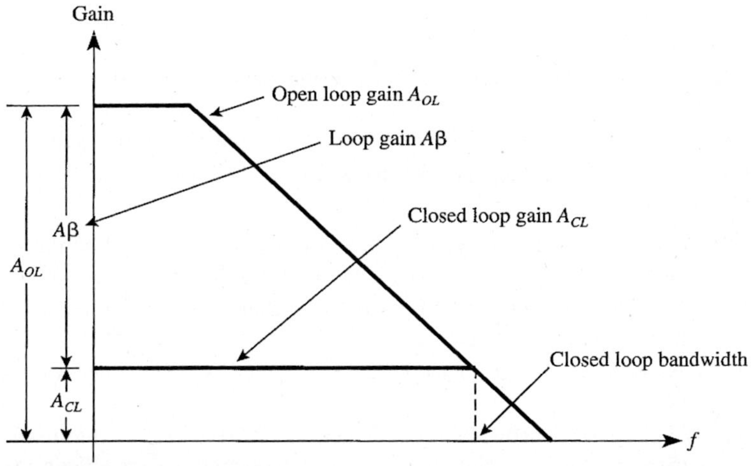

FIGURE 10-8 Loop gain.

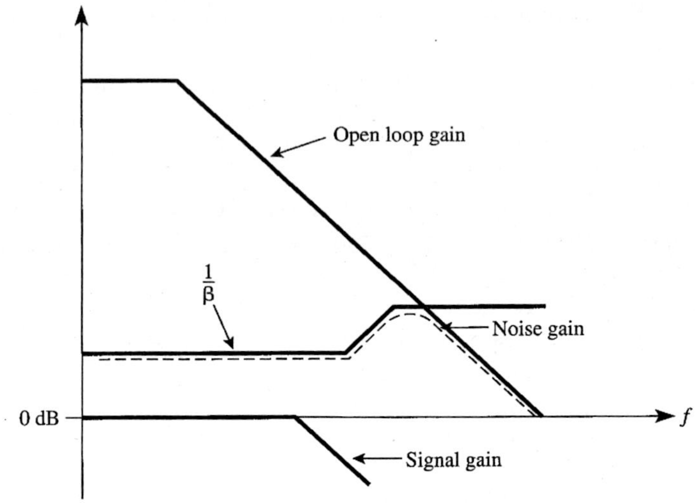

FIGURE 10-9 Noise gain.

PROBLEMS

10-1. What is the typical open-loop gain of a 741 op amp at very low frequencies?

10-2. The dc open-loop gain of an op amp is 100,000. Find the open-loop gain at its break frequency.

10-3. The transient response rise time (unity gain) of an op amp is 0.07 μs. Find the small-signal bandwidth.

10-4. An op amp has a small signal unity-gain bandwidth of 2 MHz. Find its open-loop gain at 200 kHz.

10-5. What is the difference between the open-loop and closed-loop gain of an op amp?

10-6. What is the open-loop gain for the op amp of Problem 10-4 at 2 MHz?

10-7. What is *rise time?*

10-8. An op amp has a dc open-loop gain of 100,000. It is used in an inverting amplifier circuit with $R_f = 100$ kΩ, $R_i = 10$ kΩ. Find the actual dc closed-loop gain.

10-9. The op amp of Problem 10-8 is used in a noninverting amplifier with the same R_f and R_i. Find the amplifier's actual dc closed-loop gain.

10-10. What is the small-signal bandwidth of the op amp whose frequency response is given in Fig. 10-1?

10-11. The unity-gain bandwidth of an op amp is 10 MHz. It is used to make a noninverting amplifier with an ideal closed-loop gain of 100. Find the amplifier's **(a)** small-signal bandwidth; **(b)** A_{CL} at f_H.

10-12. How fast can the output of an op amp change by 10 V if its slew rate is 1 V/μs?

10-13. Find the maximum frequency for a sine-wave output voltage of 10-V peak with an op amp whose slew rate is 1 V/μs.

10-14. Find the noise gain for an inverting amplifier with a gain of $R_f/R_i = -10$.

10-15. What is the noise gain for a five-input inverting adder?

10-16. The op amp in Example 10-9 is changed to one with a slew rate of 1 V/μs. Find its maximum full-power output frequency. Assume that $V_{o\ \max} =$ 10-V peak.

10-17. Use the Internet and compare the slew rate of an OP-177 op amp with a MAX4281.

CHAPTER 11

Active Filters

LEARNING OBJECTIVES

Upon completion of this chapter on active filters, you will be able to:

- Name the four general classifications of filters and sketch a frequency–response curve that shows the band of frequencies that they pass and stop.
- Design or analyze circuits for three types of low-pass filters: −20 dB/decade, −40 dB/decade, or −60 dB/decade roll-off.
- Design or analyze circuits for three types of high-pass filters: +20 dB, +40 dB, and +60 dB per decade of roll-off.
- Cascade a low-pass filter with a high-pass filter to make a wide bandpass filter.
- Calculate the lower and upper cutoff frequencies of either a bandpass or a notch filter if you are given (1) bandwidth and resonant frequency, (2) bandwidth and quality factor, or (3) resonant frequency and quality factor.

- Calculate the quality factor, bandwidth, and resonant frequency of a bandpass or notch filter for a given lower and upper cutoff frequency.
- Design a bandpass filter that uses only one op amp.
- Make a notch filter by (1) designing a bandpass filter circuit with the same bandwidth and a resonant frequency equal to the notch frequency, and (2) properly connecting the bandpass circuit to an inverting adder.
- Explain the operation of a stereo equalizer circuit.
- Use PSpice to simulate the performance of filter circuits.

11-0 INTRODUCTION

A *filter* is a circuit that is designed to pass a specified band of frequencies while attenuating all signals outside this band. Filter networks may be either active or passive. *Passive filter networks* contain only resistors, inductors, and capacitors. *Active filters,* which are the only type covered in this text, employ transistors or op amps plus resistors, inductors, and capacitors. Inductors are not often used in active filters, because they are bulky and costly and may have large internal resistive components.

There are four types of filters: *low-pass, high-pass, bandpass,* and *band-elimination* (also referred to as *band-reject* or *notch*) filters. Figure 11-1 illustrates frequency–

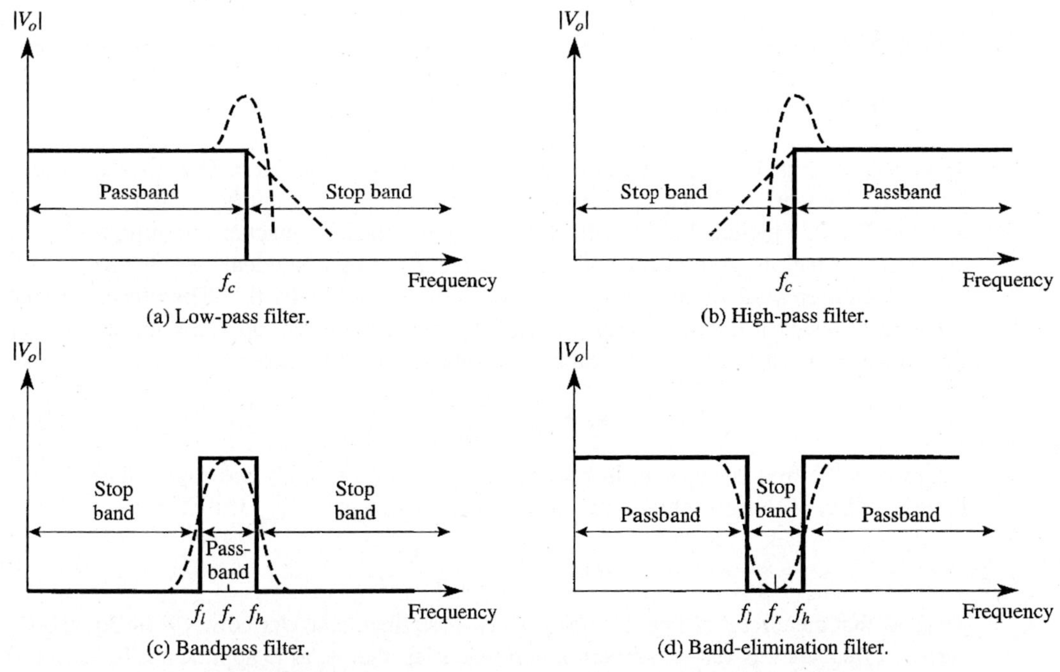

FIGURE 11-1 Frequency response for four categories of filters.

response plots for the four types of filters. A low-pass filter is a circuit that has a constant output voltage from dc up to a *cutoff frequency* f_c. As the frequency increases above f_c, the output voltage is attenuated (decreases). Figure 11-1(a) is a plot of the magnitude of the output voltage of a low-pass filter versus frequency. The solid line is a plot for the ideal low-pass filter, while the dashed lines indicate the curves for practical low-pass filters. The range of frequencies that are *transmitted* is known as the *passband.* The range of frequencies that are *attenuated* is known as the *stop band.* The cutoff frequency, f_c, is also called the 0.707 frequency, the -3-dB frequency, the corner frequency, or the break frequency.

High-pass filters attenuate the output voltage for all frequencies below the cutoff frequency f_c. Above f_c, the magnitude of the output voltage is constant. Figure 11-1(b) is the plot for ideal and practical high-pass filters. The solid line is the ideal curve, the dashed curves show how practical high-pass filters deviate from the ideal.

Bandpass filters pass only a band of frequencies while attenuating all frequencies outside the band. Band-elimination filters perform in an exactly opposite way; that is, band-elimination filters reject a specified band of frequencies while passing all frequencies outside the band. Typical frequency–response plots for bandpass and band-elimination filters are shown in Figs. 11-1(c) and (d). Once again, the solid line represents the ideal plot, while dashed lines show the practical curves.

Filters are an integral part of electronic networks and are used in applications from audio circuits to digital signal processing systems.

11-1 BASIC LOW-PASS FILTER

11-1.1 Introduction

The circuit of Fig. 11-2(a) is a commonly used low-pass active filter. The filtering is done by the *RC* network, and the op amp is used as a unity-gain amplifier. The resistor R_f is equal to R and is included for dc offset. [At dc, the capacitive reactance is infinite and the dc resistance path to ground for both input terminals should be equal (see Section 9-4).]

The differential voltage between pins 2 and 3 is essentially 0 V. Therefore, the voltage across capacitor C equals output voltage V_o, because this circuit is a voltage follower. E_i divides between R and C. The capacitor voltage equals V_o and is

$$V_o = \frac{1/j\omega C}{R + 1/j\omega C} \times E_i \tag{11-1a}$$

where ω is the frequency of E_i in radians per second ($\omega = 2\pi f$) and j is equal to $\sqrt{-1}$. Rewriting Eq. (11-1a) to obtain the closed-loop voltage gain A_{CL}, we have

$$A_{CL} = \frac{V_o}{E_i} = \frac{1}{1 + j\omega RC} \tag{11-1b}$$

To show that the circuit of Fig. 11-2(a) is a low-pass filter, consider how A_{CL} in Eq. (11-1b) varies as frequency is varied. At very low frequencies, that is, as ω approaches 0, $|A_{CL}| = 1$, and at very high frequencies, as ω approaches infinity, $|A_{CL}| = 0$. (The absolute value sign, $|\cdot|$, indicates magnitude.)

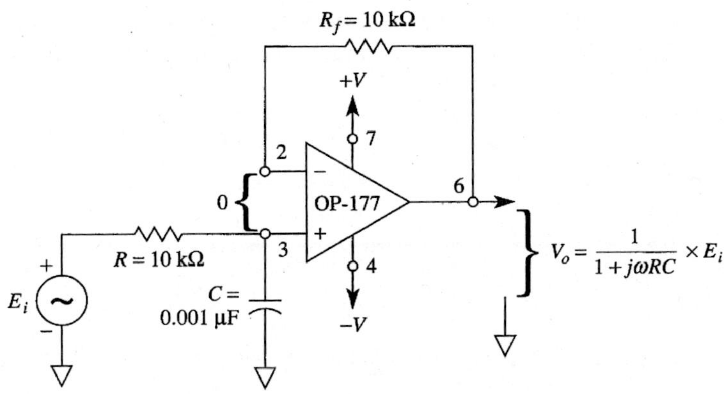

(a) Low-pass filter for a roll-off of −20 dB/decade.

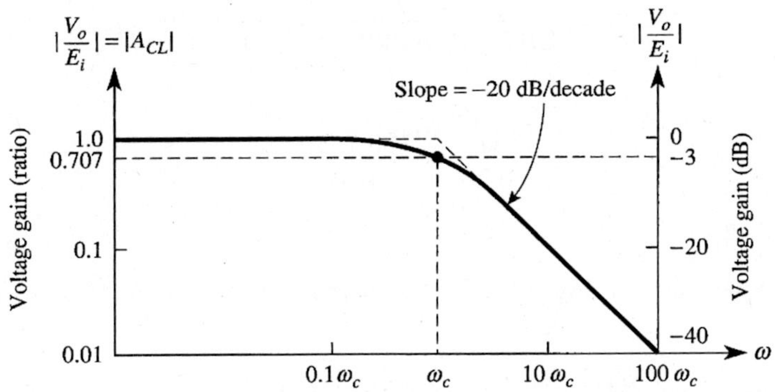

(b) Frequency–response plot for the circuit of part (a).

FIGURE 11-2 Low-pass filter and frequency–response plot for a filter with a −20-dB/decade roll-off.

Figure 11-2(b) is a plot of $|A_{CL}|$ versus ω and shows that for frequencies *greater* than the cutoff frequency ω_c, $|A_{CL}|$ decreases at a rate of 20 dB/decade*. This is the same as saying that the voltage gain is divided by 10 when the frequency of ω is increased by 10.

11-1.2 Designing the Filter

The cutoff frequency ω_c is defined as that frequency of E_i where $|A_{CL}|$ is reduced to 0.707 times its low-frequency value. This important point will be discussed further in Section 11-1.3. The cutoff frequency is evaluated from

$$\omega_c = \frac{1}{RC} = 2\pi f_c \qquad (11\text{-}2a)$$

*dB $= 20 \log_{10} \frac{Vo}{E_i}$

where ω_c is the cutoff frequency in radians per second, f_c is the cutoff frequency in hertz, R is in ohms, and C is in farads. Equation (11-2a) may be rearranged to solve for R:

$$R = \frac{1}{\omega_c\, C} = \frac{1}{2\pi f_c\, C} \tag{11-2b}$$

Example 11-1

Let $R = 10\ \text{k}\Omega$ and $C = 0.001\ \mu\text{F}$ in Fig. 11-2(a); what is the cutoff frequency?

Solution By Eq. (11-2a),

$$\omega_c = \frac{1}{(10 \times 10^3)(0.001 \times 10^{-6})} = 100\ \text{krad/s}$$

or

$$f_c = \frac{\omega_c}{6.28} = \frac{100 \times 10^3}{6.28} = 15.9\ \text{kHz}$$

Example 11-2

For the low-pass filter in Fig. 11-2(a), calculate R for a cutoff frequency of 2 kHz and $C = 0.005\ \mu\text{F}$.

Solution From Eq. (11-2b),

$$R = \frac{1}{\omega_c\, C} = \frac{1}{(6.28)(2 \times 10^3)(5 \times 10^{-9})} = 15.9\ \text{k}\Omega$$

Example 11-3

Calculate R in Fig. 11-2(a) for a cutoff frequency of 30 krad/s and $C = 0.01\ \mu\text{F}$.

Solution From Eq. (11-2b),

$$R = \frac{1}{\omega_c\, C} = \frac{1}{(30 \times 10^3)(1 \times 10^{-8})} \simeq 3.3\ \text{k}\Omega$$

Design Procedure The design of a low-pass filter similar to Fig. 11-2(a) is accomplished in three steps:

1. Choose the cutoff frequency—either ω_c or f_c.
2. Choose the capacitance C, usually between 0.001 and 0.1 μF.
3. Calculate R from Eq. (11-2b).

11-1.3 Filter Response

The value of A_{CL} at ω_c is found by letting $\omega RC = 1$ in Eq. (11-1b):

$$A_{CL} = \frac{1}{1 + j1} = \frac{1}{\sqrt{2}\ \angle 45^\circ} = 0.707\ \angle -45^\circ$$

Therefore, the magnitude of A_{CL} at ω_c is

$$|A_{CL}| = \frac{1}{\sqrt{2}} = 0.707 = -3\text{ dB}$$

and the phase angle is $-45°$.

The solid curve in Fig. 11-2(b) shows how the magnitude of the actual frequency response deviates from the straight dashed-line approximation in the vicinity of ω_c. At $0.1\omega_c$, $|A_{CL}| \simeq 1$ (0 dB), and at $10\omega_c$, $|A_{CL}| \simeq 0.1$ (-20 dB). Table 11-1 gives both the magnitude and the phase angle for different values of ω between $0.1\omega_c$ and $10\omega_c$.

Many applications require steeper roll-offs after the cutoff frequency. One common filter configuration that gives steeper roll-offs is the *Butterworth filter.*

TABLE 11-1 MAGNITUDE AND PHASE ANGLE FOR THE LOW-PASS FILTER OF FIG. 11-2(a)

ω	$\lvert A_{CL}\rvert$	Phase angle (deg)
$0.1\omega_c$	1.0	−6
$0.25\omega_c$	0.97	−14
$0.5\omega_c$	0.89	−27
ω_c	0.707	−45
$2\omega_c$	0.445	−63
$4\omega_c$	0.25	−76
$10\omega_c$	0.1	−84

11-2 INTRODUCTION TO THE BUTTERWORTH FILTER

In many low-pass filter applications, it is necessary for the closed-loop gain to be as close to 1 as possible within the passband. The *Butterworth filter* is best suited for this type of application. The Butterworth filter is also called a *maximally flat* or *flat-flat* filter, and all filters in this chapter will be of the Butterworth type. Figure 11-3 shows the ideal (solid line) and the practical (dashed lines) frequency response for three types of Butterworth filters. As the roll-offs become steeper, they approach the ideal filter more closely.

Two active filters similar to Fig. 11-2(a) could be coupled together to give a roll-off of -40 dB/decade. This would not be the most economical design, because it would require two op amps. In Section 11-3.1, it is shown how one op amp can be used to build a Butterworth filter with a single op amp to give a -40-dB/decade roll-off. Then in Section 11-4, a -40-dB/decade filter is cascaded with a -20-dB/decade filter to produce a -60-dB/decade filter.

Butterworth filters are not designed to keep a constant phase angle at the cutoff frequency. A basic low-pass filter of -20 dB/decade has a phase angle of $-45°$ at ω_c. A -40-dB/decade Butterworth filter has a phase angle of $-90°$ at ω_c, and a -60-dB/decade filter has a phase angle of $-135°$ at ω_c. Therefore, for each increase of -20 dB/decade, the phase angle will increase by $-45°$ at ω_c. We now proceed to a Butterworth filter that has a roll-off steeper than -20 dB/decade.

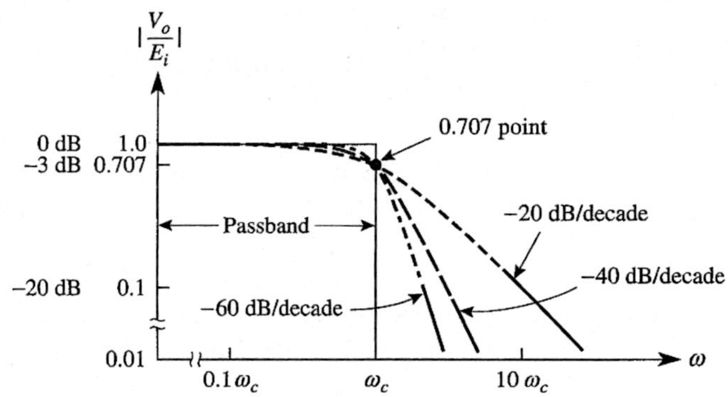

FIGURE 11-3 Frequency–response plots for three types of low-pass Butterworth filters.

11-3 −40-DB/DECADE LOW-PASS BUTTERWORTH FILTER

11-3.1 Simplified Design Procedure

The circuit of Fig. 11-4(a) is one of the most commonly used low-pass filters. It produces a roll-off of -40 dB/decade; that is, after the cutoff frequency, the magnitude of A_{CL} decreases by 40 dB as ω increases to $10\omega_c$. The solid line in Fig. 11-4(b) shows the actual frequency–response plot, which is explained in more detail in Section 11-3.2. The op amp is connected for dc unity gain. Resistor R_f is included for dc offset, as explained in Section 9-4. Since the op amp circuit is basically a voltage follower (unity-gain amplifier), the voltage across C_1 equals output voltage, V_o.

The design of the low-pass filter of Fig. 11-4(a) is greatly simplified by making resistors $R_1 = R_2 = R$. Then there are only five steps in the design procedure.

Design procedure

1. Choose the cutoff frequency, ω_c or f_c.
2. Pick C_1; choose a convenient value between 100 pF and 0.1 μF.
3. Make $C_2 = 2C_1$.
4. Calculate

$$R = \frac{0.707}{\omega_c C_1} \tag{11-3}$$

5. Choose $R_f = 2R$.

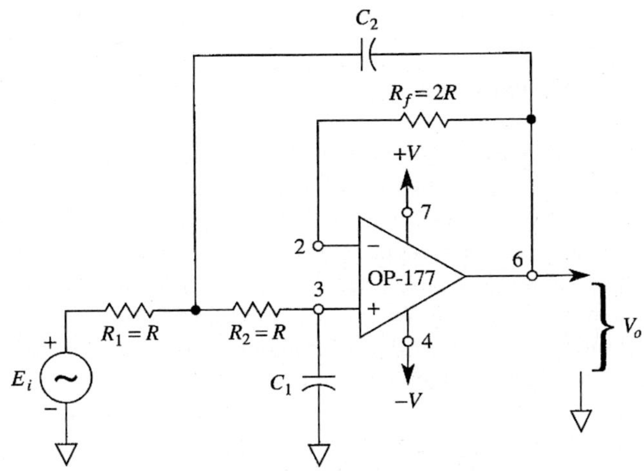

(a) Low-pass filter for a roll-off of −40 dB/decade.

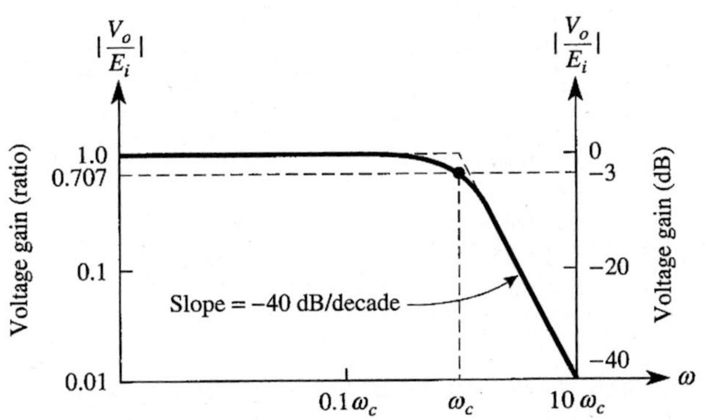

(b) Frequency–response plot for the low-pass filter of part (a).

FIGURE 11-4 Circuit and frequency plot for a low-pass filter of −40 dB/decade.

Example 11-4

Determine R_1 and R_2 in Fig. 11-4(a) for a cutoff frequency of 1 kHz. Let $C_1 = 0.01\ \mu F$.

Solution Pick $C_2 = 2C_1 = 2(0.01\ \mu F) = 0.02\ \mu F$. Select $R_1 = R_2 = R$ from Eq. (11.3):

$$R = \frac{0.707}{(6.28)(1 \times 10^3)(0.01 \times 10^{-6})} = 11{,}258\ \Omega$$

and

$$R_f = 2(11{,}258\ \Omega) = 22{,}516\ \Omega$$

11-3.2 Filter Response

The solid curve in Fig. 11-4(b) shows that the filter of Fig. 11-4(a) not only has a steeper roll-off after ω_c than does Fig. 11-2(a), but also remains at 0 dB almost up to about 0.25 ω_c. The phase angles for the circuit of Fig. 11-4(a) range from 0° at $\omega = 0$ rad/s (dc condition) to $-180°$ as ω approaches ∞ (infinity). Table 11-2 compares magnitude and phase angle for the low-pass filters of Figs. 11-2(a) and 11-4(a) from $0.1\omega_c$ to $10\omega_c$.

The next low-pass filter cascades the filter of Fig. 11-2(a) with the filter of Fig. 11-4(a) to form a roll-off of -60 dB/decade. As will be shown, the resistors are the only values that have to be calculated.

TABLE 11-2 MAGNITUDE AND PHASE ANGLE FOR FIGS. 11-2(a) AND 11-4(a)

	$\lvert A_{CL}\rvert$		Phase angle (deg)	
ω	-20 dB/decade; Fig. 11-2(a)	-40 dB/decade; Fig. 11-4(a)	Fig. 11-2(a)	Fig. 11-4(a)
$0.1\omega_c$	1.0	1.0	-6	-8
$0.25\omega_c$	0.97	0.998	-14	-21
$0.5\omega_c$	0.89	0.97	-27	-43
ω_c	0.707	0.707	-45	-90
$2\omega_c$	0.445	0.24	-63	-137
$4\omega_c$	0.25	0.053	-76	-143
$10\omega_c$	0.1	0.01	-84	-172

11-4 −60-DB/DECADE LOW-PASS BUTTERWORTH FILTER

11-4.1 Simplified Design Procedure

The low-pass filter of Fig. 11-5(a) is built using one low-pass filter of -40 dB/decade cascaded with another of -20 dB/decade to give an overall roll-off of -60 dB/decade. The overall closed-loop gain A_{CL} is the gain of the first filter times the gain of the second filter, or

$$A_{CL} = \frac{V_o}{E_i} = \frac{V_{o_1}}{E_i} \times \frac{V_o}{V_{o_1}} \tag{11-4}$$

For a Butterworth filter, the magnitude of A_{CL} must be 0.707 at ω_c. To guarantee that the frequency response is flat in the passband, use the following design steps.

Design procedure

1. Choose the cutoff frequency, ω_c or f_c.
2. Pick C_3; choose a convenient value between 0.001 and 0.1 μF.
3. Make

$$C_1 = \tfrac{1}{2}C_3 \quad \text{and} \quad C_2 = 2C_3 \tag{11-5}$$

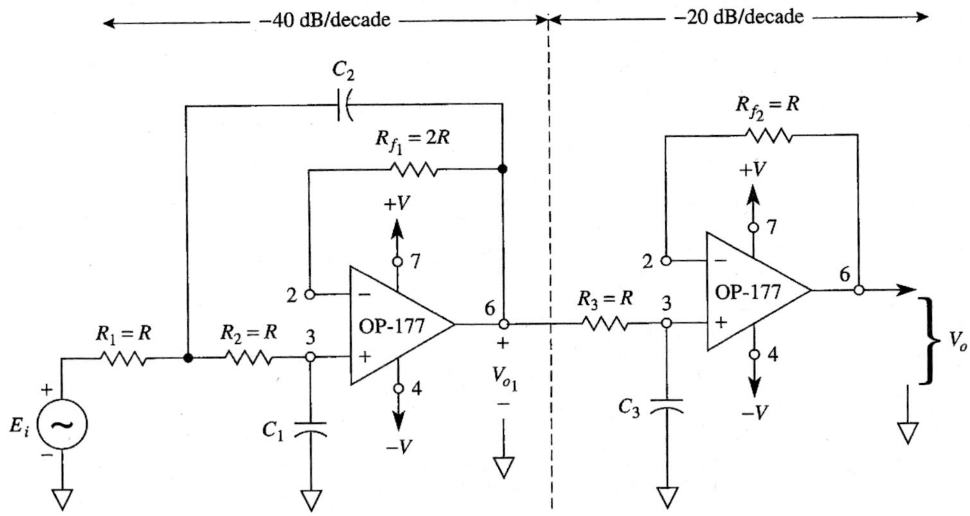

(a) Low-pass filter for a roll-off of −60 dB/decade.

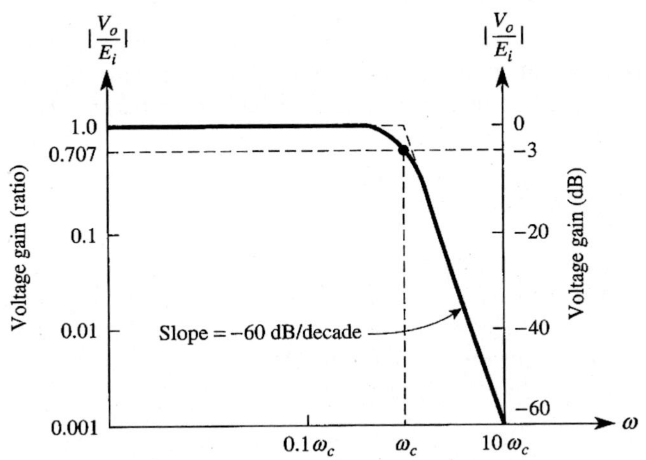

(b) Plot of frequency response for the circuit of part (a).

FIGURE 11-5 Low-pass filter designed for a roll-off of −60 dB/decade and corresponding frequency–response plot.

4. Calculate

$$R = \frac{1}{\omega_c\, C_3} \tag{11-6}$$

5. Make $R_1 = R_2 = R_3 = R$.
6. $R_{f_1} = 2R$ and $R_{f_2} = R$. For best results the value of R should be between 10 and 100 kΩ. If the value of R is outside this range, you should go back and pick a new value of C_3.

Example 11-5

For the − 60-dB/decade low-pass filter of Fig. 11-5(a), determine the values of C_1, C_2, and R for a cutoff frequency of 1 kHz. Let $C_3 = 0.01\ \mu F$.

Solution From Eq. (11-5),

$$C_1 = \tfrac{1}{2}C_3 = \tfrac{1}{2}(0.01\ \mu F) = 0.005\ \mu F$$

and

$$C_2 = 2C_3 = 2(0.01\ \mu F) = 0.02\ \mu F$$

From Eq. (11-6),

$$R = \frac{1}{(6.28)(1 \times 10^3)(0.01 \times 10^{-6})} = 15{,}915\ \Omega$$

Example 11-5 shows that the value of R in Fig. 11-5(a) is different from those of Fig. 11-4(a), although the cutoff frequency is the same. This is necessary so that $|A_{CL}|$ remains at 0 dB in the passband until the cutoff frequency is nearly reached; then $|A_{CL}| = 0.707$ at ω_c.

11-4.2 Filter Response

The solid line in Fig. 11-5(b) is the actual plot of the frequency response for Fig. 11-5(a). The dashed curve in the vicinity shows the straight-line approximation. Table 11-3 compares the magnitudes of A_{CL} for the three low-pass filters presented in this chapter. Note that the $|A_{CL}|$ for Fig. 11-5(a) remains quite close to 1 (0 dB) until the cutoff frequency, ω_c; then the steep roll-off occurs.

TABLE 11-3 $|A_{CL}|$ **FOR THE LOW-PASS FILTERS OF FIGS. 11-2(a), 11-4(a), AND 11-5(a)**

ω	−20 dB/decade; Fig. 11-2(a)	−40 dB/decade; Fig. 11-4(a)	−60 dB/decade; Fig. 11-5(a)
$0.1\omega_c$	1.0	1.0	1.0
$0.25\omega_c$	0.97	0.998	0.999
$0.5\omega_c$	0.89	0.97	0.992
ω_c	0.707	0.707	0.707
$2\omega_c$	0.445	0.24	0.124
$4\omega_c$	0.25	0.053	0.022
$10\omega_c$	0.1	0.01	0.001

The phase angles for the low-pass filter of Fig. 11-5(a) range from 0° at $\omega = 0$ (dB condition) to $-270°$ as ω approaches ∞. Table 11-4 compares the phase angles for the three low-pass filters.

All digital signal processing systems use a low-pass filter at the front end to attenuate frequencies above the Nyquist frequency, which is one-half the sampling rate.

TABLE 11-4 PHASE ANGLES FOR THE LOW-PASS FILTERS OF FIGS. 11-2(a), 11-4(a), AND 11-5(a)

ω	−20 dB/decade; Fig. 11-2(a)	−40 dB/decade; Fig. 11-4(a)	−60 dB/decade; Fig. 11-5(a)
$0.1\omega_c$	−6°	−8°	−12°
$0.25\omega_c$	− 4°	−21°	−29°
$0.5\omega_c$	−27°	−43°	−60°
ω_c	−45°	−90°	−135°
$2\omega_c$	−63°	−137°	−210°
$4\omega_c$	−76°	−143°	−226°
$10\omega_c$	−84°	−172°	−256°

11-5 HIGH-PASS BUTTERWORTH FILTERS

11-5.1 Introduction

A high-pass filter is a circuit that attenuates all signals below a specified cutoff frequency ω_c and passes all signals whose frequency is above the cutoff frequency. Thus a high-pass filter performs the opposite function of the low-pass filter.

Figure 11-6 is a plot of the magnitude of the closed-loop gain versus ω for three types of Butterworth filters. The phase angle for a circuit of 20 dB/decade is +45° at ω_c.

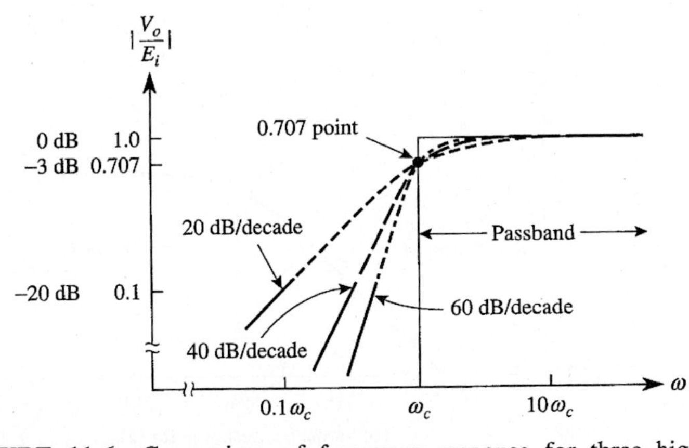

FIGURE 11-6 Comparison of frequency response for three high-pass Butterworth filters.

Phase angles at ω_c *increase* by +45° for each *increase* of 20 dB/decade. The phase angles for these three types of high-pass filters are compared in Section 11-5.5.

In this book the design of high-pass filters will be similar to that of the low-pass filters. In fact, the only difference will be the position of the filtering capacitors and resistors.

11-5.2 20-dB/Decade Filter

Compare the high-pass filter of Fig. 11-7(a) with the low-pass filter of Fig. 11-2(a) and note that C and R are interchanged. The feedback resistor R_f is included to minimize dc offset. Since the op amp is connected as a unity-gain follower in Fig. 11-7(a), the output voltage V_o equals the voltage across R and is expressed by

$$V_o = \frac{1}{1 - j(1/\omega RC)} \times E_i \tag{11-7}$$

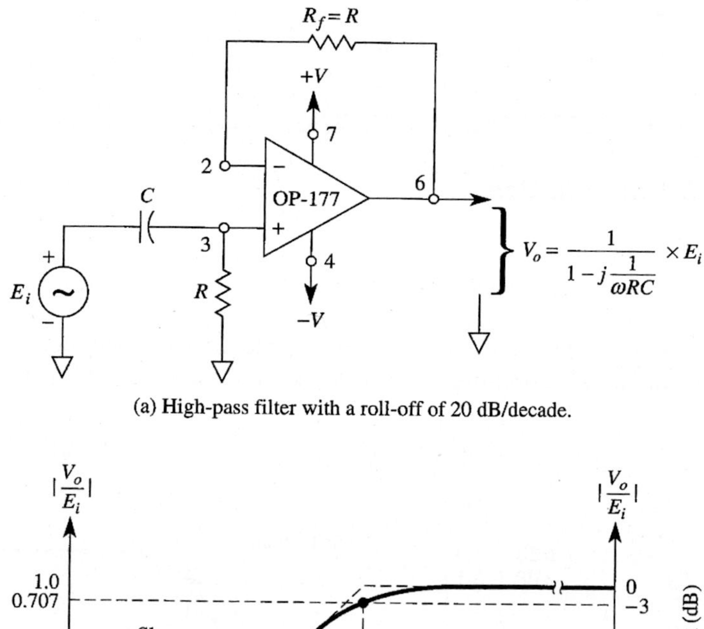

(a) High-pass filter with a roll-off of 20 dB/decade.

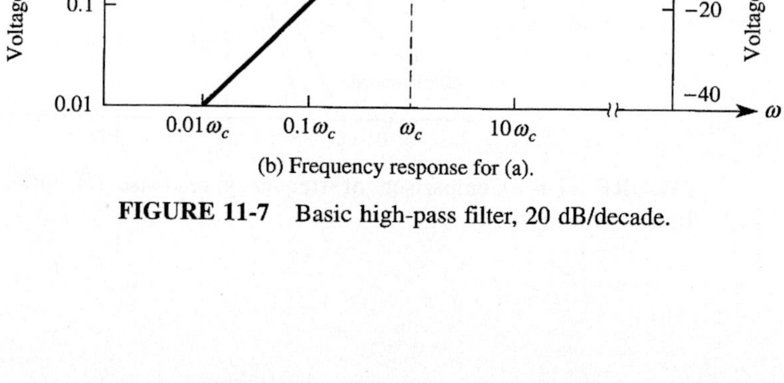

(b) Frequency response for (a).

FIGURE 11-7 Basic high-pass filter, 20 dB/decade.

When ω approaches 0 rad/s in Eq. (11-7), V_o approaches 0 V. At high frequencies, as ω approaches infinity, V_o equals E_i. Since the circuit is not an ideal filter, the frequency response is not ideal, as shown by Fig. 11-7(b). The solid line is the actual response; the dashed lines show the straight-line approximation. The magnitude of the closed-loop gain equals 0.707 when $\omega RC = 1$. Therefore, the cutoff frequency ω_c is given by

$$\omega_c = \frac{1}{RC} = 2\pi f_c \tag{11-8a}$$

or

$$R = \frac{1}{\omega_c C} = \frac{1}{2\pi f_c C} \tag{11-8b}$$

The reason for solving for R and not C in Eq. (11-8b) is that it is easier to adjust R than it is C. The steps needed in designing Fig. 11-7(a) are as follows:

Design procedure for 20-dB/decade high-pass

1. Choose the cutoff frequency, ω_c or f_c.
2. Choose a convenient value of C, usually between 0.001 and 0.1 μF.
3. Calculate R from Eq. (11-8b).
4. Choose $R_f = R$.

Example 11-6

Calculate R in Fig. 11-7(a) if $C = 0.002\ \mu\text{F}$ and $f_c = 10$ kHz.

Solution From Eq. (11-8b),

$$R = \frac{1}{(6.28)(10 \times 10^3)(0.002 \times 10^{-6})} = 8\ \text{k}\Omega$$

Example 11-7

In Fig 11-7(a) if $R = 22\ \text{k}\Omega$ and $C = 0.01\ \mu\text{F}$, calculate (a) ω_c; (b) f_c.

Solution (a) From Eq. (11-8a),

$$\omega_c = \frac{1}{(22 \times 10^3)(0.01 \times 10^{-6})} = 4.54\ \text{krad/s}$$

(b)

$$f_c = \frac{\omega_c}{2\pi} = \frac{4.54 \times 10^3}{6.28} = 724\ \text{Hz}$$

11-5.3 40-dB/Decade Filter

The circuit of Fig. 11-8(a) is to be designed as a high-pass Butterworth filter with a roll-off of 40 dB/decade below the cutoff frequency, ω_c. To satisfy the Butterworth criteria, the frequency response must be 0.707 at ω_c and be 0 dB in the pass band. These conditions will be met if the following design procedure is followed:

Design procedure for 40-dB/decade high-pass

1. Choose a cutoff frequency, ω_c or f_c.
2. Let $C_1 = C_2 = C$ and choose a convenient value.

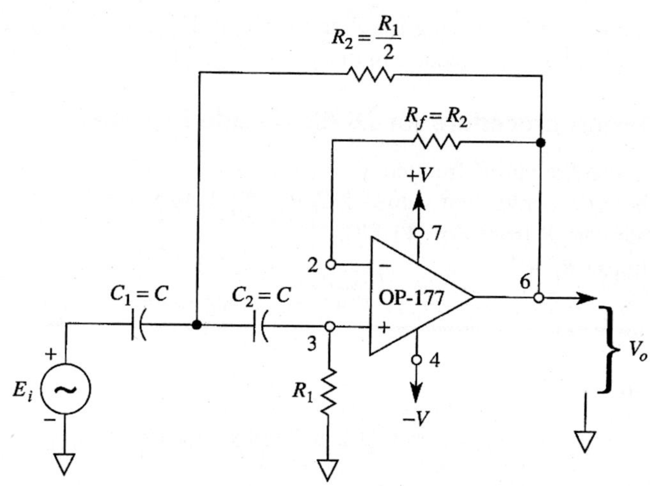

(a) High-pass filter with a roll-off of 40 dB/decade.

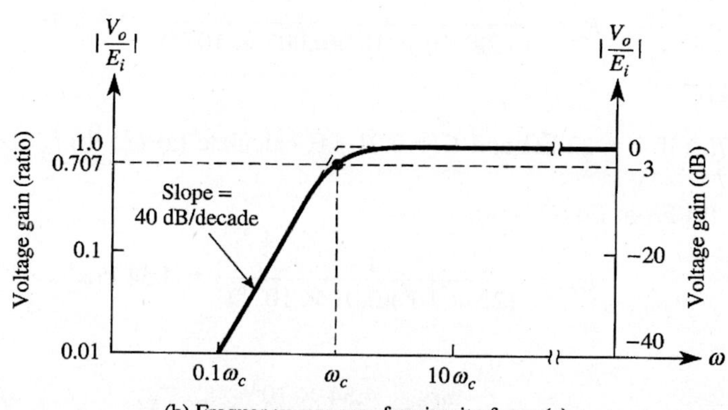

(b) Frequency response for circuit of part (a).

FIGURE 11-8 Circuit and frequency response for a 40-dB/decade high-pass Butterworth filter.

3. Calculate R_1 from

$$R_1 = \frac{1.414}{\omega_c C} \tag{11-9}$$

4. Select

$$R_2 = \tfrac{1}{2} R_1 \tag{11-10}$$

5. To minimize dc offset, let $R_f = R_1$.

Example 11-8

In Fig. 11-8(a), let $C_1 = C_2 = 0.01\ \mu$F. Calculate (a) R_1 and (b) R_2 for a cutoff frequency of 1 kHz.

Solution (a) From Eq. (11-9),

$$R_1 = \frac{1.414}{(6.28)(1 \times 10^3)(0.01 \times 10^{-6})} = 22.5\ \text{k}\Omega$$

(b) $R_2 = \frac{1}{2}(22.5\ \text{k}\Omega) = 11.3\ \text{k}\Omega$.

Example 11-9

Calculate (a) R_1 and (b) R_2 in Fig. 11-8(a) for a cutoff frequency of 80 krad/s. $C_1 = C_2 =$ 125 pF.

Solution (a) From (11-9),

$$R_1 = \frac{1.414}{(80 \times 10^3)(125 \times 10^{-12})} = 140\ \text{k}\Omega$$

(b)

$$R_2 = \tfrac{1}{2}(140\ \text{k}\Omega) = 70\ \text{k}\Omega.$$

11-5.4 60-dB/Decade Filter

As with the low-pass filter of Fig. 11-5, a high-pass filter of +60 dB/decade can be constructed by cascading a +40-dB/decade filter with a +20-dB/decade filter. This circuit (like the other high- and low-pass filters) is designed as a Butterworth filter to have the frequency response in Fig. 11-9(b). The design steps for Fig. 11-9(a) are as follows:

Design procedure for 60-dB/decade high-pass

1. Choose the cutoff frequency, ω_c or f_c.
2. Let $C_1 = C_2 = C_3 = C$ and choose a convenient value between 100 pF and 0.1 μF.

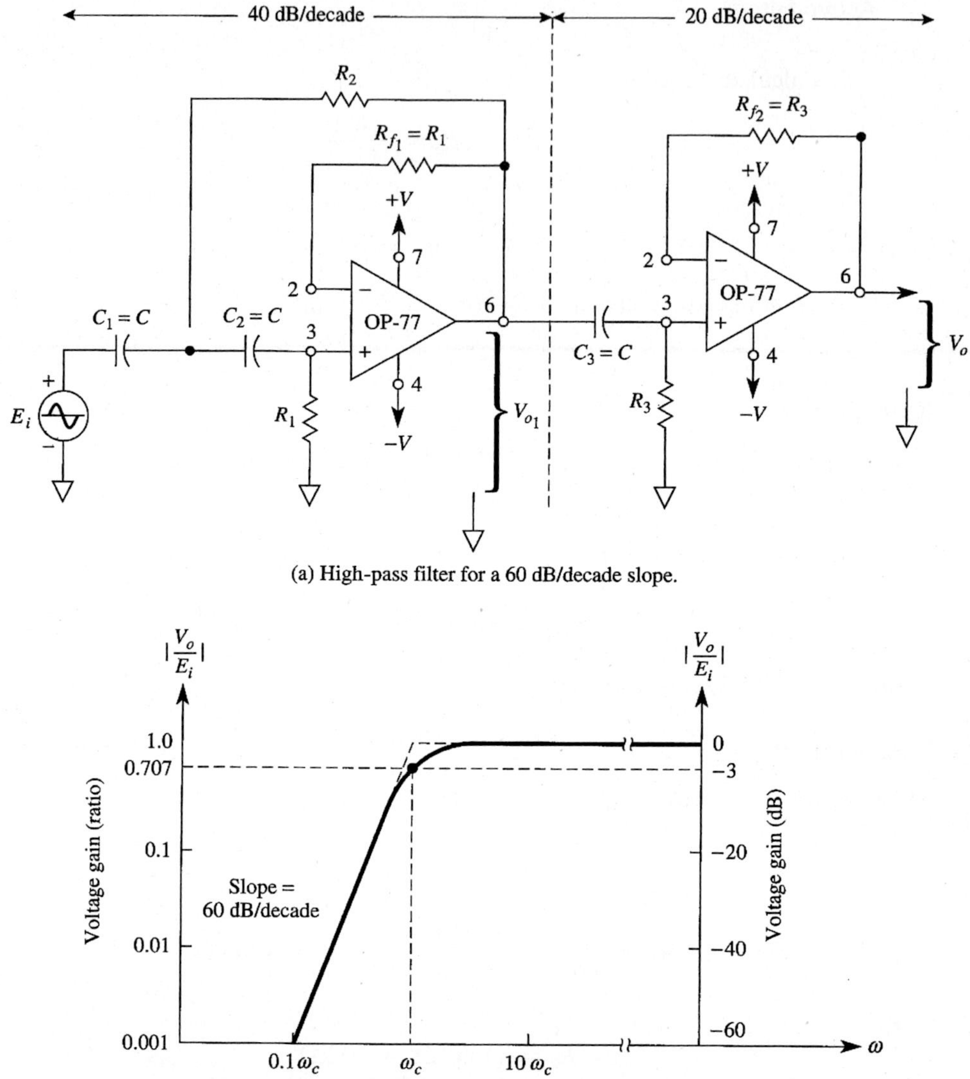

(a) High-pass filter for a 60 dB/decade slope.

(b) Frequency–response for the circuit of part (a).

FIGURE 11-9 Circuit and frequency response for a 60-dB/decade Butterworth high-pass filter.

3. Calculate R_3 from

$$R_3 = \frac{1}{\omega_c\, C} \tag{11-11}$$

4. Select

$$R_1 = 2R_3 \tag{11-12}$$

5. Select

$$R_2 = \tfrac{1}{2}R_3 \tag{11-13}$$

6. To minimize dc offset current, let $R_{f_1} = R_1$ and $R_{f_2} = R_3$.

Example 11-10

For Fig. 11-9(a), let $C_1 = C_2 = C_3 = C = 0.1\ \mu$F. Determine (a) R_3, (b) R_1, and (c) R_2 for $\omega_c = 1$ krad/s. ($f_c = 159$ Hz.)

Solution (a) By Eq. (11-11),

$$R_3 = \frac{1}{(1 \times 10^3)(0.1 \times 10^{-6})} = 10\ \text{k}\Omega$$

(b) $R_1 = 2R_3 = 2(10\ \text{k}\Omega) = 20\ \text{k}\Omega$.
(c) $R_2 = \frac{1}{2}R_3 = \frac{1}{2}(10\ \text{k}\Omega) = 5\ \text{k}\Omega$.

Example 11-11

Determine (a) R_3, (b) R_1, and (c) R_2 in Fig. 11-9(a) for a cutoff frequency of 60 kHz. Let $C_1 = C_2 = C_3 = C = 220$ pF.

Solution (a) From Eq. (11-11),

$$R_3 = \frac{1}{(6.28)(60 \times 10^3)(220 \times 10^{-12})} = 12\ \text{k}\Omega$$

(b) $R_1 = 2R_3 = 2(12\ \text{k}\Omega) = 24\ \text{k}\Omega$.
(c) $R_2 = \frac{1}{2}R_3 = \frac{1}{2}(12\ \text{k}\Omega) = 6\ \text{k}\Omega$.

If desired, the 20-dB/decade section can come before the 40-dB/decade section, because the op amps provide isolation and do not load one another.

11-5.5 Comparison of Magnitudes and Phase Angles

Table 11-5 compares the magnitudes of the closed-loop gain for the three high-pass filters. For each increase of 20 dB/decade, the circuit not only has a steeper roll-off below ω_c but also remains closer to 0 dB or a gain of 1 above ω_c.

The phase angle for a 20-dB/decade Butterworth high-pass filter is 45° at ω_c. For a 40-dB/decade filter it is 90°, and for a 60-dB/decade filter it is 135°. Other phase angles in the vicinity of ω_c for the three filters are given in Table 11-6.

TABLE 11-5 COMPARISON OF $|A_{CL}|$ FOR FIGS. 11-7(a), 11-8(a), AND 11-9(a)

ω	20 dB/decade; Fig. 11-7(a)	40 dB/decade; Fig. 11-8(a)	60 dB/decade; Fig. 11-9(a)
$0.1\omega_c$	0.1	0.01	0.001
$0.25\omega_c$	0.25	0.053	0.022
$0.5\omega_c$	0.445	0.24	0.124
ω_c	0.707	0.707	0.707
$2\omega_c$	0.89	0.97	0.992
$4\omega_c$	0.97	0.998	0.999
$10\omega_c$	1.0	1.0	1.0

TABLE 11-6 COMPARISON OF PHASE ANGLES FOR FIGS. 11-7(a), 11-8(a), AND 11-9(a)

ω	20 dB/decade; Fig. 11-7(a)	40 dB/decade; Fig. 11-8(a)	60 dB/decade; Fig. 11-9(a)
$0.1\omega_c$	84°	172°	256°
$0.25\omega_c$	76°	143°	226°
$0.5\omega_c$	63°	137°	210°
ω	45°	90°	135°
$2\omega_c$	27°	43°	60°
$4\omega_c$	14°	21°	29°
$10\omega_c$	6°	8°	12°

11-6 INTRODUCTION TO BANDPASS FILTERS

11-6.1 Frequency Response

A bandpass filter is a frequency selector. It allows one to select or pass only one particular band of frequencies from all other frequencies that may be present in a circuit. Its normalized frequency response is shown in Fig. 11-10. This type of filter has a maximum gain at a resonant frequency f_r. In this chapter all bandpass filters will have a gain of 1 or 0 dB at f_r. There is one frequency below f_r where the gain falls to 0.707. It is the *lower cutoff frequency*, f_l. At the *higher cutoff frequency*, f_h, the gain also equals 0.707, as in Fig. 11-10.

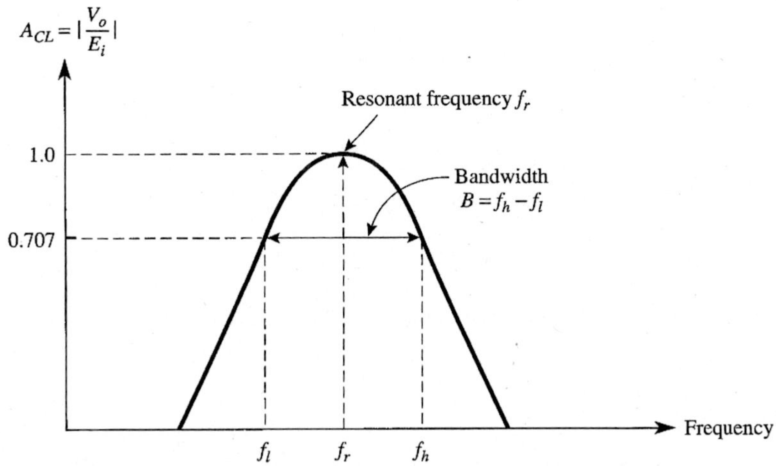

FIGURE 11-10 A bandpass filter has a maximum gain at resonant frequency f_r. The band of frequencies transmitted lies between f_l and f_h.

11-6.2 Bandwidth

The range of frequencies between f_l and f_h is called *bandwidth B,* or

$$B = f_h - f_l \tag{11-14}$$

The bandwidth is not exactly centered on the resonant frequency. (It is for this reason that we use the historical name "resonant frequency" rather than "center frequency" to describe f_r.)

If you know the values for f_l and f_h, the resonant frequency can be found from

$$f_r = \sqrt{f_l f_h} \tag{11-15}$$

If you know the resonant frequency, f_r, and bandwidth, B, cutoff frequencies can be found from

$$f_l = \sqrt{\frac{B^2}{4} + f_r^2} - \frac{B}{2} \tag{11-16a}$$

$$f_h = f_l + B \tag{11-16b}$$

Example 11-12

A bandpass voice filter has lower and upper cutoff frequencies of 300 and 3000 Hz. Find (a) the bandwidth; (b) the resonant frequency.

Solution (a) From Eq. (11-14),

$$B = f_h - f_l = (3000 - 300) = 2700 \text{ Hz}$$

(b) From Eq. (11-15),

$$f_r = \sqrt{f_l f_h} = \sqrt{(300)(3000)} = 948.7 \text{ Hz}$$

Note: f_r is always *below* the center frequency of $(3000 + 300)/2 = 1650$ Hz.

Example 11-13

A bandpass filter has a resonant frequency of 950 Hz and a bandwidth of 2700 Hz. Find its lower and upper cutoff frequencies.

Solution From Eq. (11-16a),

$$f_l = \sqrt{\frac{B^2}{4} + f_r^2} - \frac{B}{2} = \sqrt{\frac{(2700)^2}{4} + (950)^2} - \frac{2700}{2}$$

$$= 1650 - 1350 = 300 \text{ Hz}$$

From Eq. (11-16b), $f_h = 300 + 2700 = 3000$ Hz.

11-6.3 Quality Factor

The *quality factor Q* is defined as the ratio of resonant frequency to bandwidth, or

$$Q = \frac{f_r}{B} \tag{11-17}$$

Q is a measure of the bandpass filter's *selectivity.* A high Q indicates that a filter selects a smaller band of frequencies (more selective).

11-6.4 Narrowband and Wideband Filters

A *wideband* filter has a bandwidth that is two or more times the resonant frequency. That is, $Q \leq 0.5$ for wideband filters. In general, wideband filters are made by cascading a low-pass filter circuit with a high-pass filter circuit. This topic is covered in the next section. A narrowband filter ($Q > 0.5$) can usually be made with a single stage. This type of filter is presented in Section 11-8.

Example 11-14

Find the quality factor of a voice filter that has a bandwidth of 2700 Hz and a resonant frequency of 950 Hz (see Examples 11-12 and 11-13).

Solution From Eq. (11-7),

$$Q = \frac{f_r}{B} = \frac{950}{2700} = 0.35$$

This filter is classified as wideband because $Q < 0.5$.

11-7 BASIC WIDEBAND FILTER

11-7.1 Cascading

When the output of one circuit is connected in series with the input of a second circuit, the process is called *cascading* gain stages. In Fig. 11-11, the first stage is a 3000-Hz low-pass filter (Section 11-3). Its output is connected to the input of a 300-Hz high-pass filter (Section 11-5.3). The cascaded pair of active filters now form a bandpass filter from input E_i to output V_o. Note that it makes no difference if the high-pass is connected to the low-pass, or vice versa. Note: Each op amp circuit in Fig. 11-11 has unity gain.

11-7.2 Wideband Filter Circuit

In general, a wideband filter ($Q \leq 0.5$) is made by cascading a low- and a high-pass filter (see Fig. 11-11). Cutoff frequencies of the low- and high-pass sections *must not overlap,* and each must have the same passband gain. Furthermore, the low-pass filter's cutoff frequency must be 10 or more times the high-pass filter's cutoff frequency.

For cascaded low- and high-pass filters, the resulting wideband filter will have the following characteristics:

1. The lower cutoff frequency, f_l, will be determined only by the high-pass filter.
2. The high cutoff frequency, f_h, will be set only by the low-pass filter.
3. Gain will be maximum at resonant frequency, f_r, and equal to the passband gain of either filter.

These principles are illustrated next.

11-7.3 Frequency Response

In Fig. 11-11 the frequency response of a basic −40-dB/decade 3000-Hz low-pass filter is plotted as a dashed line. The frequency response of a 300-Hz high-pass filter is plotted as a solid line. The 40-dB/decade roll-off of the high-pass filter is seen to determine f_l. The −40-dB/decade roll-off of the low-pass sets f_h. Both roll-off curves make up the frequency response of the bandpass filter, V_o versus f. Observe that the resonant, low, and high cutoff frequencies plus bandwidth agree exactly with the values calculated in

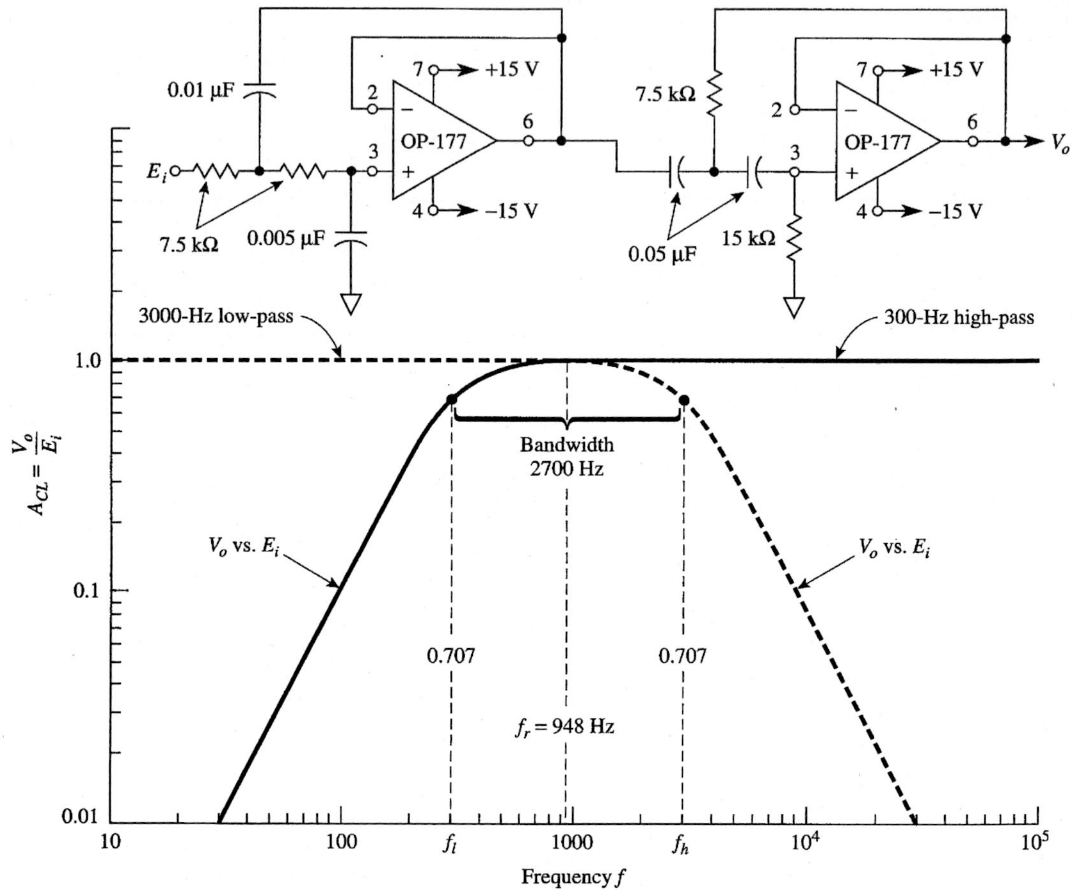

FIGURE 11-11 A 3000-Hz second-order low-pass filter is cascaded with a 300-Hz high-pass filter to form a 300- to 3000-Hz bandpass voice filter.

Examples 11-12 and 11-13. Narrow bandpass filters will be introduced in Section 11-8. Discussion of notch filters is deferred until Sections 11-9 and 11-10.

11-8 NARROWBAND BANDPASS FILTERS

Narrowband filters exhibit the typical frequency response shown in Fig. 11-12(a). The analysis and construction of narrowband filters is considerably simplified if we stipulate that the narrowband filter will have a maximum gain of 1 or 0 dB at the resonant frequency f_r. Equations (11-14) through (11-17) and bandpass terms were presented in Section 11-6. They gave an introduction to (cascaded pair) wideband filters. These equations and terms also apply to the narrowband filters that follow.

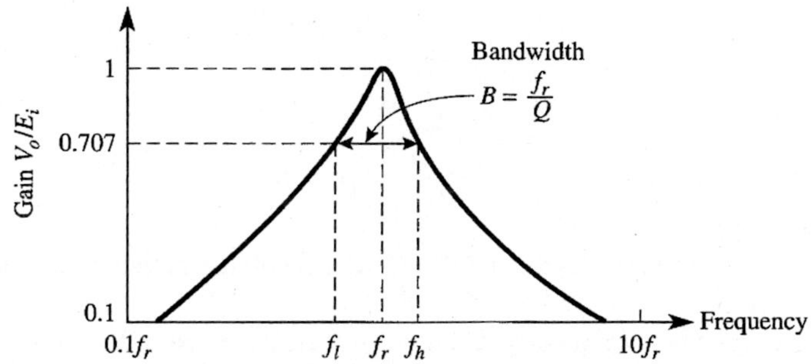

(a) Typical frequency response curve of a narrowband filter.

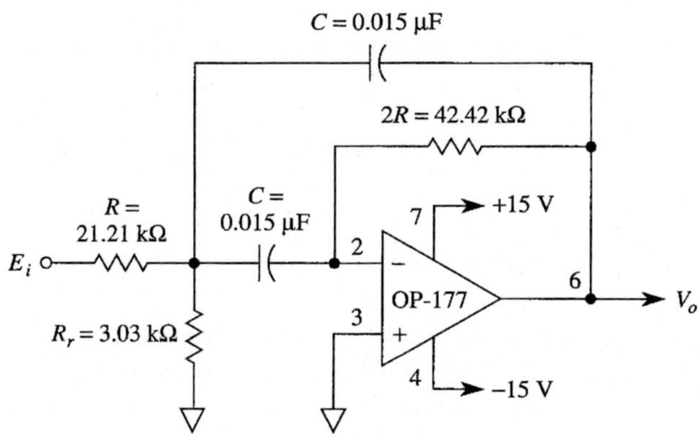

(b) Narrowband filter.

FIGURE 11-12 Narrow bandpass filter circuit and its frequency response for the component values shown; f_r = 100 Hz, B = 500 Hz, Q = 2, f_l = 780 Hz, and f_h = 1280 Hz. (a) Typical frequency response of a bandpass filter; (b) narrow bandpass filter circuit.

11-8.1 Narrowband Filter Circuit

A narrowband filter circuit uses only one op amp, as shown in Fig. 11-12. (Compare with the two-op-amp wideband filters in Fig. 11-11.) The filter's input resistance is established approximately by resistor R. If the feedback resistor ($2R$) is made two times the input resistor R, the filter's maximum gain will be 1 or 0 dB at resonant frequency f_r. By adjusting R_r one can change (or exactly trim) the resonant frequency *without changing the bandwidth or gain.*

11-8.2 Performance

The performance of the *unity*-gain narrowband filter in Fig. 11-12 is determined by only a few simple equations. The bandwidth B in hertz is determined by resistor R and the two (matched) capacitors C by

$$B = \frac{0.1591}{RC} \tag{11-18a}$$

where

$$B = \frac{f_r}{Q} \tag{11-18b}$$

Gain is a maximum of 1 at f_r provided that feedback resistor $2R$ is twice the value of input resistor R.

The resonant frequency f_r is determined by resistor R_r according to

$$R_r = \frac{R}{2Q^2 - 1} \tag{11-19}$$

If you are given component values for the circuit, its resonant frequency can be calculated from

$$f_r = \frac{0.1125}{RC} \sqrt{1 + \frac{R}{R_r}} \tag{11-20}$$

11-8.3 Stereo-Equalizer Octave Filter

A stereo equalizer has 10 bandpass filters per channel. They separate the audio spectrum from approximately 30 Hz to 16 kHz into 10 separate octaves of frequency. Each octave can then be cut or boosted with respect to the other to achieve special sound effects, equalize room response, or equalize an automotive compartment to make the radio sound like it is playing in a large hall. The construction of one such equalizer will be analyzed by an example.

Example 11-15

Octave equalizers have resonant frequencies at approximately 32, 64, 128, 250, 500, 1000, 2000, 4000, 8000, and 16,000 Hz. Q of each filter is chosen to have values between 1.4 and 2. Let's make a unity-gain narrowband filter to select the sixth octave. Specifically, make a filter with $f_r = 1000$ Hz and $Q = 2$.

Solution From Eq. (11-18b),

$$B = \frac{f_r}{Q} = \frac{1000}{2} = 500 \text{ Hz}$$

[*Note:* From Eq. (11-16), $f_l = 80$ and $f_h = 1280$ Hz.] Choose $C = 0.015\ \mu$F. Find R from Eq. (11-18a).

$$R = \frac{0.1591}{BC} = \frac{0.1591}{(500)(0.015 \times 10^{-6} \text{ F})} = 21.21 \text{ k}\Omega$$

The feedback resistor will be $2R = 42.42\ \text{k}\Omega$. Find R_r from Eq. (11-19).

$$R_r = \frac{R}{2Q^2 - 1} = \frac{21.21\ \text{k}\Omega}{2(2)^2 - 1} = \frac{21.21\ \text{k}\Omega}{7} = 3.03\ \text{k}\Omega$$

Example 11-16

Given a bandpass filter circuit with the component values in Fig. 11-12, find (a) the resonant frequency; (b) the bandwidth.

Solution (a) From Eq. (11-12),

$$f_r = \frac{0.1125}{RC}\sqrt{1 + \frac{R}{R_r}} = \frac{0.1125}{(21.21 \times 10^3)(0.015 \times 10^{-6})}\sqrt{1 + \frac{21.21\ \text{k}\Omega}{3.03\ \text{k}\Omega}}$$
$$= (353.6\ \text{Hz})\sqrt{1 + 7} = 353.6\ \text{Hz} \times 2.83 \simeq 1000\ \text{Hz}$$

(b) From Eq. (11-18a),

$$B = \frac{0.1591}{RC} = \frac{0.1591}{(21.21 \times 10^3)(0.015 \times 10^{-6})} = 500\ \text{Hz}$$

11-9 NOTCH FILTERS

11-9.1 Introduction

The notch or band-reject filter is named for the characteristic shape of its frequency–response curve in Fig. 11-13. Unwanted frequencies are attenuated in the stopband B. The desired frequencies are transmitted in the passband that lies on either side of the notch.

Notch filters usually have a passband gain of unity or 0 dB. The equations for Q, B, f_l, f_h, and f_r are identical to those of its associated bandpass filter. The reasons for this last statement are presented next.

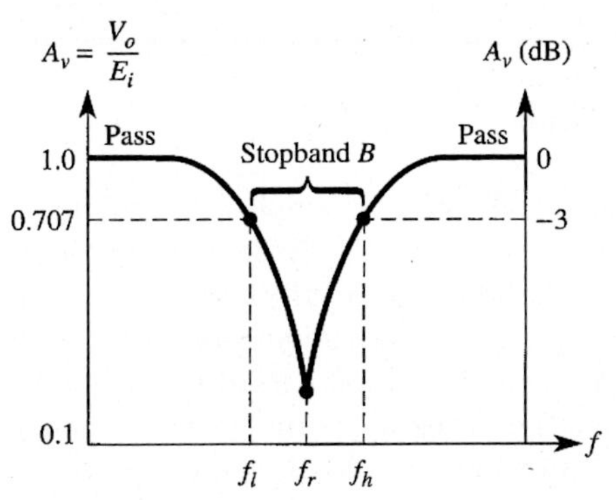

FIGURE 11-13 A notch filter transmits frequencies in the passband and rejects undesired frequencies in the stopband.

11-9.2 Notch Filter Theory

As shown in Fig. 11-14, a notch filter is made by subtracting the output of a bandpass filter from the original signal. For frequencies in the notch filter's passband, the output of the bandpass filter section approaches zero. Therefore, input E_i is transmitted via adder input resistor R_1 to drive V_o to a value equal to $-E_i$. Thus $V_o = -E_i$ in both lower and upper passbands of the notch filter.

Suppose that the frequency of E_i is adjusted to resonant frequency f_r of the narrow bandpass filter component. (*Note:* f_r of the bandpass sets the notch frequency.) E_i will exit from the bandpass as $-E_i$ and then is inverted by R_1 and R to drive V_o to $+E_i$. However, E_i is transmitted via R_2 to drive V_o to $-E_i$. Thus V_o responds to both inputs of the adder and becomes $V_o = E_i - E_i = 0$ V at f_r.

In practice, V_o approaches zero only at f_r. The depth of the notch depends on how closely the resistors and capacitors are matched in the bandpass filter and judicious fine adjustment of resistor R_1 at the inverting adder's onput. This procedure is explained in Section 11-10.3.

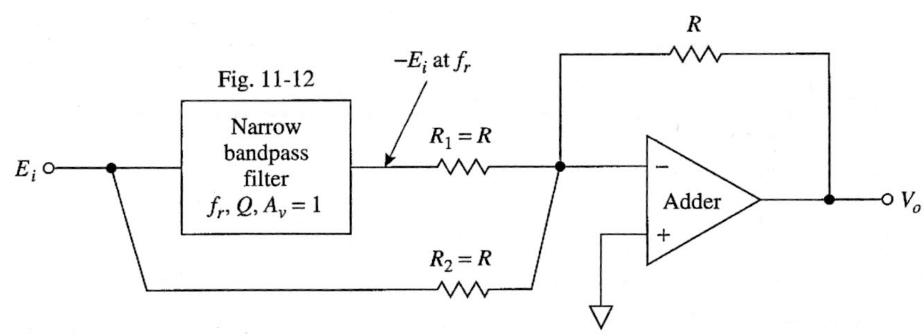

FIGURE 11-14 A notch filter is made by a circuit that subtracts the output of a bandpass filter from the original signal.

11-10 120-HZ NOTCH FILTER

11-10.1 Need for a Notch Filter

In applications where low-level signals must be amplified, there may be present one or more of an assortment of unwanted noise signals. Examples are 50-, 60-, or 400-Hz frequencies from power lines, 120-Hz ripple from full-wave rectifiers, or even higher frequencies from regulated switching-type power supplies or clock oscillators. If both signals and a signal-frequency noise component are passed through a notch filter, only the desired signals will exit from the filter. The noise frequency is "notched out." As an example, let us make a notch filter to eliminate 120-Hz hum.

11-10.2 Statement of the Problem

The problem is to make a notch filter with a notch (resonant) frequency of $f_r =$ 120 Hz. Let us select a stopband of $B = 12$ Hz. The gain of the notch filter in the passband will be unity (0 dB) so that the desired signals will be transmitted without attenuations. We use Eq. (11-17) to determine a value for Q that is required by the notch filter:

$$Q = \frac{f_r}{B} = \frac{120}{12} = 10$$

This high value of Q means that (1) the notch and component bandpass filter will have narrow bands with very sharp frequency–response curves, and (2) the bandwidth is essentially centered on the resonant frequency. Accordingly, this filter will transmit all frequencies from 0 to $(120 - 6) = 114$ Hz and all frequencies above $(120 + 6) = 126$ Hz. The notch filter will stop all frequencies between 114 and 126 Hz.

11-10.3 Procedure to Make a Notch Filter

A notch filter is made in two steps:

1. Make a bandpass filter that has the same resonant frequency, bandwidth, and consequently Q as the notch filter.
2. Connect the inverting adder of Fig. 11-15 by selecting equal resistors for R. Usually, $R = 10$ kΩ. (A practical fine-tuning procedure is presented in the next section.)

11-10.4 Bandpass Filter Components

The first step in making a 120-Hz notch filter is best illustrated by an example (see Fig. 11-15).

Design Example 11-17

Design a bandpass filter with a resonant frequency of $f_r = 120$ Hz and a bandwidth of 12 Hz so that $Q = 10$. Thus gain of the bandpass section will be 1 at f_r and approach zero at the output of the notch labeled V_o.

Solution Choose $C = 0.33$ μF. From Eq. (11-18a),

$$R = \frac{0.1591}{BC} = \frac{0.1591}{(12)(0.33 \times 10^{-6})} = 40.2 \text{ k}\Omega$$

Then the bandpass feedback resistor will be $2R$ equals 80.4 kΩ. From Eq. (11-19),

$$R_r = \frac{R}{2Q^2 - 1} = \frac{40.2\text{ k}\Omega}{2(10)^2 - 1} = \frac{40.2\text{ k}\Omega}{199} = 201\ \Omega$$

This bandpass filter component is built first and f_r is fine-tuned by adjusting R_r (see Section 11-8.2 and Fig. 11-15).

11-10.5 Final Assembly

Refer to Fig. 11-15. Simply connect an inverting adder with equal 1% input and feedback 10-kΩ resistors as shown. The resultant notch filter (from E_i to V_o) exhibits a respectable performance that is an acceptable solution to the problem. The notch depth can be increased by fine trimming R_1 or R_2.

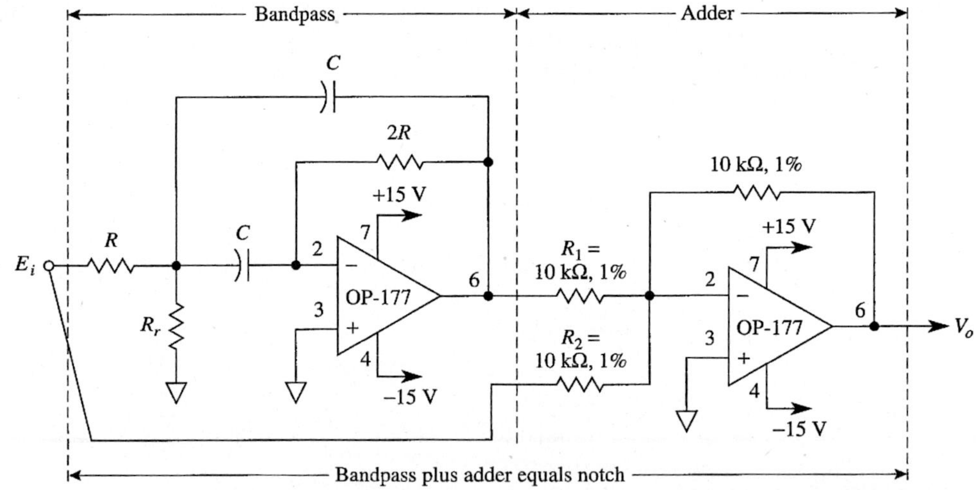

FIGURE 11-15 This two-op-amp notch filter is made from a bandpass filter plus an inverting adder. If $C = 0.33\ \mu$F, $R = 40.2$ kΩ, and $R_f = 201\ \Omega$, the notch frequency will be 120 Hz and reject a bandwidth of 12 Hz.

11-11 SIMULATION OF ACTIVE FILTER CIRCUITS USING PSPICE

We will simulate the performance of three filter circuits using PSpice: a −40 dB/decade low-pass filter, a +40 dB/decade high-pass filter, and a wide bandpass filter.

11-11.1 Low-Pass Filter

Refer to Fig. 11-4(a) and create the PSpice model of the circuit using a 741 op amp if you are using the evaluation software package. The input voltage source will be **VAC** and will be set for a 1-V magnitude. We want a plot of V_o versus frequency. To begin, place the following parts in the work area.

Draw => Get New Part

Part	Number	Library
=> **uA741**	1	**eval.slb**
=> **VAC**	1	**source.slb**
=> **VDC**	2	**source.slb**
=> **R**	4	**analog.slb**
=> **C**	2	**analog.slb**
=> **GLOBAL**	4	**port.slb**
=> **AGND**	5	**port.slb**

Note: We are using **VAC** as the input source instead of **VSIN** as we have in previous chapters. The **VAC** symbol requires only that magnitude and phase be set. The frequency range will be set in the Analysis Setup menu. Arrange the parts as shown in Fig. 11-4(a). Change the attributes of the parts to those values given in Example 11-4. Set up the **VAC** sine wave attributes by double-clicking the symbol; in the pop-up window change phase and magnitude.

ACPHASE => 0 => Save Attr
ACMAG => 1V => Save Attr => Change Display => Both name and value

Double-click on the lead from the output terminal of the op amp and label it **Vo** (see Fig. 11-16). To obtain a plot of V_o versus frequency, we must initialize the AC Sweep menu.

Analysis => Setup => Enable AC Sweep
Open AC Sweep => Decade
=> Pts/Decade => 10
=> Start Freq => 10Hz
=> End Freq => 10kHz

Save the circuit as a file with the **.SCH** extension. Run the simulation

Analysis => Simulate

In the Probe window, we need to select both Plot and Trace options from the menu bar.

Plot => Y Axis Settings => Scale => Log
Trace => Add => V[Vo]

Label the plots and obtain a printout as shown in Fig. 11-17.

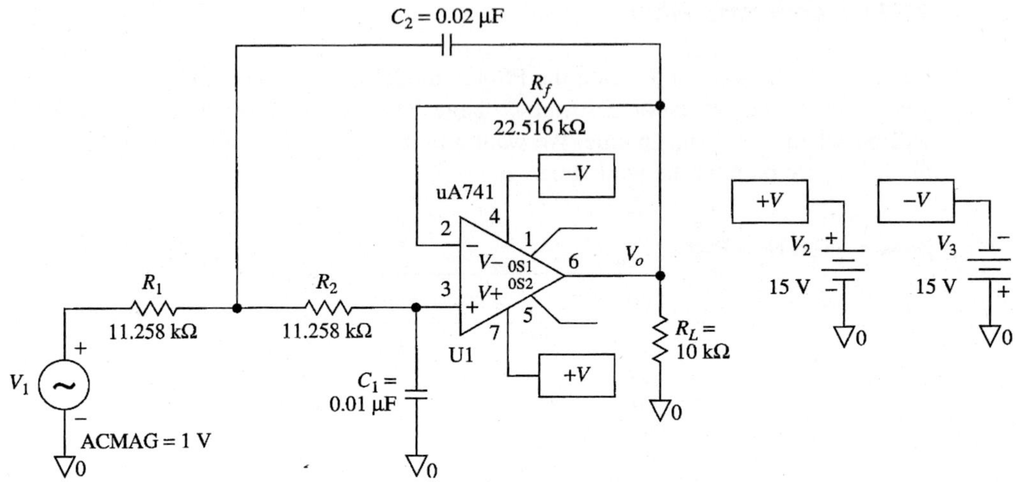

FIGURE 11-16 PSpice model of Fig. 11-4(a).

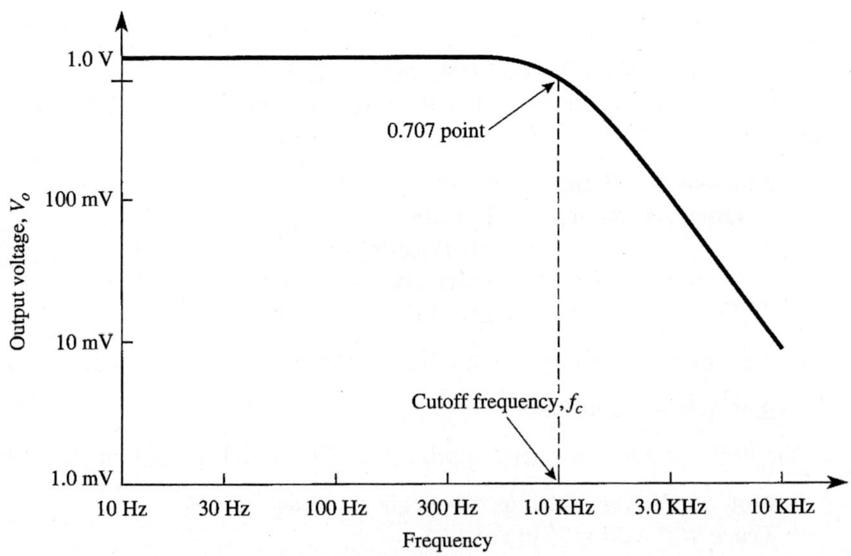

FIGURE 11-17 Frequency–response plot of a low-pass filter.

11-11.2 High-Pass Filter

The procedure for modeling and simulating a high-pass filter is similar to that for the low-pass filter previously described. Refer to Fig. 11-8(a) and create the PSpice model of the circuit using a 741 op amp. The input voltage source will be **VAC** and will be set for a 1-V magnitude. Obtain a plot of V_o versus frequency. To begin, place the following parts in the work area.

Draw => Get New Part

Part	Number	Library
=> **uA741**	1	**eval.slb**
=> **VAC**	1	**source.slb**
=> **VDC**	2	**source.slb**
=> **R**	4	**analog.slb**
=> **C**	2	**analog.slb**
=> **GLOBAL**	4	**port.slb**
=> **AGND**	5	**port.slb**

As previously mentioned we are using **VAC** as the input source instead of **VSIN** so that we may vary frequency through a range, because the **VAC** symbol requires only magnitude and phase to be set. The frequency range is set in the Analysis Setup menu. Arrange the parts as shown in Fig. 11-8(a). Change the attributes of the parts to those values given in Example 11-8. Set up the **VAC** sine wave attributes by double-clicking the symbol; in the pop-up window change phase and magnitude.

ACPHASE => 0 => Save Attr
ACMAG => 1V => Save Attr => Change Display => Both name and value

Double-click on the lead from the output terminal of the op amp and label it **Vo** (see Fig. 11-18). To obtain a plot of V_o versus frequency, we must initialize the AC Sweep menu.

Analysis => Setup => Enable AC Sweep
Open AC Sweep => Decade
=> Pts/Decade => 10
=> Start Freq => 100Hz
=> End Freq => 100kHz

Save the circuit as a file with the **.SCH** extension. Run the simulation

Analysis => Simulate

In the Probe window, we need to select both Plot and Trace options from the menu bar.

Plot => Y Axis Settings => Scale => Log
Trace => Add => V[Vo]

Label the plots and obtain a printout as shown in Fig. 11-19.

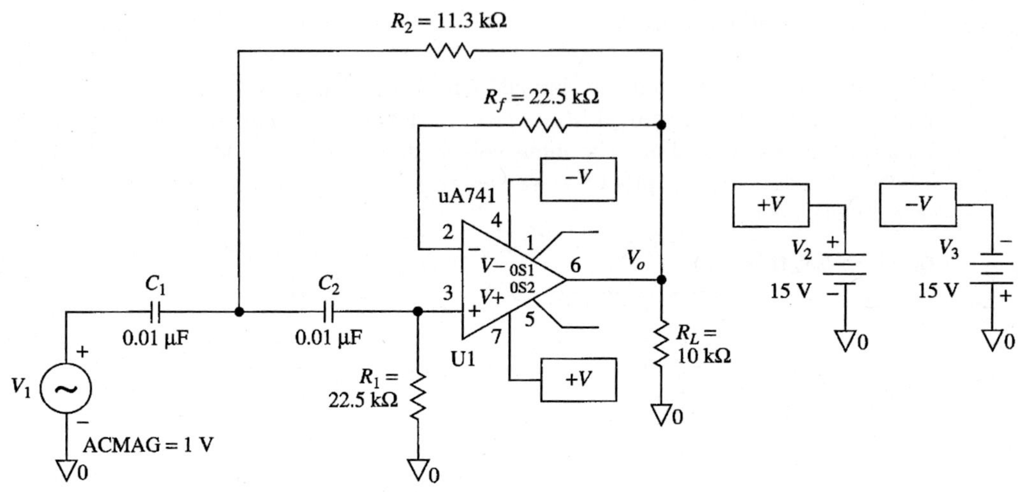

FIGURE 11-18 PSpice model of Fig. 11-8(a).

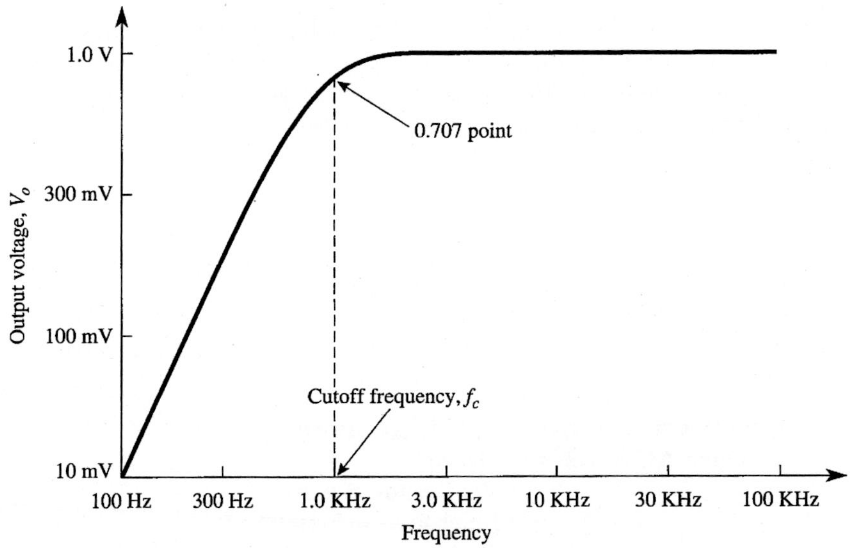

FIGURE 11-19 Frequency response of a high-pass filter.

11-11.3 Bandpass Filter

In this section, we want to model and simulate the bandpass filter shown in Fig. 11-11. This circuit is a wide bandpass filter designed by cascading a −40 dB/decade low-pass filter with a +40 dB/decade high-pass filter. Since we have already created both of these circuits in PSpice, we will create the model of the bandpass filter by copying the circuit of Fig. 11-16 and Fig. 11-18 onto a new work area, deleting some parts, changing the at-

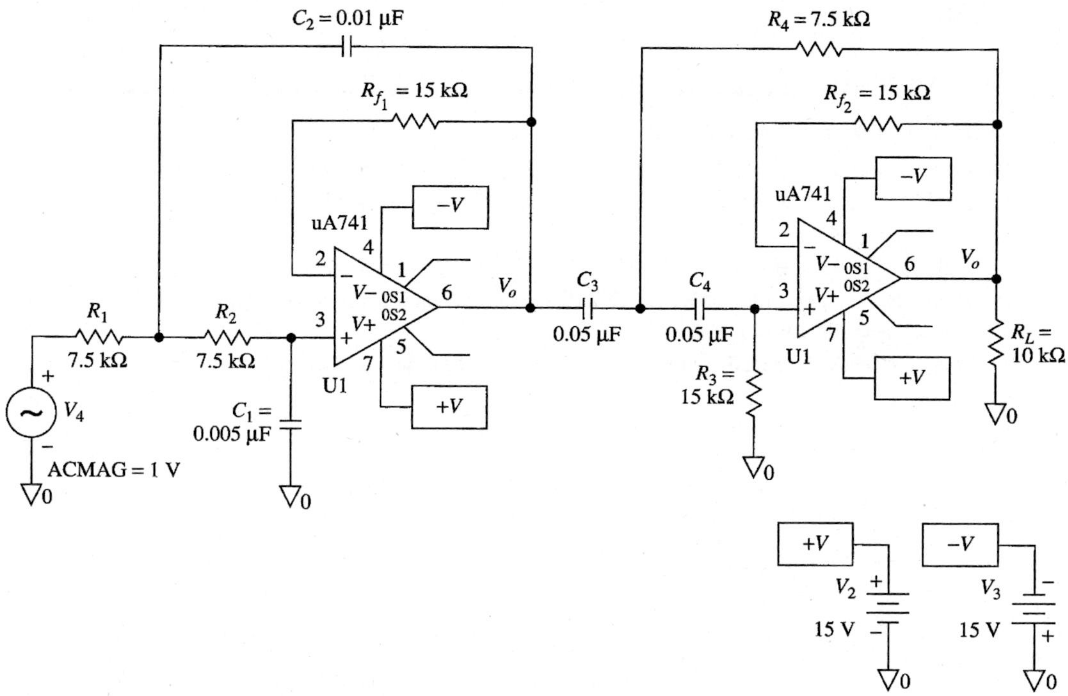

FIGURE 11-20 PSpice model of Fig. 11-11.

tributes of other parts, and saving the schematic as a new file. The model for the bandpass filter is shown in Fig. 11-20. The Analysis Setup menu has to be set as in the previous two designs. A printout from the Probe window is shown in Fig. 11-21.

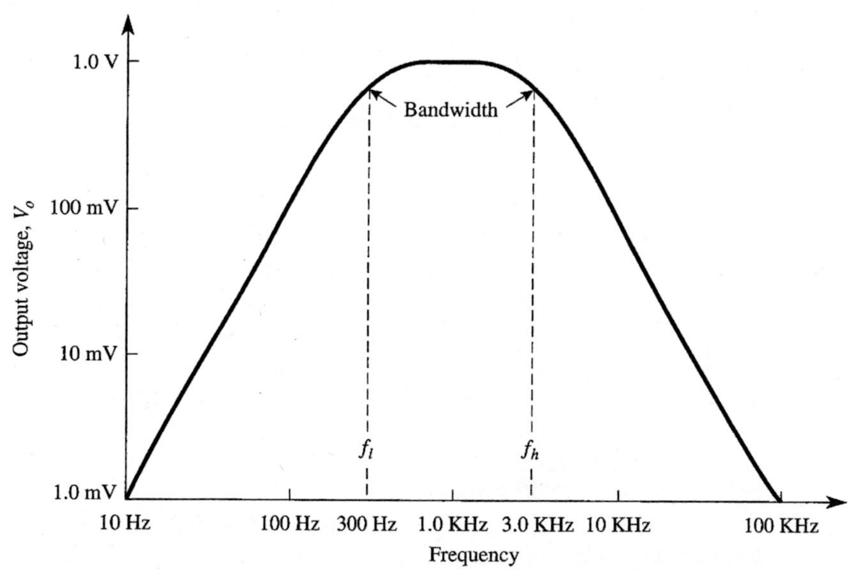

FIGURE 11-21 Frequency response of wide bandpass filter.

PROBLEMS

11-1. List the four types of filters.

11-2. What type of filter has a constant output voltage from dc up to the cutoff frequency?

11-3. What is a filter called that passes a band of frequencies while attenuating all frequencies outside the band?

11-4. In Fig. 11-2(a), if $R = 100\ \text{k}\Omega$ and $C = 0.02\ \mu\text{F}$, what is the cutoff frequency?

11-5. The low-pass filter of Fig. 11-2(a) is to be designed for a cutoff frequency of 4.5 kHz. If $C = 0.005\ \mu\text{F}$, calculate R.

11-6. Calculate the cutoff frequency for each value of C in Fig. P11-6.

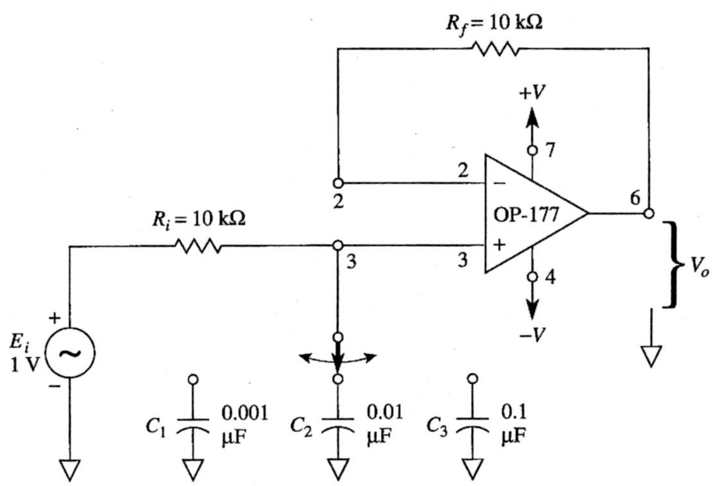

FIGURE P11-6

11-7. What are the two characteristics of a Butterworth filter?

11-8. Design a −40-dB/decade low-pass filter at a cutoff frequency of 10 krad/s. Let $C_1 = 0.02\ \mu\text{F}$.

11-9. In Fig. 11-4(a), if $R_1 = R_2 = 10\ \text{k}\Omega$, $C_1 = 0.01\ \mu\text{F}$, and $C_2 = 0.002\ \mu\text{F}$, calculate the cutoff frequency f_c.

11-10. Calculate **(a)** R_3, **(b)** R_1, and **(c)** R_2 in Fig. 11-5(a) for a cutoff frequency of 10 krad/s. Let $C_3 = 0.005\ \mu\text{F}$.

11-11. If $R_1 = R_2 = R_3 = 20\ \text{k}\Omega$, $C_1 = 0.002\ \mu\text{F}$, $C_2 = 0.008\ \mu\text{F}$, and $C_3 = 0.004\ \mu\text{F}$ in Fig. 11-5(a), determine the cutoff frequency ω_c.

11-12. In Fig. 11-5(a), $C_1 = 0.01\ \mu\text{F}$, $C_2 = 0.04\ \mu\text{F}$, and $C_3 = 0.02\ \mu\text{F}$. Calculate R for a cutoff frequency of 1 kHz.

11-13. Calculate R in Fig. 11-7(a) if $C = 0.04\ \mu\text{F}$ and $f_c = 500$ Hz.

11-14. In Fig. 11-7(a) calculate **(a)** ω_c and **(b)** f_c if $R = 10\ \text{k}\Omega$ and $C = 0.01\ \mu\text{F}$.

11-15. Design a 40-dB/decade high-pass filter for $\omega_c = 5$ krad/s. $C_1 = C_2 = 0.02\ \mu\text{F}$.

11-16. Calculate **(a)** R_1 and **(b)** R_2 in Fig. 11-8(a) for a cutoff frequency of 40 krad/s. $C_1 = C_2 = 250$ pF.

11-17. For Fig. 11-9(a), let $C_1 = C_2 = C_3 = 0.05\ \mu\text{F}$. Determine **(a)** R_3, **(b)** R_1, and **(c)** R_2 for a cutoff frequency of 500 Hz.

11-18. The circuit of Fig. 11-9(a) is designed with the values $C_1 = C_2 = C_3 = 400$ pF, $R_1 = 100\ \text{k}\Omega$, $R_2 = 25\ \text{k}\Omega$, and $R_3 = 50\ \text{k}\Omega$. Calculate the cutoff frequency f_c.

11-19. Find the **(a)** bandwidth, **(b)** resonant frequency, and **(c)** quality factor of a bandpass filter with lower and upper cutoff frequencies of 55 and 65 Hz.

11-20. A bandpass filter has a resonant frequency of 1000 Hz and a bandwidth of 2500 Hz. Find the lower and upper cutoff frequencies.

11-21. Use the capacitor and resistor values of the high-pass filter in Fig. 11-11 to prove $f_c = 3000$ Hz.

11-22. Use the capacitor and resistor values of the high-pass filter in Fig. 11-11 to prove that $f_c = 300$ Hz.

11-23. Find Q for the bandpass filter of Fig. 11-11.

11-24. Design a narrow bandpass filter using one op amp. The resonant frequency is 128 Hz and $Q = 1.5$. Select $C = 0.1\ \mu\text{F}$ in Fig. 10-12.

11-25. **(a)** How would you convert the bandpass filter of Problem 11-24 into a notch filter with the same resonant frequency and Q? **(b)** Calculate f_l and f_h for the notch filter.

CHAPTER 12

Modulating, Demodulating, and Frequency Changing with the Multiplier

LEARNING OBJECTIVES

Upon completion of this chapter on multiplier ICs, you will be able to:

- Write the output–input equation of a multiplier IC and state the value of its scale factor.
- Multiply two dc voltages or divide one dc voltage by another.
- Square the value of a dc voltage or take its square root.
- Double the frequency of any sine wave.
- Measure the phase angle between two sine waves of equal frequency.
- Show that amplitude modulation is actually a multiplication process.
- Multiply a carrier sine wave by a modulating sine wave and express the output voltage either by a product term or by a term containing sum and difference frequencies.
- Calculate the amplitude and frequency of each output frequency term.
- Make either a balanced amplitude modulator or a standard amplitude modulator.
- Show how a multiplier can be used to shift frequencies.

12-0 INTRODUCTION

Analog multipliers are arrangements of op amps and other circuit elements available as an integrated circuit. Multipliers are easy to use; some of their applications are (1) measurement of power, (2) frequency doubling and shifting, (3) detecting phase-angle difference between two signals of equal frequency, (4) multiplying two signals, (5) dividing one signal by another, (6) taking the square root of a signal, (7) squaring a signal, and (8) designing nonlinear signal conditioning circuits. Another use for multipliers is to demonstrate the principles of amplitude modulation and demodulation. The schematic of an AD633 multiplier is shown in Fig. 12-1(a). The device is a four-quadrant analog multiplier. It has high input impedance, which makes signal source loading negligible. Power supply voltages can range from ±8 V to ±18 V. No external components or user calibration are required. The output voltage is a scaled version of the x and y inputs. The *scale factor* is explained in Section 12-1.

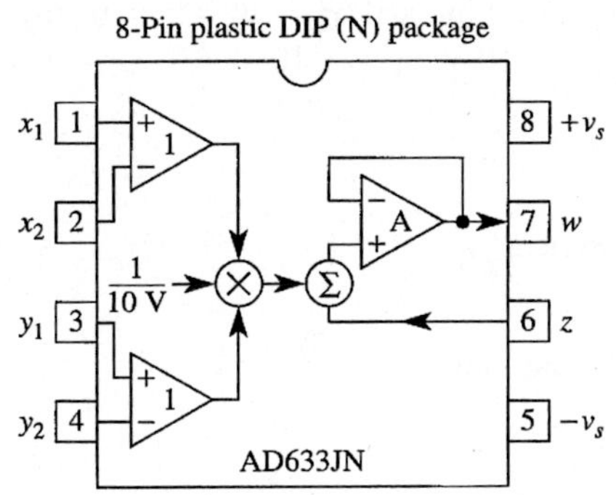

FIGURE 12-1 Introduction to the multiplier.

12-1 MULTIPLYING DC VOLTAGES

12-1.1 Multiplier Scale Factor

The 8-pin mini-DIP housing and internal schematic of the AD633 multiplier is shown in Fig. 12-1(a). In general terms, the output voltage V_o is expressed by,

$$V_o = \frac{(x_1 - x_2)(y_1 - y_2)}{10} + z \qquad (12\text{-}1)$$

where V_o is the output voltage measured at terminal W with respect to ground. The factor of $\frac{1}{10}$ is called a *scale factor* and is typical of multipliers, because multipliers are designed for the same type of power supplies used for op amps, namely ±15 V. For best results, the voltages applied to either x or y inputs should not exceed +10 V or −10 V with respect to ground. This ±10-V limit also holds for the output, so the scale factor is usually the reciprocal of the voltage limit, or 1/10 V. If both input voltages are at their positive limits of +10 V, the output will be at its positive limit of 10 V.

12-1.2 Multiplier Quadrants

Multipliers are classified by quadrants; for example, there are one-quadrant, two-quadrant, and four-quadrant multipliers. The classification is explained in two ways in Fig. 12-2. In Fig. 12-2(a), the input voltages can have four possible polarity combinations. Consider $x_2 = y_2 = z = 0$, then both x_1 and y_1 are positive and operation is in quadrant 1, since x_1 is the horizontal and y_1 the vertical axis. If x_1 is positive and y_1 is negative, quadrant 4 operation results, and so forth.

Example 12-1

Find V_o for the following combination of inputs if $x_2 = y_2 = z = 0$: (a) $x_1 = 10$ V, $y_1 = 10$ V; (b) $x_1 = -10$ V, $y_1 = 10$ V; (c) $x_1 = 10$ V, $y_1 = -10$ V; (d) $x_1 = -10$ V, $y_1 = -10$ V.

Solution From Eq. (12-1b),

$$\text{(a) } V_o = \frac{(10)(10)}{10} = 10 \text{ V, quadrant 1}$$

$$\text{(b) } V_o = \frac{(-10)(10)}{10} = -10 \text{ V, quadrant 2}$$

$$\text{(c) } V_o = \frac{(10)(-10)}{10} = -10 \text{ V, quadrant 4}$$

$$\text{(d) } V_o = \frac{(-10)(-10)}{10} = 10 \text{ V, quadrant 3}$$

In Fig. 12-2(b), V_o is plotted on the vertical axis and x_1 on the horizontal axis. If we apply 10 V to the y input and vary x from -10 V to $+10$ V, we plot the line *ab,* labeled $y = 10$ V. If y_1 is changed to -10 V, the line *cd,* labeled $y_1 = -10$ V, results. These lines can be seen on an oscilloscope by connecting V_o of the multiplier to the y_1 input of the oscilloscope and x_1 of the multiplier to the $+x_1$ input of the oscilloscope. For accuracy, V_o should be 0 V when either multiplier input is 0 V. Multiplier errors are primarily due to input and output offsets, scale factor error, and/or nonlinearity of the core multiplying unit. These errors are only fractions of a percent error, and if they need to be eliminated refer to the manufacturer's data sheet.

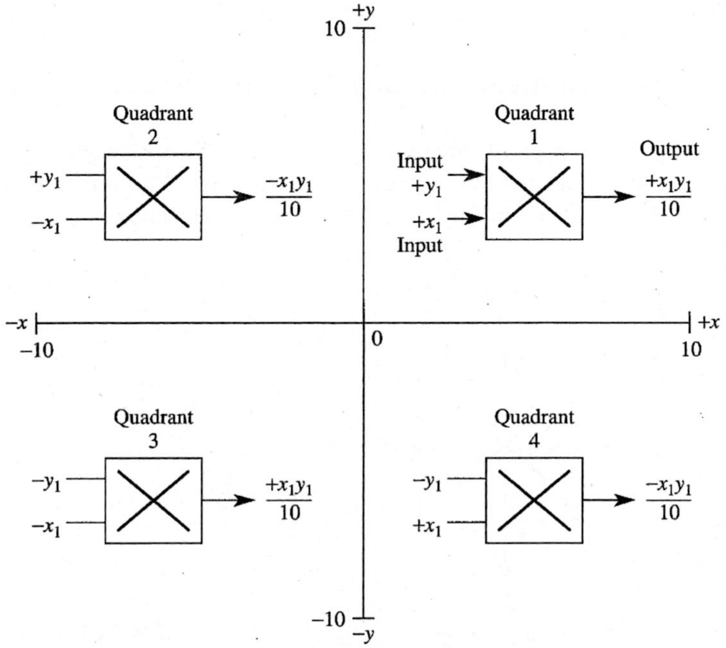

(a) y_1 vs. x_1 plot shows location of input operating point in one of four quadrants ($x_2 = y_2 = z = 0$).

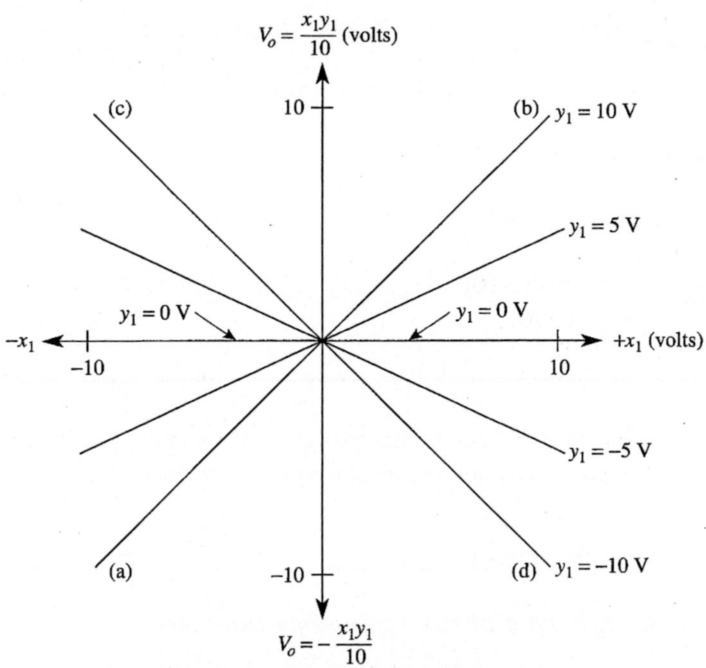

(b) Multiplier output $x_1y_1/10$ versus input x_1.

FIGURE 12-2 Multiplying two dc voltages, x_1 and y_1.

12-2 SQUARING A NUMBER OR DC VOLTAGE

Any positive or negative number can be squared by a multiplier, providing that the number can be represented by a voltage between 0 and 10 V. Simply connect the voltage E_i to *both* inputs as shown in Fig. 12-3. This type of connection is known as a *squaring circuit.*

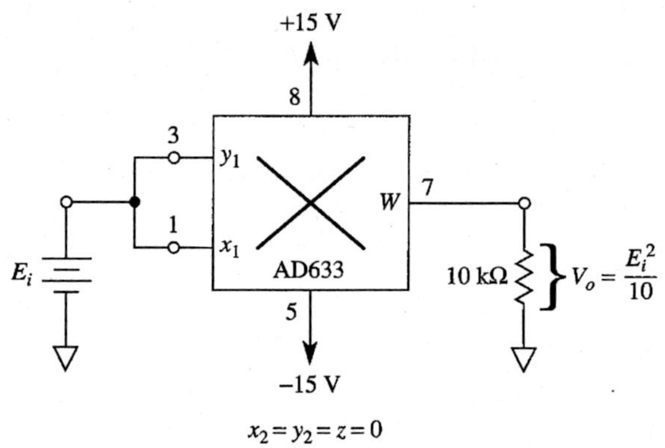

FIGURE 12-3 Squaring a dc voltage.

Example 12-2

Find V_o in Fig. 12-4 if (a) $E_i = +10$ V; (b) $E_i = -10$ V.

Solution From Fig. 12-4,

$$\text{(a) } V_o = \frac{10^2}{10} = 10 \text{ V.}$$

$$\text{(b) } V_o = \frac{(-10)(-10)}{10} = 10 \text{ V.}$$

Example 12-2 shows that the output of the multiplier follows the rules of algebra; that is, when either a positive or negative number is squared, the result is a positive number.

12-3 FREQUENCY DOUBLING

12-3.1 Principle of the Frequency Doubler

An ideal sinusoidal-wave frequency doubler would give an output voltage whose frequency is twice the frequency of the input voltage. The doubler circuit should not incorporate a tuned circuit, since the tuned circuit can be tuned only to one frequency. A true doubler should double any frequency. The multiplier is very nearly an ideal doubler if only one frequency is applied to both inputs. The output voltage for a doubler circuit is given by the trigonometric identity*

$$(\sin 2\pi ft)^2 = \frac{1}{2} - \frac{\cos 2\pi(2f)t}{2} \qquad (12\text{-}2)$$

Equation (12-2) predicts that squaring a sine wave with a frequency of (for example) $f = 10$ kHz gives a negative cosine wave with a frequency of $2f$ or 20 kHz plus a dc term of $\frac{1}{2}$. Note that *any* input frequency f will be doubled when passed through a squaring circuit.

12-3.2 Squaring a Sinusoidal Voltage

In Fig. 12-4(a), sine-wave voltage E_i is applied to both multiplier inputs x_1 and y_1. E_i has a peak value of 5 V and a frequency of 10 kHz. The output voltage V_o is predicted by the calculations shown in Example 12-3.

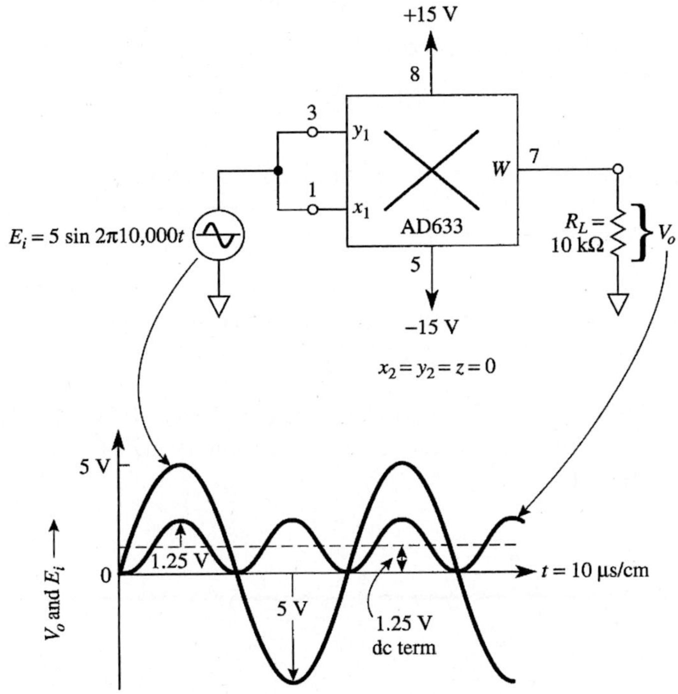

FIGURE 12-4 Squaring circuit as a frequency doubler.

*Equation (12-2) is a special case of the general equation shown in math tables:

$$(\sin A)(\sin B) = \tfrac{1}{2}[\cos(A - B) - \cos(A + B)]$$

Here $A = B = 2\pi ft$.

Example 12-3

Calculate V_o in the squaring circuit or frequency doubler of Fig. 12-4. Let $x_2 = y_2 = z = 0$.

Solution Input $E_{x_1} = E_{y_1} = E_i$ and is expressed in volts by

$$E_i = E_{x_1} = E_{y_1} = 5 \sin 2\pi 10{,}000t$$

Substituting into Eq. (12-1) yields

$$V_o = \frac{E_i^2}{10} = \frac{5^2}{10}(\sin 2\pi 10{,}000t)^2 \tag{12-3}$$

Applying Eq. (12-2), we obtain

$$V_o = 2.5\left[\frac{1}{2} - \frac{\cos 2\pi 20{,}000t}{2}\right] \text{V}$$

$$= \underbrace{1.25}_{} \quad \underbrace{-1.25 \cos 2\pi 20{,}000t}_{}$$

$$= \text{dc term of } 1.25\text{ V} - \text{frequency doubled to } 20{,}000\text{ Hz, } 1.25\text{-V peak}$$

Both E_i and V_o are shown in Fig. 12-4. If you want to remove the dc voltage, simply install a 1-μF coupling capacitor between R_L and the output terminal. If you want to measure the dc term simply connect a dc voltmeter across V_o.

Conclusion. V_o has two voltage components: (1) a dc voltage equal to $\frac{1}{20}(E_{ip})^2$, and (2) an ac sinusoidal wave whose peak value is $\frac{1}{20}(E_{ip})^2$ and whose frequency is double that of E_i.

Example 12-4

What are the dc and ac output voltage components of Fig. 12-4 if (a) E_i = 10-V peak at 1 kHz; (b) E_i = 2-V peak at 2.5 kHz?

Solution (a) dc value = $(10)^2/20$ = 5 V; peak ac value = $(10)^2/20$ = 5 V at 2 kHz. (b) dc value = $(2)^2/20$ = 0.2 V; peak ac value = $(2)^2/20$ = 0.2 V at 5 kHz.

12-4 PHASE-ANGLE DETECTION

12-4.1 Basic Theory

If two sine waves of the *same* frequency are applied to the multiplier inputs in Fig. 12-5(a), the output voltage V_o has a dc voltage component and an ac component whose frequency is twice that of the input frequency. This conclusion was developed in Section 12-3.2. The dc voltage is actually proportional to the *difference* in phase angle θ between E_{x_1} and E_{y_1}. For example, in Fig. 12-5, $\theta = 0°$, because there was no phase difference between E_{x_1} and E_{y_1}. Figure 12-5(b) shows two sine waves of identical frequency but a phase difference of 90°; therefore, $\theta = 90°$.

If one input sine wave differs in phase angle from the other, it is possible to calculate or measure the phase-angle difference from *the dc voltage component in V_o*. This dc component V_o dc is given by*

$$V_o \text{ dc} = \frac{E_{xp} E_{yp}}{20} (\cos \theta) \tag{12-4a}$$

where E_{xp} and E_{yp} are peak amplitudes of E_{x_1} and E_{y_1}. For example, if $E_{xp} = 10$ V, $E_{yp} = 5$ V, *and they are in phase,* then V_o dc would indicate 2.5 V on a dc voltmeter. This voltmeter point would be marked as a phase angle of 0° ($\cos 0° = 1$). If $\theta = 45°$ ($\cos 45° = 0.707$), the dc meter would read $0.707 \times 2.5 \text{ V} \simeq 1.75$ V. Our dc voltmeter can be calibrated as a *phase-angle meter* 0° at 2.5 V, 45° at 1.75 V, and 90° at 0 V.

Equation (12-4a) may also be expressed by*

$$\cos \theta = \frac{20V_o \text{ dc}}{E_{xp} E_{yp}} \tag{12-4b}$$

If we could arrange for the product $E_{xp} E_{yp}$ to equal 20, we could use a 0-to-1-V dc voltmeter to read $\cos \theta$ directly from the meter face and calibrate the meter face in degrees from a cosine table. That is, Eq. (12-4b) reduces to

$$V_o \text{ dc} = \cos \theta \qquad \text{for } E_{xp} = E_{yp} = 4.47 \text{ V} \tag{12-4c}$$

This point is explored further in Section 12-4.2.

*Trigonometric identity:

$$\sin A \sin B = \tfrac{1}{2}[\cos (A - B) - \cos (A + B)]$$

For equal frequencies, different phase angle:

$$A = 2\pi ft + \theta \text{ for } E_x, \qquad B = 2\pi ft \text{ for } E_y$$

Therefore,

$$[\sin (2\pi ft + \theta)][\sin 2\pi ft] = \tfrac{1}{2}[\cos \theta - \cos (4\pi ft + \theta)]$$
$$= \tfrac{1}{2}(\text{dc} - \text{double frequency term})$$

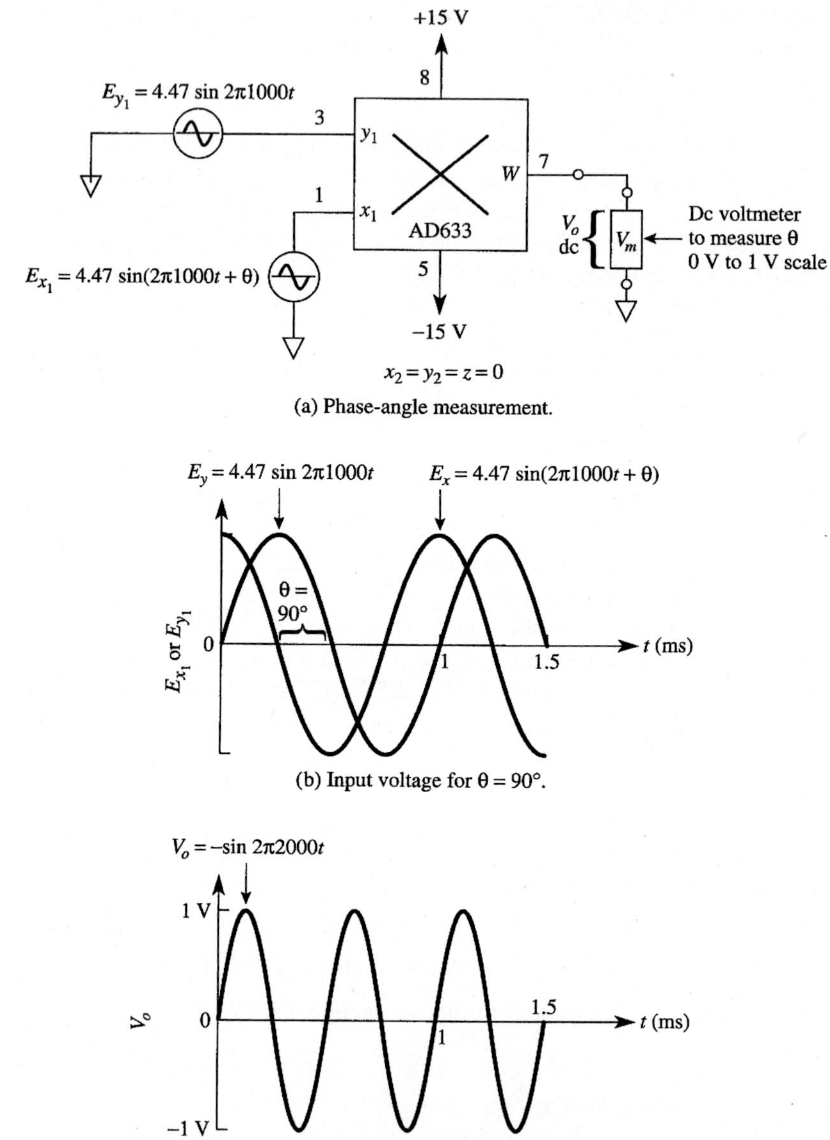

(a) Phase-angle measurement.

(b) Input voltage for θ = 90°.

(c) Output voltage for θ = 90°; dc term is 0 V.

FIGURE 12-5 Multiplier used to measure the phase-angle difference between two equal frequencies.

Example 12-5

In Fig. 12-5, $E_{xp} = E_{yp} = 5$ V and the dc component of V_o is 1.25 V from Eq. (12-4b). Prove that there is 0° phase angle between E_{x_1} and E_{y_1} (since they are the same voltage).

Solution From Eq. (12-4b),

$$\frac{20 \times 1.25}{5 \times 5} = \cos\theta = \frac{25}{25} = 1$$

Since $\cos 0° = 1$, $\theta = 0°$.

12-4.2 Phase-Angle Meter

Equation (12-4b) points the way to making a phase-angle meter. Assume that the peak values of E_{x_1} and E_{y_1} in Fig. 12-5(a) are scaled to 4.47 V by amplifiers or voltage dividers. Then a dc voltmeter is connected as shown in Fig. 12-5(a) to measure just the dc voltage component. The meter face then can be calibrated directly in degrees. The procedure is developed further in Examples 12-6 and 12-7.

Example 12-6

The average value of V_o in Fig. 12-5(c) is 0, so the dc component of $V_o = 0$ V. Calculate θ.

Solution From Eq. (12-4b), $\cos\theta = 0$, so $\theta = \pm 90°$. Note that the dc component of V_o cannot distinguish between a lead phase angle (+) or a lag phase angle (−).

Example 12-7

Calculate $V_{o\ dc}$ for phase angles of (a) $\theta = \pm 30°$; (b) $\theta = \pm 45°$; (c) $\theta = \pm 60°$.

Solution From a trigonometric table, obtain the $\cos\theta$ and apply Eq. (12-4a).

θ (deg)	cos θ	$V_{o\ dc}$ (V)
±30	0.866	0.866
±45	0.707	0.707
±60	0.500	0.500
± 0	1.000	1.000
±90	0.000	0.000

(The last two rows of this table come from Examples 12-5 and 12-6.)

The 0-to-1-V voltmeter scale can now be calibrated in degrees, 0 V for a 90° phase angle and 1.0 V for 0° phase angle. At 0.866 V, θ = 30°, and so forth. The phase-angle meter does not indicate whether θ is a leading or lagging phase angle but only the phase difference between E_{x_1} and E_{y_1}.

12-4.3 Phase Angles Greater Than ±90°

The cosine of phase angles greater than +90° or −90° is a negative value. Therefore, V_o will be negative. This extends the capability of the phase angle meter in Example 12-7.

Example 12-8

Calculate $V_{o\ dc}$ for phase angles of (a) $\theta = \pm 90°$; (b) $\theta = \pm 120°$; (c) $\theta = \pm 135°$; (d) $\theta = \pm 150°$; (e) $\theta = \pm 180°$.

Solution Using Eq. (12-4a) and tabulating results, we have

θ	±90°	±120°	±135°	±150°	±180°
$V_{o\ dc}$	0 V	−0.5 V	−0.70 V	−0.866 V	−1 V

From the results of Examples 12-7 and 12-8, a ±1-V voltmeter can be calibrated to read from 0 to ±180°.

12-5 ANALOG DIVIDER

An analog divider gives the ratio of two signals or provides gain control. It is constructed as shown in Fig. 12-6 by inserting a multiplier in the feedback loop of an op amp. Since the op amp's (−) input draws negligible current, the same value of current I flows through resistors R. Therefore, the output voltage of the multiplier V_m is equal in magnitude but opposite in polarity (with respect to ground) to E_{in} or

$$E_{in} = -V_m \tag{12-5a}$$

But V_m is also equal to one-tenth (scale factor) of the product of input E_x and output of the op amp V_o. Substituting for V_m yields

$$E_{in} = \frac{V_o E_x}{10} \tag{12-5b}$$

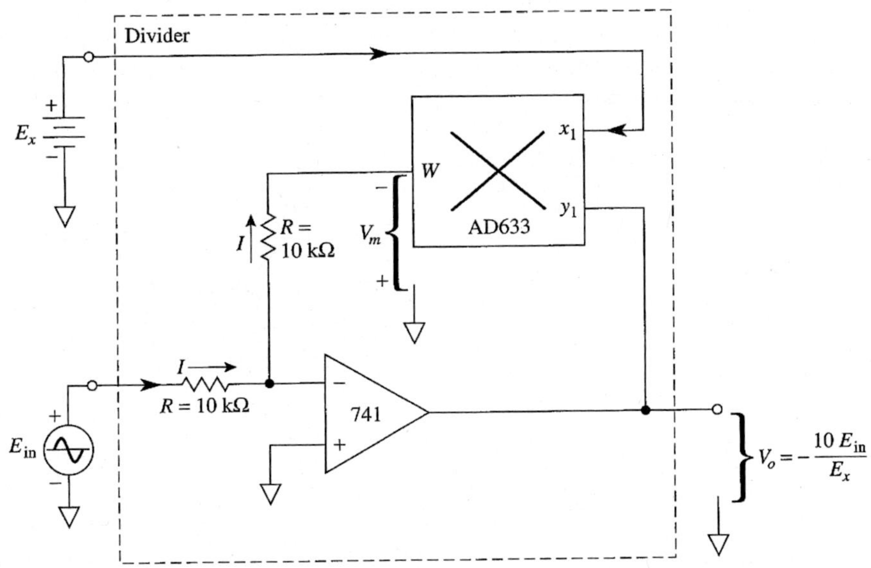

FIGURE 12-6 Division with an op amp and a multiplier ($x_2 = y_2 = z = 0$).

Solving for V_o, we obtain

$$V_o = \frac{10E_{in}}{E_x} \tag{12-5c}$$

Equation (12-5c) shows that the divider's output V_o is proportional to the ratio of inputs E_{in} and E_x. E_x should never be allowed to go to 0 V or to a negative voltage, because the op amp will saturate. E_{in} can be positive, negative, or 0 V. Note that the divider can be viewed as a voltage gain $10/E_x$ acting on E_{in}. So if E_x is changed, the gain will change. This voltage control of the gain is useful in automatic gain-control circuits.

12-6 FINDING SQUARE ROOTS

A divider can be made to find square roots by connecting both inputs of the multiplier to the output of the op amp (see Fig. 12-7). Equation (12-5a) also pertains to Fig. 12-7. But now V_m is one-tenth (scale factor) of $V_o \times V_o$ or

$$-E_{in} = V_m = \frac{V_o^2}{10} \tag{12-6a}$$

Solving for V_o (eliminate $\sqrt{-1}$) yields

$$V_o = \sqrt{10\,|E_{in}|} \tag{12-6b}$$

Equation (12-6b) states that V_o equals the square root of 10 times the *magnitude* of E_{in}. E_{in} must be a negative voltage, or else the op amp saturates. The range of E_{in} is between -1 and -10 V. Voltages smaller than -1 V will cause inaccuracies. The diode prevents $(-)$ saturation for positive E_{in}. If E_{in} has positive values, reverse the diode.

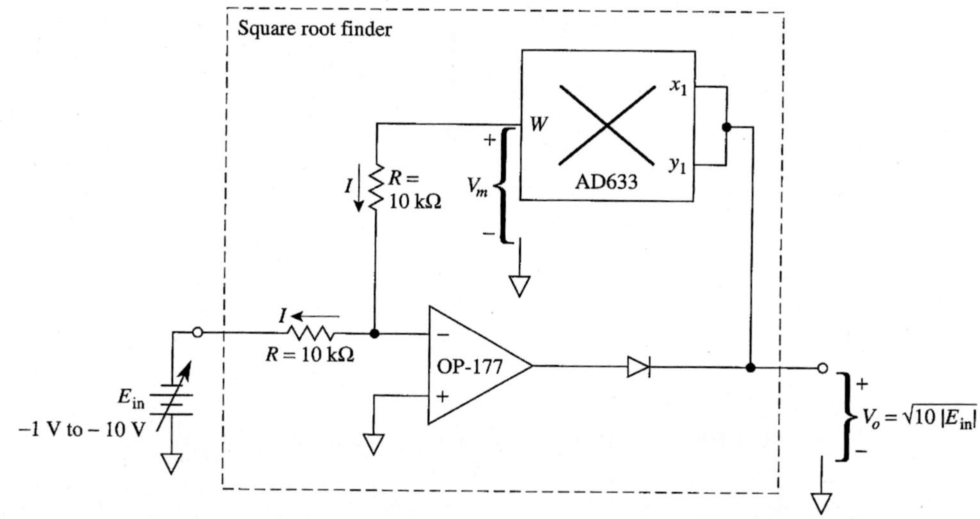

FIGURE 12-7 Square rooting with an op amp and a multiplier ($x_2 = y_2 = z = 0$).

12-7 INTRODUCTION TO AMPLITUDE MODULATION

12-7.1 Need for Amplitude Modulation

Low-frequency audio or data signals cannot be transmitted from antennas of reasonable size. Audio signals can be transmitted by changing or *modulating* some characteristic of a higher-frequency *carrier* wave. If the amplitude of the carrier wave is changed in proportion to the audio signal, the process is called *amplitude modulation* (AM). Changing the frequency or the phase angle of the carrier wave results in *frequency modulation* (FM) and *phase-angle modulation* (PM), respectively.

Of course, the original audio signal must eventually be recovered by a process called *demodulation* or detection. The remainder of this section concentrates on using the multiplier for amplitude modulation. ("Modulate" is from the ancient Greek language meaning to "change." Curiously, it is the Latin prefix "de" that converts the meaning to "change back.")

12-7.2 Defining Amplitude Modulation

The introduction to amplitude modulation begins with the amplifier in Fig. 12-8(a). The input voltage E_c is amplified by a constant gain A. Amplifier output V_o is the product gain of A and E_c. Now suppose that the amplifier's gain is varied. This concept is represented by an arrow through A in Fig. 12-8(b). Assume that A is varied from 0 to a maximum and back to 0 as shown in Fig. 12-8(b) by the plot of A versus t. This means that the amplifier multiplies the input voltage E_c by a different value (gain) over a period of time. V_o is now the amplitude of input E_c varied or multiplied by an amplitude of A. This process is an example of amplitude modulation, and the output voltage V_o is called the *amplitude modulated signal.* Therefore, to obtain an amplitude-modulated signal (V_o), the amplitude of a high-frequency carrier signal (E_c) is varied by an intelligence or data signal A.

12-7.3 The Multiplier Used as a Modulator

From Section 12-7.2 and Fig. 12-8(b), V_o equals E_c multiplied by A. Therefore, amplitude modulation is a *multiplication process.* As shown in Fig. 12-8(c), E_c is applied to a multiplier's x input. E_m [having the same shape as A in Fig. 12-8(b)] is applied to the multiplier's y input. E_c is multiplied by a voltage that varies from 0 through a maximum and back to 0. So V_o has the same envelope as E_m. The multiplier can be considered a *voltage-controlled gain device* as well as an amplitude modulator. The waveshape shown is that of a *balanced modulator.* The reason for this name will be given in Section 12-8.3.

Note carefully in Fig. 12-8(c) that V_o is not a sine wave; that is, the peak values of successive half-cycles are different. This principle is used in Section 12-12 to show how a *frequency-shifter* (*heterodyne*) circuit works, but first, we examine amplitude modulation in greater detail.

12-7.4 Mathematics of a Balanced Modulator

A high-frequency sinusoidal *carrier wave* E_c is applied to one input of a multiplier. A lower-frequency audio or data signal is applied to the second input of a modulator and will be called the *modulating wave,* E_m. For test and analysis, both E_c and E_m will be sine waves described as follows.

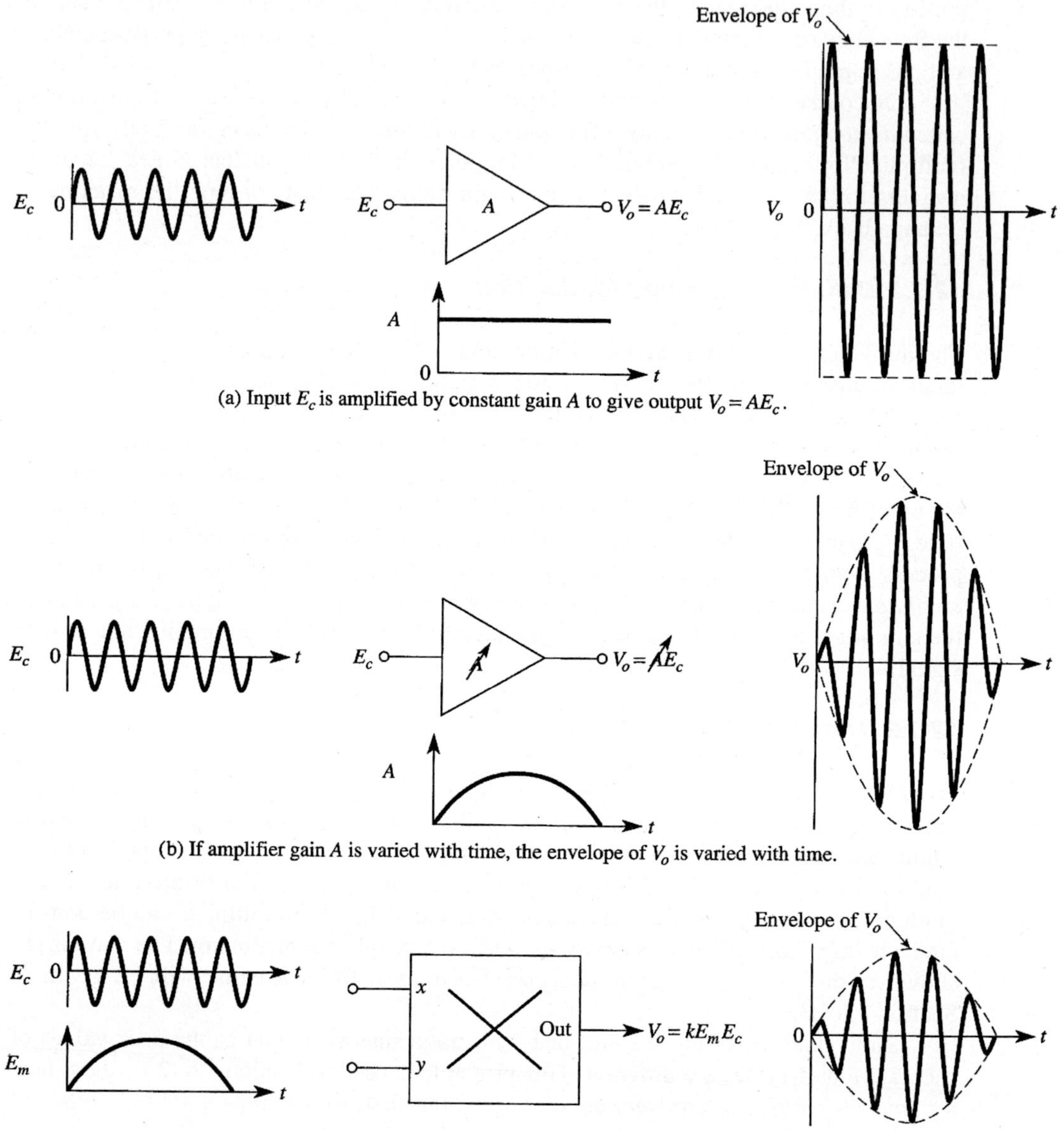

(a) Input E_c is amplified by constant gain A to give output $V_o = AE_c$.

(b) If amplifier gain A is varied with time, the envelope of V_o is varied with time.

(c) If E_m varies as A in part (b), then V_o has the same general shape as in part (b).

FIGURE 12-8 Introduction to modulation.

Carrier wave, E_c:

$$E_c = E_{cp} \sin 2\pi f_c t \tag{12-7a}$$

where E_{cp} is the peak value of the carrier wave and f_c is the carrier frequency.

Modulating wave, E_m:

$$E_m = E_{mp} \sin 2\pi f_m t \tag{12-7b}$$

where E_{mp} is the peak value of the modulating wave and f_m is the modulating frequency.

Now let the carrier voltage E_c be applied to the x input of a multiplier as E_x, and let the modulating voltage E_m be applied to the y input of a multiplier as E_y. The multiplier's output voltage V_o is expressed as a *product term* from Eq. (12-16) as

$$V_o = \frac{E_m\,E_c}{10} = \frac{E_{mp}\,E_{cp}}{10}(\sin 2\pi f_m t)(\sin 2\pi f_c t) \tag{12-8}$$

Equation (12-8) is called the *product term,* because it represents the product of two sine waves with different frequencies. However, it is not in the form used by ham radio operators or communications personnel. They prefer the form obtained by applying to Eq. (12-8) the trigonometric identity

$$(\sin A)(\sin B) = \tfrac{1}{2}[\cos (A - B) - \cos (A + B)] \tag{12-9}$$

Substituting Eq. (12-9) into Eq. (12-8), where $A = E_c$ and $B = E_m$, we have

$$V_o = \frac{E_{mp}\,E_{cp}}{20}\cos 2\pi(f_c - f_m)t - \frac{E_{mp}\,E_{cp}}{20}\cos 2\pi(f_c + f_m)t \tag{12-10}$$

Equation (12-10) is analyzed in Section 12-7.5.

12-7.5 Sum and Difference Frequencies

Recall from Section 12-7.3 that E_c is a sine wave and E_m is a sine wave, but no part of V_o is a sine wave. V_o in Fig. 12-8(c) is expressed mathematically by either Eq. (12-8) or (12-10). But Eq. (12-10) shows that V_o is made up of *two* cosine waves with frequencies *different* from either E_m or E_c. They are the *sum frequency* $f_c + f_m$ and the *difference frequency* $f_c - f_m$. The sum and difference frequencies are evaluated in Example 12-9.

Example 12-9

In Fig. 12-9, carrier signal E_c has a peak voltage of $E_{cp} = 5$ V and a frequency of $f_c =$ 10,000 Hz. The modulating signal E_m has a peak voltage of $E_m = 5$ V and a frequency of $f_m = 1000$ Hz. Calculate the peak voltage and frequency of (a) the sum frequency; (b) the difference frequency.

Solution From Eq. (12-10), the peak value of both sum and difference voltages is

$$\frac{E_{mp}\,E_{cp}}{20} = \frac{5\text{ V} \times 5\text{ V}}{20} = 1.25\text{ V}$$

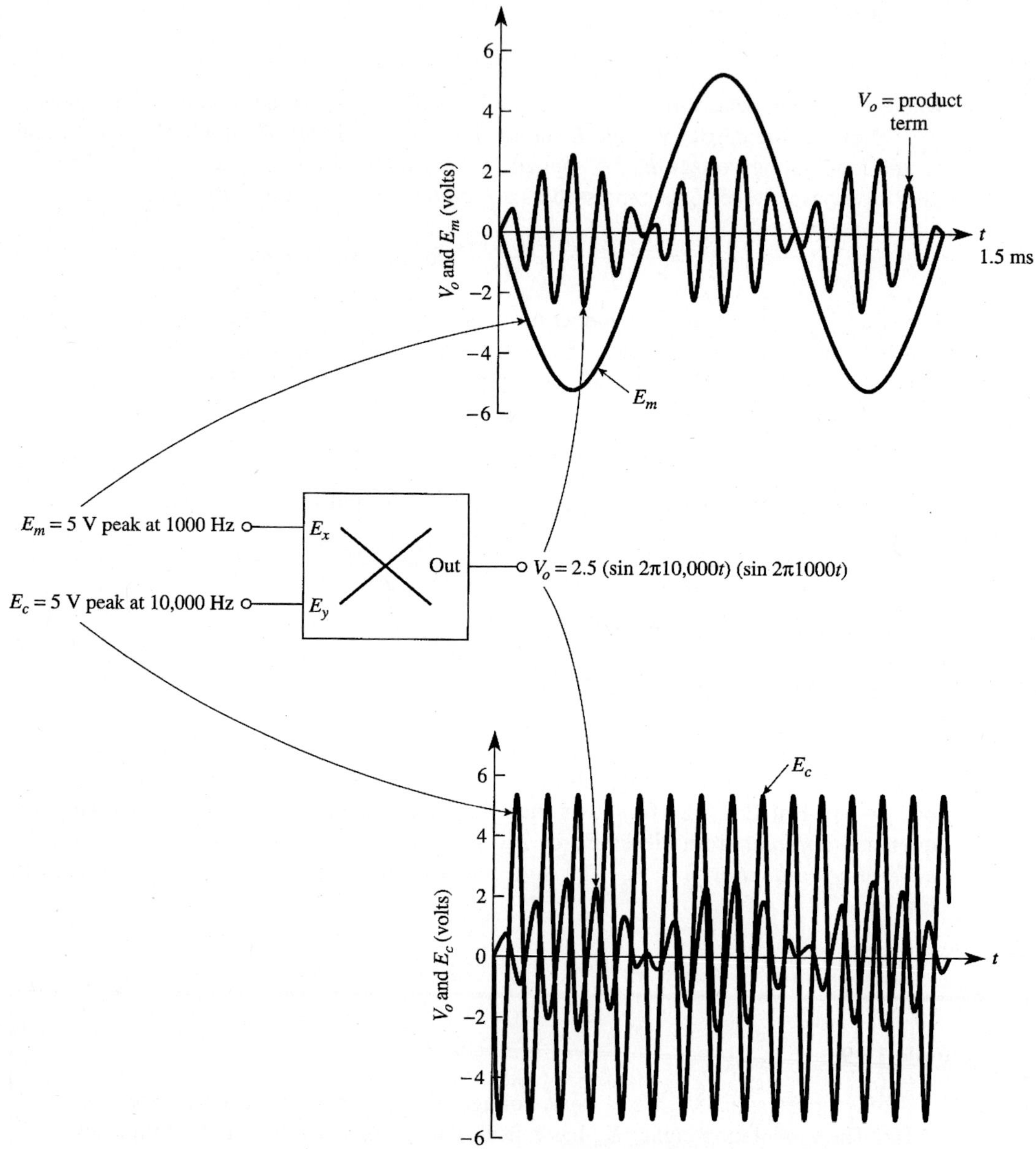

FIGURE 12-9 The multiplier as a balanced modulator.

The sum frequency is $f_c + f_m = 10{,}000\text{ Hz} + 1000\text{ Hz} = 11{,}000\text{ Hz}$; the difference frequency is $f_c - f_m = 10{,}000\text{ Hz} - 1000\text{ Hz} = 9000\text{ Hz}$. Thus V_o is made up of the difference of two cosine waves:

$$V_o = 1.25\cos 2\pi 9000t - 1.25\cos 2\pi 11{,}000t$$

This result can be verified by connecting a wave or spectrum analyzer to the multiplier's output; a 1.25-V deflection occurs at 11,000 Hz and at 9000 Hz. The original input signals of 1 kHz and 10 kHz do *not* exist at the output.

An oscilloscope can be used to show input and output voltages of the multiplier of Example 12-9. The product term for V_o is found from Eq. (12-8):

$$V_o = \underbrace{2.5\text{ V}}\underbrace{(\sin 2\pi 10{,}000t)}\underbrace{(\sin 2\pi 1000t)}$$

$$= 2.5 \times \quad E_c \quad \times \quad E_m$$

V_o is shown with E_m in the top drawing and with E_c in the bottom drawing of Fig. 12-9. Observe that E_m and E_c have peak voltages of 5 V. The peak value of V_o is 2.5 V. Note that the upper and lower envelopes of V_o are *not* the same shape as E_m. Therefore, we *cannot* rectify and filter V_o to recover E_m. This characteristic distinguishes the balance modulator.

12-7.6 Side Frequencies and Sidebands

Another way of displaying the output of a modulator is by a graph showing the peak amplitude as a vertical line for each frequency. The resulting *frequency spectrum* is shown in Fig. 12-10(a). The sum and difference frequencies in V_o are called *upper* and *lower side* frequencies because they are above and below the carrier frequency on the graph. When more than one modulating signal is applied to the modulator (y input) in Fig. 12-9, each generates a sum and difference frequency in the output. Thus, there will be two side frequencies for each y input frequency, placed symmetrically on either side of the carrier. If the expected range of modulating frequencies is known, the resulting range of side frequencies can be predicted. For example, if the modulating frequencies range between 1 and 4 kHz, the lower side frequencies fall in a band between $(10 - 4)\text{ kHz} = 6\text{ kHz}$ and $(10 - 1)\text{ kHz} = 9\text{ kHz}$. The band between 6 and 9 kHz is called the *lower sideband.* For this same example, the *upper sideband* ranges from $(10 + 1)\text{ kHz} = 11\text{ kHz}$ to $(10 + 4)\text{ kHz} = 14\text{ kHz}$. Both upper and lower sidebands are shown in Fig. 12-10(b).

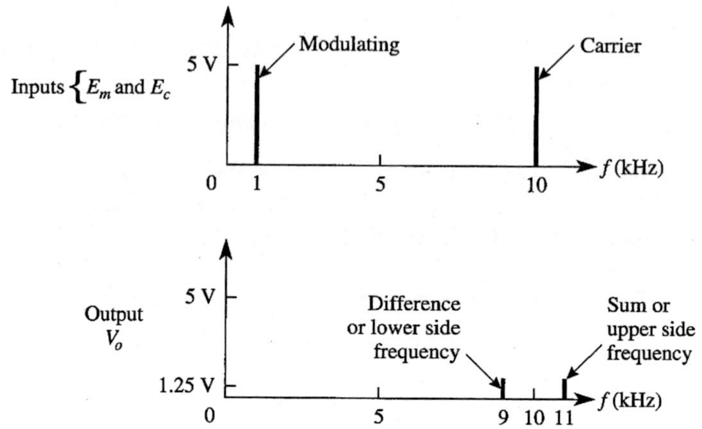

(a) Frequency spectrum for $f_c = 10$ kHz and $f_m = 1$ kHz in Example 12-9.

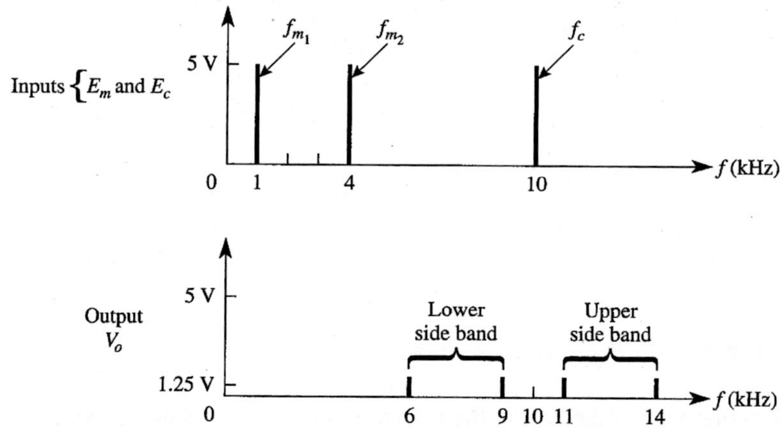

(b) Frequency spectrum for $f_c = 10$ kHz and $f_{m_1} = 1$ kHz, $f_{m_2} = 4$ kHz.

FIGURE 12-10 Frequency spectrum for a balanced modulator.

12-8 STANDARD AMPLITUDE MODULATION

12-8.1 Amplitude Modulator Circuit

The circuits of Section 12-7 multiplied the carrier and modulating signals to generate a balanced output that is expressed either as (1) a product term, or (2) a sum and difference frequency. The term *balanced modulator* originated in the days of vacuum tube–transformer technology. It was very difficult to "balance out" the carrier. With today's modern multipliers, the absence of carrier E_c in the output is a zero-cost bonus. The classical or standard amplitude modulator (AM) adds the carrier term to the output. The AM car radio uses standard AM. One way of adding the carrier term to generate a standard AM output is shown in Fig. 12-11(a). The modulating signal is fed into one input of an adder. A dc voltage equal to the peak value of the carrier voltage E_{cp} is fed into the other input. The output of the adder is then fed into the y input of a multiplier, as shown in Fig. 12-11(b). The carrier signal is fed into the x input. The multiplier multiplies E_x by E_y, and its output voltage is the standard AM voltage given by either of the following equations:

$$V_o = \begin{cases} \dfrac{E^2_{cp}}{10} \sin 2\pi f_c t & \text{(carrier term)} \\ \qquad + \\ \dfrac{E_{cp}\,E_{mp}}{10} (\sin 2\pi f_c t)(\sin 2\,\pi f_m t) & \text{(product term)} \end{cases} \tag{12-11}$$

or

$$V_o = \begin{cases} \dfrac{E^2_{cp}}{10} \sin 2\pi f_c t & \text{(carrier term)} \\ \qquad + \\ \dfrac{E_{cp}\,E_{mp}}{20} \cos 2\pi(f_c - f_m)t & \text{(lower side frequency)} \\ \qquad - \\ \dfrac{E_{cp}\,E_{mp}}{20} \cos 2\pi(f_c + f_m)t & \text{(upper side frequency)} \end{cases} \tag{12-12}$$

The output voltage V_o is shown in Fig. 12-11(b). The voltage levels are worked out in the following example.

Example 12-10

In Fig. 12-11, $E_{cp} = E_{mp} = 5$ V. The carrier frequency $f_c = 10$ kHz, and the modulating frequency is $f_m = 1$ kHz. Evaluate the peak amplitudes of the output carrier and product terms.

Solution From Eq. (12-11), the carrier term peak voltage is

$$\frac{(5\text{ V})(5\text{ V})}{10} = 2.5\text{ V}$$

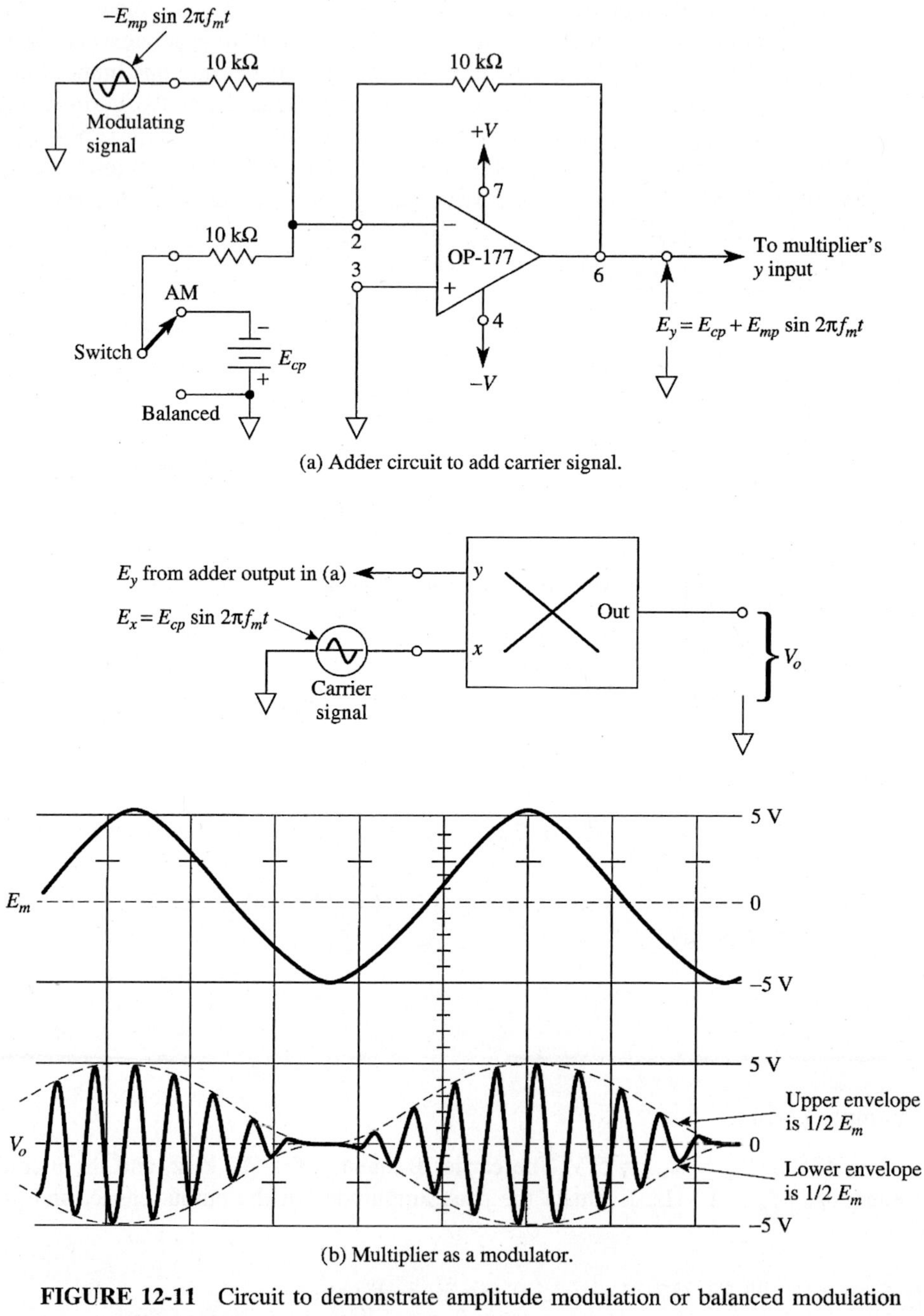

(a) Adder circuit to add carrier signal.

(b) Multiplier as a modulator.

FIGURE 12-11 Circuit to demonstrate amplitude modulation or balanced modulation (see also Fig. 12-13).

The product term peak voltage is

$$\frac{(5\text{ V})(5\text{ V})}{10} = 2.5\text{ V}$$

The side frequencies peak voltages are

$$\frac{(5\text{ V})(5\text{ V})}{20} = 1.25\text{ V}$$

The waveshape of V_o is shown in Fig. 12-11(b). Observe that the upper and lower envelopes of V_o are the same shape as E_m. This is characteristic of a standard amplitude modulator (AM), *not* of the balanced modulator. It allows easy recovery of audio signal E_m by a half-wave rectifier and suitable filter capacitor.

12-8.2 Frequency Spectrum of a Standard AM Modulator

The signal frequencies present in V_o for the standard AM output of Fig. 12-11 are found from Eq. (12-11). Using the voltage values in Example 12-10, we have

$$\text{Carrier term} = 2.5\text{ V peak at } 10{,}000\text{ Hz}$$

$$\text{Lower side frequency} = 1.25\text{ V peak at } 9000\text{ Hz}$$

$$\text{Upper side frequency} = 1.25\text{ V peak at } 11{,}000\text{ Hz}$$

These frequencies are plotted in Fig. 12-12 and should be compared with the balanced modulator of Fig. 12-10.

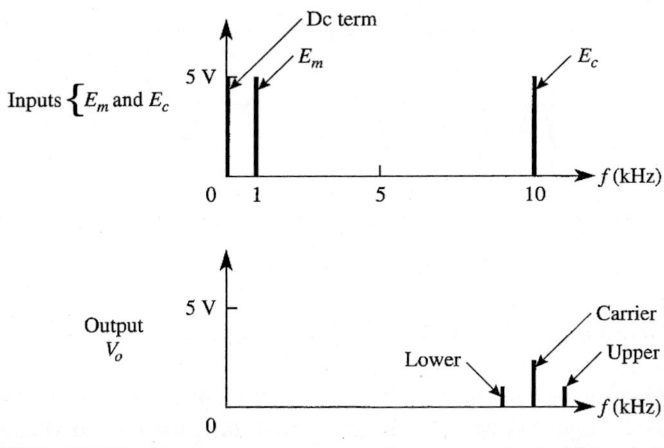

FIGURE 12-12 Frequency spectrum for a standard AM modulator, $f_c = 10$ kHz, $f_m = 1$ kHz.

12-8.3 Comparison of Standard AM Modulators and Balanced Modulators

If the switch in Fig. 12-11(a) is positioned to AM, V_o will contain three frequencies—f_c, $f_c + f_m$, and $f_c - f_m$—carrier, plus sum and difference frequencies. Observe that the envelopes of V_o have the same shape as the intelligence signal E_m. This observation can be used to recover E_m from the AM signal, as stated in Example 12-10. Note that if there is no signal frequency, the station still transmits carrier f_c. Radio receivers use this fact to activate signal-strength meters, tuning lights, and automatic volume control (AVC).

If the switch in Fig. 12-11(a) is positioned to "Balanced," V_o will contain only the product term with only two frequencies, $f_c + f_m$ and $f_c - f_m$. The envelope of V_o does *not* follow E_m. Since V_o does *not* contain f_c, this type of modulation is called *balanced modulation* in the sense that the carrier has been balanced out. It is also called *suppressed carrier modulation,* since the carrier is suppressed in the output. If no modulating frequency is present, the radio station does not transmit. This is a good system for clandestine operation. For a comparison of balanced and standard AM modulation, both outputs are shown together in Fig. 12-13.

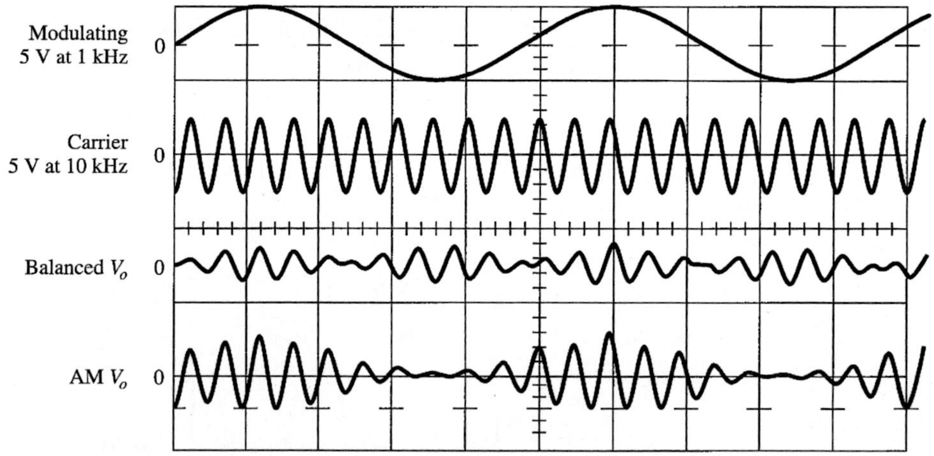

FIGURE 12-13 Comparison of balanced modulation and standard AM from Fig. 12-11.

12-9 DEMODULATING AN AM VOLTAGE

Demodulation, or *detection,* is the process of recovering a modulating signal E_m from the modulated output voltage V_o. To accomplish this, the AM modulated wave is applied to the *y* input of a multiplier as shown in Fig. 12-14. Each *y* input frequency is multiplied by the *x* input carrier frequency and generates a sum and difference frequency as shown

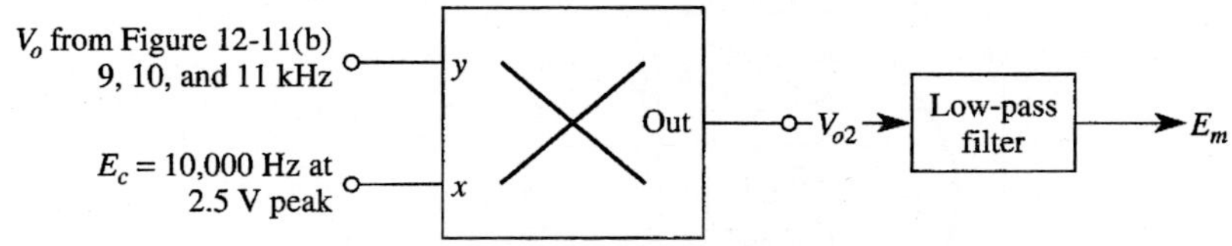

(a) Multiplier used as a demodualtor.

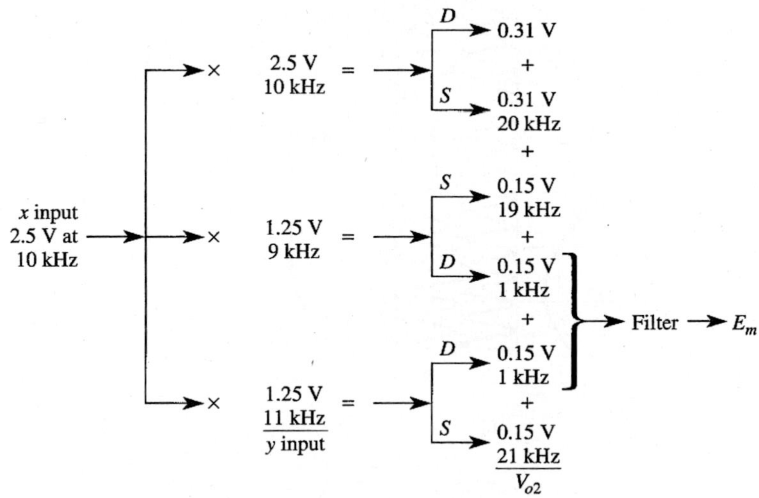

(b) Frequency and peak amplitude of signal components at x input, y input, multiplier output, and filter.

FIGURE 12-14 The demodulator is a multiplier plus a low-pass filter.

in Fig. 12-14(b). Since only the 1-kHz frequency is the modulating signal, use a low-pass filter to extract E_m. Thus the demodulator is simply a multiplier with the carrier frequency applied to one input, and the AM signal to be demodulated is fed into the other input. The multiplier's output is fed into a low-pass filter whose output is the original modulating data signal E_m. Thus a multiplier plus a low-pass filter and carrier signal equals a demodulator.

Waveshapes at inputs and outputs of both the AM modulator and demodulator are shown in Fig. 12-15. Note the unusual shape of V_{o_2} because it contains six components [detailed in Fig. 12-14(b)].

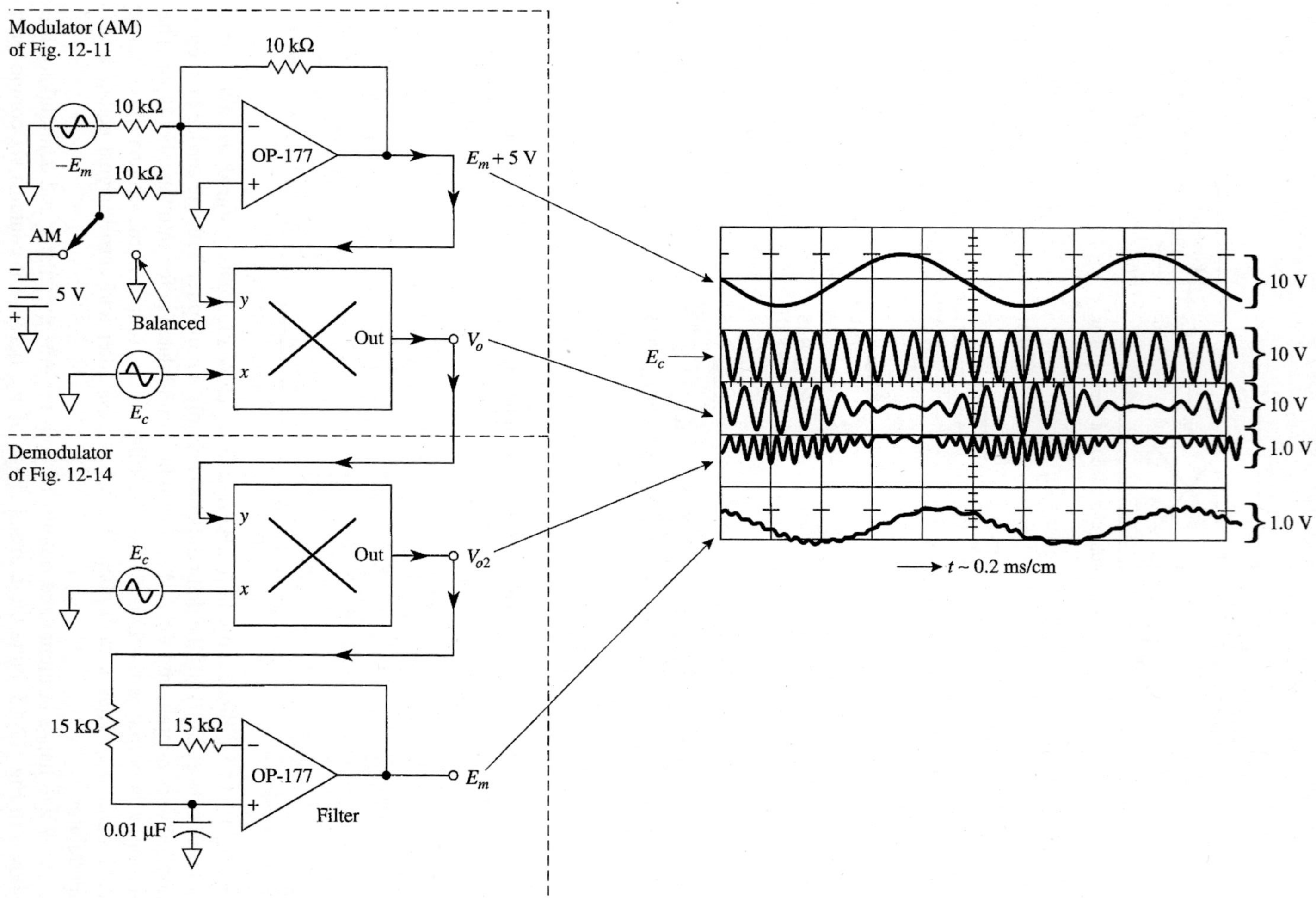

FIGURE 12-15 Voltage waveforms in an amplitude modulator and demodulator ($f_c = 10$ kHz, $f_m = 1$ kHz).

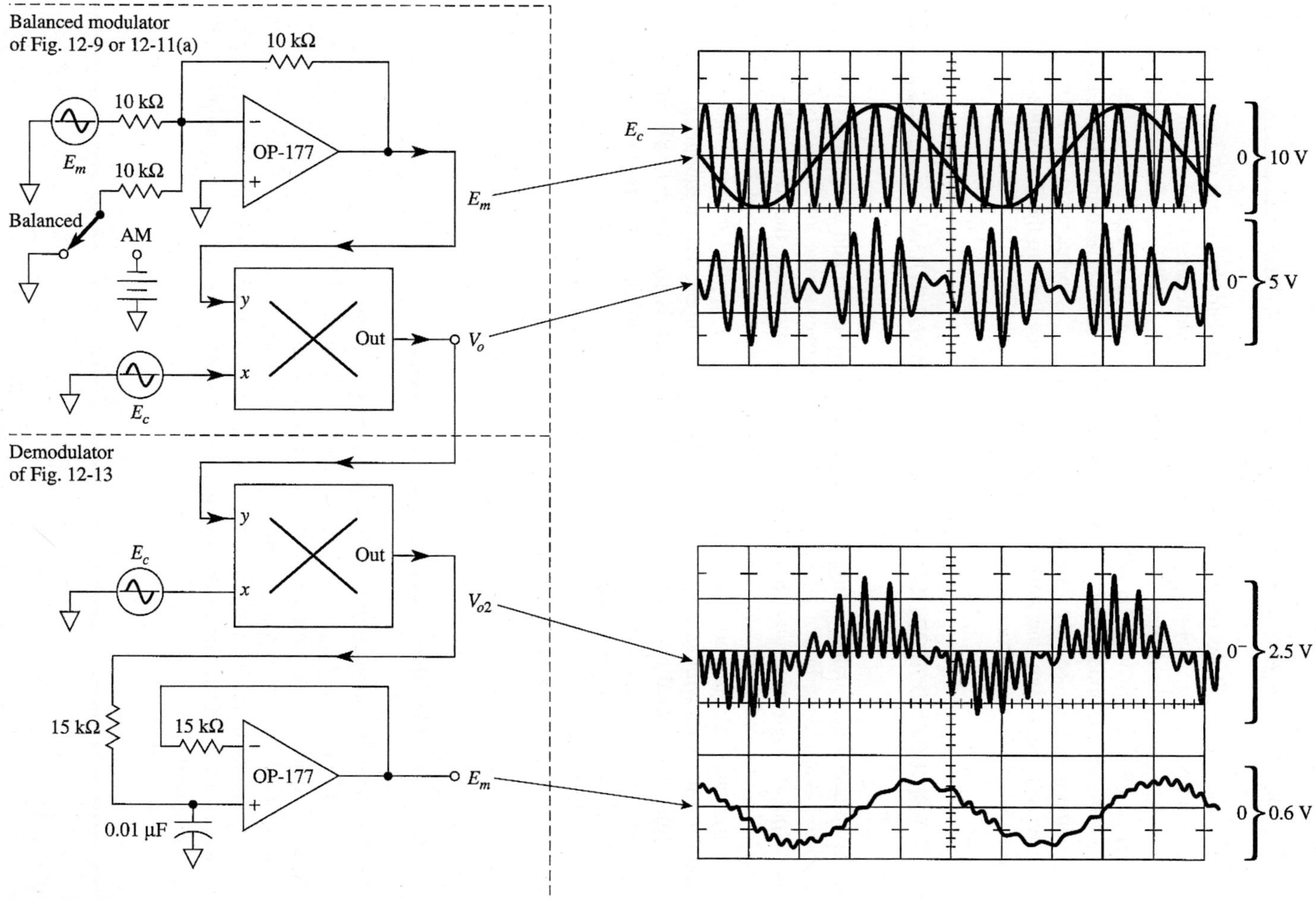

FIGURE 12-16 Demonstration of balanced modulator and demodulator with waveshapes.

12-10 DEMODULATING A BALANCED MODULATOR VOLTAGE

Modulating signal E_m is recovered from a balanced modulator by means of the same technique employed in Fig. 12-14 and Section 12-9. The only difference is the absent carrier frequency of 10 kHz at the demodulator's *y* input. This missing 10 kHz also eliminates both the dc and 20-kHz term in V_{o_2}. The circuit arrangement in Fig. 12-16 was built to demonstrate the demodulating technique and show the resulting waveshapes. The demodulated E_m is not a pure sine wave because only a simple filter was used. If f_c is increased to 100 kHz, E_m will be closer to being a pure sine wave. The carrier's frequency fed into the demodulator should be *exactly* equal to the carrier frequency driving the modulator.

12-11 SINGLE-SIDEBAND MODULATION AND DEMODULATION

In the balanced modulator of Figs. 12-9 and 12-10, we could add a high-pass filter (see Chapter 11) to the modulator's output. If the filter removes all the lower side frequencies, the output is *single sideband* (SSB). If the filter only attenuates the lower side frequencies (to leave a *vestige* of the lower sideband), we have a *vestigial* sideband modulator.

Assume that only one modulating frequency f_m is applied to our single-sideband modulator together with carrier f_c. Its output is a single, upper side frequency $f_c + f_m$. To demodulate this signal and recover f_m, all we have to do is connect the SSB signal $f_c + f_m$ to one multiplier input and f_c to the other input. According to the principles set forth in Section 12-7.4, the demodulator's output would have a sum frequency of $(f_c + f_m) + f_c$ and a difference frequency of $(f_c + f_m) - f_c = f_m$. A low-pass filter would recover the modulating signal f_m and easily eliminate the high-frequency signal, whose frequency is $2f_c + f_m$.

12-12 FREQUENCY SHIFTING

In radio communication circuits, it is often necessary to shift a carrier frequency f_c with its accompanying side frequencies down to a lower intermediate frequency f_{IF}. This shift of each frequency is accomplished with the multiplier connections of Fig. 12-17(a). The modulated carrier signals are applied to the *y* input. A local oscillator is adjusted to a frequency f_o equal to the sum of the carrier and desired intermediate frequency and applied to the *x* input. The frequencies present in the output of the multiplier are calculated in the following example.

Example 12-11

In Fig. 12-17(a) amplitudes and frequencies of an AM modulated wave are present at each input as follows:

y input:

Peak amplitude (V)	Frequency (kHz)
1	$(f_c + f_m) = 1005$
4	$f_c = 1000$
1	$(f_c - f_m) = 995$

where f_c is the broadcasting station's carrier frequency and $(f_c + f_m)$ and $(f_c - f_m)$ are the upper and lower side frequencies due to a 5-kHz modulating frequency, and 1000-kHz carrier frequency.

x input: The local oscillator is set for a 5-V-peak sine wave at 1445 kHz, because the desired intermediate frequency is 455 kHz.

Find the peak value and frequency of each signal component in the output of the multiplier.

Solution From Eq. (12-12), the peak amplitude of each *y* input frequency is multiplied by the peak amplitude of the local oscillator frequency. This product is multiplied by $\frac{1}{20}$ ($\frac{1}{10}$ for the scale factor $\times \frac{1}{2}$ from the trigonometric identity) to obtain the peak amplitude of the resulting sum and difference frequencies at the multiplier's output. The results are tabulated in Fig. 12-17.

All frequencies present in the multiplier's output are plotted on the frequency spectrum of Fig. 12-17(c). A low-pass filter or band-pass filter is used to pass only the three lower intermediate frequencies of 450, 455, and 460 kHz. The upper intermediate frequencies of 2450, 2455, and 2460 kHz may be used if desired, but they are usually filtered out.

We conclude from Example 12-11 that each frequency present at the *y* input is shifted down and up to new intermediate frequencies. The lower set of intermediate frequencies can be extracted by a filter. Thus, the information contained in the carrier f_c has been preserved and shifted to another subcarrier or intermediate frequency. The process of frequency shifting is also called *heterodyne.* The heterodyne principle will be used in Section 12-13 to construct a universal AM receiver that will demodulate standard AM, and balance modulator and single-sideband signals.

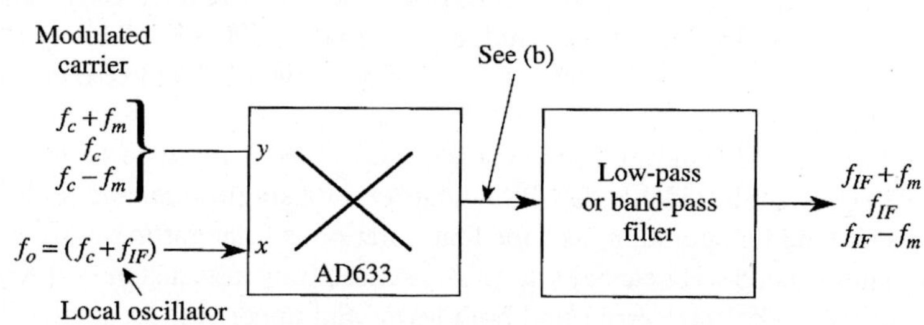

(a) Circuit for a frequency shifter.

FIGURE 12-17 The multiplier as a frequency shifter.

Frequency at y input (kHz)	Multiplier output	
	Peak (V)	Frequency (kHz)
1005	$\frac{1 \times 5}{20} = 0.25$	1455 + 1005 = 2460 1455 − 1005 = 450
1000	$\frac{4 \times 5}{20} = 1.0$	1455 + 1000 = 2455 1455 − 1000 = 455
995	$\frac{1 \times 5}{20} = 0.25$	1455 + 995 = 2450 1455 − 995 = 460

(b) Frequencies present in multiplier output.

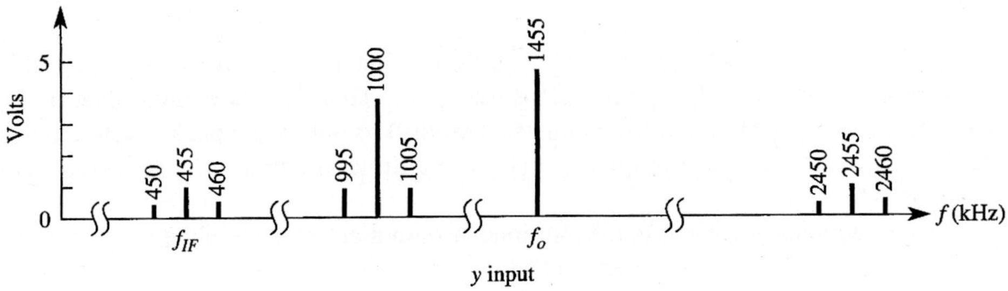

(c) The y input frequencies are shifted to the intermediate frequency.

FIGURE 12-17 *(cont.)*

12-13 UNIVERSAL AMPLITUDE MODULATION RECEIVER

12-13.1 Tuning and Mixing

The ordinary automobile or household AM radio can receive only standard AM signals that occupy the AM broadcast band, about 500 to 1500 kHz. This type of radio receiver cannot extract the audio or data signals from single-sideband (CB) or suppressed carrier transmission.

Figure 12-18 shows a receiver that will receive any type of AM transmission, carrier plus sidebands, sidebands without carrier, or a single sideband (either upper or lower). To understand its operation, assume that a station is transmitting a 5-kHz audio signal that modulates a 1005-kHz carrier wave. The station transmits a standard AM frequency spectrum of the 1005-kHz carrier and both lower and upper side frequencies of 1000 kHz and 1010 kHz [see Fig. 12-18(a)].

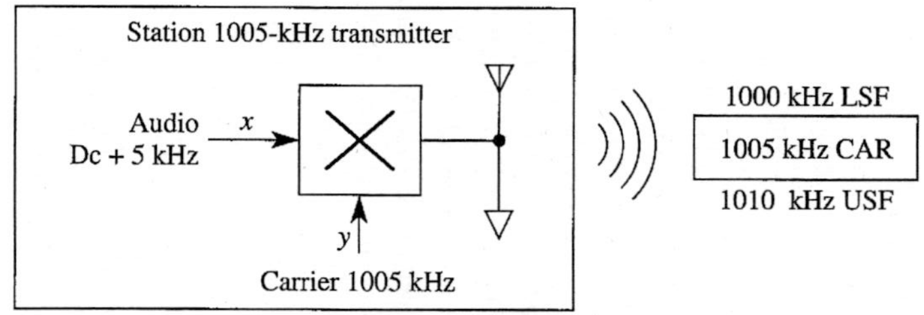

(a) Standard AM station broadcasting a 5-kHz signal on a carrier of 1005 kHz.

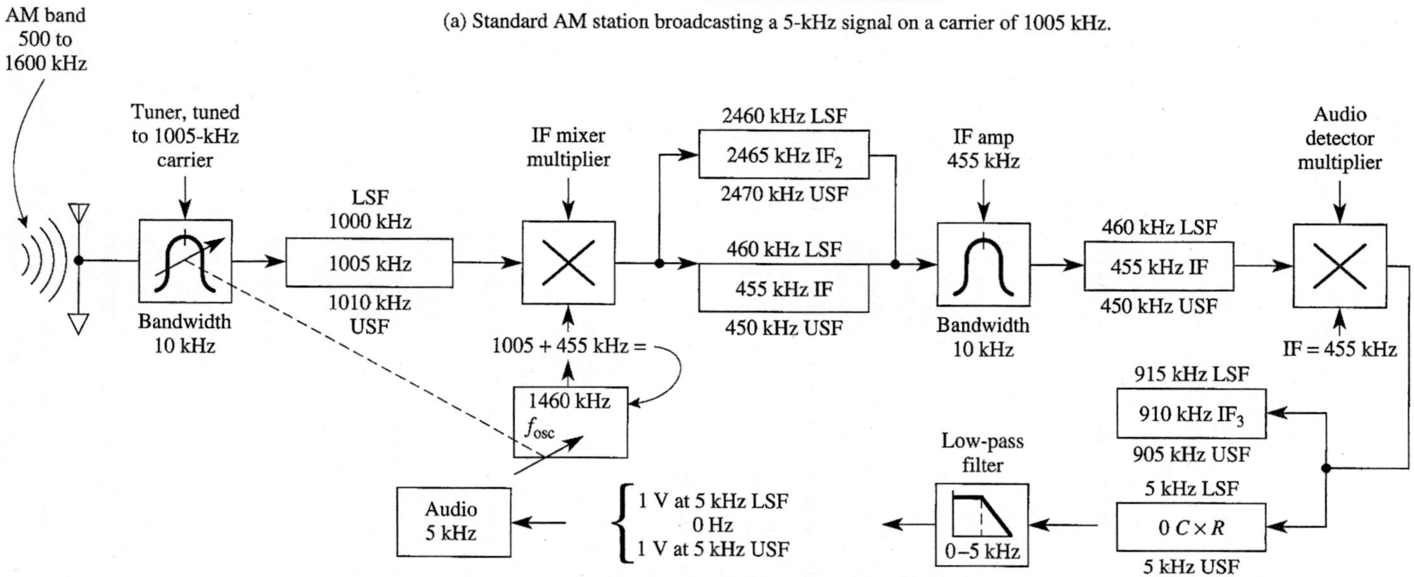

(b) Universal AM receiver frequency distribution.

FIGURE 12-18 The superheterodyne receiver in (b) can demodulate or detect audio signals from the standard AM transmission in (a) and also single-sideband or suppressed carrier AM.

In Fig. 12-18(b) the receiver's *tuner* is tuned to select this one station's 10-kHz band of frequencies out of the entire broadcast band of frequencies that are present on the receiver's antenna. A *local oscillator* in the radio is designed to produce a signal that *tracks* the tuner and is always 455 kHz higher than the tuner frequency. The oscillator and tuner output frequencies are multiplied by the *IF mixer.* The IF mixer acts as a frequency shifter to shift the incoming radio-frequency carrier down to an *intermediate-frequency* (IF) carrier of 455 kHz.

12-13.2 Intermediate-Frequency Amplifier

The output of the IF mixer contains both sum frequencies (2465-kHz IF_2 carrier) and difference frequencies (455-kHz IF carrier). Only the difference frequencies are amplified by the tuned high-gain *IF amplifier.* This first frequency shift (heterodyning) is performed so that most of the signal amplification is done by a single narrowband tuned-IF amplifier that usually has three stages of gain. *Any* station carrier that is selected by the tuner is shifted by the local oscillator and mixer multiplier down to the IF frequency for amplification. This frequency downshift scheme is used because it is much easier to build a reliable narrowband IF amplifier (10-kHz bandwidth centered on a 455-kHz carrier) than it is to build an amplifier that can provide equal amplification of, *and select* 10-kHz bandwidths over, the *entire* AM broadcast band.

12-13.3 Detection Process

The output of the IF amplifier is multiplied by the IF frequency in the *audio detector multiplier.* The term *detection* means that we are going to *detect* or *demodulate* the audio signal from the 455-kHz IF carrier. The audio detector shifts the incoming IF carrier and side frequencies up and down as sum and difference frequencies. Only the difference frequencies are transmitted through the low-pass filter in Fig. 12-18(b). The astute reader will note that the *low-pass filter* output frequencies are *not* labeled (+) 5 kHz for the upper side frequency and (−) 5 kHz for the lower side frequency. If you work the mathematics out using sine waves for audio, carrier, local oscillator, and IF signals, it turns out that both audio 5-kHz signals are in phase (as negative cosine waves). The output of the low-pass filter is applied to an audio amplifier and finally to a speaker.

12-13.4 Universal AM Receiver

Why will this receiver do what most other receivers cannot do? The previous sections dealt with a standard AM transmission carrier plus *both* upper and lower side frequencies. Suppose that you eliminated the 1005-kHz carrier at the transmitter in Fig. 12-18(a). Note how the carrier is identified by enclosure in a rectangle as it progresses through the receiver in Fig. 12-18(b). If no carrier enters the receiver, but only the side frequencies, both audio signals (USF and LSF) will still enter the audio amplifier. Thus this receiver can recover audio information from either standard AM or balanced AM modulation. An AM car radio will *not* recover the audio signal from balanced AM transmissions.

Next suppose that you eliminated (by filters) the carrier *and* upper side frequency from the transmitter in Fig. 12-18(a); only the lower side frequency of 1000 kHz would be broadcast. The entire lower sideband would occupy 1000 to 1005 kHz. This is *single-sideband transmission.* At the receiver, the tuner would select 1000 kHz (to 1005 kHz). The IF amplifier would output 460 kHz (to 455 kHz). Finally, the low-pass filter would output 5 kHz (to 0 kHz); thus this receiver can also receive single-sideband transmission. The versatility of this type of receiver is inherent in the design. It requires no switches to activate circuit changes for different types of AM modulation.

PROBLEMS

12-1. Find V_o in Fig. 12-1 for the following combination of inputs: **(a)** $x = 5$ V, $y = 5$ V; **(b)** $x = -5$ V, $y = 5$ V; **(c)** $x = 5$ V, $y = -5$ V; **(d)** $x = -5$ V, $y = -5$ V.

12-2. State the operating point quadrant for each combination in Problem 12-1 [see Fig. 12-2(a)].

12-3. What is the name of the procedure used to make $V_o = 0$ when both x and y inputs are at 0 V?

12-4. Find V_o in Fig. 12-3 if $E_i = -3$ V.

12-5. The peak value of E_i in Fig. 12-4 is 8 V, and its frequency is 400 Hz. Evaluate the output's **(a)** dc terms; **(b)** ac term.

12-6. In Fig. 12-5, $E_{xp} = 10$ V, $E_{yp} = 10$ V, and $\theta = 30°$. Find V_o.

12-7. Repeat Problem 12-6 for $\theta = -30°$.

12-8. $E_x = 10$ V and $E_{in} = -1$ V in Fig. 12-6. Find V_o.

12-9. In the balanced modulator of Fig. 12-9, E_x is a 15-kHz sine wave at 8-V peak and E_y is a 3-kHz sine wave at 5-V peak. Find the peak voltage of each frequency in the output.

12-10. In Fig. 12-9, the carrier frequency is 15 kHz. The modulating frequencies range between 1 and 2 kHz. Find the upper and lower side bands.

12-11. The switch is on AM in Fig. 12-11. The modulating frequency is 10 kHz at 5-V peak. The carrier is 100 kHz at 8-V peak. Identify the peak value and each frequency contained in the output.

12-12. If the switch is thrown to "Balanced" in Problem 12-11, what changes result in the output?

12-13. The x input of Fig. 12-14 is three sine waves of 5 V at 20 kHz, 2.0 V at 21 kHz, and 2.0 V at 19 kHz. The y input is 5 V at 20 kHz. What are the output signal frequency components?

12-14. You need to shift a 550-kHz signal to a 455-kHz intermediate frequency. What frequency should be generated by the local oscillator?

CHAPTER 13

Integrated-Circuit Timers

LEARNING OBJECTIVES

Upon completion of this chapter on integrated-circuit timers, you will be able to:

- Name three operating states of a 555 timer and tell how they are controlled by the trigger and threshold terminals.
- Draw circuits that produce a time delay or an initializing pulse upon application of power.
- Connect the 555 to make an oscillator for any desired frequency.
- Use 555 oscillators to make a tone-burst oscillator or voltage-controlled frequency shifter.
- Explain the operation of a 555 when it is wired to perform as a one-shot or monostable multivibrator.
- Use the 555 one-shot as a touch switch, frequency divider, or missing pulse detector.
- Describe the operation of an XR2240 programmable timer/counter.
- Connect the XR2240 as a long-interval timer, free-running oscillator, binary pattern generator, or frequency synthesizer.

- Build a switch programmable timer.
- Use PSpice to model and simulate performance of 555 timer circuits.

13-0 INTRODUCTION

Applications such as oscillators, pulse generators, ramp or square-wave generators, one-shot multivibrators, burglar alarms, and voltage monitors all require a circuit capable of producing timing intervals. The most popular integrated-circuit timer is the 555, first introduced by Signetics Corporation. Like general-purpose op amps, the 555 is reliable, easy to use in a variety of applications, and low in cost. The 555 can also operate from supply voltages of +5 V to +18 V, making it compatible with both TTL (transistor–transistor logic) circuits and op amp circuits. The 555 timer can be considered a functional block that contains two comparators, two transistors, three equal resistors, a flip-flop, and an output stage. These are shown in Fig. 13-1.

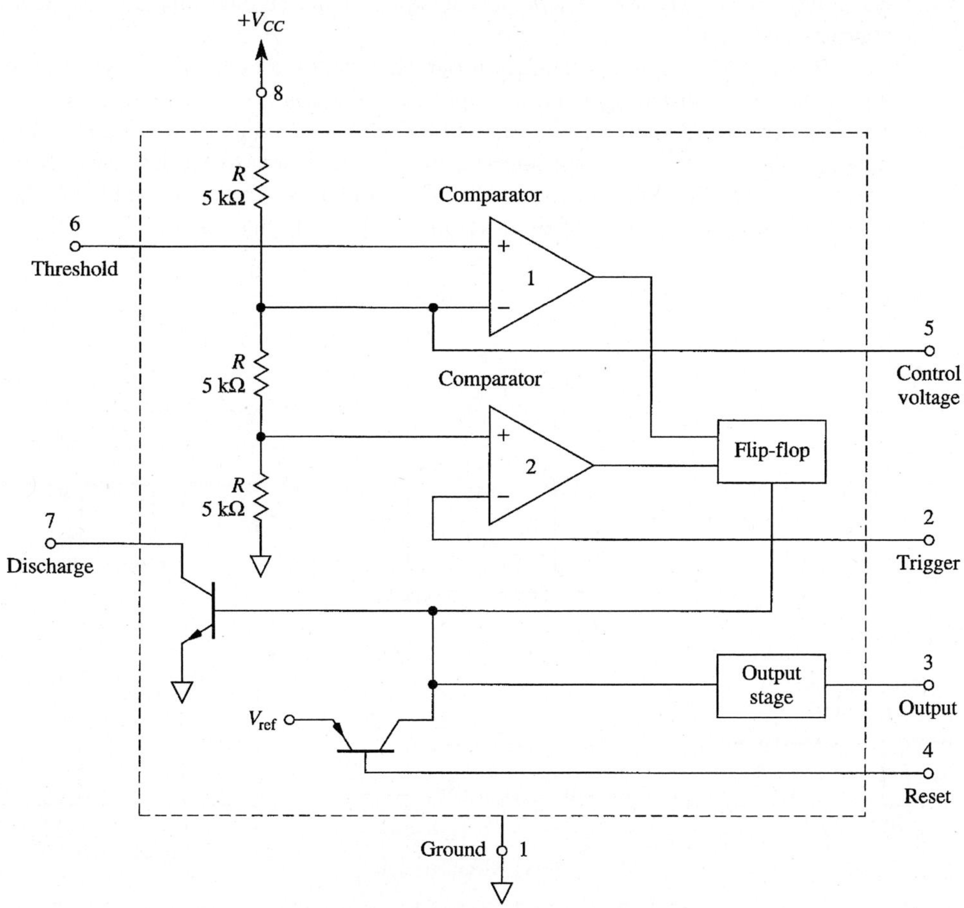

FIGURE 13-1 A 555 integrated-circuit timer.

Along with the 555 timer, there are also available counter timers such as Exar's XR 2240. The 2240 contains a 555 timer plus a programmable binary counter in a single 16-pin package. A single 555 has a maximum timing range of approximately 15 minutes. Counter timers have a maximum timing range of days. The timing range of both can be extended to months or even years by cascading. Our study of timers will begin with the 555 and its applications and then proceed to the counter timers.

13-1 OPERATING MODES OF THE 555 TIMER

The 555 IC timer has two modes of operation, either as an astable (free-running) multivibrator or as a monostable (one-shot) multivibrator. Free-running operation of the 555 is shown in Fig. 13-2(a). The output voltage switches from a high to a low state and back again. The time the output is either high or low is determined by a resistor–capacitor network connected externally to the 555 timer (see Section 13-2). The value of the high output voltage is slightly less than V_{CC}. The value of the output voltage in the low state is approximately 0.1 V.

When the timer is operated as a one-shot multivibrator, the output voltage is low until a negative-going trigger pulse is applied to the timer; then the output switches high. The time the output is high is determined by a resistor and capacitor connected to the IC timer. At the end of the timing interval, the output returns to the low state. Monostable operation is examined further in Sections 13-5 and 13-6. To understand how a 555 timer operates, a brief description of each terminal is given in Section 13-2.

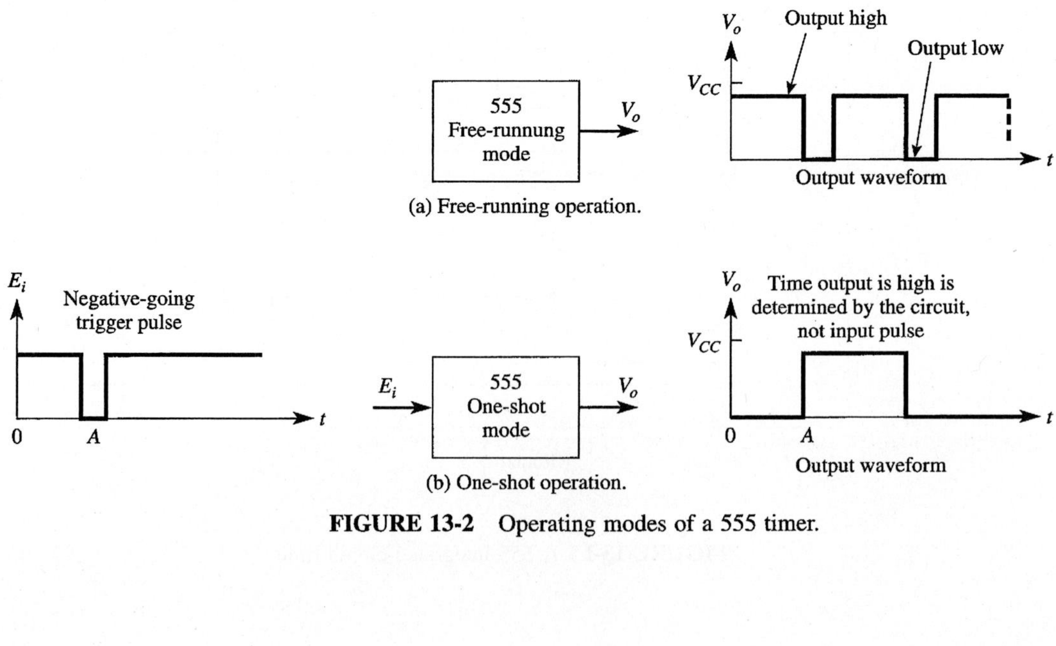

FIGURE 13-2 Operating modes of a 555 timer.

13-2 TERMINALS OF THE 555

13-2.1 Packaging and Power Supply Terminals

The 555 timer is available in two package styles, TO-99 and DIP, as shown in Fig. 13-3(a). Pin 1 is the common, or ground terminal, and pin 8 is the positive voltage supply terminal V_{CC}. V_{CC} can be any voltage between +5 V and +18 V. Thus the 555 can be powered by existing digital logic supplies (+5 V), linear IC supplies (+15 V), and automobile or dry cell batteries. Internal circuitry requires about 0.7 mA per supply volt (10 mA for V_{CC} = +15) to set up internal bias currents. Maximum power dissipation for the package is 600 mW.

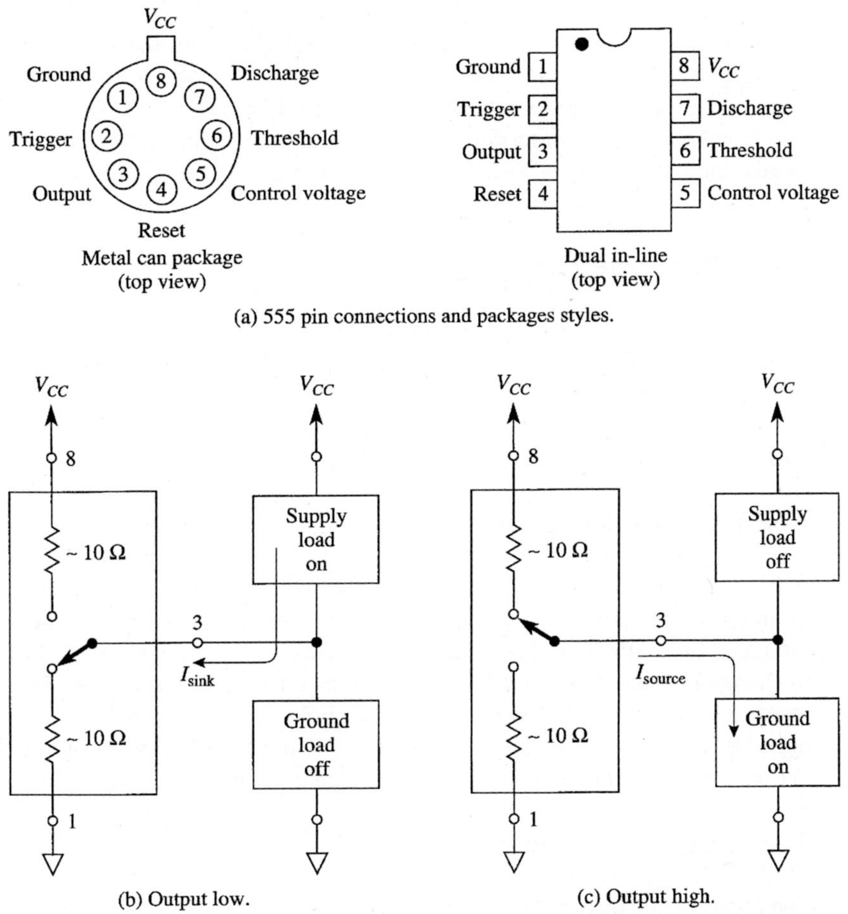

FIGURE 13-3 The 555 timer output operation and package terminals. Either a grounded or a supply load can be connected, although usually not simultaneously.

13-2.2 Output Terminal

As shown in Figs. 13-3(b) and (c), the output terminal, pin 3, can either source or sink current. A *floating* supply load is *on* when the *output* is *low,* and *off* when the output is *high.* A *grounded* load is *on* when the *output* is *high,* and *off* when the output is *low.* In normal operation either a supply load or a grounded load is connected to pin 3. Most applications do not require both types of loads at the same time.

The maximum sink or source current is technically 200 mA, but more realistically is 40 mA. The high output voltage [Fig. 13-3(c)] is about 0.5 V below V_{CC}, and the low output voltage [Fig. 13-3(b)] is about 0.1 V above ground, for load currents below 25 mA.

13-2.3 Reset Terminal

The reset terminal, pin 4, allows the 555 to be disabled and override command signals on the trigger input. When not used, the reset terminal should be wired to $+V_{CC}$. If the reset terminal is grounded or its potential reduced below 0.4 V, both the output terminal, pin 3, and the discharge terminal, pin 7, are at approximately ground potential. In other words, the output is held low. If the output was high, a ground on the reset terminal immediately forces the output low.

13-2.4 Discharge Terminal

The discharge terminal, pin 7, is usually used to discharge an external timing capacitor during the time the output is low. When the output is high, pin 7 acts as an open circuit and allows the capacitor to charge at a rate determined by an external resistor or resistors and capacitor. Figure 13-4 shows a model of the discharge terminal for when C is discharging and for when C is charging.

13-2.5 Control Voltage Terminal

A 0.01-μF filter capacitor is usually connected from the control voltage terminal, pin 5, to ground. The capacitor bypasses noise and/or ripple voltages from the power supply to minimize their effect on threshold voltage. The control voltage terminal may also be used to change both the threshold and trigger voltage levels. For example, connecting a 10-kΩ resistor between pins 5 and 8 changes threshold voltage to 0.5 V_{CC} and the trigger voltage to 0.25 V_{CC}. An external voltage applied to pin 5 will change both threshold and trigger voltages and can also be used to modulate the output waveform.

13-2.6 Trigger and Threshold Terminals

The 555 has two possible operating states and one memory state. They are determined by *both* the *trigger* input, pin 2, and the *threshold* input, pin 6. The trigger input is compared by comparator 1 in Fig. 13-1, with a lower threshold voltage V_{LT} that is equal to $V_{CC}/3$. The threshold input is compared by comparator 2 with a higher threshold voltage V_{UT} that

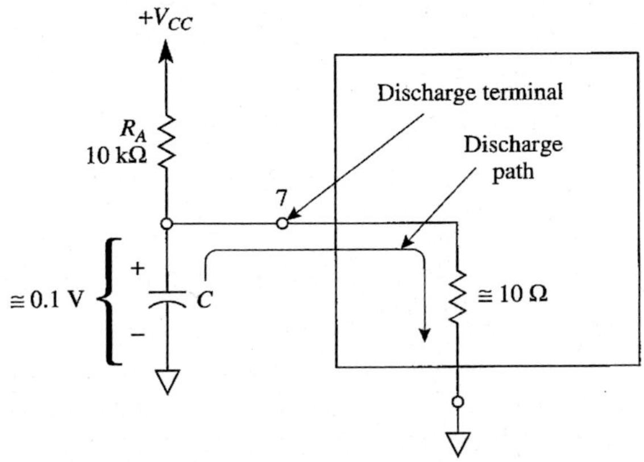

(a) Model of the discharge terminal when the output is low and the capacitor is discharging.

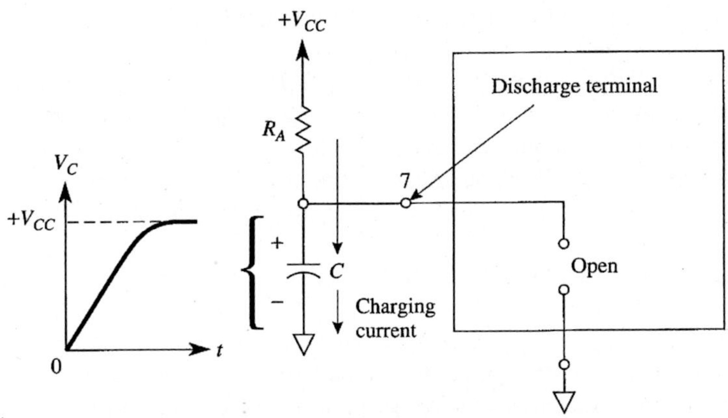

(b) Model of the discharge terminal when the output is high and the capacitor is charging.

FIGURE 13-4 Operation of discharge terminal.

is equal to $2V_{CC}/3$. Each input has two possible voltage levels, either above or below its reference voltage. Thus with two inputs there are four possible combinations that will cause four possible operating states.

The four possible input combinations and corresponding states of the 555 are given in Table 13-1. In operating state *A, both* trigger and threshold are *below* their respective threshold voltages and the output terminal (pin 3) is *high.* In operating state *D, both* inputs are *above* their threshold voltages and the output terminal is *low.*

The observation that low inputs give a high output, and high inputs give a low output, might lead you to conclude that the 555 acts as an inverter. However, as shown in Table 13-1, the 555 also has a *memory* state. Memory state *C* occurs when the trigger input is *above, and* the threshold input is *below* their respective reference voltages.

TABLE 13-1 OPERATING STATES OF A 555 TIMER: $V_{UT} = 2V_{CC}/3$, $V_{LT} = V_{CC}/3$; HIGH $\simeq V_{CC}$, LOW OR GROUND $\simeq$ 0 V

Operating state	Trigger pin 2	Threshold pin 6	State of terminals	
			Output 3	Discharge 7
A	Below V_{LT}	Below V_{UT}	High	Open
B	Below V_{LT}	Above V_{UT}	High	Open
C	Above V_{LT}	Below V_{UT}	Remembers last state	
D	Above V_{LT}	Above V_{UT}	Low	Ground

A visual aid in understanding how these operating states occur is presented in Fig. 13-5. An input voltage E_i is applied to *both* trigger and threshold input terminals. When E_i is below V_{LT} during time intervals *A–B* and *E–F*, state *A* operation results, so that output V_{o_3} is high. When E_i lies above V_{LT} but below V_{UT}, within time *B–C*, the 555 enters state *C* and remembers its last *A* state. When E_i exceeds V_{UT}, state *D* operation sends the output low. When E_i drops between V_{UT} and V_{LT} during time *D–E*, the 555 remembers the last *D* state and its output stays low. Finally, when E_i drops below V_{LT} during time *E–F*, the *A* state sends the output high.

By plotting output V_{o_3} against E_i in Fig. 13-5, we see a hysteresis characteristic. Recall from Chapter 4 that a *hysteresis loop* means that the circuit has *memory.* This also means that if the inputs are in one of the memory states, you cannot tell what state the output is now in, unless you know the previous state. Two *power-on* applications will now be given to show how to analyze circuit operation from Table 13-1.

13-2.7 Power-on Time Delays

There are two types of timing events that may be required during a power-on application. You may wish to apply power to one part of a system and wait for a short interval before starting some other part of a system. For example, you may need to reset all counters to zero before starting a personal computer at the beginning of a business day. A circuit that solves this problem is shown in Fig. 13-6(a). When the power switch is thrown to on at $t = 0$, the initial capacitor voltage is zero. Therefore, both pins 2 and 6 are above their respective thresholds and the output stays low in operating state *D*. As capacitor *C* charges, the threshold drops below V_{UT} while the trigger is still above V_{LT}, forcing the 555 into memory state *C*. Finally, both trigger and threshold drop just below V_{LT}, where the 555 enters state *A* and forces the output high at time *T*.

The net result is that an output from pin 3 of the 555 is delayed for a time interval *T* after the switch closure at $t = 0$. The time delay is found from $T = 1.1\ R_AC$.

By interchanging R_A and *C*, a time delay with a high output can be generated. In the circuit of Fig. 13-6(b), power is applied to a system when the switch is closed. The 555's output goes high for a period of time *T* and then goes low. *T* is found from Eq. (13-9). This type of startup pulse is typically used to reset counters and initialize computer sequences after a power failure. It also can allow time before arming an alarm system so that an operator can exit after the system has been turned on.

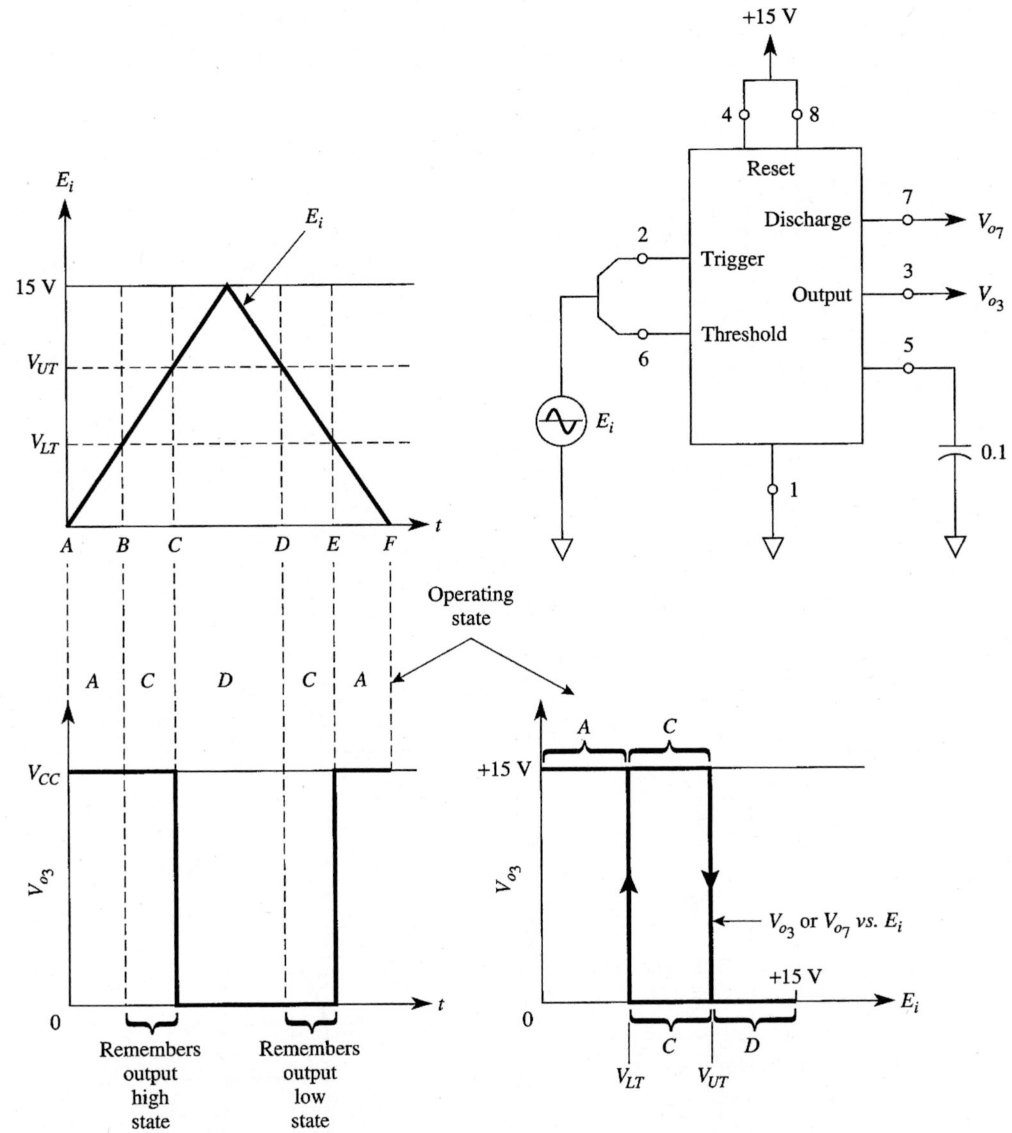

FIGURE 13-5 Three of the four operating states of a 555 timer are shown by a test circuit to measure E_i and V_{o_3} versus time and V_{o_3} versus E_i.

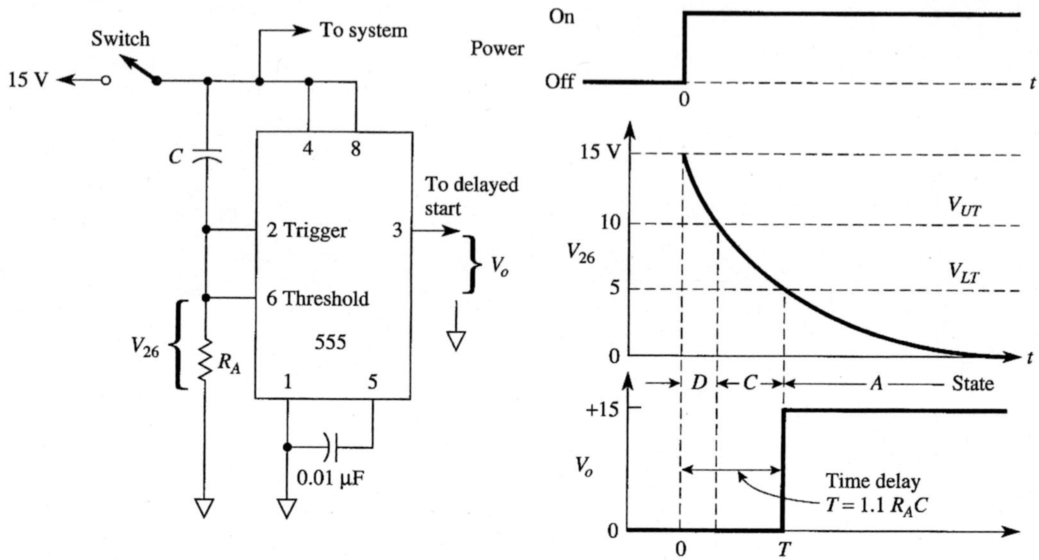

(a) Output V_o does not go high until a time interval T elapses after application of power at $t = 0$.

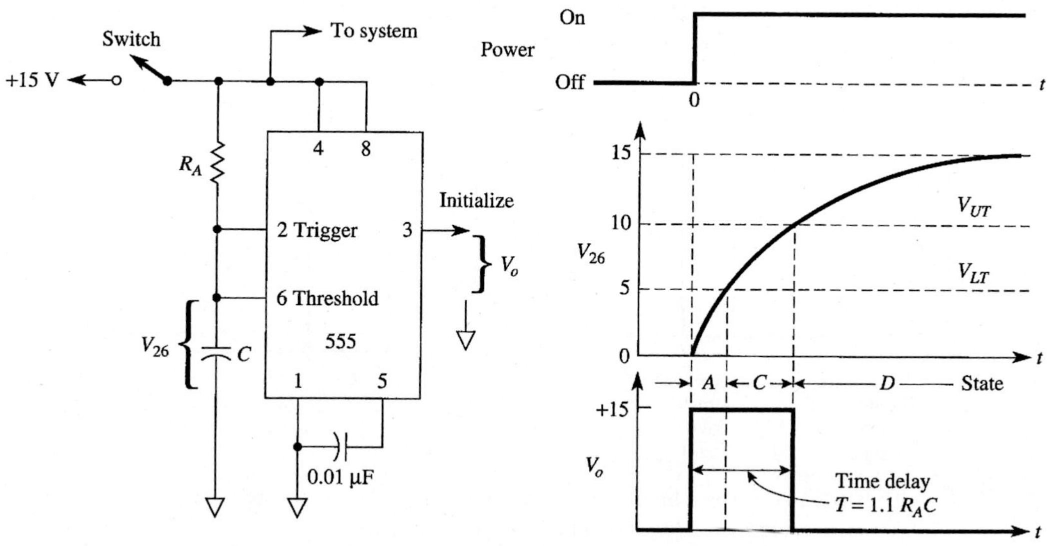

(b) Output V_o goes high for a time interval T after power is applied.

FIGURE 13-6 Power-on time-delay applications are analyzed by reference to Table 13-1.

13-3 FREE-RUNNING OR ASTABLE OPERATION

13-3.1 Circuit Operation

The 555 is connected as a free-running multivibrator in Fig. 13-7(a). Refer to the wave-shapes in Fig. 13-7(b) to follow the circuit's operation. At time *A* both pins 2 and 6 go just below $V_{LT} = \frac{1}{3} V_{CC}$ and output pin 3 goes high (state *A*). Pin 7 also becomes an open, so capacitor *C* charges through $R_A + R_B$. During output high time *A–B*, the 555 is in memory state *C*, remembering the previous *A* state. When V_C goes just above $V_{UT} = \frac{2}{3} V_{CC}$ at time *B*, the 555 enters state *D* and sends the output low. Pin 7 also goes low and capacitor *C* discharges through resistor R_B. During output low time *B–C*, the 555 is in memory state *C*, remembering the previous state *D*. When V_C drops just below V_{LT}, the sequence repeats.

13-3.2 Frequency of Oscillation

The output stays high during the time interval that *C* charges from $\frac{1}{3} V_{CC}$ to $\frac{2}{3} V_{CC}$ as shown in Fig. 13-7(b). This time interval is given by

$$t_{\text{high}} = 0.695(R_A + R_B)C \tag{13-1}$$

The output is low during the time interval that *C* discharges from $\frac{2}{3} V_{CC}$ to $\frac{1}{3} V_{CC}$ and is given by

$$t_{\text{low}} = 0.695 R_B C \tag{13-2}$$

Thus the total period of oscillation *T* is

$$T = t_{\text{high}} + t_{\text{low}} = 0.695(R_A + 2R_B)C \tag{13-3}$$

The free-running frequency of oscillation *f* is

$$f = \frac{1}{T} = \frac{1.44}{(R_A + 2R_B)C} \tag{13-4}$$

Figure 13-7(c) is a plot of Eq. (13-4) for different values of $(R_A + 2R_B)$ and quickly shows what combinations of resistance and capacitance are needed to design an astable multivibrator.

Example 13-1

Calculate (a) t_{high}, (b) t_{low}, and (c) the free-running frequency for the timer circuit of Fig. 13-7(a).

Solution (a) By Eq. (13-1),

$$t_{\text{high}} = 0.695(6.8\text{ k}\Omega + 3.3\text{ k}\Omega)(0.1\ \mu\text{F}) = 0.7\text{ ms}$$

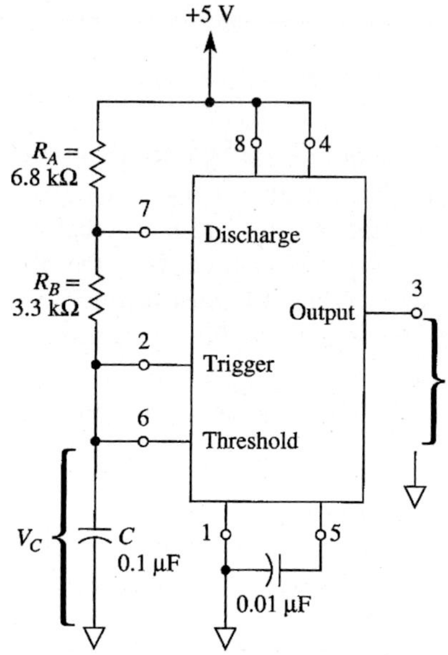

(a) Astable operation.

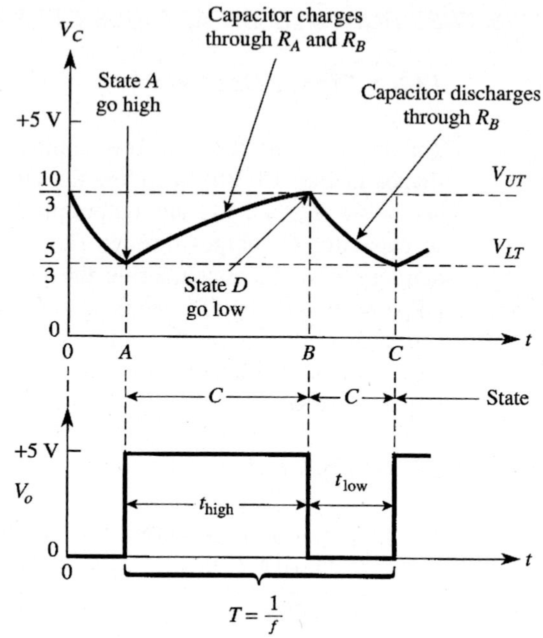

(b) Waveshapes.

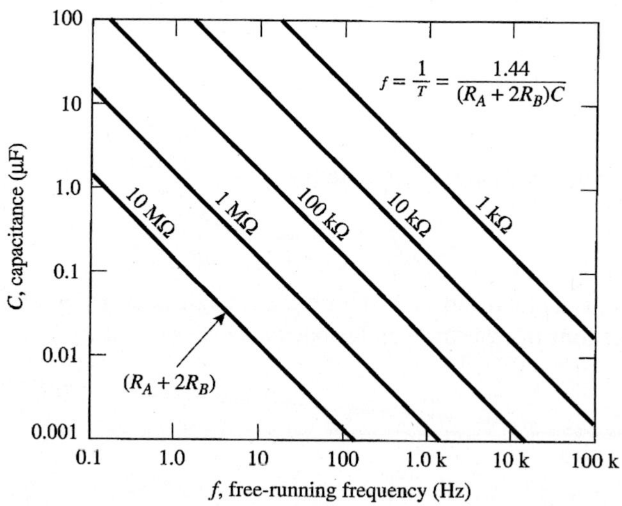

FIGURE 13-7 Waveshapes are shown in (b) for the free-running astable multivibrator in (a). Frequency of operation is determined by the resistance and capacitance values in (c).

(b) By Eq. (13-2),

$$t_{low} = 0.695(3.3\ \text{k}\Omega)(0.1\ \mu\text{F}) = 0.23\ \text{ms}$$

(c) By Eq. (13-4),

$$f = \frac{1.44}{[6.8\ \text{k}\Omega + (2)(3.3\ \text{k}\Omega)][0.1\ \mu\text{F}]} = 1.07\ \text{kHz}$$

The answer to part (c) agrees with results obtainable from Fig. 13-7(c).

13-3.3 Duty Cycle

The ratio of time when the output is low, t_{low}, to the total period T is called the *duty cycle D*. In equation form,*

$$D = \frac{t_{low}}{T} = \frac{R_B}{R_A + 2R_B} \qquad (13\text{-}5)$$

The next section presents an inexpensive way of extending the duty cycle.

Example 13-2

Calculate the duty cycle for the values given in Fig. 13-7(a).

Solution By Eq. (13-5),

$$D = \frac{3.3\ \text{k}\Omega}{6.8\ \text{k}\Omega + 2(3.3\ \text{k}\Omega)} = 0.25$$

This checks with Fig. 13-6(b), which shows that the timer's output is low for approximately 25% of the total period T. Equation (13-5) shows that it is impossible to obtain a duty cycle of $\frac{1}{2}$, or 50%. As presented, the circuit of Fig. 13-7(a) is not capable of producing a square wave. The only way D in Eq. (13-5) can equal $\frac{1}{2}$ is for R_A to equal 0. Then there would be a short circuit between V_{CC} and pin 7. However, R_A must be large enough so that when the discharge transistor is "on," current through it is limited to 0.2 A. Thus the minimum value of R_A in ohms is given by

$$\text{minimum } R_A \simeq \frac{V_{CC}}{0.2\ \text{A}} \qquad (13\text{-}6)$$

In practice, keep R_A equal to or greater than 1 kΩ.

*The original literature published by Signetics (the maker of the 555) defined duty cycle as shown. We go along with the originator in this chapter. In most other texts and papers, duty cycle is expressed, in percent, as the ratio of high time to period.

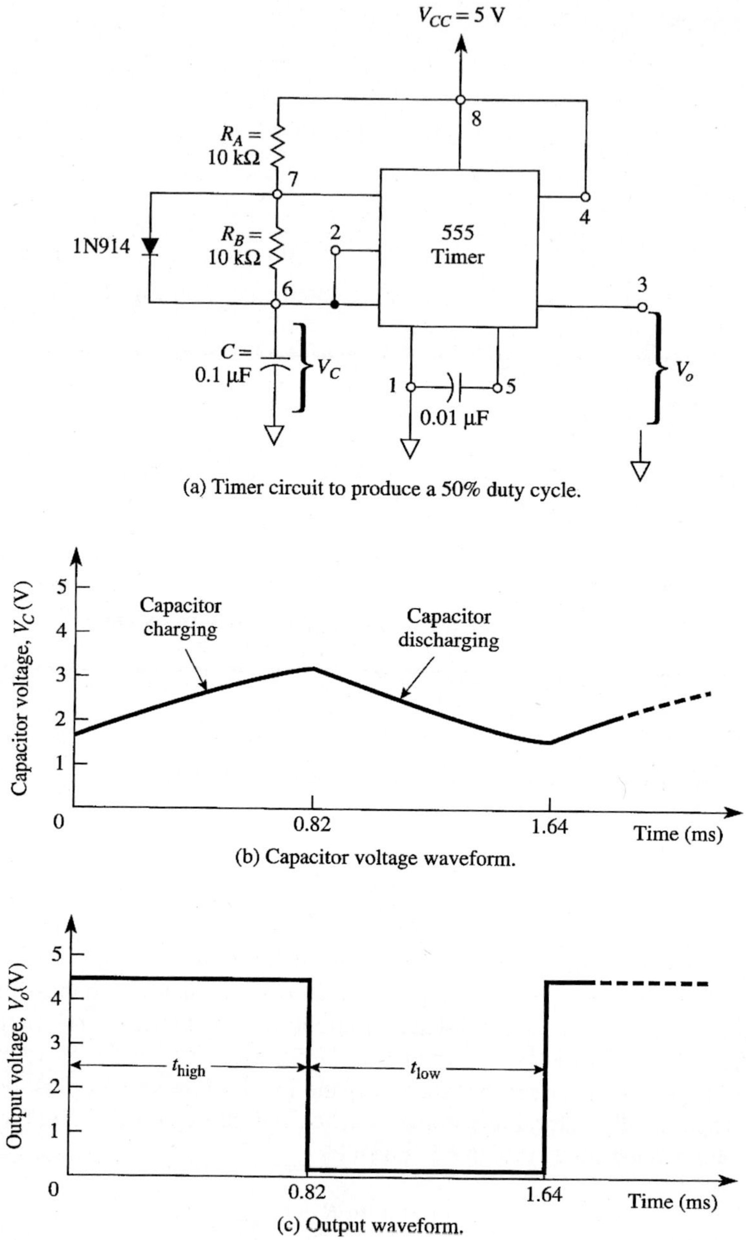

FIGURE 13-8 Connecting a diode across R_B to produce duty cycles of 50%.

13-3.4 Extending the Duty Cycle

The duty cycle for the circuit of Fig. 13-7(a) can never be equal to or greater than 50%, as discussed in Section 13-3.3. By connecting a diode in parallel with R_B in Fig. 13-8(a), a duty cycle of 50% or greater can be obtained. Now the capacitor charges through R_A and the diode, but discharges through R_B. The times for the output waveform are

$$t_{\text{high}} = 0.695 R_A C \tag{13-7a}$$

$$t_{\text{low}} = 0.695 R_B C \tag{13-7b}$$

$$T = 0.695(R_A + R_B)C \tag{13-7c}$$

Equations (13-7a) and (13-7b) show that if $R_A = R_B$, then the duty cycle is 50%, as shown in Figs. 13-8(b) and (c).

13-4 APPLICATIONS OF THE 555 AS AN ASTABLE MULTIVIBRATOR

13-4.1 Tone-Burst Oscillator

With the switch in Fig. 13-9 set to the "continuous" position, the *B* 555 timer functions as a free-running multivibrator. The frequency can be varied from about 1.3 kHz to 14 kHz by the 10-kΩ potentiometer. If the potentiometer is replaced by a thermistor or photoconductive cell, the oscillating frequency will be proportional to temperature or light intensity, respectively.

The *A* 555 timer oscillates at a slower frequency. The 1-MΩ potentiometer sets the lowest frequency at about 1.5 Hz. Lower frequencies are possible by replacing the 1-μF capacitor with a larger value. When the connecting switch is thrown to the "burst" position, output pin 3 of the *A* timer alternately places a ground or high voltage on reset pin 4 of the *B* 555 timer. When pin 4 of the *B* timer is grounded, it cannot oscillate, and when ungrounded the timer oscillates. This causes the *B* timer to oscillate in bursts. The output of the tone-burst generator is V_o and is taken from pin 3 of timer *B*. V_o can drive either an audio amplifier or a stepdown transformer directly to a speaker. The 556 IC timer contains two 555 timers in a single 14-pin dual-in-line package. Thus the tone-burst generator can be made with one 556. This circuit is modeled and its performance is simulated in Section 13-11.

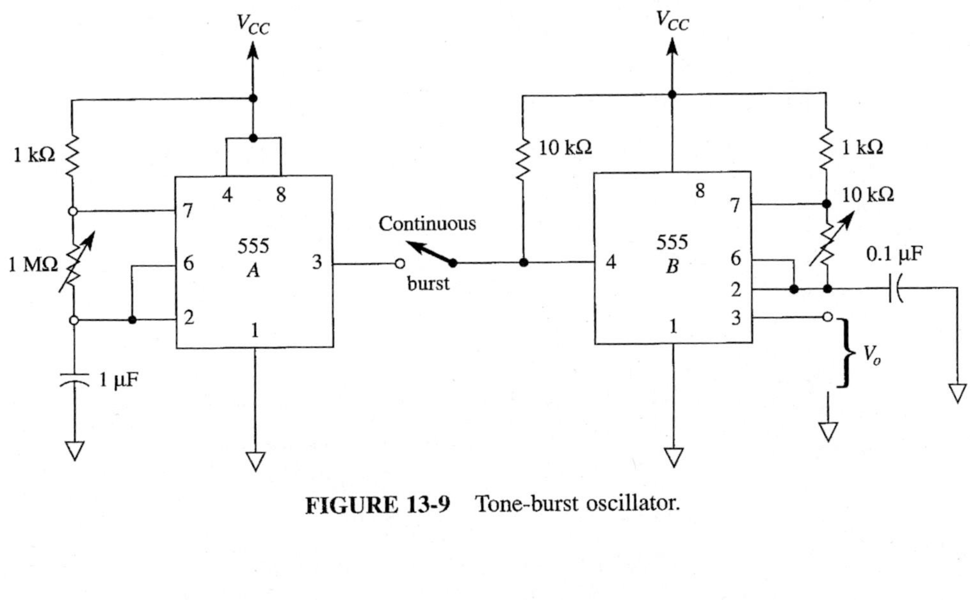

FIGURE 13-9 Tone-burst oscillator.

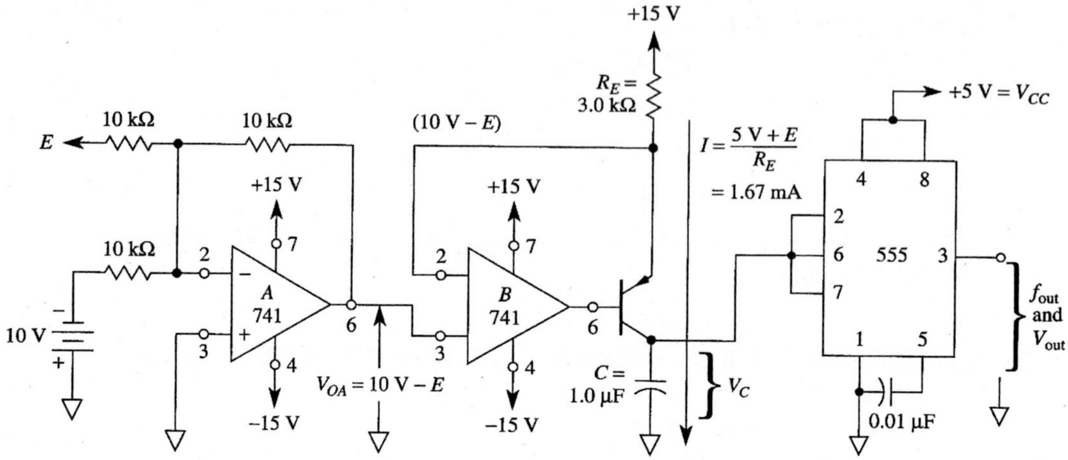

(a) Circuit for voltage-controlled frequency shifter.

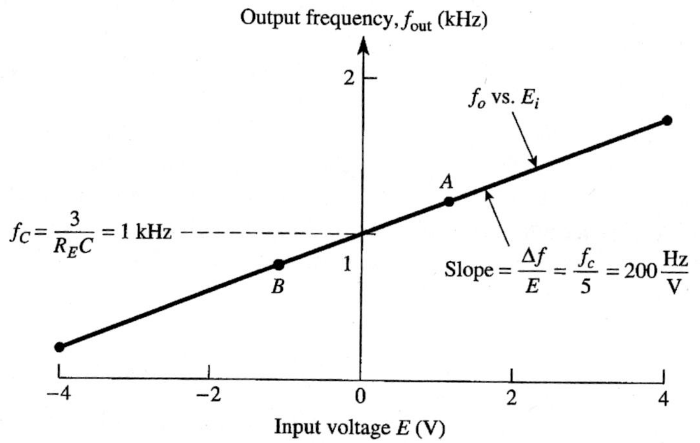

(b) Output–input characteristic of the frequency shifter: when $E = 0$, the output voltage is a 10-μs negative spike with a frequency of 1 kHz; output frequency f_{out} increases by 200 Hz for every positive-going 1 V of E. If E goes negative, f_{out} decreases at the same rate.

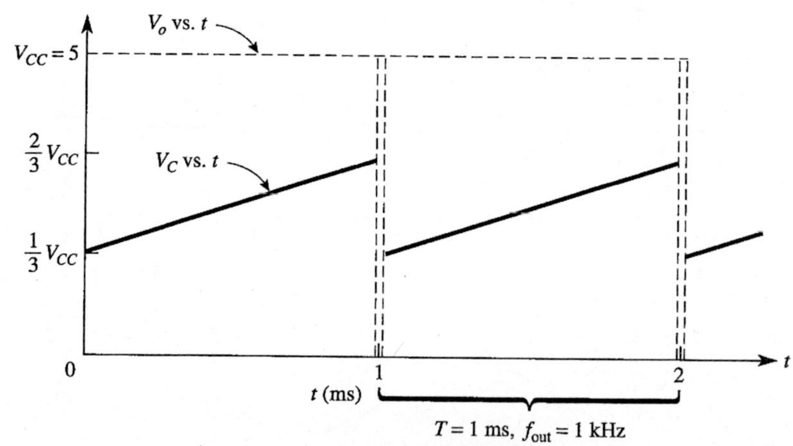

(c) Waveshapes for $E = 0$ V, $f_{out} = 1$ kHz.

FIGURE 13-10 Voltage-controlled frequency shifter.

13-4.2 Voltage-Controlled Frequency Shifter

A low-cost, low-frequency voltage-controlled frequency shifter is presented in Fig. 13-10(a). Since the 555 timer is powered by $V_{CC} = 5$ V, $V_{UT} = 5\text{ V}(\frac{2}{3})$ and $V_{LT} = 5\text{ V}(\frac{1}{3})$. Capacitor voltage V_C will charge to V_{UT} at which time the 555 will ground pin 7 to rapidly discharge C to V_{LT}. Then the discharge path for C is disconnected. The voltage waveshape described for C is shown in Fig. 13-10(c).

Capacitor C is charged by a constant current I. I is set by the voltage across R_E and the value of R_E. The voltage across R_E is determined by the difference between the 15-V supply and the voltage at pin 2 (10 V − E) of voltage follower 741B. Thus $V_{RE} = 15 - (10 - E) = 5\text{ V} + E$. Pin 2 follows pin 3 of op amp B, or V_{oA}. The inverting adder, 741A, has an output voltage of $V_{oA} = 10 - E$.

The charge lost by C for each cycle equals $C(\Delta V_C)$ where $\Delta V_C = 5\text{ V}(\frac{2}{3}) - 5\text{ V}(\frac{1}{3}) = 5\text{ V}/3$. Charge stored by C equals charge current I times period T (the charge time). For equilibrium

$$\text{charge stored} = \text{charge lost} \tag{13-8a}$$

$$IT = C\,\Delta V \tag{13-8b}$$

$$\frac{5\text{ V} - E}{R_E}T = C\frac{5\text{ V}}{3} \tag{13-8c}$$

Since period $T = 1/f_{\text{out}}$, we can rewrite Eq. (13-8c) as

$$f_{\text{out}} = \text{center frequency } f_c + \text{shift frequency } \Delta f \tag{13-8d}$$

where

$$f_c = \frac{3}{R_E\,C} \text{ when } E = 0\text{ V} \tag{13-8e}$$

and

$$\Delta f = 0.2 f_c E \tag{13-8f}$$

This reasoning will be clarified by an example.

Example 13-3

For the frequency shifter of Fig. 13-10 calculate (a) the charge current I for $E = 0$ V; (b) the center frequency f_c when $E = 0$ V; (c) the frequency shift for $E = \pm 1$ V and f_{out}. (d) Point out the positive and negative limits for E_i.

Solution (a)

$$I = \frac{5\text{ V} - E}{R_E} = \frac{5\text{ V}}{3\text{ k}\Omega} = 1.67\text{ mA}$$

(b) From Eq. (13-8e), when $E = 0$,

$$f_c = \frac{3}{R_E\,C} = \frac{3}{(3\text{ k}\Omega)(1 \times 10^{-6}\text{ F})} = \frac{3}{3\text{ ms}} = 1\text{ kHz}$$

Therefore, $f_{out} = 1$ kHz when $E = 0$.
(c) From Eq. (13-8f),

$$\Delta f = 0.2 f_c E = 0.2(1000)1 = 200\text{ Hz}$$

$f_{out} = f_c + \Delta f = 1000 + 200\text{ Hz} = 1200$ Hz [see Fig. 13-10(b), point A]. For $E = -1$ V, then, $\Delta f = -200$ Hz and $f_{out} = 800$ Hz [see Fig. 13-10(b), point B].
(d) Pin 2 of 741B cannot go closer to the 15-V supply voltage than about 1 V. This restricts the lower limit of E to about -4 V and f_{out} to 200 Hz. V_{CE} of the transistor needs about 2 V headroom above $V_{UT} = 3.3$ V, so the upper limit of E is about $+4$ V, where

In summary, the 555 oscillates at a center frequency f_c determined by Eq. (13-8e). E increases or decreases this center frequency by an amount of $0.2 f_c$ per volt for positive and negative values of E, respectively.

13-5 ONE-SHOT OR MONOSTABLE OPERATION

13-5.1 Introduction

Not all applications require a continuous repetitive wave such as that obtained from a free-running multivibrator. Many applications need to operate only for a specified length of time. These circuits require a one-shot or monostable multivibrator. Figure 13-11(a) is a circuit diagram using the 555 for monostable operation. When a negative-going pulse is applied to pin 2, the output goes high and terminal 7 removes a short circuit from capacitor C. The voltage across C rises from 0 volts at a rate determined by R_A and C. When capacitor voltage reaches $\frac{2}{3}V_{CC}$, comparator 1 in Fig. 13-1 causes the output to switch from high to low. The input and output voltage waveforms are shown in Fig. 13-11(a). The output is high for a time given by

$$t_{high} = 1.1 R_A\,C \tag{13-9}$$

Figure 13-11(b) is a plot of Eq. (13-9) and quickly shows the wide range of output pulses that are obtainable and the required values of R_A and C. Figure 13-11(a) gives the idea of a one-shot. In practice we must add more parts to make a workable circuit (see Section 13-5.2).

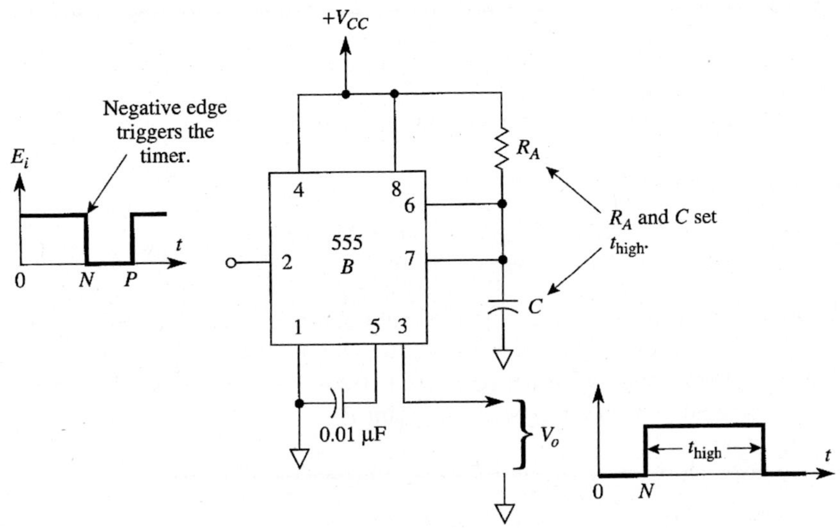

(a) A 555 timer wired for monostable operation.
Peak value of E_i must be $\geq \frac{2}{3}V_{CC}$.

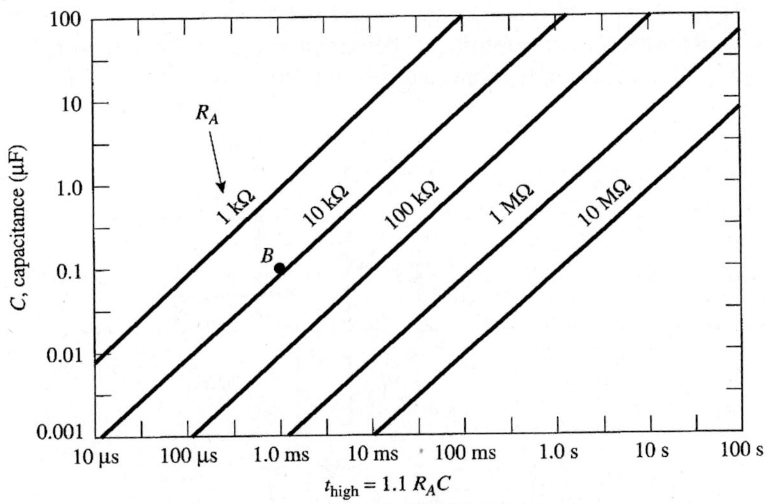

(b) Design aid to determine output pulse duration.

FIGURE 13-11 Monostable operation.

Example 13-4

If $R_A = 9.1\ \text{k}\Omega$, find C for an output pulse duration of 1 ms.

Solution Rearrange Eq. (13-9):

$$C = \frac{t_{\text{high}}}{1.1R_A} = \frac{1 \times 10^{-3}\ \text{s}}{1.1(9.1 \times 10^3)\ \Omega} = 0.1\ \mu\text{F}$$

This answer checks with that obtainable at point B in Fig. 13-11(b). For the 555 timer to trigger properly in this type of operation, the width of the trigger pulse must be less than t_{high} and a trigger input pulse network is needed so that the output does not switch on the positive-going edge of the trigger pulse (point P).

13-5.2 Input Pulse Circuit

Figure 13-12 shows the multivibrator wired for monostable operation in contrast with Fig. 13-11(a). R_i, C_i, and diode D are needed to generate a single output pulse for one input pulse. Resistor R_A and capacitor C determine the time that the output is high, as given by Eq. (13-9). Resistor R_i is connected between V_{CC} and pin 2 to ensure that the output is nor-

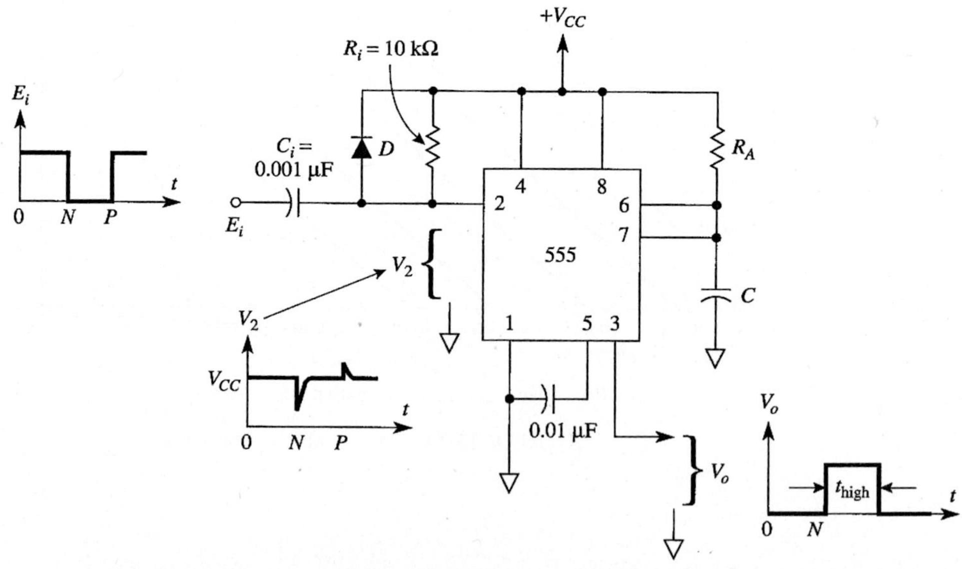

FIGURE 13-12 For satisfactory monostable operation, the input pulse network of R_i, C_i, and D is needed.

mally low. C_i is charged to $(V_{CC} - E_i)$ until the negative-going trigger pulse occurs. The time constant of R_i and C_i should be small with respect to the output timing interval t_{high}. Diode D prevents the 555 timer from triggering on positive-going edges of E_i. Waveforms for the input pulse, E_i, the pulse at pin 2, V_2, and the output pulse, V_o, are all shown in Fig. 13-12.

Example 13-5

(a) If $R_A = 10\ \text{k}\Omega$ and $C = 0.2\ \mu\text{F}$ in Fig. 13-12, find t_{high}. (b) What is the time constant of R_i and C_i in Fig. 13-12?

Solution (a) By Eq. (13-19),

$$t_{\text{high}} = 1.1(10 \times 10^3)(0.2 \times 10^{-6}) = 2.2\ \text{ms}$$

(b) Time constant $= R_iC_i = (10 \times 10^3)(0.001 \times 10^{-6}) = 0.01$ ms.

Just as with astable operation, reset terminal pin 4 is normally tied to supply voltage V_{CC}. If pin 4 is grounded at any time, the timing cycle is stopped. When the reset terminal is grounded, both output pin 3 and discharge terminal 7 go to ground potential. Thus the output goes low and any charge accumulated by the timing capacitor C is removed. As long as the reset terminal is grounded, these conditions remain.

13-6 APPLICATIONS OF THE 555 AS A ONE-SHOT MULTIVIBRATOR

13-6.1 Water-Level Fill Control

In Fig. 13-13(a), the start switch is closed and the output of the 555 is low. When the start switch is opened, the output goes high to actuate the pump. The output is high for a time interval given by Eq. (13-9). Upon completion of the timing interval, the output of the 555 returns to its low state, turning the pump off. The height of the water level is set by the timing interval, which is set by R_A and C. In the event of a potential overflow, the overfill switch must place a ground on reset pin 4, which causes the timer's output to go low and stops the pump.

13-6.2 Touch Switch

The 555 is wired as a one-shot multivibrator in Fig. 13-13(b) to perform as a touch switch. A 22-MΩ resistor to pin 2 holds the 555 in its idle state. If you scuff your feet to build up a static charge, the 555 will produce a single-shot output pulse when you touch the finger plate. If the electrical noise level is high (due, for example, to fluorescent lights) the 555 may oscillate when you touch the finger plate. Reliable and consistent triggering

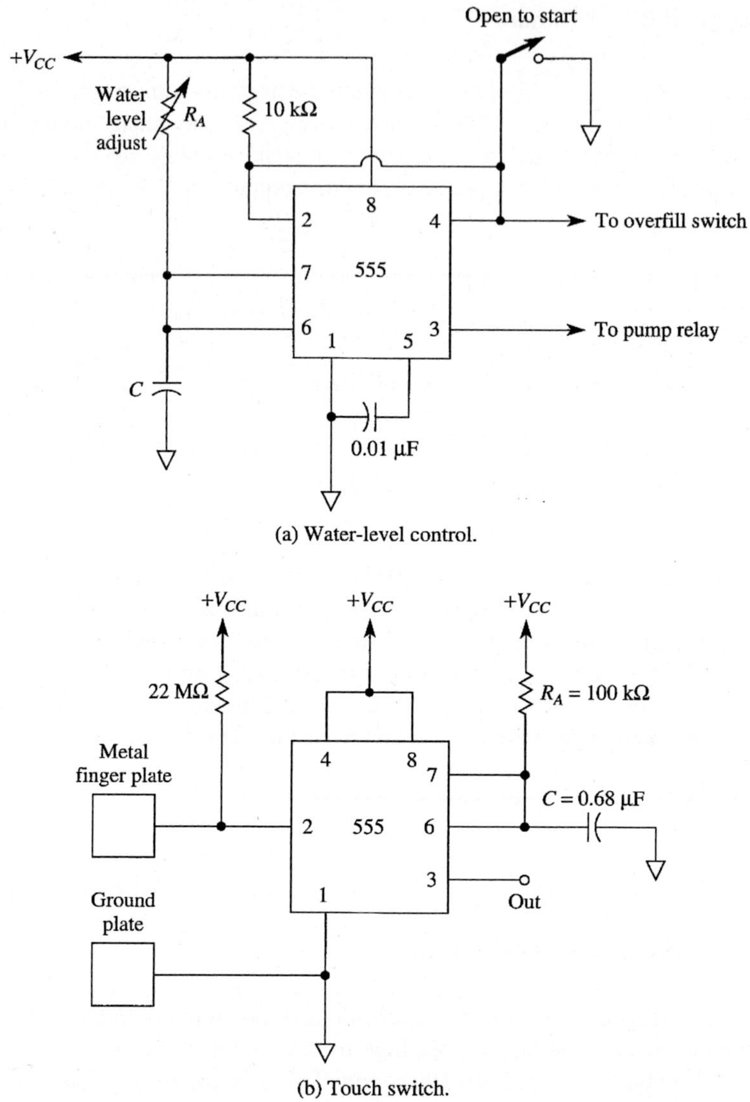

FIGURE 13-13 Basic one-shot application of the 555.

will occur if a thumb is placed on a ground plate and fingers of the same hand tap the finger plate. An isolated power supply or batteries should be used for safety.

13-6.3 Frequency Divider

Figure 13-12 also can be used as a frequency divider if the timing interval is adjusted to be longer than the period of the input signal E_i. For example, suppose that the frequency of E_i is 1 kHz, so that its period is 1 ms. If $R_A = 10\ \text{k}\Omega$ and $C = 0.1\ \mu\text{F}$, the timing interval given by Eq. (13-9) is $t_{high} = 1.1$ ms. Therefore, the one shot will be triggered by the first negative-going pulse of E_i, but the output will still be high when the second negative-going pulse occurs. The one-shot will, however, be re-triggered on the third negative-going pulse. In this example, the one-shot triggers on every other pulse of E_i, so there is only one output for every two input pulses; thus E_i is divided by 2.

Example 13-6

(a) Calculate the timing interval in Fig. 13-12 if $R_A = 10\ k\Omega$ and $C = 0.1\ \mu F$.
(b) What value of R_A should be installed to divide a 1-kHz input signal by 3?

Solution (a) By Eq. (13-9), $t_{high} = 1.1(10 \times 10^3)(0.1 \times 10^{-6}) = 1.1$ ms. (b) t_{high} should exceed two periods of E_i, or 2 ms, and be less than three periods (3 ms). Choose $t_{high} = 2.2$ ms; then $2.2\ ms = 1.1 R_A \times 0.1 \times 10^{-6}$ F; $R_A = 20\ k\Omega$.

13-6.4 Missing Pulse Detector

Transistor Q is added to the 555 one-shot in Fig. 13-14(a) to make a missing pulse detector. When E_i is at ground potential (0 V), the emitter diode of transistor Q clamps capacitor voltage V_C to a few tenths of a volt above ground. The 555 is forced into its idle state

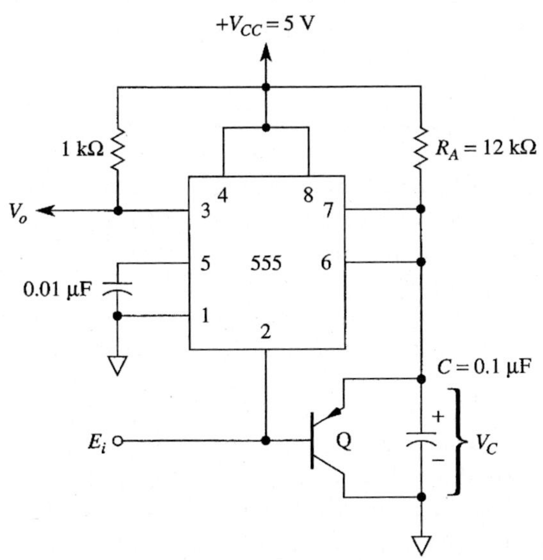

(a) Circuit.

FIGURE 13-14 Missing pulse detector.

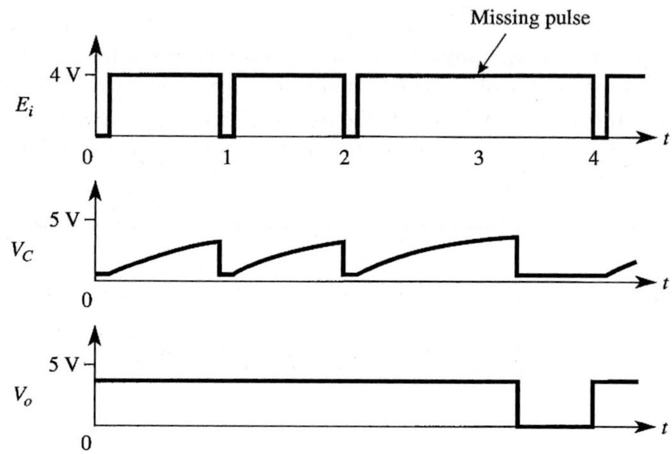

(b) Voltage waveforms.

FIGURE 13-14 (*cont.*)

with a high output voltage V_o at pin 3. When E_i goes high, the transistor cuts off and capacitor C begins to charge. This action is shown by the waveshapes in Fig. 13-14(b). If E_i again goes low before the 555 completes its timing cycle, the voltage across C is reset to about 0 V. If, however, E_i does *not* go low before the 555 completes its timing cycle, the 555 enters its normal state and output V_o goes low. This is exactly what happens if the $R_A\,C$ timing interval is slightly longer than the period of E_i and E_i suddenly misses a pulse. This type of circuit can detect a missing heartbeat. If E_i pulses are generated from a rotating wheel, this circuit tells when the wheel speed drops below a predetermined value. Thus the missing pulse detector circuit also performs speed control and measurement.

13-7 INTRODUCTION TO COUNTER TIMERS

When a timer circuit is connected as an oscillator and is used to drive a counter, the resultant circuit is a *counter timer.* Typically, the counter has many separate output terminals. One output terminal gives one pulse for each period T of the oscillator. A second output terminal gives one output pulse for every two periods ($2T$) of the oscillator. A third output terminal gives one output pulse for every four oscillator periods ($4T$), and so on, depending on the design of the counter. Thus each output terminal is rated in terms of the basic oscillator period T.

Some counters are designed so that their outputs can be connected together. The resultant output pulse is the *sum* of the individual output pulses. For example, if the first, second, and third output terminals are wired together, the result is one output pulse for every $1T + 2T + 4T = 7T$ oscillator periods. A counter with this capability is said to be *programmable,* because the user can program the counter to give one output pulse for any combination of timer outputs. One such programmable timer/counter is Exar's XR 2240.

This integrated-circuit device is representative of the timer/counter family, and some of its features will be studied next. Another timer/counter is Exar's XR 2242 or its pin replacement from Harris Semiconductor, the ICM 7242. This device is a long-range fixed timer. The timing can be set from microseconds to days and is cascadable. The device is housed in an 8-pin mini-DIP.

13-8 THE XR 2240 PROGRAMMABLE TIMER/COUNTER

13-8.1 Circuit Description

As shown in Fig. 13-15, the XR 2240 consists of one modified 555 timer, one 8-bit binary counter, and a control circuit. They are all contained in a single 16-pin dual-in-line package.

A positive-going pulse applied to *trigger* input 11 starts the 555 time base oscillator. A positive-going pulse on *reset* pin 10 stops the 555 time base oscillator. The threshold voltage for both trigger and reset terminals is about +1.4 V.

The time base period T for one cycle of the 555 oscillator is set by an external RC network connected to the *timing* pin 13. T is calculated from

$$T = RC \tag{13-10}$$

where R is in ohms, C in farads, and T in seconds. R can range from 1 kΩ to 10 MΩ and C from 0.05 to 1000 μF. Thus the period of the 555 can range from microseconds to hours.

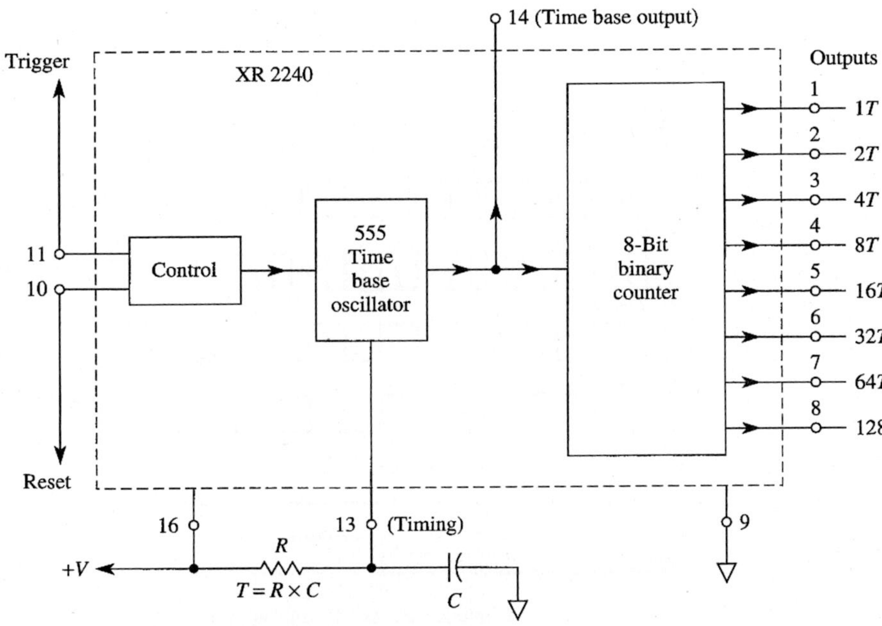

FIGURE 13-15 Block diagram of the XR 2240 programmable timer/counter.

The output of the 555 time base oscillator is available for measurement at pin 14 and also drives the 8-bit binary counter. Operation of the counter is discussed in Section 13-8.2.

13-8.2 Counter Operation

A simplified schematic of the 8-bit binary counter is shown in Fig. 13-16. Output of the 555 time base oscillator is shown as a switch. One side of the switch is connected to

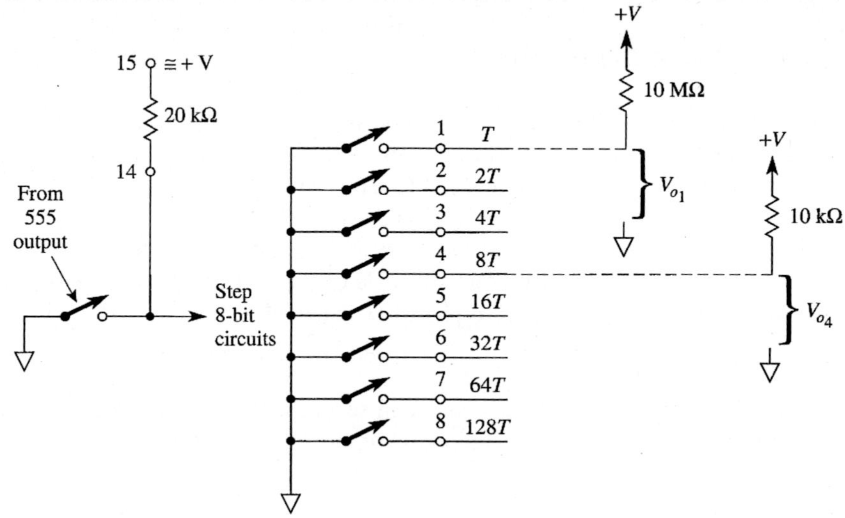

(a) Simplified outputs of the 2240.

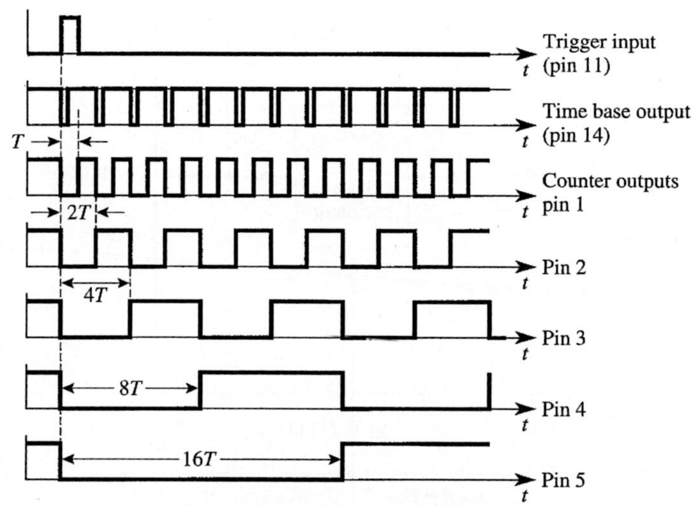

(b) Timing chart for XR 2240 outputs.

FIGURE 13-16 Counter operation.

ground while the other side is wired to a 20-kΩ pull-up resistor. A regulated positive voltage is available at pin 15. Each negative-going edge from the 555 steps the 8-bit counter up by one count.

Normally, the 2240 is in its *reset* position. That is, all 8 output pins (pins 1 to 8) act like open circuits, as shown by the output switch models in Fig. 13-16. Pull-up resistors (10 kΩ) should be installed, as shown, to those terminals that are going to be used. Outputs 1 and 4 will then be high in the reset condition.

When the 2240 is triggered (pulse applied to pin 11), all output switches of the counter are closed by the control circuit and outputs 1 to 8 go low. Thus the counter begins its count with all outputs essentially grounded. At the end of every time base period, the 555 steps the counter once. The counter's T switch on terminal 1 opens after the first time base period (output 1 goes high) and closes after the second time base period. This counting action of the timer is shown in Fig. 13-16(b).

Output pin 2 is labeled $2T$ in Fig. 13-16(a). It is seen from Fig. 13-16(b) that the output on pin 2 has stayed low for two time base periods ($2T$). Thus the second output stays low for twice the time interval of the first output. This conclusion may be generalized to all outputs of the binary counter; that is, each output stays low for twice the time interval of the preceding output. Time intervals for pins 1 to 5 are shown in Fig. 13-16(b) and are given for all outputs in Table 13-2.

TABLE 13-2 OUTPUT TERMINAL TIME CHART

Terminal number	Time output stays low after trigger pulse
1	T
2	$2T$
3	$4T$
4	$8T$
5	$16T$
6	$32T$
7	$64T$
8	$128T$

Example 13-7

After triggering, how long will the following output terminals stay low? (a) pin 3; (b) pin 4; (c) pin 7; (d) pin 8. $R = 100$ kΩ and $C = 0.01$ μF.

Solution By Eq. (13-10), the time base period is

$$T = (100 \times 10^3)(0.01 \times 10^{-6}) = 1 \text{ ms}$$

From Table 13-2, (a) $t_{low} = 4(1 \text{ ms}) = 4$ ms; (b) $t_{low} = 8(1 \text{ ms}) = 8$ ms; (c) $t_{low} = 64(1 \text{ ms}) = 64$ ms; (d) $t_{low} = 128(1 \text{ ms}) = 128$ ms.

The conclusion to be drawn from Example 13-7 is that after triggering, there are eight pulses of different time intervals available from the counter timer.

13-8.3 Programming the Outputs

The output circuits are designed to be used either individually or wired together, which is called *wire-or.* The term *wire-or* indicates that two or more output terminals can be jumpered together with a common wire (output bus) to a single pull-up resistor, as shown in Fig. 13-17(a). The resultant timing cycle for V_o is found by redrawing the individual timing of pins 4 and 5 in Fig. 13-17(b). Here we see that as long as either pin 4 *or* pin 5

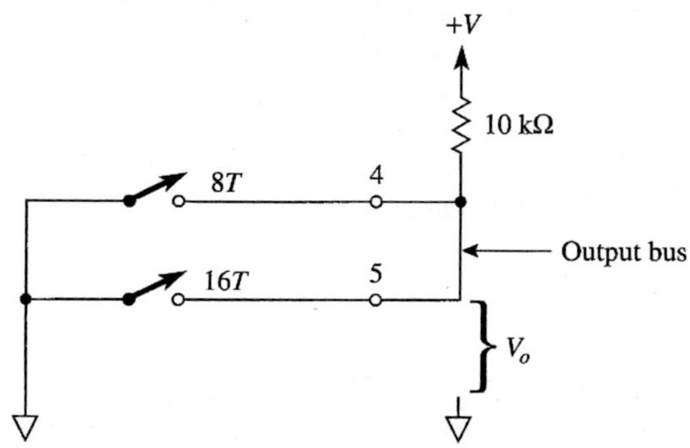

(a) Pins 4 and 5 are wired together to program 24*T*.

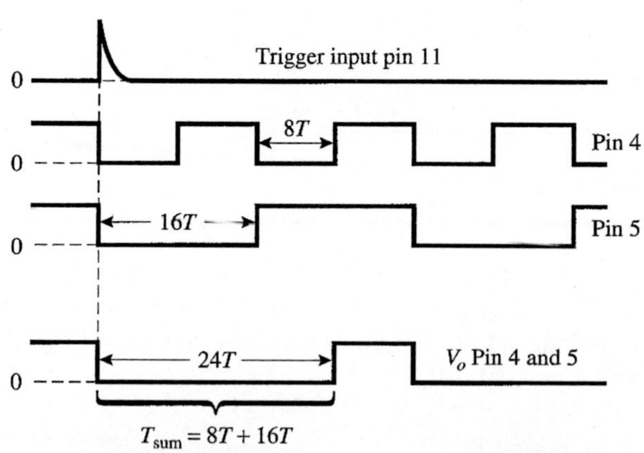

(b) Common bus V_o stays low as long as either pin 4 or pin 5 stays low.

FIGURE 13-17 Programming the outputs.

is low, V_o will be low. Only when both outputs go high (output switches open) will the output go high. Thus the timing cycle for the output bus is found simply by calculating the sum T_{sum} of the individual outputs.

Example 13-8

Calculate the timing cycle for (a) Fig. 13-17(a); (b) a circuit where pins 3, 6, and 7 are jumpered to a common bus. Let $T = 1$ s.

Solution (a) $T_{sum} = 8T + 16T = 24T = 24 \times 1\ \text{s} = 24\ \text{s}$; (b) $T_{sum} = 4T + 32T + 64T = 100T = 100 \times 1\ \text{s} = 100\ \text{s}$.

By using switches instead of jumper wires, T_{sum} can be easily changed or *programmed* for any desired timing cycle from T to $255T$.

13-9 TIMER/COUNTER APPLICATIONS

13-9.1 Timing Applications

The 2240 is wired for monostable operation in the programmable timer application in Fig. 13-18. When the trigger input goes high, the output bus goes low for a timing cycle period equal to T_{sum} (see Section 13-8.3). At the end of the timing cycle, the output bus goes high. The connection from output bus via a 51-kΩ resistor to reset pin 10 forces the timer to reset itself when the output bus goes high. Thus after each trigger pulse, the 2240 generates a timing interval selected by the program switches.

Example 13-9

In Fig. 13-18, $C = 1.0\ \mu\text{F}$ and $R = 5\ \text{M}\Omega$ to establish a time base period given by Eq. (13-10) to be 5 s. What is (a) the timing cycle for switch positions shown in Fig. 13-18; (b) the minimum programmable timing cycle; (c) the maximum programmable timing cycle?

Solution (a) $T_{sum} = 4T + 8T = 12T = 12 \times 5\ \text{s} = 60\ \text{s} = 1$ min; (b) minimum timing cycle is $1T = 5$ s; (c) with all program switches closed.

$$T_{sum} = T + 2T + 4T + 8T + 16T + 32T + 64T + 128T = 255T$$

$$255T = 255 \times 5\ \text{s} = 1275\ \text{s} = 21\ \text{min}\ 15\ \text{s}$$

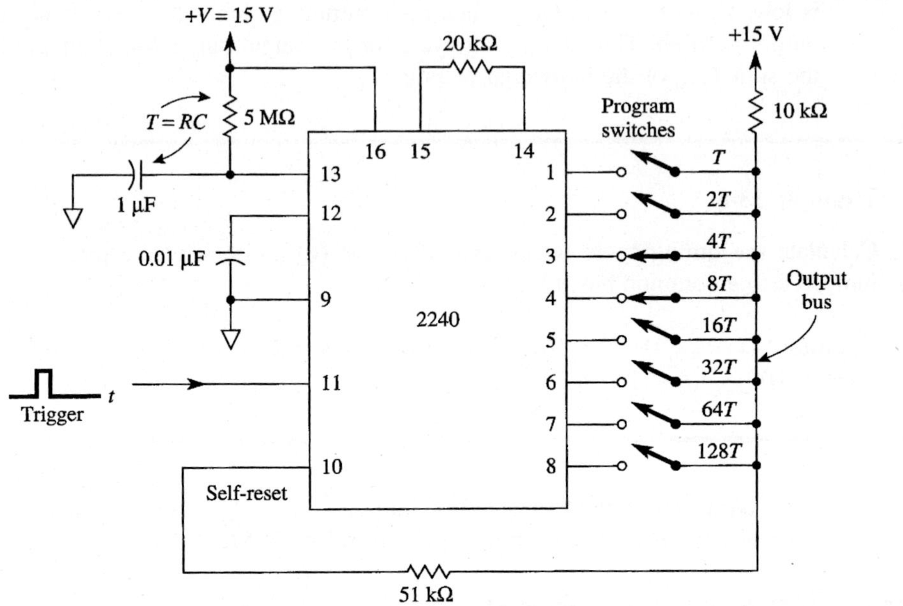

FIGURE 13-18 Programmable timer: 5 s to 21 min, 15 s in 5-s intervals.

13-9.2 Free-Running Oscillator, Synchronized Outputs

The 2240 operates as a free-running oscillator in the circuit in Fig. 13-19. The reset terminal is grounded so that the 2240 will stay in its timing cycle once it is started. When

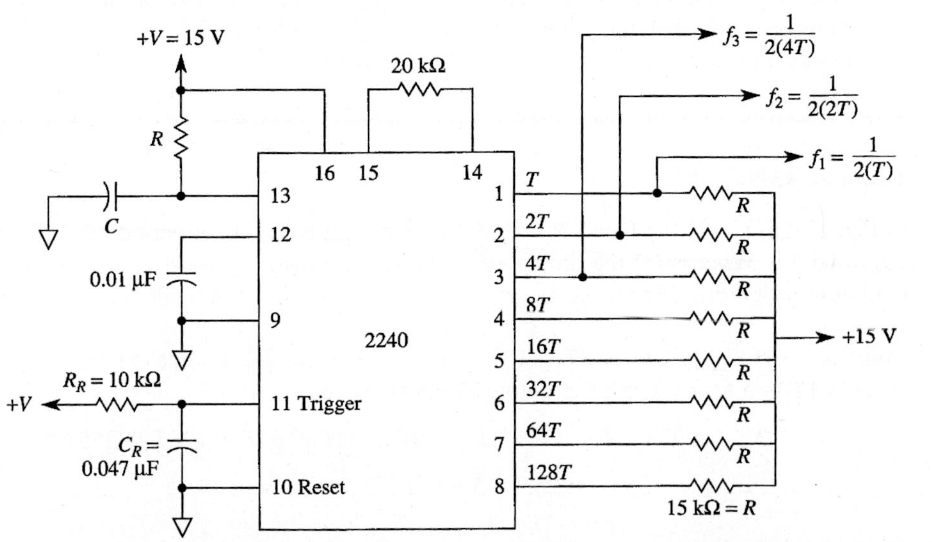

FIGURE 13-19 Free-running oscillator with synchronized outputs.

power is applied, R_R and C_R couple a positive-going pulse into trigger input 11 to start the internal time base oscillator running.

Each output is wired through an external control switch to an individual pull-up resistor. A square-wave output voltage is available at each counter output. Their frequencies have a binary relationship. That is, the frequency available at each pin is one-half the frequency present at the preceding pin. The waveshapes are identical to those in Fig. 13-16(b). Observe that the *period* of the f_1 frequency at pin 1 is twice the time base period rating T or $2(T)$. Thus $f_1 = 1/2T$. At pin 4, the period is $2(8T)$ and $f_4 = 1/16T$.

Example 13-10

In Fig. 13-19, $T = 2.5$ ms; what frequencies are present at (a) output 1; (b) output 2; (c) output 3; (d) output 4?

Solution Tabulating calculations, we obtain:

Pin number	Time base rating	Period	Frequency (Hz)
1	T	$2T = 5$ ms	200
2	$2T$	$4T = 10$ ms	100
3	$4T$	$8T = 20$ ms	50
4	$8T$	$16T = 40$ ms	25

The connections to pins 10 and 11 may be removed to allow the oscillator to be started with a positive-going trigger pulse at pin 11. To stop oscillation, apply a positive-going pulse to reset pin 10.

13-9.3 Binary Pattern Signal Generator

Pulse patterns similar to those shown in Fig. 13-20 are generated by a modified version of Fig. 13-19. The modification requires the eight output resistors to be replaced by program switches and a single 10-kΩ resistor similar to that shown in Fig. 13-18, and it eliminates the 51-kΩ resistor between the output bus and the self-reset terminal.

The output is a train of pulses that depends on which program switches are closed. The period of the pulse pattern is set by the highest program switch that is closed, and the pulse width is set by the lowest program switch that is closed. For example, if the $4T$ (pin 3) and $1T$ (pin 1) switches are closed, the pulse pattern is repeated every $2 \times 4T = 8T$ seconds (see Fig. 13-20). The minimum pulse width is $1T$. To determine the actual pulse pattern, refer to the timing chart in Fig. 13-16(b). If switches $1T$ and $4T$ are closed,

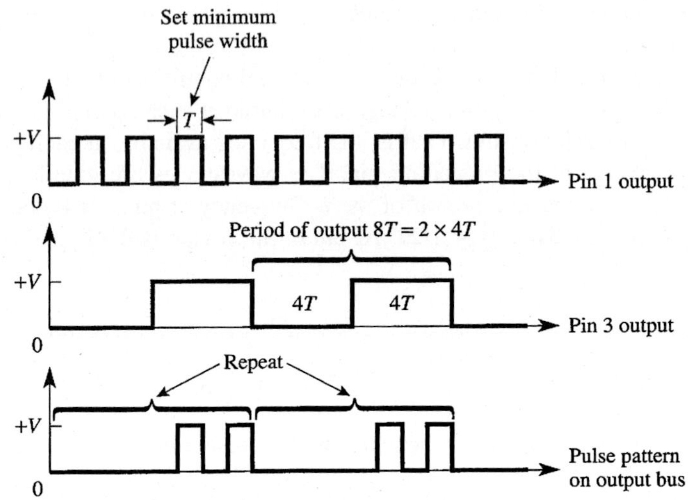

FIGURE 13-20 Binary pattern signal generator with outputs T and $4T$ connected to output bus.

there is an output pulse only when there are high output pulses from each line. The repeating pulse patterns are shown in Fig. 13-20.

13-9.4 Frequency Synthesizer

The output bus in Fig. 13-21(a) is capable of generating any one of 255 related frequencies. Each frequency is selected by closing the desired program switches to program a particular frequency at output V_o.

To understand circuit operation, assume that the output bus goes high. This will drive reset pin 10 high and couple a positive-going pulse into trigger pin 11. The reset terminal going positive resets the 2240 (all outputs low). The positive pulse on pin 10 retriggers the 2240 time base oscillator, to begin generation of a time period that depends on which program switches are closed. For example, assume that switches T and $4T$ are closed in Fig. 13-21(a). The timing for these switches is shown in Fig. 13-21(b). The output bus stays low for $4T$ from pin 3 plus $1T$ from pin 1 before going high (to initiate a reset–retrigger sequence as noted above). The time period and frequency of the output signal V_o are thus expressed by

$$\text{period} = T_{\text{sum}} + T \tag{13-11a}$$

and

$$f = \frac{1}{\text{period}} \tag{13-11b}$$

where T_{sum} is found by adding the time base rating for each output terminal connected to the output bus.

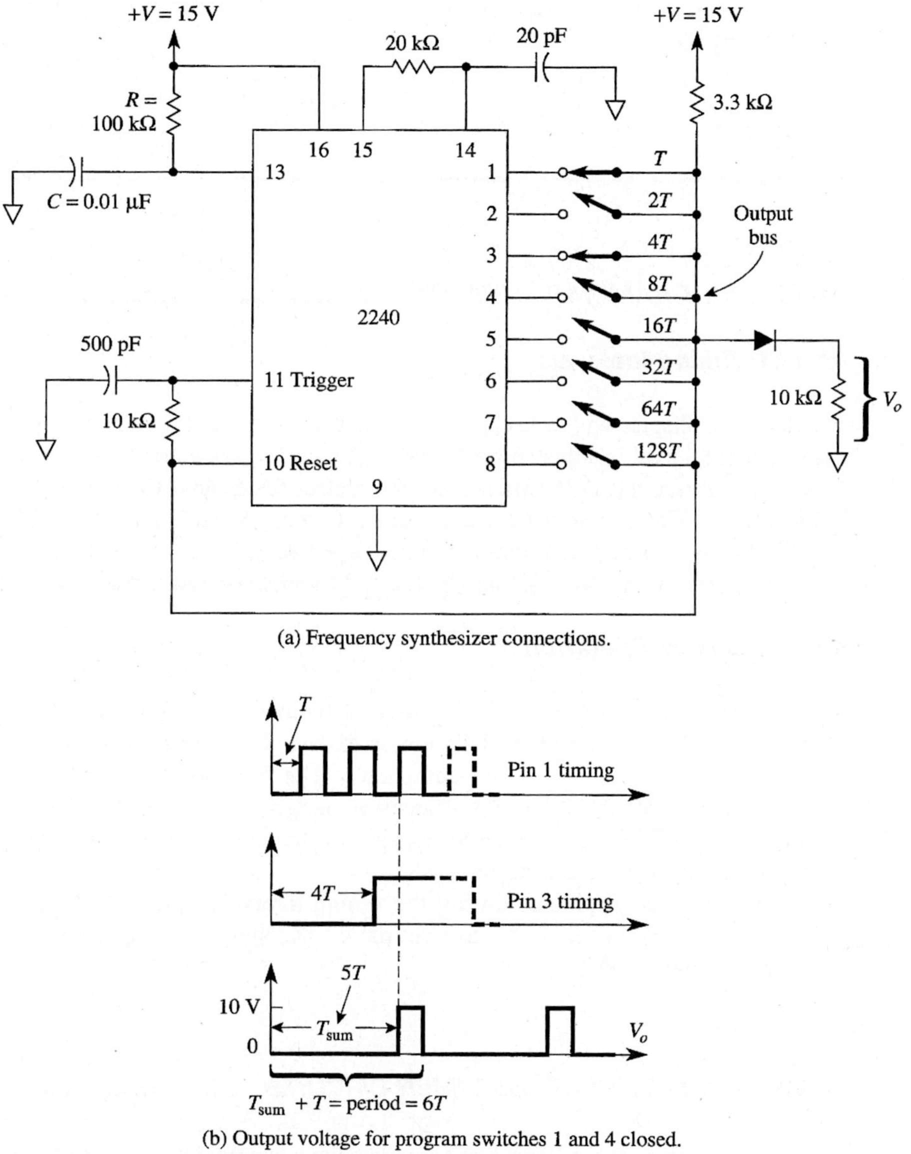

FIGURE 13-21 Frequency synthesizer $T = 1$ ms, $f = 166$ Hz.

Example 13-11

Find the output frequency for Fig. 13-21(a).

Solution From Eq. (13-11a), $T_{sum} = 1T + 4T = 5T$, and period $= (T_{sum} + T) = 6T = 6 \times 1$ ms $= 6$ ms. By Eq. (13-11b),

$$f = \frac{1}{6 \times 10^{-3}\ \text{s}} = 166\ \text{Hz}$$

13-10 SWITCH PROGRAMMABLE TIMER

13-10.1 Timing Intervals

We close this chapter with a useful programmable timer. In the version of Fig. 13-22, the basic timing interval is adjusted for $T = 5.0$ s by timing capacitor C and timing resistor R. A 16-pin, eight-circuit DIP switch is used to select the desired time interval, as shown in Example 13-9. Close switch 1 for a 5-s timer. Closing S_2 adds 10 s, S_3 adds 20 s, and so on. With all switches closed, the maximum time interval is $2^8 \times 5$ s $= 1280$ s or 21 min 20 s. The basic timing interval can be changed by picking new values for R and/or C.

13-10.2 Circuit Operation

Start switch S_s is opened momentarily, driving the trigger pin high to start the timer. All outputs of the XR 2240 go low. This low is extended by any closed select switch(es) S_1 through S_8, to (−) pin 2 of the 301 comparator. The 301's output goes high to light LED D_1 and turn on transistor Q_1. The on-transistor energizes the relay, thus switching the contacts from the NC position to the NO position. This completes the power circuit to turn on any appliance for the timing interval.

To turn *off* an appliance during the timing interval, simply move the ac common wire from terminal A of the relay to terminal C. The supply voltage must be regulated to obtain repeatable results.

13-11 PSPICE SIMULATION OF 555 TIMER CIRCUITS

In this section, we will use PSpice and simulate the performance of two timing circuits—the astable multivibrator shown in Fig. 13-7(a) and the tone-burst circuit shown in Fig. 13-9.

13-11.1 Astable or Free-Running Multivibrator

Refer to Fig. 13-7(a) and create the PSpice model of the 555 timing circuit. Place the following parts on the right-hand side of the work area.

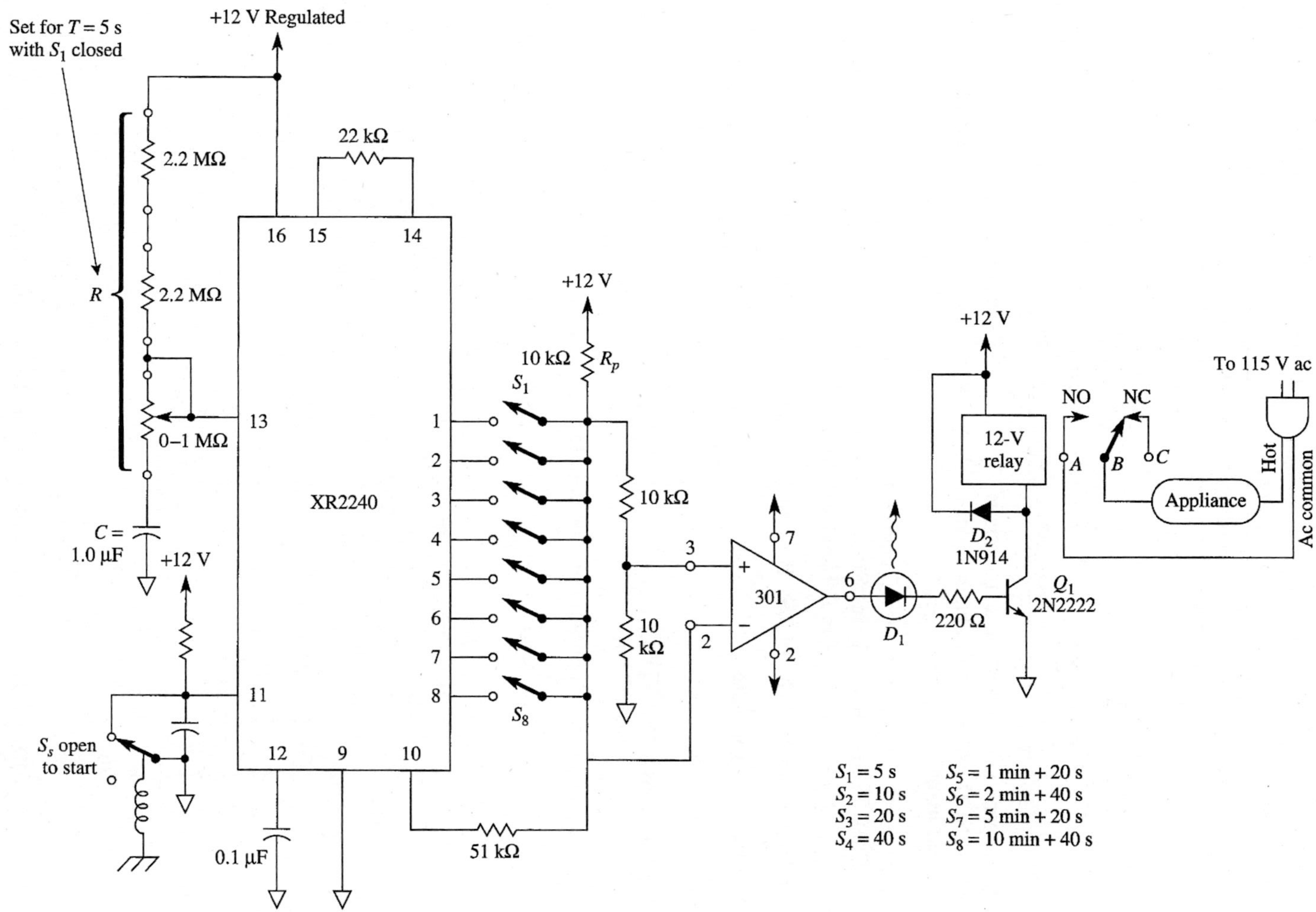

FIGURE 13-22 Switch programmable timer. A basic time interval is set to $T = 5$ s by R and C. Time select switches S_1 through S_8 determine how long the appliance load will be turned on after momentary-open start switch S_s is depressed. $S_1 = 5$ s. S_2 adds 10 s. The remaining switches extend the time interval as shown above.

Draw => Get New Part

Part	Number	Library
=> 555D	1	**eval.slb**
=> VDC	1	**source.slb**
=> R	3	**analog.slb**
=> C	2	**analog.slb**
=> GLOBAL	2	**port.slb**
=> AGND	4	**port.slb**

Arrange the parts as shown in the schematic of Fig. 13-23 and wire the network. Save the file with an **.SCH** extension. To obtain a plot of V_C and V_o versus time, we must initialize the Transient menu.

Analysis => Setup => Select **Transient**
Click **Transient => Print Step:=> 10μs**
=> Final Time: => 2ms

Save the circuit as a file with the **.SCH** extension. Run the simulation

Analysis => Simulate

In the Probe window, select

Trace => Add => V[Vc]
=> V[Vo]

Label the plots and obtain a printout as shown in Fig. 13-24.

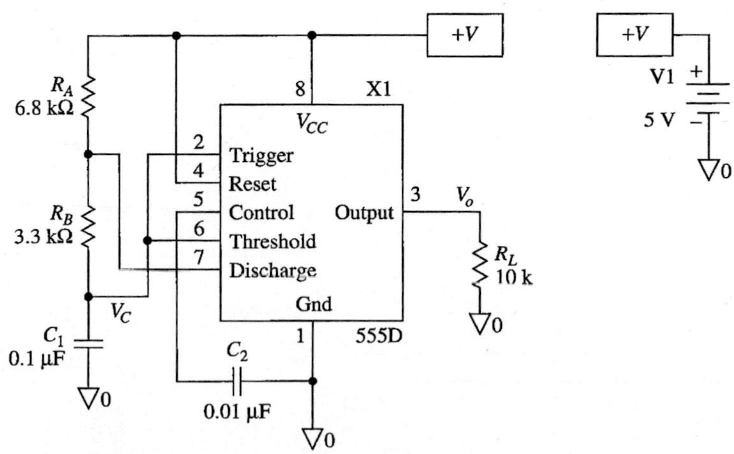

FIGURE 13-23 PSpice model of the astable multivibrator.

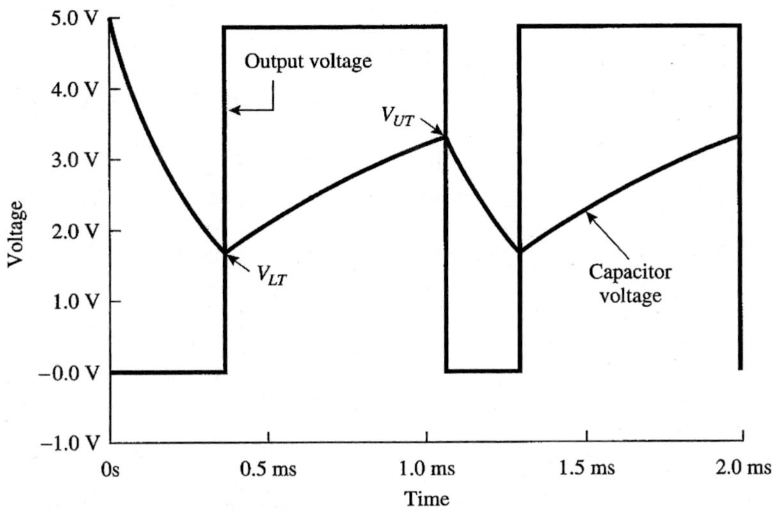

FIGURE 13-24 Capacitor voltage, V_c, and output voltage, V_o, from PSpice model of an astable multivibrator.

13-11.2 Tone-Burst-Control Circuit

Create the PSpice model of the two 555 timers cascaded as a tone-burst-control circuit shown in Fig. 13-9. Use a 220-kΩ resistor in place of the 1-MΩ resistor to set the timing for timer *A*. Place the following parts on the right-hand side of the work area.

Draw => Get New Part

Part	Number	Library
=> **555D**	2	**eval.slb**
=> **VDC**	1	**source.slb**
=> **R**	5	**analog.slb**
=> **C**	4	**analog.slb**
=> **GLOBAL**	3	**port.slb**
=> **AGND**	6	**port.slb**

Arrange the parts and wire the network as shown in the schematic of Fig. 13-25. Label the outputs of the timers *A* and *B* as **Vo1** and **Vo2,** respectively. Save the file with an **.SCH** extension. In order to obtain a plot of V_{o_1} and V_{o_2} versus time, we must initialize the Transient menu.

Analysis => **Setup** => Select **Transient**
Click **Transient** => **Print Step:** => **0.5s**
=> **Final Time:** => **1s**

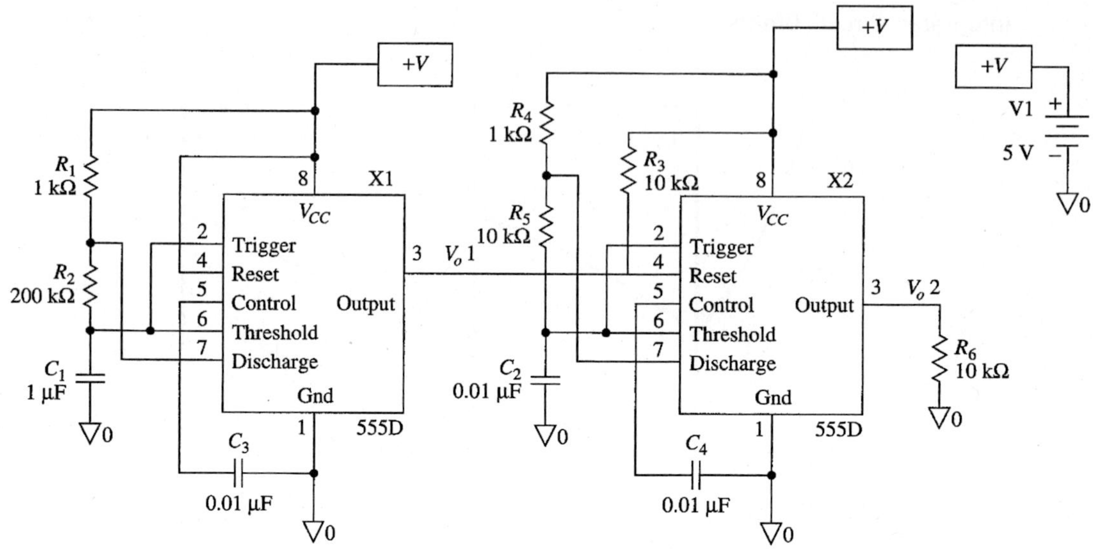

FIGURE 13-25 PSpice model of the tone-burst circuit shown in Fig. 13-9.

Save the circuit as a file with the **.SCH** extension. Run the simulation

Analysis => Simulate

This simulation takes several minutes to execute. In the Probe window, select the Plot and the Trace menus

Plot => Y Axis Settings => Data Range => User Defined => 0V to 6V
Trace => Add => V[Vo_1]
=> V[Vo_2]

Label the plots and obtain a printout as shown in Fig. 13-26.

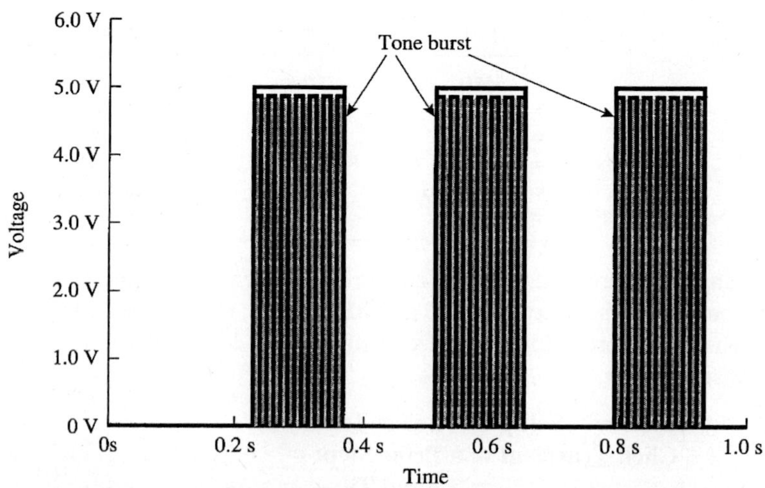

FIGURE 13-26 Tone-burst output from the model shown in Fig. 13-25.

PROBLEMS

13-1. What are the operating modes of the 555 timer?

13-2. In Fig. 13-6(a), $R_A = R_B = 10$ kΩ and $C = 0.1$ μF. Find **(a)** t_{high}; **(b)** t_{low}; **(c)** frequency of oscillation.

13-3. Using the graph of Fig. 13-7, estimate the free-running frequency of oscillation f if $(R_A + 2R_B) = 1$ MΩ and $C = 0.02$ μF.

13-4. What is the duty cycle in Problem 13-2?

13-5. In Example 13-1, R_A and R_B are increased by a factor of 10 to 68 kΩ and 33 kΩ. Find the new frequency of oscillation.

13-6. In Fig. 13-8, R_A and R_B are each reduced to 5 kΩ. What is the effect on **(a)** the duty cycle; **(b)** the period T of the output?

13-7. In Fig. 13-9, at what value should the 10-kΩ resistor be set for a 2-kHz output from the B 555?

13-8. Capacitor C is changed to 0.1 μF in Fig. 13-10. Calculate **(a)** the center frequency f_c when $E = 0$ V; **(b)** the frequency shift for $E = \pm 2$ V.

13-9. In Fig. 13-11(a), $R_A = 100$ kΩ and $C = 0.1$ μF. Find t_{high}.

13-10. R_A is changed to 20 kΩ in Example 13-5. Find t_{high}.

13-11. In Example 13-6(b), what value of R_A is required to divide a 1-kHz signal by 2?

13-12. Refer to Example 13-7. How long will the following output terminals stay low? **(a)** pin 1; **(b)** pin 2; **(c)** pin 5; **(d)** pin 6.

13-13. In Fig. 13-16(a), T is set for 1 ms and pins 2, 4, 6, and 8 are connected to the output bus. Find the timing interval.

13-14. In Problem 13-13, the odd-numbered pins, 1, 3, 5, and 7, are connected to the output bus. Find the timing interval.

13-15. In Example 13-9, C is changed to 0.1 μF and R to 500 kΩ. Find **(a)** the time base period; **(b)** the timing cycle for switch positions shown in Fig. 13-18; **(c)** the maximum timing cycle.

13-16. In Example 13-10, what frequencies are present at pins **(a)** 5; **(b)** 6; **(c)** 7; **(d)** 8?

13-17. In Fig. 13-21, only switches to pins 1, 2, 3, and 4 are closed. Find the output frequency.

CHAPTER 14

Digital-to-Analog Converters

LEARNING OBJECTIVES

Upon completion of this chapter on digital-to-analog converters, you will be able to:

- Write the general input–output equation for a digital-to-analog converter (DAC) and calculate the output for any given input.
- Understand basic DAC specifications.
- Distinguish between a DAC and an MDAC.
- Tell what features must be present to make a DAC compatible with a microprocessor.
- Explain how a microprocessor selects only one DAC out of all peripheral devices and sends data to it.
- Dynamically test a microprocessor-compatible DAC, the AD558.
- Explain the operation of a serial DAC and how to program a microprocessor to send data to it.

14-0 INTRODUCTION

Real-world processes produce analog signals that vary continuously. The rate may be very slow, such as a change in temperature variations, or very fast, such as in an audio system. Analog processes are best described by decimal numbers and letters of the alphabet. Microprocessors and computers, however, use binary patterns to represent numbers, letters, or symbols.

It is not easy to store, manipulate, compare, calculate, or retrieve data with accuracy using analog technology. However, computers can perform these tasks quickly and on an almost unlimited mass of data with precision, using digital techniques. Thus the need for converters to interface between the analog and digital worlds emerged. Analog-to-digital converters (ADCs) allow the analog world to communicate with computers. Computers communicate with people and physical processes via digital-to-analog converters (DACs). This chapter and the next study this interconnection between the analog and digital worlds. This chapter concentrates on digital-to-analog converters while Chapter 15 covers the analog-to-digital interface. The reason for studying digital-to-analog converters first is that some analog-to-digital converters use DAC circuitry as you will see in Chapter 15.

14-1 DAC CHARACTERISTICS

A digital-to-analog converter converts an input binary pattern of logic 1s and 0s to a discrete analog output, either voltage or current. The input–output relationship for a DAC is a set of points corresponding to the input digital code and its equivalent analog output. This relationship is the DAC's *transfer function*. Depending on how the DAC is configured, the transfer can be unipolar (only positive output values) or bipolar (positive and negative values). DACs can be purchased with an analog output voltage or with a current output. Current output DACs require the user to add an op amp to the DAC's output terminal. However, this allows the user to select the best device and set the circuit gain for the application. This section describes some of a DAC's specifications.

14-1.1 Resolution

The circuit symbol and the input–output characteristics of a 3-bit DAC are shown in Figure 14-1(a). Each digital input requires an electrical signal representing either a logic 1 or a logic 0. D_0 is the least significant bit, LSB, and for this example, D_2 is the most significant bit, MSB.

Resolution is the number of distinct analog outputs (voltage or current) that can be produced by a DAC. For an n-bit DAC,

$$\text{resolution} = 2^n \tag{14-1}$$

where n is the number of digital inputs. Therefore, for the 3-input DAC of Fig. 14-1(a), there are $2^3 = 8$ distinct outputs. Figure 14-1(b) is a plot of the discrete analog outputs for the corresponding digital input codes. This plot is referred to as the DAC's transfer

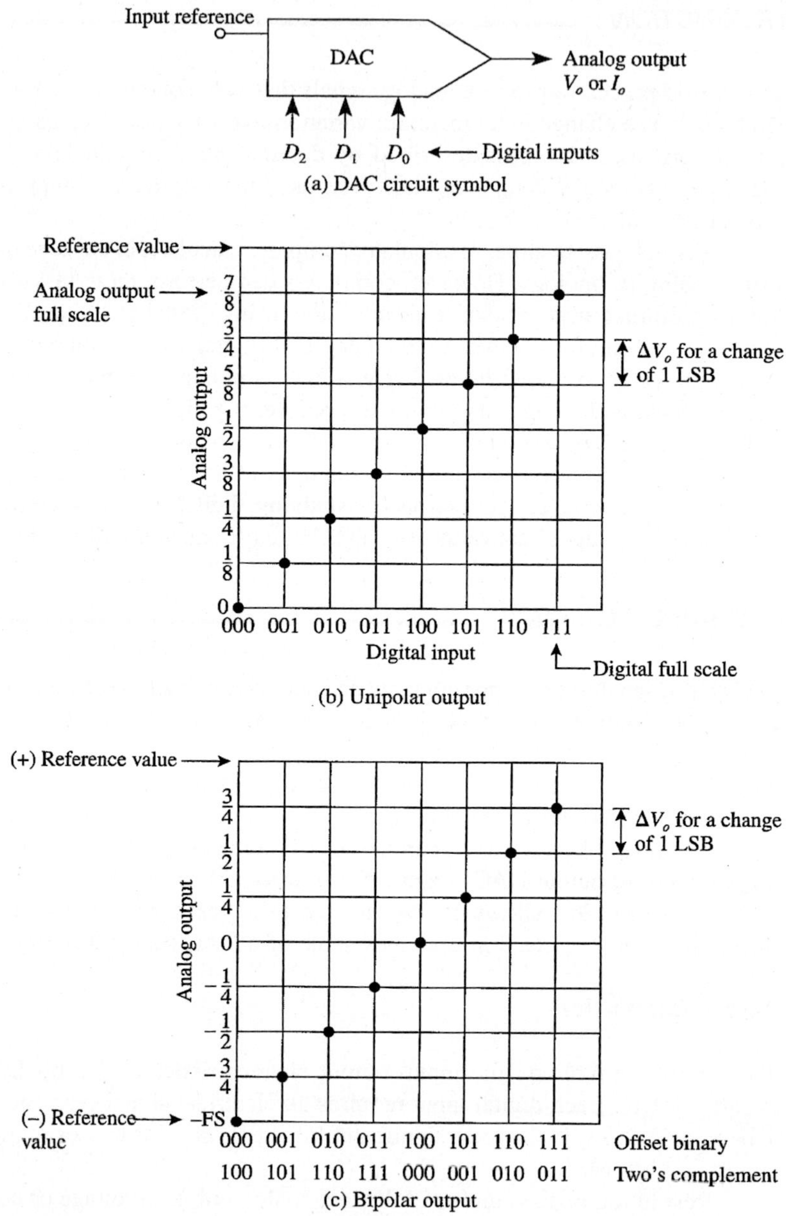

FIGURE 14-1 (a) Circuit symbol; (b) and (c) plot of a 3-bit DAC's transfer function for unipolar and bipolar output.

function. Figure 14-1(b) is a plot for unipolar operation (all ouput values are positive). Figure 14-1(c) is a plot of a transfer function for a 3-bit DAC wired for bipolar operation. The digital input code for bipolar operation may be either offset binary or two's complement. Both codes are shown in Fig. 14-1(c).

The resolution value is used to determine the change in analog output voltage, ΔV_o, resulting from a change of 1 LSB at the digital input. See Figs. 14-1 (b) and (c). In order to determine this change, we need to know the DAC's full-scale output range and its resolution. In equation form,

$$\Delta V_o = \frac{\text{full-scale range}}{2^n} \tag{14-2}$$

Consider the 3-bit DAC of Fig. 14-1(a) with the input reference set at 1.2 V. Then from Eq. (14-2), the change of the output voltage for a change of 1 LSB is 150 mV/bit (1.2 V/8).

Note: A digital code equal to the full-scale range (FSR) does not exist. However, the industry standard positive full scale is the full-scale range less 1 LSB and is given by

$$V_{fs} = V_{ref}\left(1 - \frac{1}{2^n}\right) \tag{14-3}$$

where V_{fs} is the ideal full-scale output voltage when the digital inputs are all 1s. V_{ref} is the output reference value, which is usually equal to a multiple of the input reference. For our applications, V_{ref} equals the input reference voltage. This value may be set by a voltage reference IC as described in Section 2-6.

Example 14-1

An 8-bit DAC wired for unipolar operation has a full-scale voltage range from 0 to 5.12 V. What is (a) the DAC's resolution, (b) the output voltage change per bit, and (c) the ideal full-scale output voltage when all logic 1s are applied at the input?

Solution

(a) Applying Eq. (14-1),

$$\text{resolution} = 2^8 = 256$$

(b) The output change per bit is the output full-scale range (the reference value) divided by the resolution as given by Eq. (14-2).

$$\Delta V_o = \frac{5.12\text{ V}}{256} = 20\text{ mV/bit}$$

(c) Applying Eq. (14-3),

$$V_{fs} = 5.12 \text{ V} \times \left(1 - \frac{1}{2^8}\right) = 5.1 \text{ V}$$

Note: The full-scale output is always 1 LSB less than the maximum full-scale voltage range (the reference value).

Example 14-2

An 8-bit DAC has a resolution of 20 mV/bit. What is the analog output voltage for the following digital input codes (the MSB is the leftmost bit)?

(a) 00010110

(b) 10000000

Solution One method is to obtain the decimal equivalent value of the binary pattern and then multiply the decimal value by 20 mV/bit. The binary weighted code for unipolar operation is

$$D = b_7 2^7 + b_6 2^6 + b_5 2^5 + b_4 2^4 + b_3 2^3 + b_2 2^2 + b_1 2^1 + b_0 2^0$$

where D is the decimal equivalent value.

(a) Applying the binary weighted code yields

$$D = 0 + 0 + 0 + (1)2^4 + 0 + (1)2^2 + (1)2^1 + 0 = 22$$

Then

$$V_o = 22 \times 20 \text{ mV/bit} = 0.44 \text{ V}$$

(b) Applying the binary weighted code yields

$$D = (1)2^7 + 0 + 0 + 0 + 0 + 0 + 0 + 0 = 128$$

Then

$$V_o = 128 \times 20 \text{ mV/bit} = 2.56 \text{ V}$$

Example 14-3

If the DAC of Example 14-1 is wired for bipolar operation, what is the change in output voltage per bit?

Solution For bipolar operation, the full-scale output voltage range is twice the unipolar reference value. Hence,

$$\Delta V_o = \frac{2 \times V_{ref}}{2^8} = \frac{2 \times 5.12\ \text{V}}{256} = 40\ \text{mV/bit}$$

For bipolar operation, each bit change causes twice the output change as unipolar operation. Compare Fig. 14-1(c) with Fig. 14-1(b).

Typical DAC resolutions are 8, 10, 12, 14, 16, 18, 20 bits. However, DACs with higher resolutions are available. A DAC's digital input circuitry is designed for TTL, ECL, or CMOS voltage levels and this specification should be checked to match your application.

14-1.2 Offset Error

The transfer functions of Fig. 14-1 are ideal. They do not show errors that occur in practical devices. Figure 14-2 shows a plot of a transfer function due to an offset error of one LSB. Although this plot is for a positive offset error, the error could have been negative. For most devices, the offset error is usually measured by applying an input digital code of all logic 0s and measuring the analog output—the deviation from zero output is the offset error. *Note:* An offset error does not always occur just at zero. The reason that the offset error is measured at zero is because there usually are no other errors at zero. However, in some DAC this is not so, and the user should check with the manufacturer's data sheet to see if other techniques have been used. Figure 14-2(b) shows an offset error range, positive and negative, for a DAC wired for bipolar operation.

Example 14-4

A 12-bit unipolar DAC has a maximum offset error specified as 0.05% of FSR. The input reference voltage is set at 10.24 V. (a) What is the offset voltage? (b) What is the offset value in terms of least significant bit(s)?

Solution (a)

$$\text{Offset voltage} = 0.05\% \times 10.24\ \text{V} = 5\ \text{mV}$$

(b) In terms of least significant bit(s)

$$\Delta V_o = \frac{10.24\ \text{V}}{2^{12}} = 2.5\ \text{mV/bit}$$

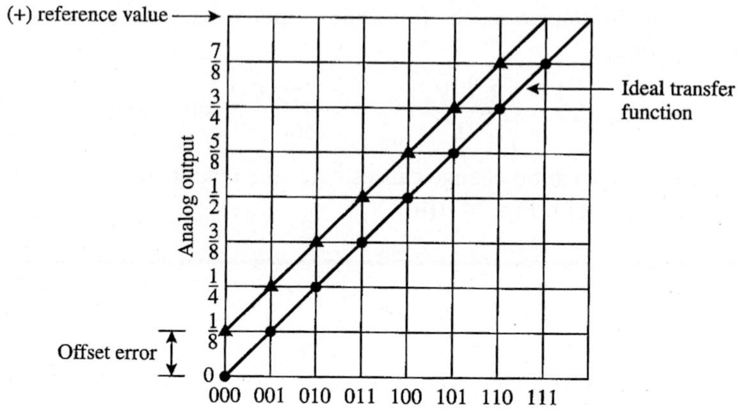

(a) An offset range of +1 LSB.

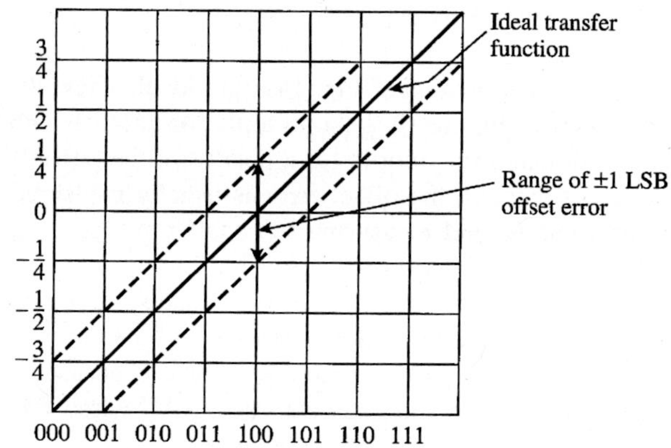

(b) An offset range of ±1 LSB.

FIGURE 14-2 Offset errors.

$$\therefore \frac{\text{Offset voltage}}{\Delta V_o} = \frac{5\text{ mV}}{2.5\text{ mV/bit}} = 2\text{ LSBs}$$

Although this example specified only a positive offset error, offset errors are usually specified as ± percentages of FSR.

14-1.3 Gain Error

Another error that can exist in the output of a DAC is gain error. Figure 14-3 shows the result of gain error. The slope of the transfer function differs from the ideal case. In the ideal case, the slope can be determined by drawing a straight line through the end points.

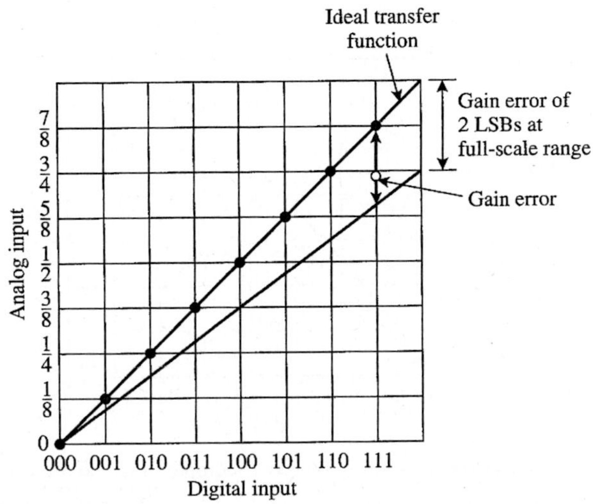

FIGURE 14-3 A gain error of 2 LSBs at full-scale digital input.

The slope is the gain of the transfer function. Figure 14-3 shows the effects of gain error and how it differs from the ideal transfer function. Gain error is given as a percentage of the full-scale range because it usually affects the output by the same percentage. To measure gain error, first determine the offset error, then set the digital inputs to all logic 1s, and finally measure the analog output. Ideally this output should be 1 LSB away from reference value. In terms of voltage,

$$\text{Gain error (\%)} = \left(\frac{V_{11} - V_{os}}{V_{ref}\left(1 - \dfrac{1}{2^n}\right)} - 1 \right)(100\%) \tag{14-4}$$

where V_{11} is the output voltage when the input digital code is all logic 1s, V_{os} is the offset error, and V_{ref} is the full-scale range (reference) value. Note the denominator is the ideal full-scale output voltage [see Eq. (14-3)].

Example 14-5

An 8-bit DAC wired for unipolar operation has a maximum gain error of ±0.2% of FSR. If the input reference is set at 5.12 V, what is the minimum output voltage when the input is all logic 1s? Assume the offset error is zero.

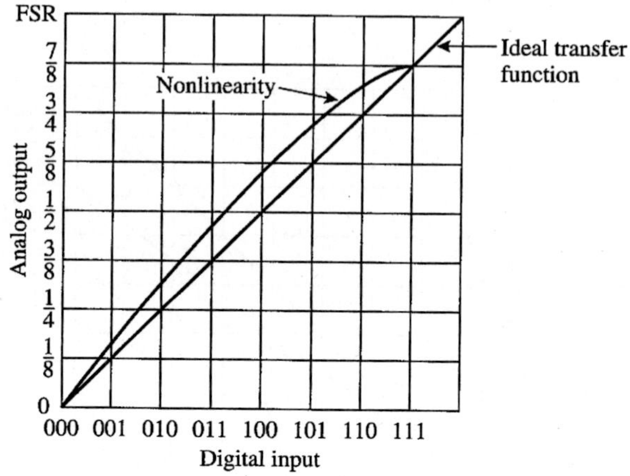

FIGURE 14-4 Nonlinearity error compared to the ideal transfer function.

Solution The denominator of Eq. (14-4) is 5.1 V [refer to Example 14-1 part (c)]. Substitute the values into Eq. (14-4).

$$-0.2\% = \left(\frac{V_{11} - 0}{5.1\ \text{V}} - 1\right) \times 100\%$$

Solving for V_{11} yields V_{11} = 5.0898 V. Note the output voltage cannot exceed 5.1 V.

14-1.4 Monotonic

A DAC is said to be monotonic if the analog output increases or remains the same as the digital input increases. This results in the output always being single-valued (an input digital code always produces one output value).

14-1.5 Relative Accuracy

Consider Fig. 14-1(a). The ideal transfer function is a straight line drawn from zero through full scale. Relative accuracy or integral nonlinearity (INL) is the maximum deviation at any input digital code from the ideal. See Fig. 14-4.

14-2 DIGITAL-TO-ANALOG CONVERSION PROCESS

14-2.1 Block Diagram

The block diagram for a resistive type of DAC is drawn in Fig. 14-5. Reference voltage V_{ref} is connected to a resistance network. A digital input code, via control circuitry, flips

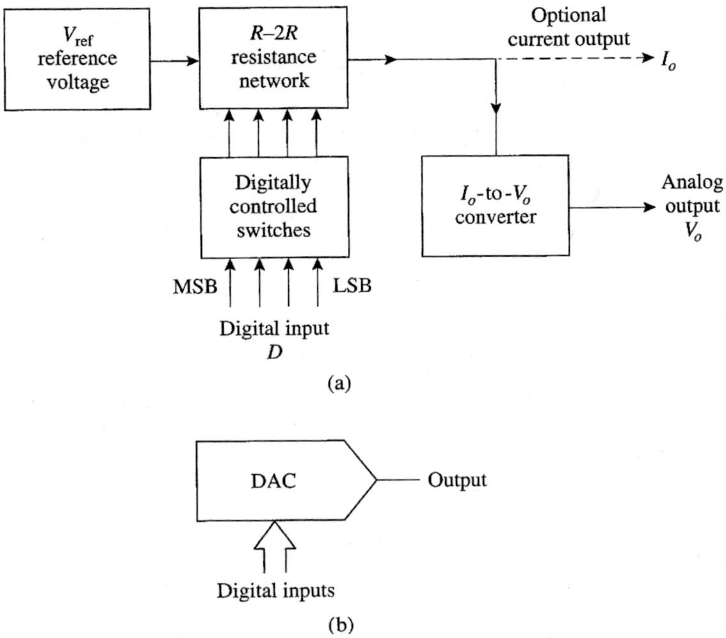

FIGURE 14-5 Block diagram and circuit symbol for a basic DAC.

switches (one for each bit) connected to the resistance network. The output of the resistance network is in the form of a current. This current may then be converted to a voltage. As previously mentioned, both current and voltage outputs are analog representations of the digital input code.

The actual digital-to-analog conversion takes place within the resistance network. Accordingly, we begin our study of DAC circuitry by looking next at the standard resistance network, called an R–$2R$ ladder network.

14-2.2 R–2R Ladder Network

A 4-bit R–$2R$ ladder network is drawn in Fig. 14-6. Each digital input controls the position of its corresponding current switch. A current switch steers its ladder current either into a real ground (position 0) or a virtual ground (position 1). Thus the wiper of each switch is always at ground potential, so that the rung currents are constant except for the brief transition time of each switch.

In Fig. 14-6, rail currents flow horizontally; rung currents flow down through the bit switches. The rail current I_1 enters node 0, where it sees a resistance R_0. R_0 is the equivalent resistance of a $2R$ resistor via switch D_0 to ground in parallel with a $2R$ terminate resistor. Thus $R_0 = 2R \parallel 2R = R$. As rail current I_1 leaves node 1, it sees R in series with $R_0 = R$ or $2R$. If we work our way back from the termination end to the voltage source, the value of resistance "looking" into a node is R. As shown in Fig. 14-6,

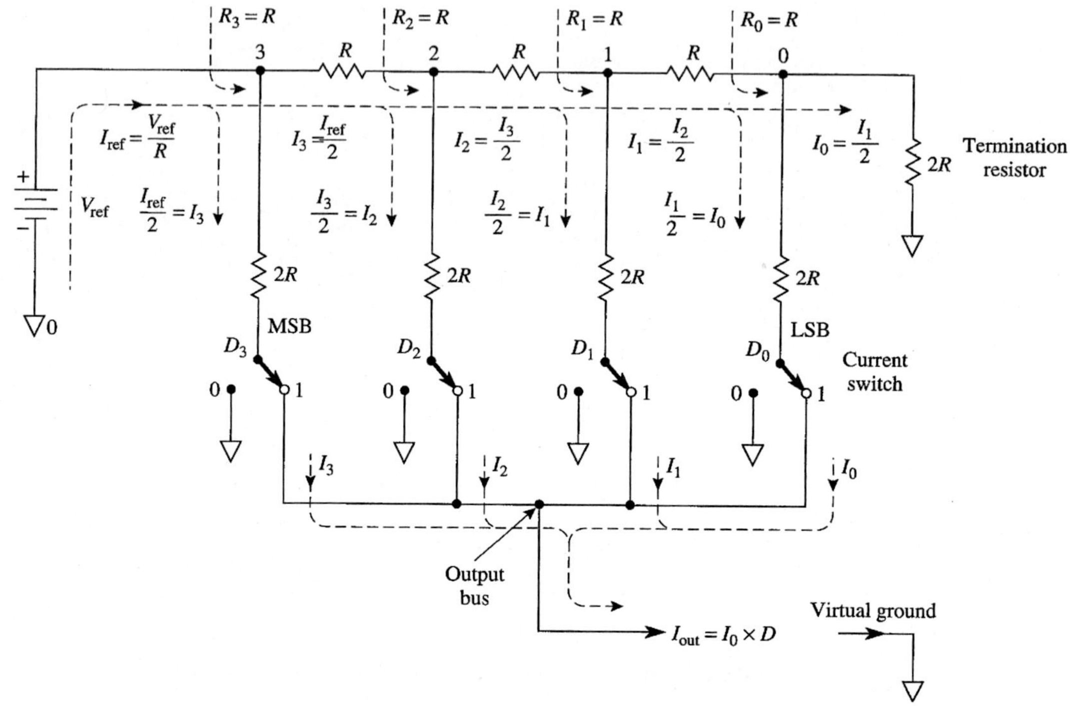

FIGURE 14-6 This R–$2R$ ladder converts the digital input code into an analog output current I_{out}.

$R_3 = R_2 = R_1 = R_0 = R$. R is called the *characteristic resistance* of the ladder network. In other words, V_{ref} sees the entire ladder network as a single resistor equal to R.

14-2.3 Ladder Currents

Since V_{ref} sees the ladder network as a resistance R, rail current I_{ref} is

$$I_{ref} = \frac{V_{ref}}{R} \tag{14-5}$$

The current pattern of the R–$2R$ network in Fig. 14-6 is analyzed as follows: The current I_{ref} splits into two equal parts at node 3: the rung current $I_3 = I_{ref}/2$ and the rail current $I_3 = I_{ref}/2$. Each rail current divides equally again at each node as it proceeds down the ladder. Rung currents are evaluated from

$$I_3 = \frac{I_{ref}}{2} \qquad I_2 = \frac{I_3}{2} = \frac{I_{ref}}{4}$$
$$I_1 = \frac{I_2}{2} = \frac{I_{ref}}{8} \qquad I_0 = \frac{I_1}{2} = \frac{I_{ref}}{16} \tag{14-6}$$

I_0 is the current controlled by the LSB switch.

14-2.4 Ladder Equation

The output current bus receives current from a rung if the bit switch is in position 1. To write the output–input equation for the ladder network, we observe that I_{out} is the sum of all rung currents steered into the output bus by the bit switches. In equation form,

$$I_{out} = I_0 \times D \qquad (14\text{-}7)$$

where D equals the decimal value of the digital input and I_0 is the smallest value of current in the ladder network. If we define I_0 as the resolution of the ladder, the output–input equation (14-7) can be expressed as

$$I_{out} = \text{resolution of 1 LSB} \times D \qquad (14\text{-}8)$$

where

$$\text{current resolution of 1 LSB} = I_0 = \frac{I_{ref}}{2^n} = \frac{1}{2^n} \times \frac{V_{ref}}{R} \qquad (14\text{-}9)$$

Example 14-6

The 4-bit resistance ladder of Fig. 14-6 has resistor values of $R = 10\ \text{k}\Omega$ and $2R = 20\ \text{k}\Omega$. V_{ref} equals 10 V. Find (a) the resolution of the ladder; (b) its output–input equation; (c) I_{out} for a digital input of 1111.

Solution (a) From Eq. (14-9),

$$\text{resolution of 1 LSB} = \frac{1}{2^n} \times \frac{V_{ref}}{R}$$

$$= \frac{1}{2^4} \times \frac{10\ \text{V}}{10\ \text{k}\Omega}$$

$$= \frac{1}{16} \times 1\ \text{mA} = 62.5\ \mu\text{A}$$

(b) Applying Eq. (14-8) yields

$$I_{out} = 62.5\ \mu\text{A} \times D$$

(c) The decimal value of binary 1111 is 15; therefore, $D = 15$ and

$$I_{out} = 62.5\ \mu\text{A} \times 15 = 0.9375\ \mu\text{A}$$

14-3 VOLTAGE OUTPUT DACs

As shown in Fig. 14-7, the output ladder current can be converted into a voltage by adding an op amp and a feedback resistor. Output voltage V_o is given by

$$V_o = -I_{out} R_f \tag{14-10a}$$

Substituting for I_{out} from Eq. (14-8) gives

$$V_o = -(\text{current resolution} \times D) \times R_f \tag{14-10b}$$

Rewrite Eq. (14-10b) as

$$V_o = -(\text{current resolution} \times R_f) \times D \tag{14-10c}$$

The coefficient of D is the voltage resolution and is given as

$$\text{voltage resolution} = I_0 R_f \tag{14-10d}$$

and V_o can be written very simply as

$$V_o = -\text{voltage resolution} \times D \tag{14-10e}$$

In terms of the actual hardware, V_o is expressed as

$$V_o = -\left(\frac{V_{ref}}{R} \times \frac{1}{2^n} R_f\right) \times D \tag{14-11}$$

Example 14-7

For the voltage output DAC of Fig. 14-7, find (a) its voltage resolution of 1 LSB and (b) V_o when the digital input is 1111.

Solution (a) From Example 14-6, the value of $I_0 = 62.5\ \mu\text{A}$. From Eq. (14-10d),

$$\text{voltage resolution} = I_0 R_f = 62.5\ \mu\text{A} \times 10\ \text{k}\Omega = 0.625\ \text{V}$$

A 1-bit input change causes a 0.625-V output voltage change.

(b) Calculate the DAC's performance equation from Eq. (14-11).

$$V_o = -\left(\frac{10\ \text{V}}{10\ \text{k}\Omega} \times \frac{1}{2^4} \times 10\ \text{k}\Omega\right) \times D = -0.625\ \text{V} \times D$$

For a digital input of 1111, $D = 15$. Therefore,

$$V_o = -0.625\ \text{V} \times 15 = -9.375\ \text{V}$$

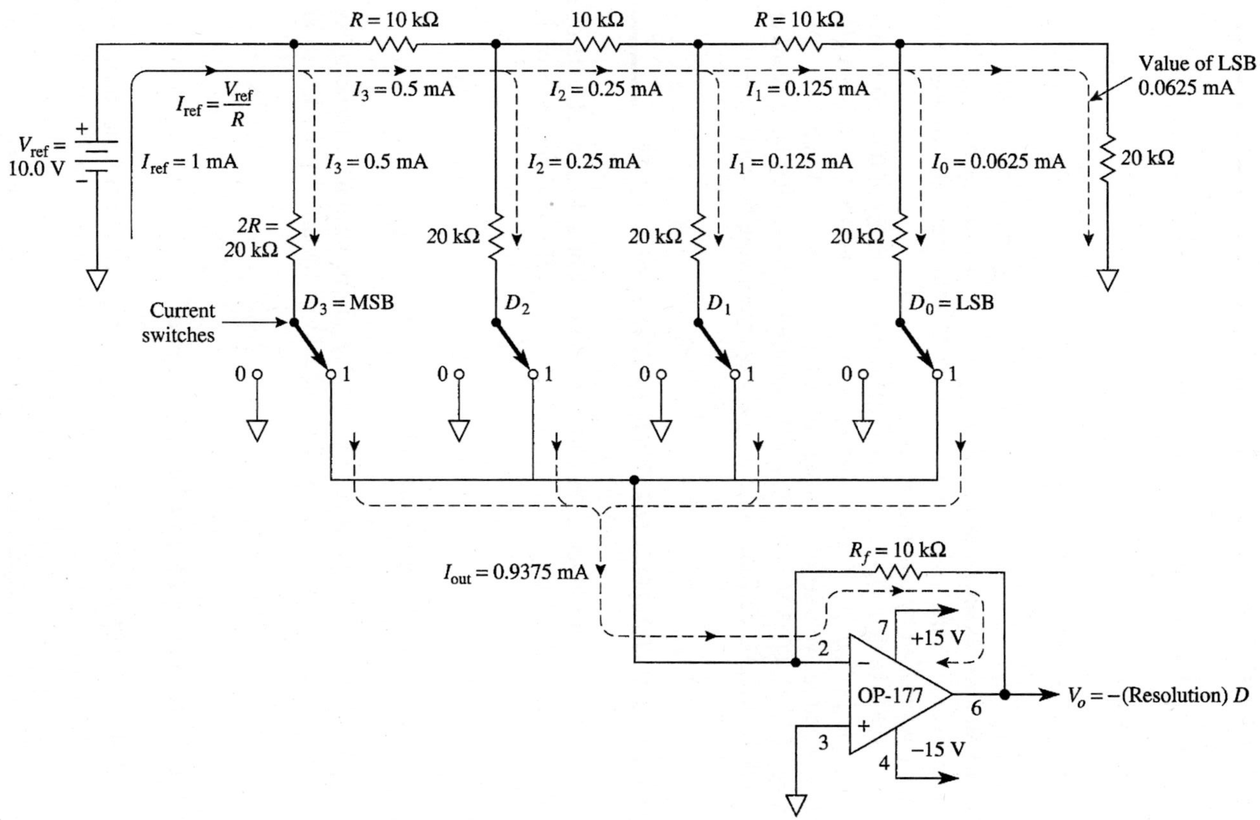

FIGURE 14-7 I_{out} of the R–$2R$ ladder is converted into a voltage by the op amp and feedback resistor R_f.

14-4 MULTIPLYING DAC

Equation (14-11) can be rewritten to show how an MDAC or multiplying DAC operates.

$$V_o = (\text{constant}) \times V_{\text{ref}} \times D \tag{14-12}$$

where

$$\text{constant} = -\frac{R_f}{2^n\, R}$$

Equation (14-12) shows that V_o is the *product* of two input signals, V_{ref} and D, and *both* signals can be variables. One example for MDAC use is the volume control of an audio signal controlled by a microprocessor.

Suppose that V_{ref} is an audio signal that varies from 0 to 10 V in Fig. 14-7. From Example 14-7, if $D = 0001$, V_o would vary from 0 to 0.625 V. If the digital input is 1000, $D = 8$ and the volume of V_o would increase from 0 to $8 \times 0.625\ \text{V} = 5\ \text{V}$. Maximum volume of $15 \times 0.625\ \text{V} = 9.375\ \text{V}$ will occur when the digital input word is 1111, $D = 15$. Thus the MDAC performs as a digitally operated volume control.

14-5 8-BIT DIGITAL-TO-ANALOG CONVERTER: THE DAC-08

The DAC-08 is a low-cost, fast MDAC, housed in a 16-pin DIP. Its operating principles are examined by reference to the task performed by each of its terminals in Fig. 14-8.

14-5.1 Power Supply Terminals

Pins 13 and 3 are the positive and negative supply terminals, respectively, and can have any value from ± 4.5 to ± 18 V. They should be bypassed with 0.1-μF and 0.01-μF capacitors, as shown in Fig. 14-8a.

14-5.2 Reference (Multiplying) Terminal

Flexibility of the DAC-08 is enhanced by having *two* rather than one reference input. Pins 14 and 15 allow positive or negative reference voltages, respectively. A positive reference voltage input is shown in Fig. 14-8a.

The user can adjust the DAC-08's input ladder current I_{ref} quite easily from 4 μA to 4 mA with a typical value of 2 mA.

$$I_{\text{ref}} = \frac{V_{\text{ref}}}{R_{\text{ref}}} \tag{14-13}$$

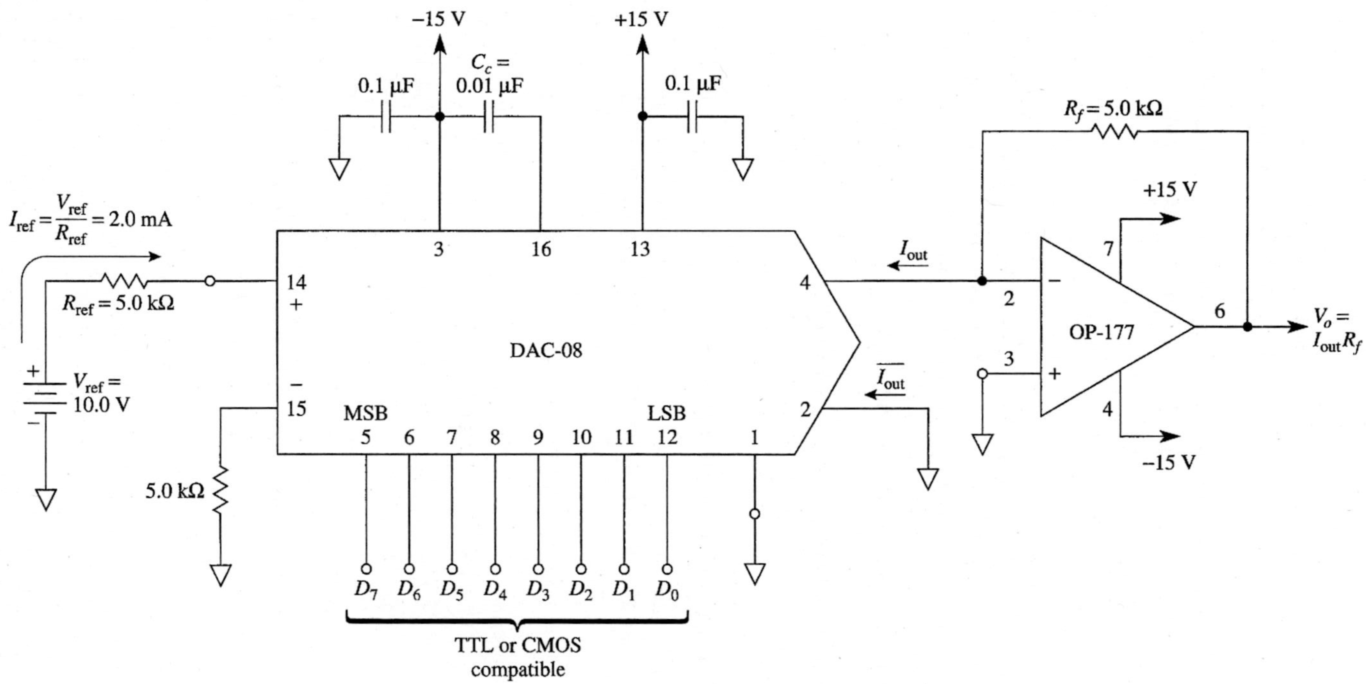

(a) DAC-08 wired for positive output voltages.

	Digital inputs								Analog output	
	D_7	D_6	D_5	D_4	D_3	D_2	D_1	D_0	I_{out}	V_o
LSB	0	0	0	0	0	0	0	1	7.812 μA	39 mV
Half scale	1	0	0	0	0	0	0	0	1.000 mA	5.0 V
Full scale	1	1	1	1	1	1	1	1	1.992 mA	9.96 V

(b) Summary of Examples 14-7 and 14-8.

FIGURE 14-8 An 8-bit DAC-08 is wired for unipolar voltage in (a). I_{out} has values given in (b) for three digital input words. The op amp converts I_{out} to voltage V_o.

14-5.3 Digital Input Terminals

Pins 5 through 12 identify the digital input terminals. Pin 5 is the most significant bit (MSB), D_7. Pin 12 is the LSB, D_0. The terminals are TTL or CMOS compatible. Logic input "0" is 0.8 V or less. Logic "1" is 2.0 V or greater, regardless of the power supply voltages. Usually, pin 1, V_{LC}, is grounded. However, it can be used to adjust the *logic input threshold* voltage V_{TH} according to $V_{TH} = V_{LC} + 1.4$ V. These digital inputs control eight internal current switches.

14-5.4 Analog Output Currents

Two current output terminals are provided in Fig. 14-8 to increase the DAC-08's versatility. Pin 4 conducts output current I_{out} and pin 2 conducts its complement, $\overline{I_{out}}$. If an internal switch is positioned to "1," its ladder rung current flows in the I_{out} bus. If positioned to "0," ladder rung current flows in the $\overline{I_{out}}$ bus.

The current value of 1 LSB (resolution) is found from

$$\text{current resolution} = (\text{value of 1 LSB}) = \frac{V_{ref}}{R_{ref}} \times \frac{1}{2^n} \tag{14-14a}$$

I_{out} is calculated from

$$I_{out} = (\text{value of 1 LSB}) \times D \tag{14-14b}$$

where D is the decimal value of the digital input word. The full-scale output current in the pin 4 output bus occurs when the digital input is 11111111, so that $D = 255$. Let's define this current as I_{fs}, where

$$I_{fs} = (\text{value of 1 LSB}) \times 255 \tag{14-15a}$$

The sum of all ladder rung currents in the DAC-08 equals I_{fs}. Since this sum always divides between the I_{out} and $\overline{I_{out}}$, the value of $\overline{I_{out}}$ is given by

$$\overline{I_{out}} = I_{fs} - I_{out} \tag{14-15b}$$

Example 14-8

Calculate (a) the ladder input current I_{ref} of the DAC-08 in Fig. 14-8; (b) the current value of 1 LSB.

Solution (a) From Eq. (14-13),

$$I_{ref} = \frac{10\text{ V}}{5\text{ k}\Omega} = 2\text{ mA}$$

(b) From Eq. (14-14a),

$$\text{current value of 1 LSB} = \frac{10\text{ V}}{5\text{ k}\Omega} \times \frac{1}{2^8} = 7.812\ \mu\text{A}$$

Example 14-9

For the DAC-08 circuit in Fig. 14-8, find the values of I_{out} and $\overline{I_{out}}$ when the digital input words are (a) 00000001; (b) 10000000; (c) 11111111.

Solution Example 14-8 showed that current output resolution is 7.812 μA/bit. From Eq. (14-15a), evaluate I_{fs}.

$$I_{fs} = (\text{resolution})255 = 7.812\ \mu\text{A} \times 255 = 1.992\text{ mA}$$

The value of D is 1 for (a), 128 for (b), and 255 for (c). I_{out} can now be found from Eq. (14-14b):

(a) $I_{out} = 7.812\ \mu\text{A} \times 1 = 7.812\ \mu\text{A}$ for 00000001 input
(b) $I_{out} = 7.812\ \mu\text{A} \times 128 = 1.000\text{ mA}$ for 10000000 input
(c) $I_{out} = 7.812\ \mu\text{A} \times 255 = 1.992\text{ mA}$ for 11111111 input

From Eq. (14-15b),

(a) $\overline{I_{out}} = 1.992\text{ mA} - 7.812\ \mu\text{A} = 1.984\text{ mA}$
(b) $\overline{I_{out}} = 1.992\text{ mA} - 1.0\text{ mA} = 0.992\text{ mA}$
(c) $\overline{I_{out}} = 1.992\text{ mA} - 1.992\text{ mA} = 0$

The results of Examples 14-8 and 14-9 are tabulated in Fig. 14-8(b).

14-5.5 Unipolar Output Voltage

In Fig. 14-8(a), the DAC-08's current output I_{out} is converted to an output voltage V_o by an external op amp and resistor R_f. The voltage output has a resolution of

$$\text{voltage resolution of 1 LSB} = \frac{V_{ref}}{R_{ref}} \times R_f \times \frac{1}{2^n} \tag{14-16a}$$

and V_o is given by

$$V_o = \text{voltage resolution} \times D = I_{out}R_f \tag{14-16b}$$

Example 14-10

For the DAC-08 circuit of Fig. 14-8(a), find V_o for digital inputs of (a) 00000001; (b) 11111111.

Solution From Eq. (14-16a),

$$\text{voltage resolution} = (10\text{ V})\,\frac{5\text{ k}\Omega}{5\text{ k}\Omega} \times \frac{1}{256} = 39.0\text{ mV/bit}$$

(a) From Eq. (14-16b) with the value of $D = 1$,

$$V_o = 39.0\text{ mV} \times 1 = 39.0\text{ mV for 00000001 input}$$

(b) The value of $D = 255$. From Eq. (14-16b),

$$V_o = 39.0\text{ mV} \times 255 = 9.961\text{ V for 11111111 input}$$

14-5.6 Bipolar Analog Output Voltage

The versatility of the DAC-08 is shown by wiring it to give a bipolar analog output voltage in response to a digital input word [Fig. 14-9(a)]. The op amp and two resistors convert the *difference* between I_{out} and $\overline{I_{out}}$ into a voltage V_o:

$$V_o = (I_{out} - \overline{I_{out}})R_f \tag{14-17}$$

I_{out} drives V_o positive and $\overline{I_{out}}$ drives V_o negative. If the digital input word increases by 1 bit, I_{out} increases by 1 LSB. However, $\overline{I_{out}}$ must therefore *decrease* by 1 LSB. Therefore, the differential output current changes by 2 LSBs; thus we would expect the bipolar output voltage span to be twice that of a unipolar output (see Example 14-3).

V_{ref} has been increased slightly in Fig. 14-9(a) so that I_{ref} increases to 2.048 mA [Eq. (14-13)]. This increases the current value of 1 LSB to an even 8 μA [Eq. (14-14a)]. We show how the output voltage responds to digital inputs by an example.

Example 14-11

For the circuit of Fig. 14-9(a), calculate V_o for digital inputs of (a) 00000000; (b) 01111111; (c) 10000000; (d) 11111111.

Solution The current value of 1 LSB equals 8 μA. From Eq. (14-15a), $I_{fs} = (8\ \mu\text{A})255 = 2.040$ mA.

(a) From Eq. (14-14b), $I_{out} = (8\ \mu\text{A}) \times 0 = 0$. Then from Eq. (14-15b), $\overline{I_{out}} = 2.040\text{ mA} - 0 = 2.04$ mA. Find V_o from Eq. (14-17):

$$V_o = (0 - 2.04\text{ mA})(5\text{ k}\Omega) = -10.20\text{ V}$$

The values of I_{out}, $\overline{I_{out}}$, and V_o are calculated for (b), (c), and (d) and are summarized in Fig. 14-9(b).

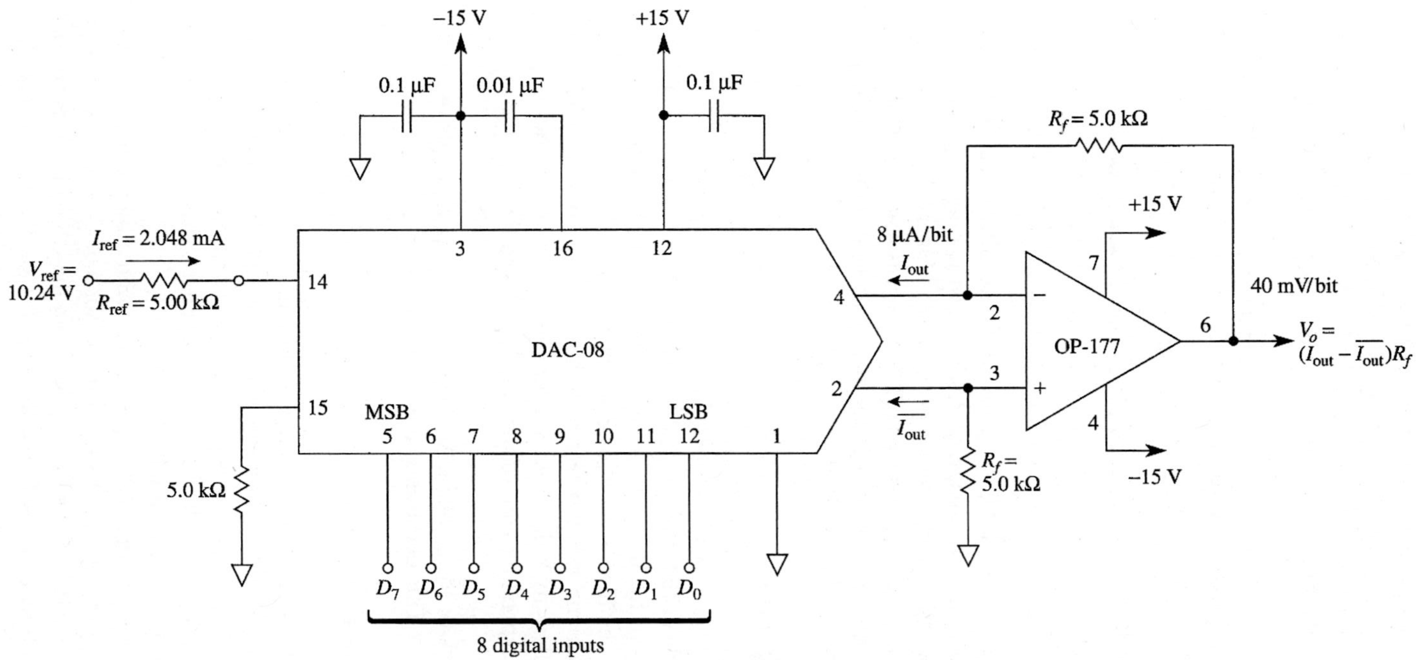

(a) DAC-08 wired for bipolar output voltage.

	Digital inputs								Analog outputs		
	D_7	D_6	D_5	D_4	D_3	D_2	D_1	D_0	I_{out} (mA)	$\overline{I_{out}}$ (mA)	V_o (V)
Negative full scale	0	0	0	0	0	0	0	0	0	2.040	−10.20
Negative zero	0	1	1	1	1	1	1	1	1.016	1.024	−0.040
Positive zero	1	0	0	0	0	0	0	0	1.024	1.016	0.040
Positive full scale	1	1	1	1	1	1	1	1	2.040	0	10.20

(b) Tabulated solutions for Example 14-9.

FIGURE 14-9 The op amp converts complementary output currents of the DAC-08 into bipolar output voltages. The op amp is wired as a differential current-to-voltage converter.

Note that full-scale negative output voltage of -10.20 V occurs for an all-zero digital input. All ones give a positive full-scale output of plus 10.20 V. Note also that V_o never goes to precisely zero volts. When I_{out} is less than $\overline{I_{out}}$ by 8 μA (01111111), V_o equals -40 mV. Since this is the closest that V_o approaches to 0 V from negative full-scale, $V_o = -40$ mV is called negative zero.

Many DACs must operate under the control of a microprocessor or computer. We therefore next present a microprocessor-compatible DAC.

14-6 MICROPROCESSOR COMPATIBILITY

14-6.1 Interfacing Principles

The programmer views the location of a DAC's register(s) or any other peripheral chip's registers as an address in the total memory space. DACs have "write only" registers. This means that a DAC has a register that the microprocessor can send binary digits to via the data bus. An ADC's registers are "read only" registers. These devices have registers whose contents can be "read" by the microprocessor via the data bus. Both DACs and ADCs have logic that permits selection via the address bus.

14-6.2 Memory Buffer Registers

"Read only" or "write only" registers have two operating states: *transparent* and *latching*. An idle register always remembers (latches) the last digital word written into it, and the register can be disconnected from the data bus. More specifically, the interface between data bus and register can be in the high-*Z* state, or essentially an open circuit.

When a register is *transparent,* it is connected to the data bus. For example, an 8-bit register in a DAC would allow its 8 data bits, D_7 through D_0, to "read" the logic 1s and/or 0s present on each corresponding wire of the data bus placed there by the microprocessor (see Fig. 14-10).

How does the microprocessor locate one DAC, out of all other peripherals or memory addresses, that it is selected? This question is answered next.

14-6.3 The Selection Process

Every DAC register has an address just like any memory location of the microprocessor. To write to one particular DAC, the microprocessor places the address of that DAC on the address bus (see Fig. 14-10). One output of a local decoder goes low to enable a chip select, $\overline{CS}$, terminal on the selected DAC. The DAC's digital input buffer registers do not yet become transparent. The DAC is only partially selected.

To fully select the DAC, the microprocessor places a low on a line to the *chip enable,* $\overline{CE}$, terminal. This line is controlled by the microprocessor's read/write line, which may also be referred to as MRW, MEMW, or R/$\overline{W}$. When both the $\overline{CS}$ and $\overline{CE}$ terminals are low, only one DAC can communicate with the microprocessor. Its internal register

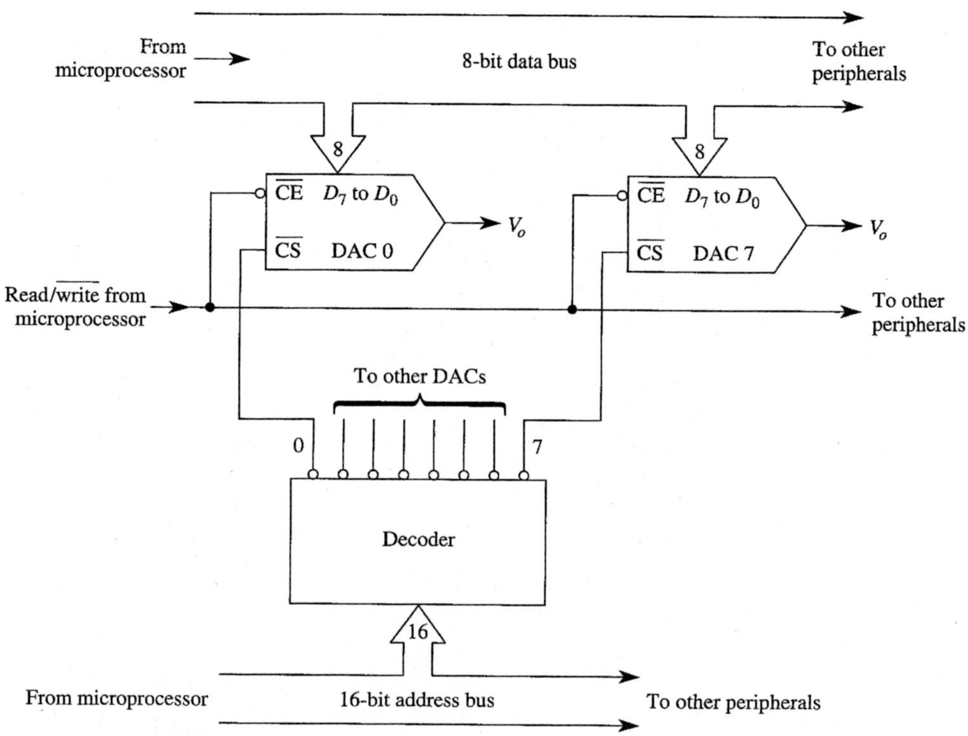

FIGURE 14-10 To select one DAC, the microprocessor places its address on the address bus. One output of the decoder goes low in response to its corresponding address code and enables the chip select terminal, $\overline{CS}$, of the DAC selected.

becomes transparent and accepts data from the data bus. The DAC immediately converts this digital data to an analog output voltage, V_o. This completes a "write only" operation. When either $\overline{CS}$ or $\overline{CE}$ goes high, the DAC's register enters the latching state and remembers the last data written into it. Thus V_o is held at the analog equivalent voltage.

14-7 AD558 MICROPROCESSOR-COMPATIBLE DAC

14-7.1 Introduction

An example of a complete 8-bit microprocessor-compatible D/A converter is introduced in Fig. 14-11. The AD558 can operate continuously or it can be controlled by a microprocessor. It is complete with an on-board precision reference voltage, latching digital inputs, and select terminals. It also contains an op amp to give an analog output voltage that

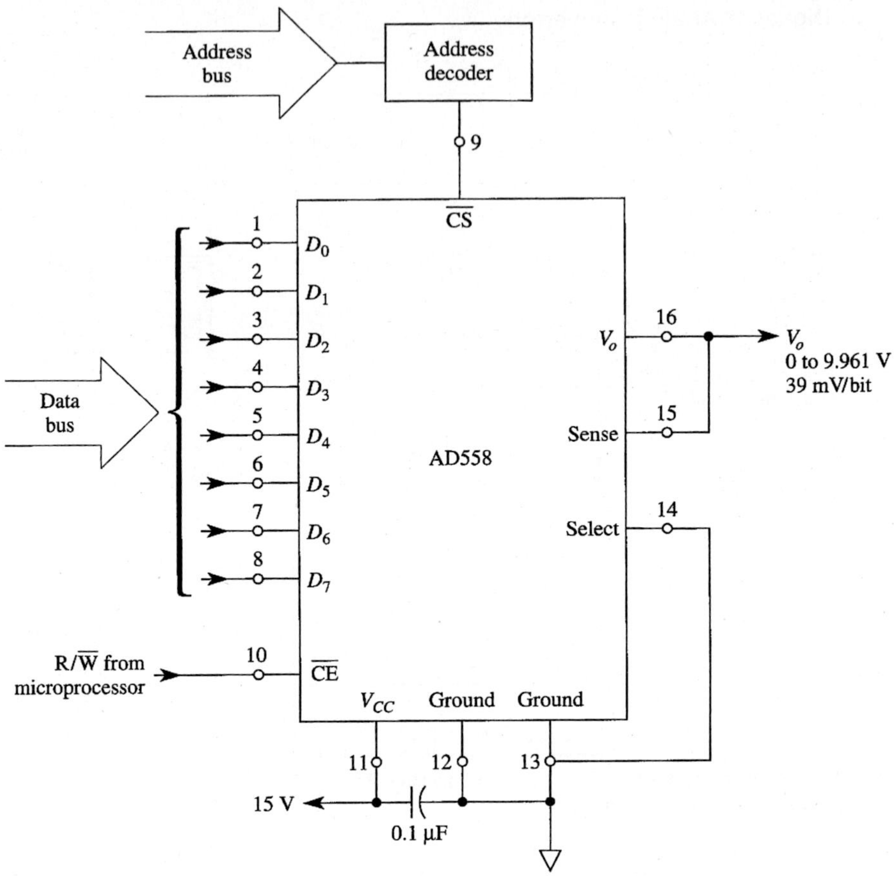

(a) AD558 terminals pin programmed for $V_o = 0$ to 9.961 V.

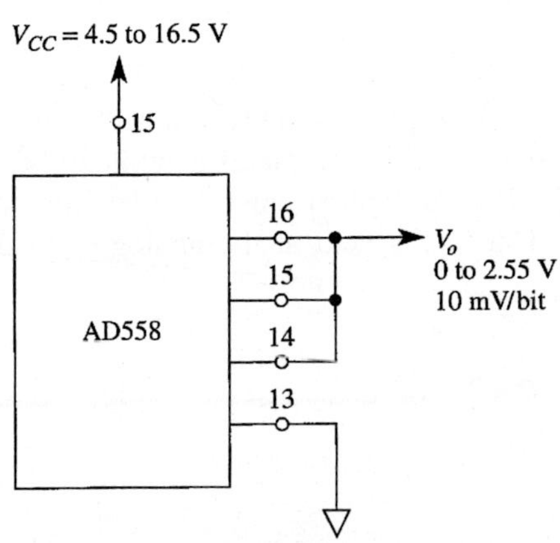

(b) $V_o = 0$ to 2.55 V.

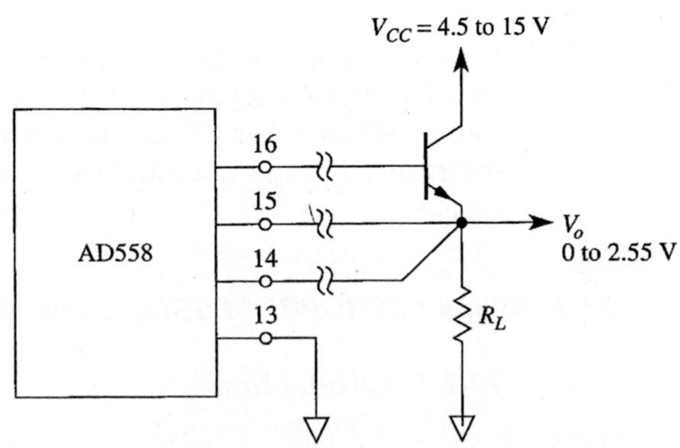

(c) Terminals 14 and 15 are extended to load R_L to eliminate the V_{BE} drop in the current boost transistor.

FIGURE 14-11 The AD588 is an 8-bit microprocessor-compatible D/A converter whose pinouts are shown in (a). It may be pin-programmed for 0 to 2.55 V as shown in (b). Terminal 15 and gain select terminal 14 may be extended to a load for current boost as shown in (c).

is pin-programmable for output ranges of 0 to 2.56 V or 0 to 10.0 V. Operation of the AD558 is studied by analyzing those tasks performed by its terminals.

14-7.2 Power Supply

Pin 11 is the power supply terminal V_{CC} in Fig. 14-11(a). It requires a minimum +4.5 V and has a maximum rating of +16.5 V. Pins 12 and 13 are the digital and analog grounds, respectively. This allows the user to maintain *separate* analog and digital grounds throughout a system, joining them at *only* one point. Usually, pins 12 and 13 are wired together and a 0.1-μF bypass capacitor *must* be connected between V_{CC} and pin 12 or 13.

14-7.3 Digital Inputs

Pins 1 through 8 are the digital inputs D_0 to D_7, with D_0 the LSB and D_7 the MSB. They are compatible with standard TTL or *low*-voltage CMOS. Logic 1 is 2.0 V minimum for a "1" bit. Logic 0 is 0.8 V maximum for a "0" bit.

The digital input pins connect the data bus to the AD558's internal *memory latching register,* when the AD558 is selected. This condition is called *transparent.* When unselected, the latching register is essentially disconnected from the data bus and remembers the last word written into the latching register. This condition is called "latching."

14-7.4 Logic Circuitry

The microprocessor executes a *write* command over the address bus via an address decoder and write line to the AD558's logic control pins 9 and 10. They are called chip select ($\overline{CS}$) and chip enable ($\overline{CE}$), respectively.

If a "1" is present on either $\overline{CS}$ or $\overline{CE}$, the digital inputs are in the "latching" mode. They are disconnected from the data bus. The input latches remember the last word written by the microprocessor over the data bus. If both $\overline{CS}$ and $\overline{CE}$ are "0," the AD558's inputs are "transparent" and connect the input memory latch register to the data bus. The microprocessor can now write data into the DAC. Digital-to-analog conversion takes place immediately and is completed in about 200 ns.

14-7.5 Analog Output

As shown in Fig. 14-11, the analog output voltage (V_o) appears between pins 16 and 13 (analog ground). Pin 14 is called "select" (V_o gain). It is wired to pins 15 and 16 to set the output voltage range at 0 to 2.56 V, as in Fig. 14-11(b). The actual analog output range is 0 to 2.55 V, or 10 mV/bit for a digital input of 00000000 to 11111111. A 0-to-10-V output range connection is shown in Fig. 14-11a. The actual range is 0 to 9.961 V or 38.9 mV/bit (power supply voltage must *exceed* maximum V_o by 2 V

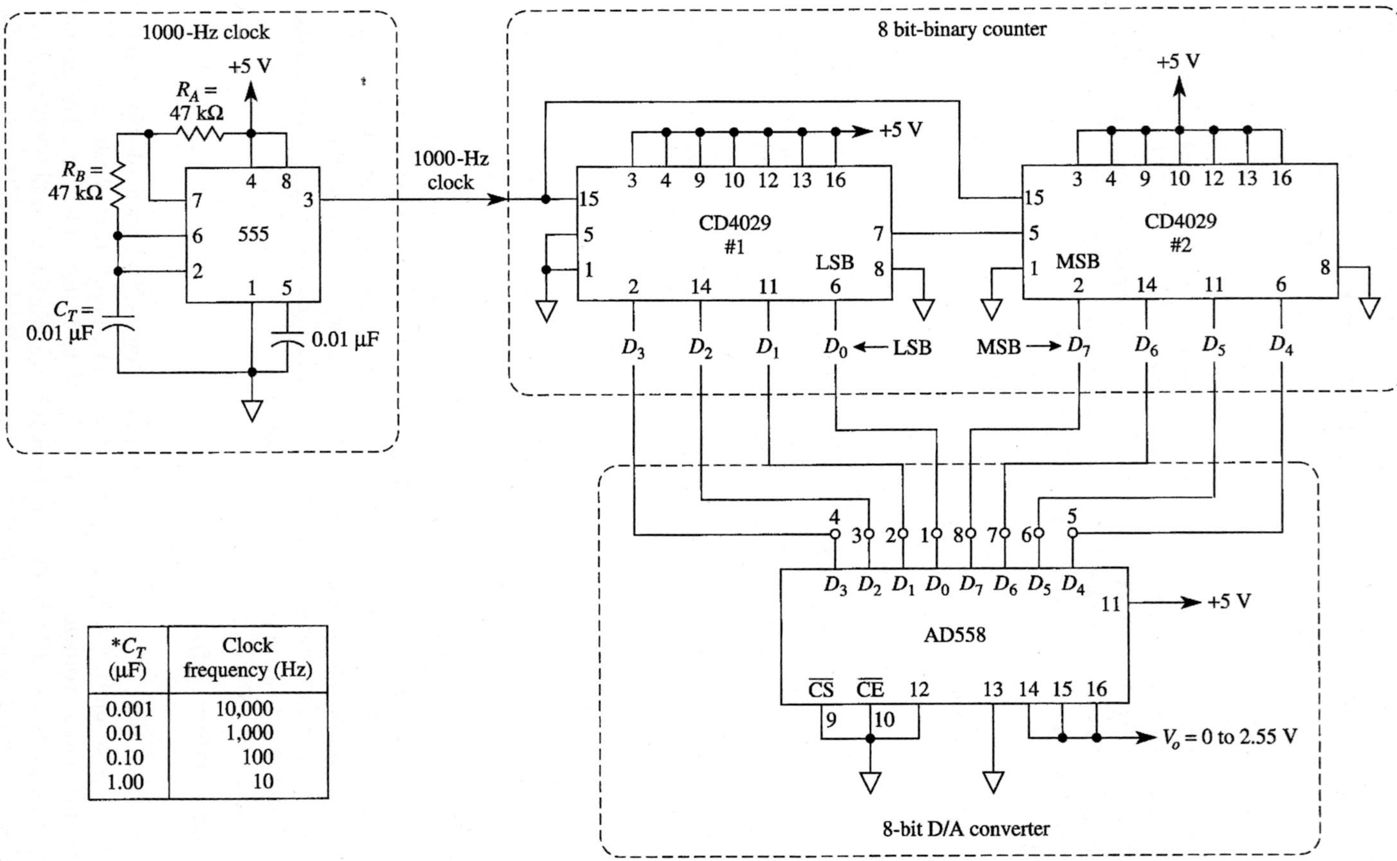

*C_T (μF)	Clock frequency (Hz)
0.001	10,000
0.01	1,000
0.10	100
1.00	10

FIGURE 14-12 The 555 clock drives an 8-bit binary counter made from two CD4049 ICs. The outputs count in binary from 00000000 to 11111111 and then repeat. The digital count is converted by the DAC into an analog voltage that resembles a staircase.

minimum). Sense terminal 15 allows remote load-voltage sensing to eliminate the effects of *IR* drops in long leads to the load. It can also be used for current boost as in Fig. 14-11(c).

14-7.6 Dynamic Test Circuit

A single AD558 can be tested dynamically *without* a microprocessor by the stand-alone, low-parts-count test circuit of Fig. 14-12. Pins 9 and 10 of the AD558 are grounded. This connects the AD558's input register (transparent) to an 8-bit synchronous counter that simulates a data bus.

The test circuit consists of three ICs. One 555 timer is wired as a 1-kHz clock. It steps an 8-bit synchronous binary counter made from two CD4029s. The counter's outputs are wired to the digital inputs of the DAC. An oscilloscope is connected (dc coupled) to display V_o. It should be externally triggered from the negative edge of the MSB (pin 8 of the AD558 or pin 2 of the right CD4029 in Fig. 14-12). The analog voltage waveshape appearing at V_o will resemble a staircase.

Each clock pulse steps the counter up by one count and increases V_o by 10 mV. Thus the risers will equal 10 mV and the staircase will have 256 treads from 0 to 2.55 V. The tread on each step will occupy about 1 ms. Thus one staircase waveshape is generated every 256 ms. Any glitch, nonlinearity, or other abnormality will be quite apparent. The capacitor C_T can be changed to give faster or slower clock frequencies, as indicated in Fig. 14-12. Any visible glitches can be minimized but *not* eliminated from V_o. However, a *sample-and-hold* or *follow-and-hold amplifier* can be connected to V_o. It waits until the glitch settles down, samples V_o, and holds this correct value. The principle of sample-and-hold is presented in Chapter 15.

14-8 SERIAL DACs

14-8.0 Introduction

This section describes a 12-bit serial D/A converter. The data is shifted into the DAC one bit at a time usually from a microprocessor or a microcontroller. The authors selected the DAC-8043 from Analog Devices to show typical operation of serial DACs. This DAC is housed in an 8-pin mini-DIP. Figure 14-13 shows the pin designation as well as the functional block diagram. The device consists of a 12-bit serial-in to parallel-out shift register, a 12-bit DAC register, and a 12-bit digital-to-analog converter. It also contains the necessary control logic. Serial data is clocked into the input shift register, one bit on each rising edge of a clock signal. After 12 bits have been shifted in, the data is transferred to the DAC register by an active low state on the $\overline{LD}$ pin. Data in the DAC register is converted to an output current by the D/A converter. Since this particular device produces an output current, an external op amp must be connected to the DAC's output as shown in Fig. 14-14.

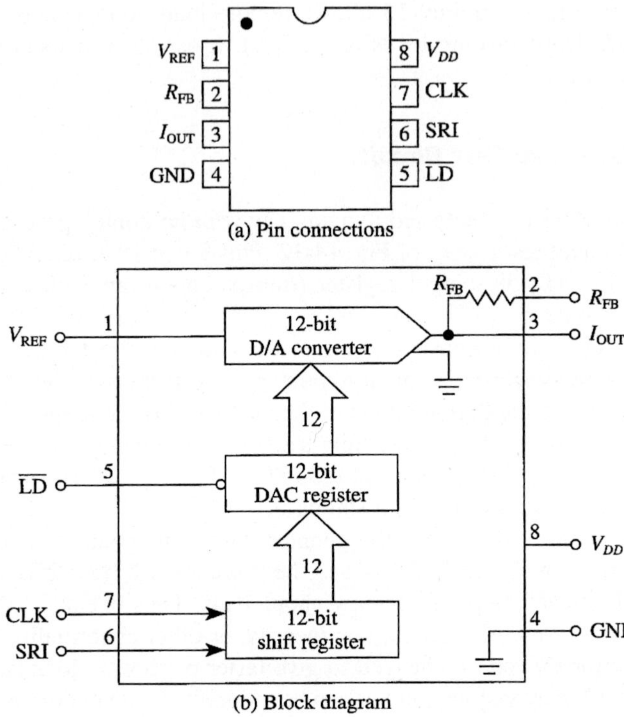

FIGURE 14-13 Pin designation and functional block diagram for a 12-bit serial DAC.

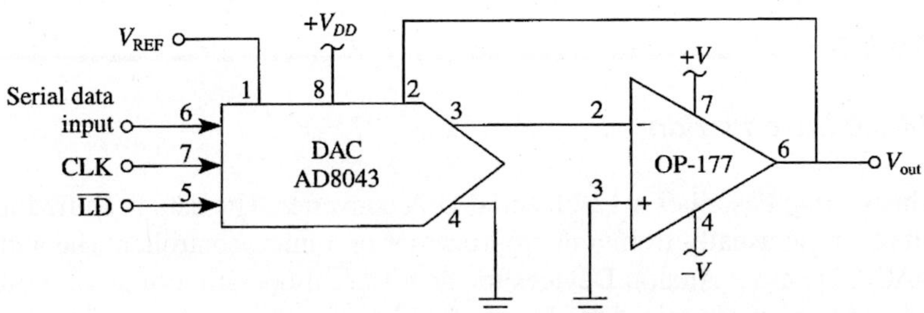

FIGURE 14-14 An op amp connected to the output of a serial DAC.

14-8.1 Interfacing a Serial DAC to a Microprocessor

Figure 14-15(a) shows a DAC-8043 interfaced to an MC6800 microprocessor. For this configuration the DAC-8043 is wired so that the input shift register is at address 0000_{hex}

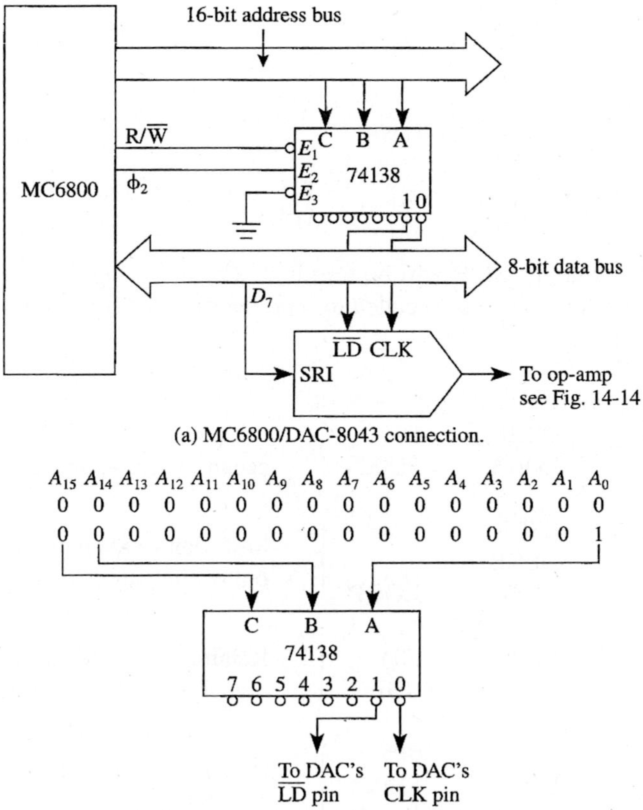

(a) MC6800/DAC-8043 connection.

(b) Connections for the 74138 3-to-8 decoder.

FIGURE 14-15 Interfacing the DAC-8043 serial D/A converter to the MC6800 microprocessor.

and the data is transferred to the DAC register when address 0001_{hex} is sent by the μP. A more detailed schematic of the decoder chip (74138) and the microprocessor's address lines are shown in Fig. 14-15(b).

14-8.2 Assembly Language Programming

Note that the 12 bits of data to be sent to the DAC are stored at memory locations $D000_{hex}$ and $D001_{hex}$ as shown in Fig. 14-16. For this example, the data to be sent are all logic 1s. The Xs in Fig. 14-16 indicate don't care bits and are not considered valid data for this application; they are not to be sent. The following subroutine will rotate the data at location

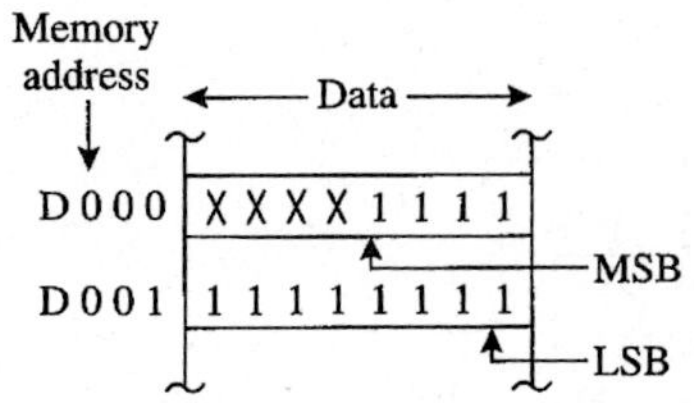

FIGURE 14-16 Data to be sent is stored in memory.

D000 so that the MSB will be sent first. The following assembly language code is written as a subroutine that is called to send the data one bit at a time to the DAC on data bus line D_7.

Subroutine Code

```
        LDAB    #04     }  Initialize
        LDAA    $D000   }  accumulators A and B

LOOP    LSLA            }
        DECB            }  Shift don't care data, Xs,
        BNE     LOOP    }  out of the way

        LDAB    #04     }  Reinitialize count value,
        BSR     SHIFT   }  branch to SHIFT subroutine

        LDAB    #08     }
        LDAA    $D001   }  Reinitialize to send
        BSR     SHIFT   }  next 8 bits

        STAA    $0001   }  Transfer data from shift reg. to DAC reg.
        RTS             }  Return from subroutine.

SHIFT   STAA    $0000   }
        LSLA            }  Subroutine to shift
        DECB            }  data on data bus line
        BNE     SHIFT   }  D7.
        RTS             }
```

The 74138 IC in Fig. 14-15 could be used to interface three more serial DACs to the microprocessor.

PROBLEMS

14-1. What is the resolution for **(a)** 10-bit DAC and **(b)** 14-bit DAC?

14-2. If the input reference voltage is 10 V for each of the DACs of Problem 14-1, what is the equivalent least significant voltage value?

14-3. What is the maximum output voltage for a 10-bit DAC if the reference voltage is set at 10.24 V?

14-4. A 12-bit DAC has a least significant bit resolution of 2.5 mV/bit, what is the output voltage if the input binary patterns are:
(a) 100011001000
(b) 001100100101
The most significant bit is the leftmost bit.

14-5. The output voltage of an 8-bit DAC is 1.9 V. What is the input digital code if the reference voltage is 5.12 V?

14-6. A 12-bit DAC has an offset error of −1/2 LSB and a gain error of −1/2 LSB. If the input reference voltage is 10.24 V, what is the output voltage when the input digital code is all 1s?

14-7. Refer to the R–2R ladder network of Fig. 14-7. Let $V_{ref} = 10$ V, $R_f = 5$ kΩ, $R = 5$ kΩ, and $2R = 10$ kΩ. Find **(a)** characteristic ladder resistance, **(b)** I_0; **(c)** voltage resolution; **(d)** output–input equation; **(e)** V_o.

14-8. What is the voltage at pins 2 and 3 of the op amp in Fig. 14-9(a) when the digital inputs are 1000000? Also calculate the voltage across the feedback resistor to get V_o.

14-9. A basic DAC consists of a reference voltage, ladder network, current switches, and op amp. Name two additional features required to make the DAC microprocessor compatible.

14-10. Can the AD558 and DAC-08 be used as multiplying DACs?

14-11. These questions refer to the AD558. **(a)** Name the terminals that allow this DAC to be selected. **(b)** Describe the digital input register's latching mode and **(c)** transparent mode of operation.

14-12. **(a)** What is the output–input equation for an AD558 DAC?
(b) Find V_o for an input code of 10000000.

14-13. If the 74138 decoder chip shown in Fig. 14-15 is used to interface three more serial DACs, at what address would their registers be located?

CHAPTER 15

Analog-to-Digital Converters

LEARNING OBJECTIVES

Upon completion of this chapter on analog-to-digital converters, you will be able to:

- Write the general output–input equations of an analog-to-digital converter (ADC) and calculate its output for any given input.
- Define and understand basic ADC specifications and terms.
- Name the most common types of ADCs and tell how each one operates.
- Explain how a microprocessor selects only one ADC out of all peripherals and reads its data.
- Calculate the maximum sine wave frequency that can be digitized to an accuracy of $\pm\frac{1}{2}$ LSB by an ADC or sample-and-hold amplifier.
- Operate a microprocessor-compatible ADC, the AD670.

15-0 INTRODUCTION

The previous chapter dealt with converting digital patterns of logic 1s and 0s to an analog signal—digital-to-analog converters (DACs). These devices are part of the output interface circuitry of a microprocessor-based system. They transform stored data or the results of digital processing back to real-world signals for control applications, for sound systems, or further analog processing, and so forth. In this chapter, we shall study part of a computer's input interface circuitry—analog-to-digital converters. ADCs are required to convert real-world signals produced by sensors and their signal-conditioning circuits to a digital pattern used by a computer. Digital data is easily stored, processed, and transmitted; hence, most modern control systems use some form of digital processing. Real-world parameters such as temperature, pressure, humidity, and so on are first converted by a sensor to an electrical quantity and then signal-conditioned analog circuitry before being applied to an A/D converter. These analog signals may vary slowly with time (for example, temperature measurements) or quickly with time (for example, audio recording). The latter systems require sample-and-hold circuitry before the signal can be applied to an A/D converter.

15-1 ADC CHARACTERISTICS

15-1.1 Resolution

The digital output of an ideal 3-bit ADC is plotted against a normalized analog input in Fig. 15-1. Similar to a DAC, the resolution of an ADC is defined as

$$\text{resolution} = 2^n \tag{15-1}$$

where n is the number of digital outputs. Since the input is an analog voltage and assumed to be continuous, resolution refers to number of discrete outputs. The ratio of the device's full-scale input voltage range, FSR, to the resolution gives the minimum change of input voltage to cause a change of one digital count at the output. In equation form

$$\Delta V_i \text{ for 1 LSB} = \frac{\text{FSR}}{2^n} \tag{15-2}$$

where FSR is the full-scale input voltage range specified on the manufacturer's data sheet and n is the number of digital outputs. The ratio of Eq. (15-2) is simply referred to as 1 LSB as shown in Fig. 15-1. *Note:* For an A/D converter 1 LSB is a voltage value as we will see in Example 15-1.

Since the digital count begins at 0, then the maximum full-scale input voltage to cause the output to be all logic 1s is 1 LSB less than the full-scale range value.

$$V_{\text{ifs}} = \text{FSR} - 1\text{ LSB} \tag{15-3}$$

where V_{ifs} is the maximum input voltage to cause all logic 1s at the ADC's output. In general, the output–input equation of an ADC for any input voltage is given by

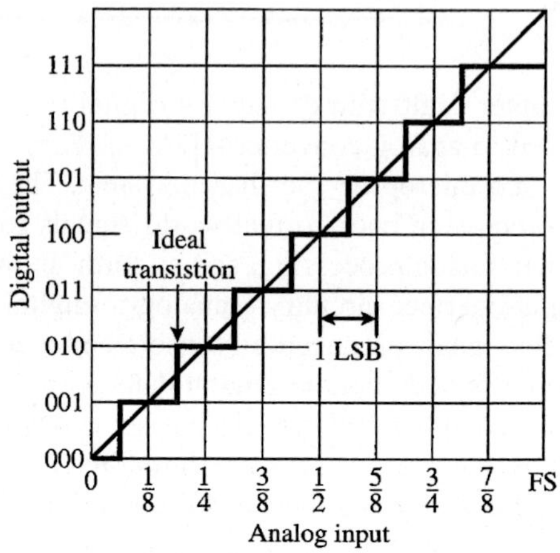

FIGURE 15-1 Ideal input–output relationship for a 3-bit A/D converter.

$$\text{digital output code} = \text{binary equivalent of } D \tag{15-4}$$

where D is the number of LSBs and is found from

$$D = \frac{V_{\text{in}}}{1 \text{ LSB}} \tag{15-5}$$

V_{in} is the applied input voltage and 1 LSB is given by Eq. (15-2).

Example 15-1

An 8-bit ADC is capable of accepting an input unipolar (positive values only) voltage 0 to 10 V. (a) What is the minimum value of input voltage to cause a digital output change of 1 LSB? (b) What input voltage will cause all 1s at the ADC's output? (c) What is the digital output code if the applied input voltage is 5.2 V?

Solution (a) From Eq. (15-2)

$$1 \text{ LSB} = \frac{10 \text{ V}}{2^8} = 39.1 \text{ mV}$$

(b) From Eq. (15-3)

$$V_{\text{ifs}} = 10 \text{ V} - 39.1 \text{ mV} = 9.961 \text{ V}$$

(c) Applying Eq. (15-5)

$$D = \frac{5.2 \text{ V}}{39.1 \text{ mV}} = 132.99 \cong 133$$

Converting this value to binary yields 10000101.

Figure 15-1 shows the graph for an ideal 3-bit A/D converter. Since there are only 8 discrete outputs for this ADC, the analog input range from 0 to full scale must be *quantized* into these 8 digital outputs. Hence, voltages within the same range have the same digital code. The ideal transition from one digital code to the next is taken to be at the mid-range value as shown. A/D converters like DACs, have offset, gain, and linearity errors but they also have a quantization error.

15-1.2 Quantization Error

Figure 15-1 shows that the digital output is 100 for all values of $\frac{1}{2}$ FSR $\pm \frac{1}{2}$ LSB. Hence, there is an unavoidable uncertainty about the exact value of V_{in} when the output is 100. This uncertainty is specified as quantization error. Its value is $\pm \frac{1}{2}$ LSB. Increasing the number of bits results in a finer resolution and therefore there is a smaller quantization error.

15-1.3 Offset Error

Similar to DACs, analog-to-digital converters also have an offset error specification. It is usually given as a fraction of 1 LSB such as $\pm \frac{1}{2}$ LSB. Figure 15-2 shows how an A/D

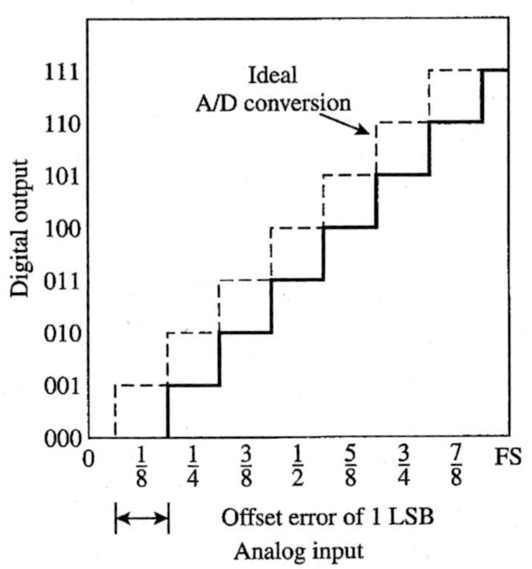

FIGURE 15-2 Comparison of the ideal A/D conversion with a conversion having offset error.

converter's output–input plot can be viewed with an offset error. This offset error modifies Eq. (15-3) as shown in the next example.

Example 15-2

Consider that the 8-bit A/D converter of Problem 15-1 has an offset error of $\pm \frac{1}{2}$ LSB. What analog input voltage will cause a digital output code of all 1s?

Solution From the solution of Problem 15-1, 1 LSB is 39.1 mV. Therefore, $\frac{1}{2}$ LSB is 19.5 mV. Modifying Eq. (15-3) by the offset error yields

$$\begin{aligned} V_{\text{ifs}} &= \text{FSR} - 1\text{ LSB} + \tfrac{1}{2}\text{ LSB} \\ &= 10\text{ V} - 39.1\text{ mV} + 19.5\text{ mV} = 9.9804\text{ V} \end{aligned}$$

15-1.4 Gain Error

Refer to Fig. 15-1. In the ideal input–output relationship the difference between the first transition and the last transition is FSR − 2 LSB. If this expression is not true, there is a gain error. See Fig. 15-3 to compare the ideal relationship with an A/D converter that has a gain error. Manufacturers specify gain error as a percentage of full-scale voltage (% FSR).

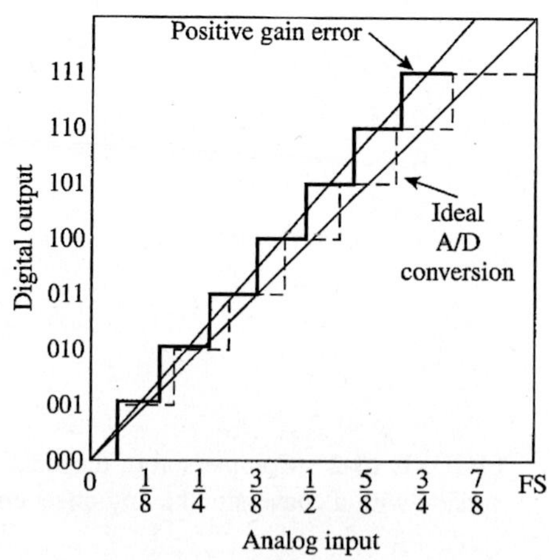

FIGURE 15-3 Gain error.

Example 15-3

The A/D converter of Problem 15-1 has a gain error of 0.5% of FSR. What value of V_{in} will cause the output digital code to be all 1s?

Solution In terms of voltage, the gain error is 50 mV (0.5% × 10 V). The ideal full-scale input voltage to cause the output to be all logic 1s is given by Eq. (15-3) and from Problem 15-1 part (c) V_{ifs} is 9.961 V. This value is reduced by a positive gain error. Therefore the new value of input voltage is V_{in} = 9.961 V − 50 mV = 9.911 V.

15-1.5 Linearity Error

Linearity error in an A/D converter exists if the differences between all of the transitions are not equal. Figure 15-4 shows that the difference between transitions are not equal. The range difference is specified as a fraction of 1 LSB such as 0.75 LSB.

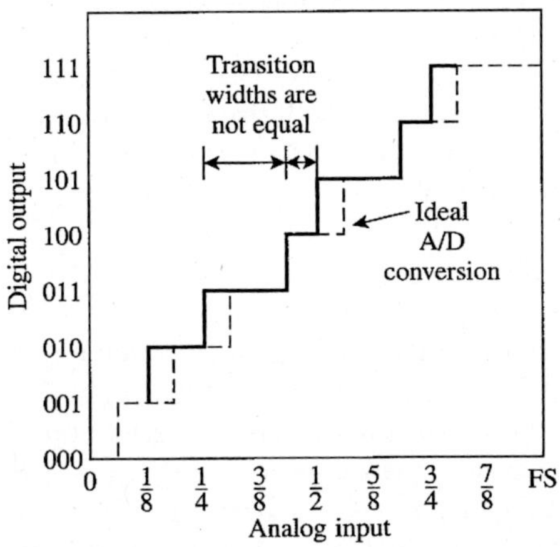

FIGURE 15-4 Linearity error.

15-2 INTEGRATING ADC

15-2.1 Types of ADCs

General characteristics of ADCs were introduced in Section 15-1. There are three standard types classified according to their conversion times. The *slow integrating ADC* typically requires 300 ms to perform a conversion; it is the best choice for measuring slowly varying dc voltages. The faster *successive approximation ADC's* conversion time is a few microseconds and can digitize audio signals. Fastest of all are the more costly *flash* converters, which can digitize video signals.

15-2.2 Principles of Operation

The block diagram of a typical dual-slope integrating A/D converter is shown in Fig. 15-5. An on-board divide-by-4 counter drives the control logic at a rate of 12 kHz. This frequency is set by the user via external timing resistor R_T and C_T. It must be a multiple of the local line frequency (50 or 60 Hz) to render the ADC immune to line-frequency noise.

The control logic unit activates a complex network of logic circuits and analog switches to convert analog input voltage V_{in} into a digital output. The analog-to-digital conversion is performed in three phases and requires about one-third of a second. These operating phases are called *signal integrate phase* T_1, *reference integrate phase* T_2, and *auto-zero phase* T_z. These will be discussed in sequence.

15-2.3 Signal Integrate Phase, T_1

The control logic unit of Fig. 15-5(b) connects V_{in} to an integrator to begin phase T_1. The integrator or ramp generator's output V_o ramps up or down depending on the polarity of V_{in} and at a rate set by V_{in}, R_{int}, and C_{int}. If V_{in} is negative, V_o ramps up, as shown in Fig. 15-5(a). Time T_1 is set by the logic unit for 1000 clock pulses. Since the 12-kHz clock has a period of 83.3 μs per count, T_1 lasts 83.33 ms.

If $V_{in} = -100$ mV, V_o will ramp from 0 V to 833 mV. The maximum allowed full-scale value of V_{in} is ± 200 mV. When $V_{in} = -200$ mV, V_o will rise to a maximum 1666 mV. Clearly, V_o is directly proportional to V_{in}. At the end of 1000 counts, the logic unit disconnects V_{in} and connects V_{ref} to the integrator. This action ends T_1 and begins T_2.

15-2.4 Reference Integrate Phase, T_2

During phase T_1, the logic unit determined the polarity of V_{in} and charged a reference capacitor, C_{ref} (not shown), to a reference voltage $V_{ref} = 100$ mV. At the beginning of phase T_2, the logic unit connects C_{ref} to the integrator so that V_{ref} has a polarity opposite to V_{in}. Consequently, V_{ref} will ramp the integrator back toward zero. Since V_{ref} is constant, the integrator's output V_o will ramp down at a *constant* rate, as shown in Fig. 15-5a.

When V_o reaches zero, a comparator "tells" the logic unit to terminate phase T_2 and begin the next auto-zero phase. T_z is thus proportional to V_o and consequently, V_{in}. The exact relationship is

$$T_2 = T_1 \frac{V_{in}}{V_{ref}} \tag{15-6a}$$

Since $T_1 = 83.33$ ms and $V_{ref} = 100$ mV,

$$T_2 = \left(0.833 \frac{\text{ms}}{\text{mV}}\right) V_{in} \tag{15-6b}$$

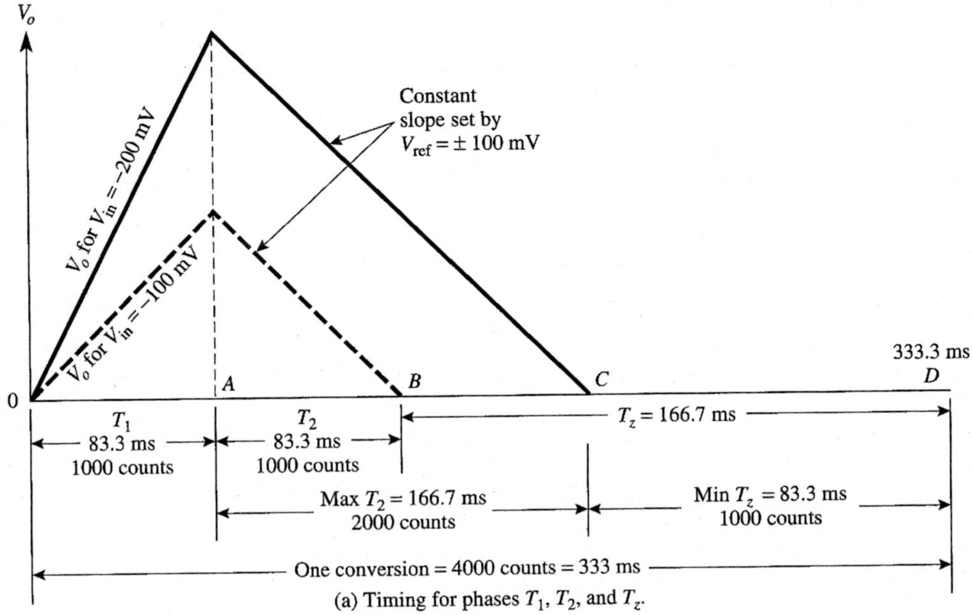

(a) Timing for phases T_1, T_2, and T_z.

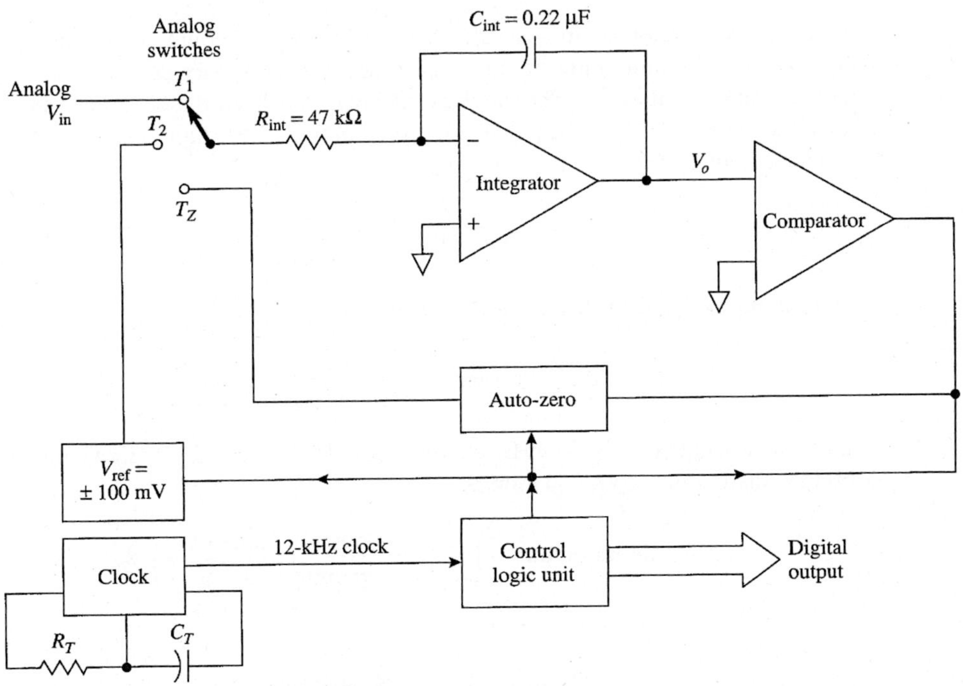

(b) Simplified block diagram.

FIGURE 15-5 (a) Timing diagram of a typical dual-slope integrating ADC; (b) simplified block diagram of a dual-slope integrating ADC. In (a), one A/D conversion takes place in three phases: signal integrating phase T_1, reference T_2, and auto-zero T_z.

Example 15-4

For the ADC of Fig. 15-5, calculate T_2 if (a) $V_{in} = \pm 100$ mV; (b) $V_{in} = \pm 200$ mV.

Solution (a) From Eq. (15-6b),

$$T_2 = \left(0.833 \frac{\text{ms}}{\text{mV}}\right)(100 \text{ mV}) = 83.33 \text{ ms}$$

(b)

$$T_2 = \left(0.833 \frac{\text{ms}}{\text{mV}}\right)(200 \text{ mV}) = 166.6 \text{ ms}$$

15-2.5 The Conversion

The actual conversion of analog voltage V_{in} into a digital count occurs during T_2 as follows. The control unit connects the clock to an internal binary-coded-decimal counter at the beginning of phase T_2. The clock is disconnected from the counter at the end of T_2. Thus the counter's content becomes the digital output. This digital output is set by T_2 and the clock frequency:

$$\text{digital output} = \left(\frac{\text{counts}}{\text{second}}\right)T_2 \tag{15-7a}$$

but T_2 is set by V_{in} from Eq. (15-6a) and therefore

$$\text{digital output} = \left(\frac{\text{counts}}{\text{second}}\right)(T_1)\left(\frac{V_{in}}{V_{ref}}\right) \tag{15-7b}$$

Since clock frequency is 12 kHz for the 7106/7107 ADC, $T_1 = 83.33$ ms, and $V_{ref} =$ 100 mV, the output–input equation is

$$\text{digital output} = \left(12{,}000 \frac{\text{counts}}{\text{second}}\right)\left(\frac{83.33 \text{ ms}}{100 \text{ mV}}\right)V_{in}$$

or

$$\text{digital output} = \left(10 \frac{\text{counts}}{\text{mV}}\right)V_{in} \tag{15-7c}$$

The counter's output is connected to an appropriate $3\frac{1}{2}$-digit display.

Example 15-5

V_{in} equals + 100 mV in the ADC of Fig. 15-5. Find the digital output.

Solution From Eq. (15-7c),

$$\text{digital output} = \left(10\ \frac{\text{counts}}{\text{mV}}\right)(100\text{ mV}) = 1000\text{ counts}$$

Example 15-5 shows the need for some human engineering. The display reads 1000, but it *means* that V_{in} equals 100 mV. *You* must wire in a decimal point to display 100.0 and paste an "mV" sign beside the display.

15-2.6 Auto-Zero

The block diagram of Fig. 15-5(b) contains a section labeled "Auto-zero." During the third and final phase of conversion, T_z, the logic unit activates several analog switches and connects an auto-zero capacitor C_{AZ} (not shown).

The auto-zero capacitor is connected across the integrating capacitor, C_{int}, and any input offset voltages of both integrating and comparator op amps. C_{AZ} charges to a voltage approximately equal to the average error voltage due to C_{int} and the offset voltages. During the following phases, T_1 and T_2, the error voltage stored on C_{AZ} is connected to cancel any error voltage on C_{ref}. Thus the ADC is automatically zeroed for every conversion.

15-2.7 Summary

Refer to the timing diagram in Fig. 15-5(a). The logic unit allocates 4000 counts for one conversion. At 83.33 μs per count, the conversion takes 333 ms. The control unit always allocates 1000 counts or 83.3 ms to phase T_1.

The number of counts required for T_2 depends on V_{in}. Zero counts are used for V_{in} = 0 V and a maximum of 2000 counts, or 166.7 ms are used when V_{in} is at its maximum limit of $\pm$200 mV.

T_2 and T_z always share a total of 3000 counts for a total of 250 ms. For V_{in} 0 V, $T_2 = 0$ counts and $T_z = 3000$ counts. For $V_{in} = \pm 200$ mV, $T_2 = 2000$ counts and $T_z =$ 1000 counts.

Some manufacturers of ADCs market a complete $3\frac{1}{2}$-digit digital voltmeter kit. The kit contains a 40-pin dual-slope integrating A/D converter, all necessary parts, printed circuit board, and instructions. The instructions make it easy to construct and easy to use, and form an excellent tutorial on integrating ADCs.

15-3 SUCCESSIVE APPROXIMATION ADC

The block diagram of a successive approximation register (ADC) is shown in Fig. 15-6. It consists of a DAC, a comparator, and a *successive approximation register* (SAR). One terminal is required for analog input voltage V_{in}. The digital output is available in either serial or parallel form. A minimum of three control terminals are required. *Start conversion* initiates an A/D conversion sequence and *end of conversion* tells when the conversion is completed. An external clock terminal sets the time to complete each conversion.

15-3.1 Circuit Operation

Refer to Fig. 15-6. An input *start conversion* command initiates one analog-to-digital conversion cycle. The successive approximation register (SAR) connects a sequence of digital numbers, one number for each bit to the inputs of a DAC. This process was explained in Chapter 14.

The DAC converts each digital number into an analog output V_o. Analog input voltage, V_{in}, is compared to V_o by a comparator. The comparator tells the SAR whether V_{in} is greater or less than DAC output V_o, once for each bit. For a 3-bit output, three comparisons would be made.

Comparisons are made beginning with the MSB and ending with the LSB, as will be explained. At the end of the LSB comparison, the SAR sends an end-of-conversion signal. The digital equivalent of V_{in} is now present at the SAR's digital output.

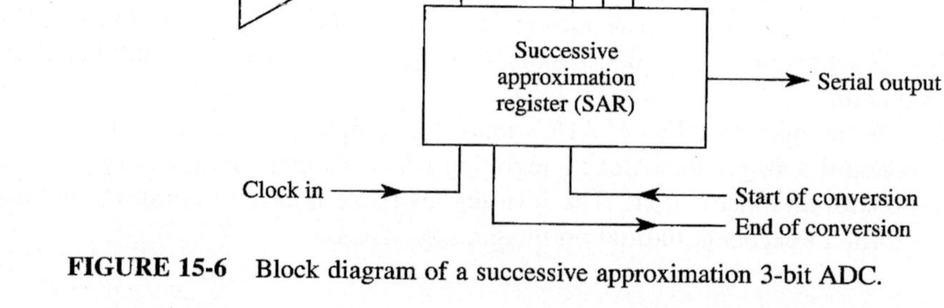

FIGURE 15-6 Block diagram of a successive approximation 3-bit ADC.

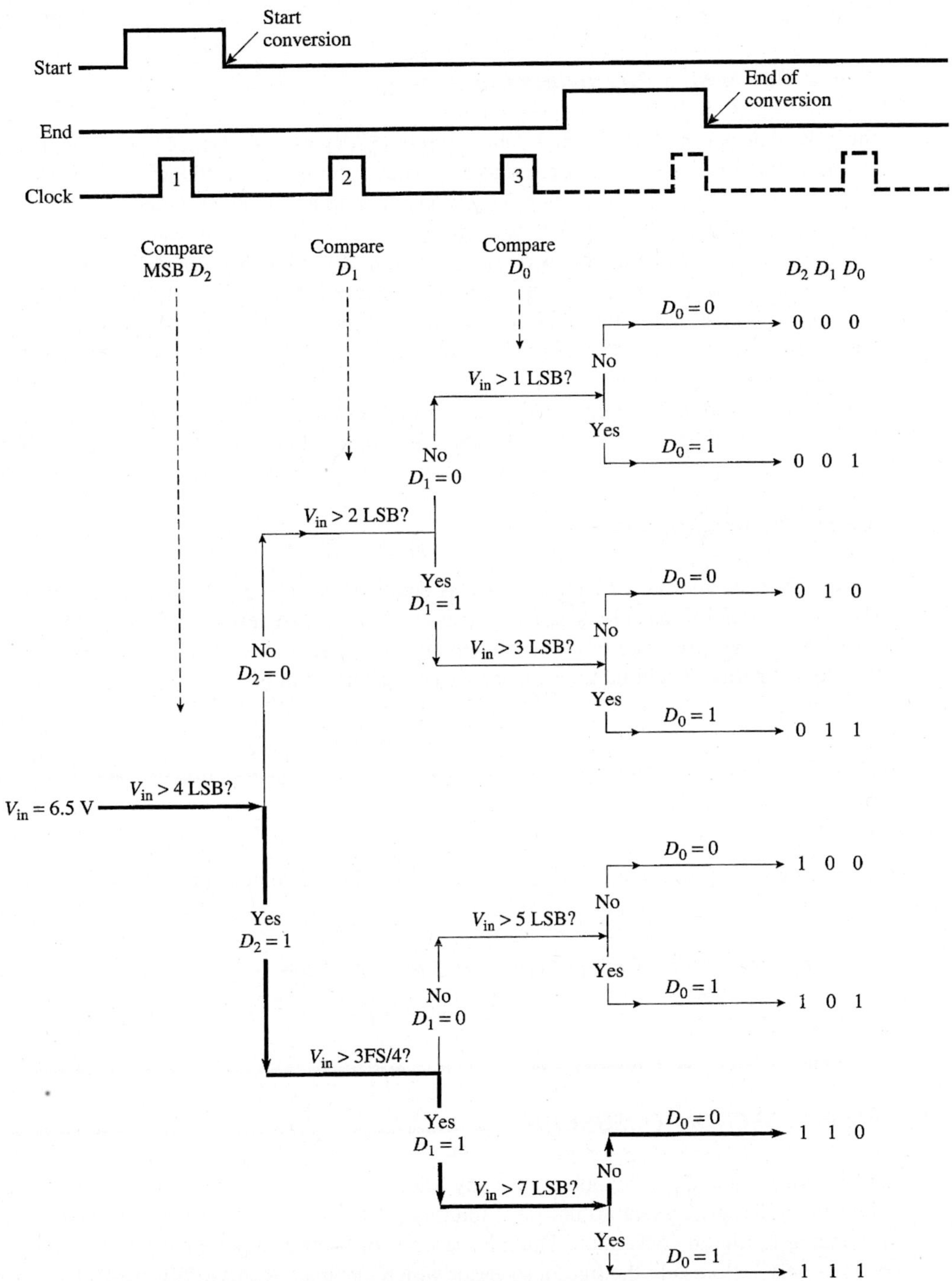

FIGURE 15-7 "Start" begins operation of this 3-bit successive approximation register. Beginning with the MSB, the weight of each bit is compared with V_{in} by the comparator of Fig. 15-6. If V_{in} is greater, the SAR's output is set to 1, or to 0 if V_{in} is smaller. Dark lines show the conversion for V_{in} = 6.5 V.

15-3.2 Successive Approximation Analogy

Suppose that you had 1-, 2-, and 4-lb weights (SAR) plus a balance scale (comparator and DAC). Think of the 1-lb weight as 1 LSB and the most significant 4-lb weight as 4 LSB. Refer to Figs. 15-6 and 15-7. V_{in} corresponds to an unknown weight.

Let us convert $V_{in} = 6.5$ V to a digital output (unknown weight = 6.5 lb). You would place the unknown weight on one platform of the balance, the 4-lb weight on the other, and compare if the unknown weight (V_{in}) exceeded the 4-lb weight. The SAR uses one clock pulse to apply the MSB 100 to the DAC in Fig. 15-7. Its output, $V_o = 4$ V, is compared with V_{in}. The MSB (D_2) is set to 1 if $V_{in} > V_o$. This is analogous to your leaving the 4-lb weight on the scale.

The SAR then applies 110 (add a 2-lb weight) to the DAC, D_1 is set to 1 since $V_{in} =$ 6.5 V is greater than $V_o = 6$ V. Finally, the SAR applies 111 to the DAC (add 1 lb). Since $V_{in} = 6.5$ V is less than 7 V, D_0 is set to zero (1-lb weight removed).

15-3.3 Conversion Time

Figure 15-7 shows that one clock pulse is required for the SAR to compare each bit. However, an additional clock pulse is usually required to reset the SAR prior to performing a conversion. The time for one analog-to-digital conversion must depend on both the clock's period T and number of bits n. The relationship is

$$T_C = T(n + 1) \tag{15-8}$$

Example 15-6

An 8-bit successive approximation ADC is driven by a 1-MHz clock. Find its conversion time.

Solution The time for one clock pulse is 1 μs. From Eq. (15-8),

$$T_C = 1\ \mu\text{s}(8 + 1) = 9\ \mu\text{s}$$

15-4 ADCs FOR MICROPROCESSORS

One of the principal selling points of many microcontrollers is an internal A/D converter. However, microprocessors do not have internal A/Ds, so we must use an external device for analog-to-digital conversion. The microprocessor "views" a peripheral ADC simply as a "read only" address in the microprocessor's memory map. Refer to Fig. 15-8. The ADC must have a tristate *memory buffer register* (MBR). In the idle state, the MBR will contain a digital code resulting from the ADC's *last* conversion. Also, the MBR will be disconnected from the data bus.

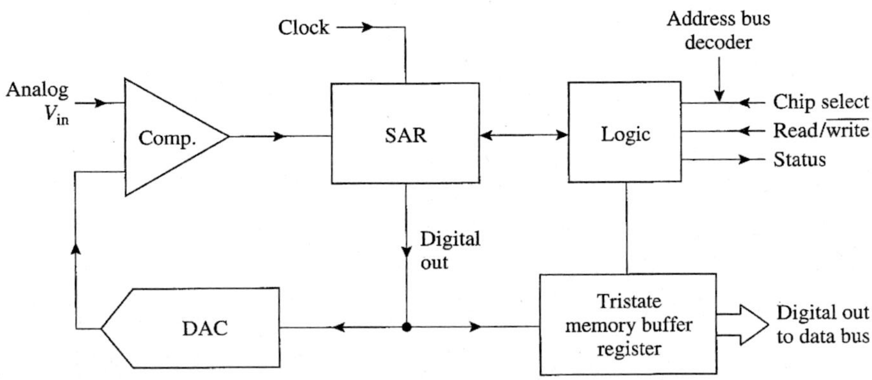

FIGURE 15-8 To be compatible with microprocessors, the ADC of Fig. 15-6 requires selection logic and a memory buffer register.

The microprocessor uses the address bus and decoders to select one ADC out of all the others by bringing its *chip select* terminal low. A low on the *chip select* terminal in Fig. 15-8 tells the ADC that a command is coming to its read/$\overline{write}$ terminal. If read/$\overline{write}$ is brought low by the microprocessor, the ADC converts V_{in} into a digital code and loads or writes it into its own MBR. When read/$\overline{write}$ is high *and* chip select is low, the ADC's memory buffer register is connected (transparent) to the data bus.

It is important to look at this operation from the microprocessor's viewpoint. A *read* command means that the microprocessor is going to read data stored in the ADC's memory buffer register. The ADC's digital tristate outputs must go from high-Z (high impedance) to transparent and connect the digital word to the data bus. A *write command is actually a start conversion command* to the ADC. The microprocessor thus tells the ADC: (1) perform a conversion; (2) store (and *write*) it in your memory; and (3) don't tell me the result until I want to *read* it.

Finally, the microprocessor-compatible ADC must tell the microprocessor via its *status* line when a conversion is in progress; the status line goes high. If a conversion is completed, the status line goes low to signal the microprocessor that data is valid and ready for reading. We select the Analog Devices AD670 to learn how all the foregoing features are available in a single 20-pin integrated circuit.

15-5 AD670 MICROPROCESSOR-COMPATIBLE ADC

The AD670 is an 8-bit microprocessor-compatible successive approximation analog-to-digital converter. The 20-pin package of Fig. 15-9 contains all the features described in Section 15-3 and Fig. 15-8. In addition, it contains an on-board clock, voltage reference, and instrumentation amplifier, and needs only a single 5-V supply. To understand how the AD670 operates, we examine the tasks performed by each of its terminals and associated circuit blocks.

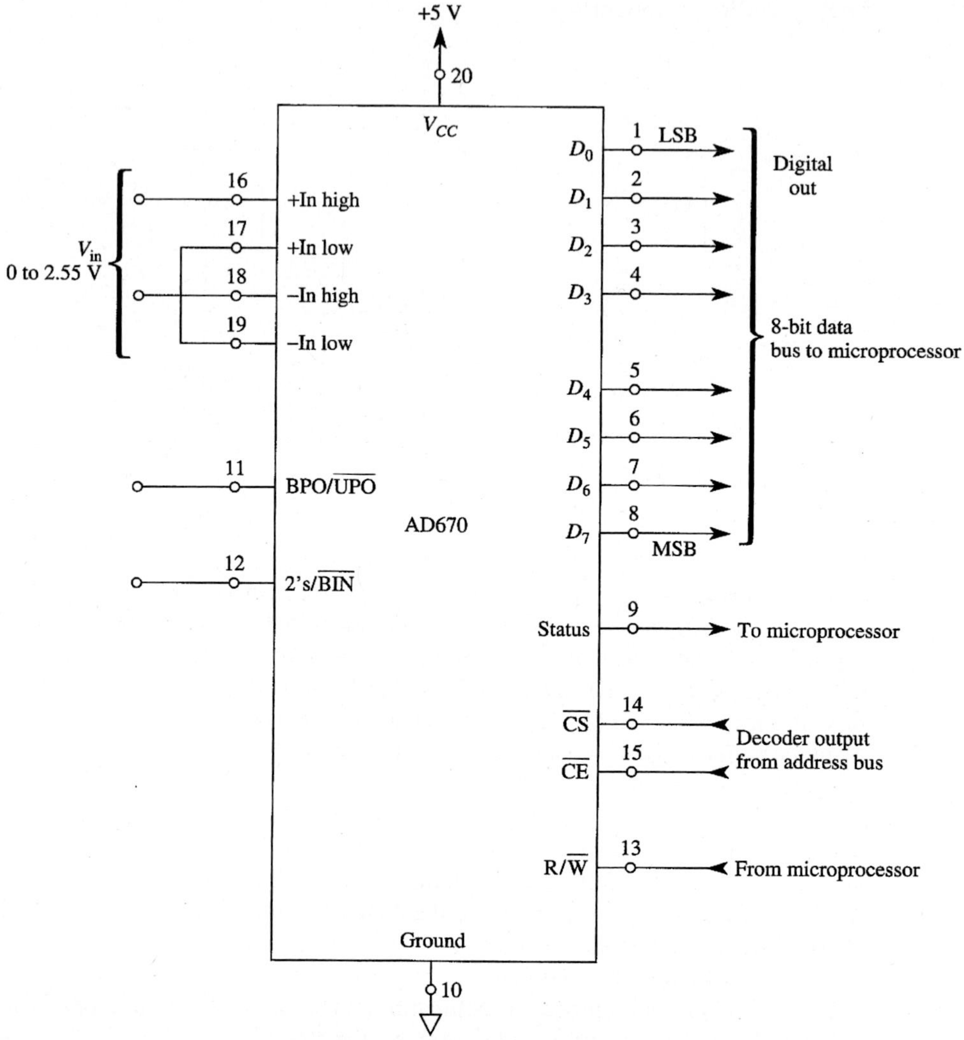

(a) Inputs wired for a resolution of 10 mV/bit.

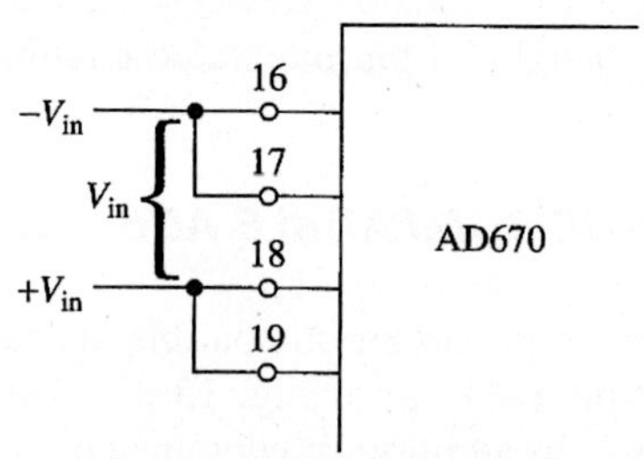

(b) Inputs wired for 1 mV/bit.

FIGURE 15-9 (a) AD670 ADC pin connections. Full-scale analog input voltages are 0 to 2.55 V or 0 to ±1.28 V and 0 to 255 mV or 0 to ±128 mV in (b).

15-5.1 Analog Input Voltage Terminals

Four analog input terminals are pins 16, 17, 18, and 19 in Fig. 15-9. They are inputs to an instrumentation amplifier configured to handle unipolar or bipolar analog input voltages. They are also pin-programmable to make it easy for the user to select resolution. Figure 15-9(a) shows operation for an analog input of 0 to 2.55 V, resolution = 10 mV/LSB. Figure 15-9(b) shows operation for 0 to 255 mV or 1 mV/LSB.

15-5.2 Digital Output Terminals

Pins 1 through 8 are tristate, buffered, latching digital outputs for the data bus digits, D_0 through D_7, respectively. When a microprocessor tells the AD670 to perform a conversion (write), the result is latched into its memory buffer register. Tristate output switches are held in the *high-impedance* (high-*Z*) state until the microprocessor sends a read command. Thus the ADC's memory register is normally disconnected from the data bus.

15-5.3 Input Option Terminal

Pin 11 is called BPO/$\overline{\text{UPO}}$ and allows the microprocessor to tell the AD670 whether to accept a bipolar analog input voltage range or a unipolar input range. A low on pin 11 selects unipolar operation. A range of 0 to 2.55 V or 0 to 255 mV is set by the user as in Figs. 15-9(a) and (b). A high sent to pin 11 selects bipolar operation. The V_{in} range is then ± 1.28 V [Fig. 15-9(a)] or ± 128 mV [Fig. 15-9(b)].

15-5.4 Output Option Terminal

In Fig. 15-9, pin 12 is labeled "2's/$\overline{\text{BIN}}$." It allows the microprocessor to tell the AD670 to present an *output format* in either 2's-complement code or binary code. A binary output code format will be *straight binary* if V_{in} is unipolar (pin 11 = low) or *offset binary* if V_{in} is bipolar (pin 11 = high). The four possible options are shown in Fig. 15-10(a).

The digital output responses to analog input V_{in} are shown in Figs. 15-10(b) and (c). V_{in} is the *differential* input voltage and is defined by

$$V_{in} = (+V_{in}) - (-V_{in}) \tag{15-9}$$

where $+V_{in}$ and $-V_{in}$ are measured with respect to ground.

15-5.5 Microprocessor Control Terminals

As shown in Fig. 15-9, pins 13, 14, and 15 are used by a microprocessor to control the AD670. Terminal 14 is called *chip select* ($\overline{\text{CS}}$) and terminal 15 is called *chip enable* ($\overline{\text{CE}}$). Pin 13 is called read/$\overline{\text{write}}$ (R/$\overline{\text{W}}$).

If $\overline{\text{CS}}$, $\overline{\text{CE}}$, and R/$\overline{\text{W}}$ are all brought low, the ADC converts continuously. It performs one conversion every 10 μs or less. The result of each conversion is latched into the output buffer register. However, the digital output code is *not* connected to the data bus

Pin 11 BPO/$\overline{UPO}$	Input range	Pin 12 2's/$\overline{BIN}$	Output format
0	Unipolar	0	Straight binary
1	Bipolar	0	Offset binary
0	Unipolar	1	2's complement
1	Bipolar	1	2's complement

(a)

Differential V_{in}	Unipolar/straight binary, pin 11 = 0, 12 = 0
0	0000 0000
1 mV	0000 0001
128 mV	1000 0000
255 mV	1111 1111

(b)

Differential V_{in}	Bipolar/offset binary, pin 11 = 1, 12 = 0	Bipolar/2's complement, pin 11 = 1, 12 = 1
−128 mV	0000 0000	1000 0000
−1 mV	0111 1111	1111 1111
0	1000 0000	0000 0000
1 mV	1000 0001	0000 0001
127 mV	1111 1111	0111 1111

(c)

FIGURE 15-10 Input ranges are unipolar or bipolar, and output formats are determined by pins 11 and 12 in (a). Output codes are given for unipolar inputs in (b) and for bipolar inputs in (c). (a) Input range and output format are controlled by pins 12 and 11, respectively; (b) digital output codes for unipolar V_{in} inputs wired as in Fig. 15-9a; (c) digital output codes for bipolar V_{in} inputs wired as in Fig. 15-9b.

because the outputs are high impedance. This condition is called a *write and convert* command; that is, the microprocessor tells the AD670 to write converted data into its own buffer register. If $\overline{CS}$ or R/$\overline{W}$ or $\overline{CE}$ is high, the AD670 is unselected (high impedance) and retains the last conversion in its register.

The status terminal, pin 9, stays high during a conversion. When a conversion is completed, pin 9 outputs a low to "tell" the microprocessor that data is valid in the AD670's buffer register. To read data out of the AD670, the microprocessor brings R/$\overline{W}$ high while status and $\overline{CS}$ *and* $\overline{CE}$ are low. This is a *read* command from the microprocessor.

The AD670's buffer becomes transparent and connects the eight digital outputs (D_7 through D_0) to the data bus. Data will remain on the bus until the AD670 is disconnected by bringing $\overline{CS}$ high, $\overline{CE}$ high, or R/$\overline{W}$ low.

Summary

1. A low on $\overline{CE}$ and $\overline{CS}$ selects the AD670. What happens next depends on R/$\overline{W}$.
2. If R/$\overline{W}$ is low (for at least 0.3 μs), a conversion is performed and the result is written into the buffer register. Outputs are high impedance. The conversion requires 10 μs.

3. If R/$\overline{W}$ is high, the last conversion is stored in the buffer and the outputs are transparent. No further conversions are performed. The contents of the register can now be read by the microprocessor via the data bus.
4. Status tells the microprocessor what is going on within the AD670. Status = high means that conversion is being performed. Status = low tells the microprocessor that data are valid. The microprocessor is free to read the selected AD670's data by placing a high to R/$\overline{W}$.

15-6 TESTING THE AD670

Figure 15-11 shows how to wire an AD670 to perform continuous conversions *without* a microprocessor. This circuit can be used as a laboratory exercise to gain experience operating ADCs. Each data output, D_0 through D_7, is connected to an inverter, resistor, and LED. These components simulate a data bus. An LED lights to signify that a logic 1 is present on its associated data bus wire.

Pins 14 and 15 are wired so that $\overline{CS}$ and $\overline{CE}$ are low. This causes continuous conversion. The 555 timer drives R/$\overline{W}$ low for 5 μs to simulate a write command. R/$\overline{W}$ thus returns high before a conversion is completed in the 10-μs conversion time. At the end of 10 μs, the high on R/$\overline{W}$ simulates a read command and data are displayed on the LEDs. If R_T = 1.5 MΩ, the AD670 makes one conversion and one readout 1000 times per second. Reduce R_T to 120 kΩ for convert/reads of 10,000 times per second.

15-7 FLASH CONVERTERS

15-7.1 Principles of Operation

Fastest of all A/D converters is the *flash* converter, shown in Fig. 15-12(a). A reference voltage and resistor divider network establishes a resolution of 1 V/LSB. Analog input voltage V_{in} is applied to the + inputs of all comparators. Their outputs drive an 8-line-to-3-line priority encoder. The encoder logic outputs a binary code that represents the analog input.

For example, suppose that V_{in} = 5.0 V. The outputs of comparators 1 through 5 would go high and 6 through 8 would go low. As shown in Fig. 15-12(b), the digital output would be 101.

15-7.2 Conversion Time

The conversion time of the flash converter is limited only by the response times of comparators and logic gates. They can digitize video or radar signals. The flash converter's high speed becomes more expensive as resolution is increased. Figure 15-12 shows that the flash converter requires seven comparators (or $2^3 - 1$) to perform a 3-bit conversion. The number of comparators required for n-bit resolution is

$$\text{number of comparators} = 2^n - 1 \tag{15-10}$$

R	V_{in}
10 kΩ	0–440 mV
10 kΩ ‖820 Ω	0–2.60 V

R_T	Conversions per second
1.5 MΩ	1,000
120 kΩ	10,000

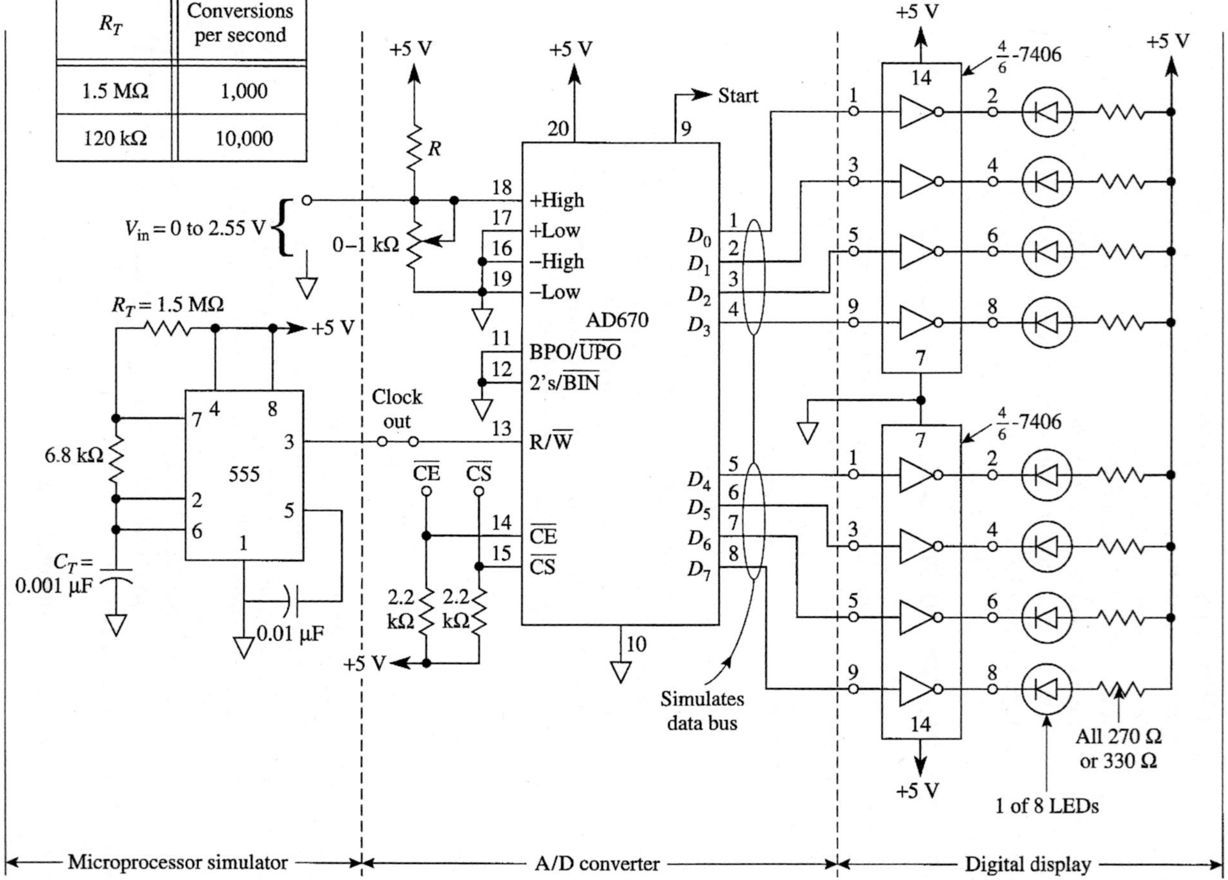

FIGURE 15-11 The operation of the AD670 can be studied without the need for a microprocessor. Pins 14 and 15 can be grounded to simulate a microprocessor selection via an address bus. The 555 timer simulates continuous

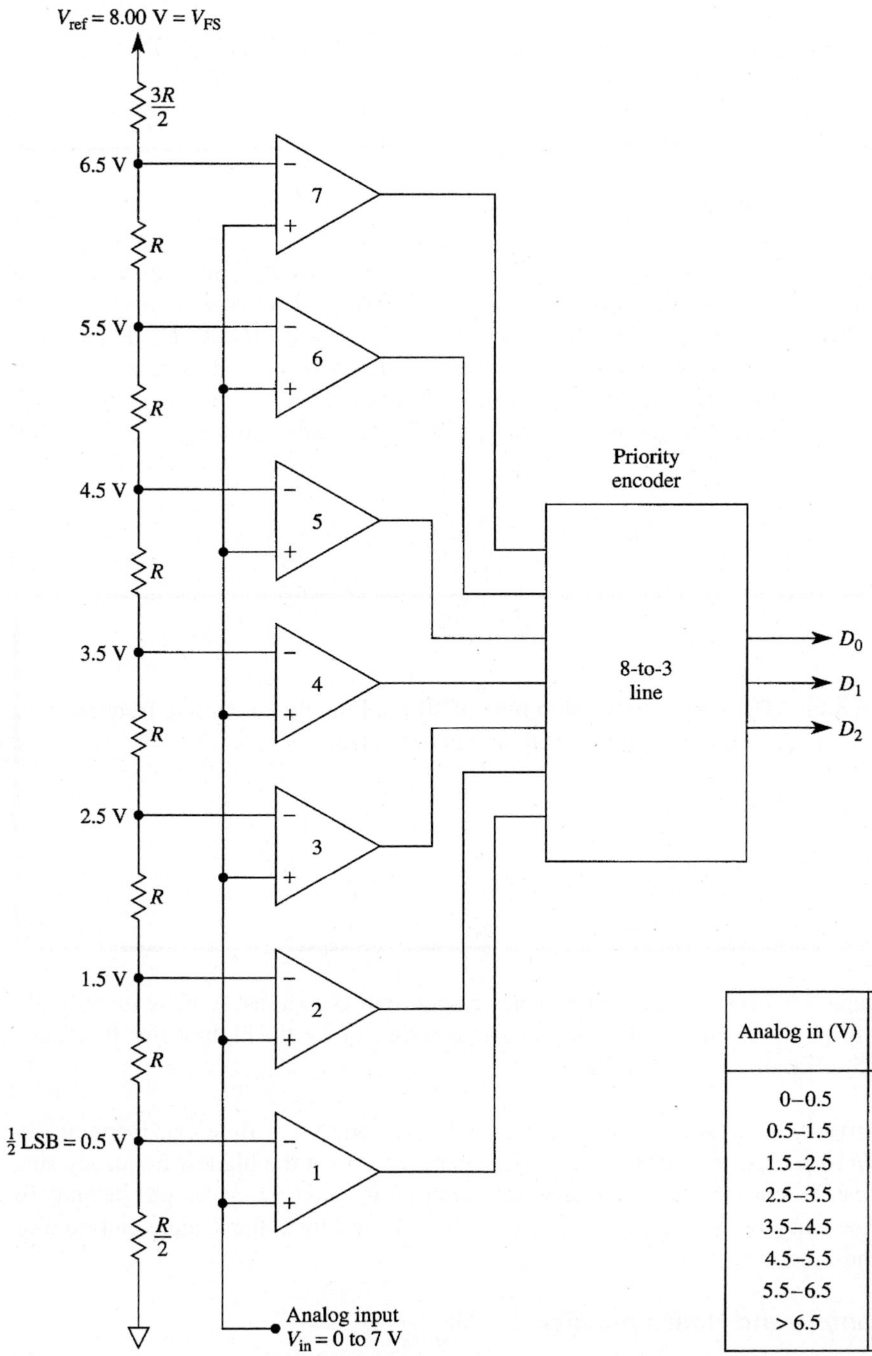

(a)

Analog in (V)	Digital out		
	D_2	D_1	D_0
0–0.5	0	0	0
0.5–1.5	0	0	1
1.5–2.5	0	1	0
2.5–3.5	0	1	1
3.5–4.5	1	0	0
4.5–5.5	1	0	1
5.5–6.5	1	1	0
> 6.5	1	1	1

(b)

FIGURE 15-12 (a) Three-bit flash (parallel) A/D converter; (b) output versus input.

For example, an 8-bit flash converter requires $(2^8 - 1)$ or 255 comparators. Encoder logic would be more complex, requiring a 256-line-to-8-line priority encoder.

15-8 FREQUENCY RESPONSE OF ADCs

15-8.1 Aperture Error

During conversion time, T_C, the analog input voltage must not change by more than $\pm\frac{1}{2}$ LSB (total 1 LSB), or the conversion will be incorrect. This type of inaccuracy is called *aperture error.* The rate of change of V_{in} with respect to time is called *slew rate.* If V_{in} is a sine wave, its slew rate is maximum at its zero crossings. The sine wave's slew rate is determined by both its peak voltage and frequency.

For an A/D converter, the maximum frequency for a sine wave V_{in} to be digitized within an accuracy of $\pm\frac{1}{2}$ LSB is

$$f_{max} \simeq \frac{1}{2\pi\,(T_C)2^n} \qquad (15\text{-}11)$$

Example 15-7

The AD670 is an 8-bit ADC with a conversion time of 10 μs. Find the maximum frequency of an input sine wave that can be digitized without aperture error.

Solution From Eq. (15-11),

$$f_{max} \simeq \frac{1}{2\pi\,(2^8)10\ \mu\text{s}} = \frac{1}{2\pi\,(256)10 \times 10^{-6}\ \text{s}} = 62\ \text{Hz}$$

Example 15-7 shows that the frequency response of even a fast ADC is surprisingly low. For a 10-bit integrating ADC with a conversion time of $\frac{1}{3}$ s, the highest sine frequency is about 0.5 mHz, or 1 cycle per 2000 s.

Summary. An 8-bit converter with a 10-μs conversion time can theoretically perform $[1/(10\ \mu\text{s})]\mu\text{s} = 100{,}000$ conversions per second, yet the highest frequency sine wave that can be converted without slew-rate limiting is about 62 cycles per second. To raise the frequency response, we must add another circuit block, the sample-and-hold or follower-and-hold amplifier.

15-8.2 Sample-and-Hold Amplifier

The sample-and-hold (S/H) or follow-and-hold amplifier of Fig. 15-13 is made from two op amps, a hold capacitor (C_H), and a high-speed analog switch. This amplifier is connected between an analog input signal and the input to an ADC.

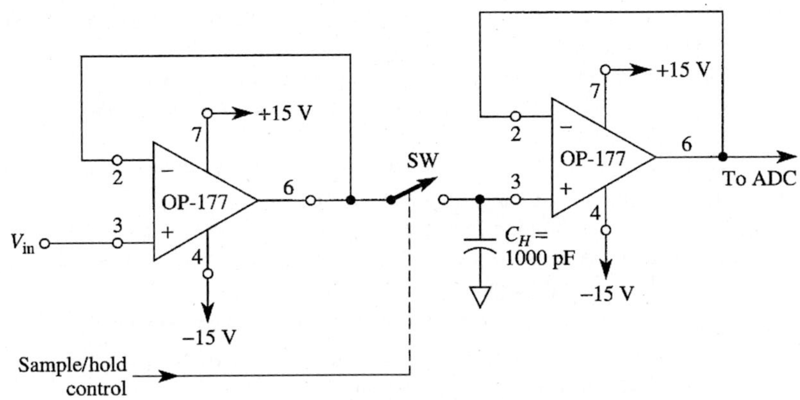

FIGURE 15-13 Sample-and-hold amplifier.

When the S/H amplifier is in the *sample* mode, the switch is closed and hold capacitor (C_H) voltage *follows* V_{in}. A *hold* command opens the switch and C_H retains a charge equal to V_{in} at the moment of switching. The S/H amplifier thus acts to hold V_{in} (stored on C_H) constant, while the ADC performs a conversion.

Conversion time of the ADC no longer limits frequency response. Instead, the limit is the *aperture time* of the S/H amplifier, which can be made much less than the conversion time. Aperture time is the time elapsed between a hold command and a switch opening. If the hold command is advanced by a time equal to the aperture time, C_H will hold the desired sample of V_{in}. Then the only remaining error is *aperture time uncertainty,* the switch jitter variation for each hold command.

Commercial S/H amplifiers have aperture time uncertainties lower than 50 ns. An example shows the improvement in frequency response due to an added S/H amplifier.

Example 15-8

An S/H amplifier with an aperture time uncertainty of 50 ns is connected to an 8-bit ADC. Find the highest-frequency sine wave that can be digitized within an error of 1 LSB.

Solution Replace conversion time by aperture uncertainty time in Eq. (15-11):

$$f_{max} \simeq \frac{1}{2\pi\,(2^8)50 \times 10^{-9}\ \text{s}} = 12.4\ \text{kHz}$$

PROBLEMS

15-1. A 10-bit A/D converter is capable of accepting an input voltage 0 to 5.12 V. What is the digital output code if the input voltage is 2.4 V?

15-2. If the 8-bit A/D converter of Example 15-2 has an offset error of $-\frac{1}{2}$ LSB, what analog input voltage will cause the digital output to be all 1s?

15-3. Note that the A/D converter of Example 15-1 has an offset error of $+\frac{1}{2}$ LSB and a gain error of -0.4% of FSR. What is the analog input voltage needed to cause the digital output to be all 1s?

15-4. Name three types of ADCs and indicate their relative conversion speeds (slow or fast).

15-5. $V_{in} = 50$ mV in the integrating ADC circuit of Fig. 15-5.
(a) What is the duration for integrating phase T_1 and the value of V_o?
(b) What is the name of phase T_2, the value of V_{ref}, and the duration of T_2?
(c) Find the circuit output.

15-6. Name the three components of a successive approximation 8-bit ADC.

15-7. A microprocessor issues a write command to an ADC. Does the ADC send data to the microprocessor or perform a conversion?

15-8. An input voltage with a range of 0 to 2.55 V is applied to pin 16 of an AD670 and pin 18 is grounded. Which other input pins should be jumpered or grounded to select this range?

15-9. How do you pin-program the AD670 for straight binary output?

15-10. How does a microprocessor "tell" an AD670 to **(a)** perform a conversion; **(b)** place the result on the data bus? **(c)** How does the microprocessor know when the AD670 has finished a conversion and its data are valid?

15-11. **(a)** What is the conversion time for an AD670?
(b) How many conversions can it perform per second?
(c) What is the maximum sine wave frequency that it can convert without adding a sample-and-hold amplifier?

15-12. If the sample-and-hold amplifier of an 8-bit ADC has an aperture uncertainty time of 10 ns, what maximum sine wave frequency can it convert within $\pm\frac{1}{2}$ LSB?

15-13. How many comparators are required to make an 8-bit flash converter?

CHAPTER 16

Power Supplies

LEARNING OBJECTIVES

Upon completion of this chapter on power supplies, you will be able to:

- Draw the schematic for a full-wave bridge (FWB) rectifier unregulated power supply.
- Identify the components of an FWB and tell what each component does in the circuit.
- Design an FWB rectifier; choose the specifications for the transformer, diodes, and capacitor; purchase these components from standard stock; build the rectifier; test it; and document its performance.
- Measure the percent regulation and percent ripple, draw the load voltage waveshapes at no load or full load, and plot the regulation curve for an FWB unregulated power supply.
- Design or analyze a bipolar or two-value unregulated power supply.
- Explain the need for voltage regulators.
- Connect an IC voltage regulator to an unregulated FWB rectifier circuit to make a voltage-regulated power supply.

- Design, build, and test a ±15-V regulated power supply for analog ICs.
- Build a regulated 5-V supply for TTL logic.
- Connect an LM317 to an unregulated supply to obtain a laboratory-type voltage regulator that can be adjusted precisely to a required voltage.

16-0 INTRODUCTION

Most electronic devices require dc voltages to operate. Batteries are useful in low-power or portable devices, but operating time is limited unless the batteries are recharged or replaced. The most readily available source of power is the 60-Hz 110-V ac wall outlet. The circuit that converts this ac voltage to a dc voltage is called a *dc power supply.*

The most economical dc power supply is some type of rectifier circuit. Unfortunately, some ac ripple voltage rides on the dc voltage, so the rectifier circuit does not deliver pure dc. An equally undesirable characteristic is a reduction in dc voltage as more load current is drawn from the supply. Since dc voltage is *not* regulated (that is, constant with changing load current), this type of power supply is classified as *unregulated.* Unregulated power supplies are introduced in Sections 16-1 and 16-2. It is necessary to know their limitations before such limitations can be minimized or overcome by adding regulation. It is also necessary to build an unregulated supply before you connect a voltage regulator to it.

Without a good regulated voltage supply, none of the circuits in this text (or any other text for that matter) will work. Therefore, this chapter shows the simplest way to analyze or design power supplies for linear or digital ICs.

It is possible to make a good voltage regulator with an op amp plus a zener diode, resistors, and a few transistors. However, it is wiser to use a modern integrated circuit voltage regulator. The types of superb regulators are so vast there is no problem in finding one that will suit your needs.

We will present an op amp regulator to illustrate the workings of a few of the features within an IC regulator. Then we will proceed to a representative sampling of some of the widely used IC voltage regulators. But we begin with the unregulated supply.

16-1 INTRODUCTION TO THE UNREGULATED POWER SUPPLY

16-1.1 Power Transformer

A transformer is required for reducing the nominally 115-V ac wall outlet voltage to the lower ac value required by transistors, ICs, and other electronic devices. Transformer voltages are given in terms of rms values. In Fig. 16-1, the transformer is rated as 115 to 24 V center tap. With the 115-V rms connected to the primary, 24-V rms is developed between secondary terminals 1 and 2. A third lead, brought out from the center of the secondary, is called a center tap, CT. Between terminals CT and 1 or CT and 2, the rms voltage is 12 V.

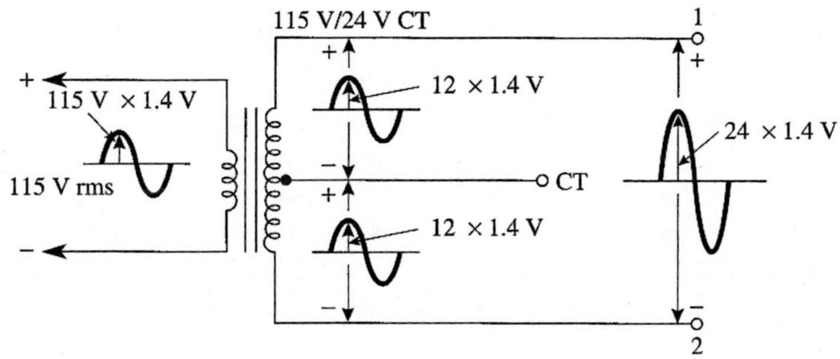

(a) Peak voltages for the positive half-cycle.

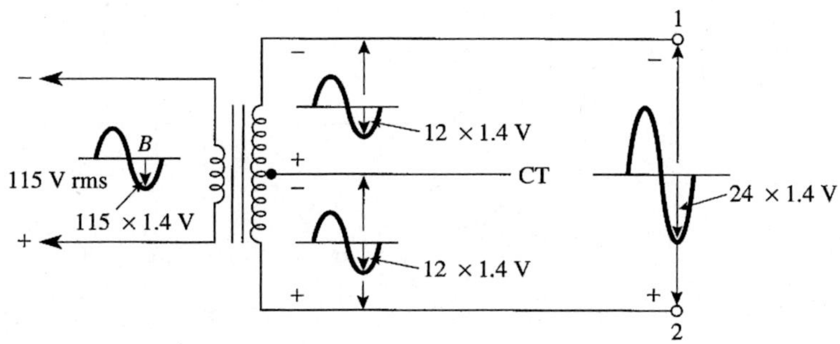

(b) Peak voltages for the negative half-cycle.

FIGURE 16-1 Power transformer 115 V/24 VCT.

An oscilloscope would show the sinusoidal voltages shown in Fig. 16-1. The maximum instantaneous voltage E_m is related to the rms value E_{rms} by

$$E_m = 1.4(E_{rms}) \tag{16-1}$$

In Fig. 16-1(a), voltage polarities are shown for the positive primary half-cycle; those for the negative half-cycle are shown in Fig. 16-1(b).

Example 16-1

Find E_m in Fig. 16-1 between terminals 1 and 2.

Solution By Eq. (16-1), $E_m = 1.4(24\ \text{V}) = 34$ V.

16-1.2 Rectifier Diodes

The next step in building a dc power supply is to convert the lower secondary ac voltage of the transformer to a pulsated dc voltage. This is accomplished by silicon diodes.

In Fig. 16-2(a), four diodes are arranged in a diamond configuration called a full-wave bridge rectifier. They are connected to terminals 1 and 2 in the transformer of Fig. 16-1. When terminal 1 is positive with respect to terminal 2, diodes D_1 and D_2 conduct. When terminal 2 is positive with respect to terminal 1, diodes D_3 and D_4 conduct. The result is a pulsating dc voltage between the output terminals.

16-1.3 Positive versus Negative Supplies

Note that the bridge has two input terminals labeled ac. The output terminals are labeled (+) and (−), respectively. Also note that the output dc voltage *cannot* as yet be desig-

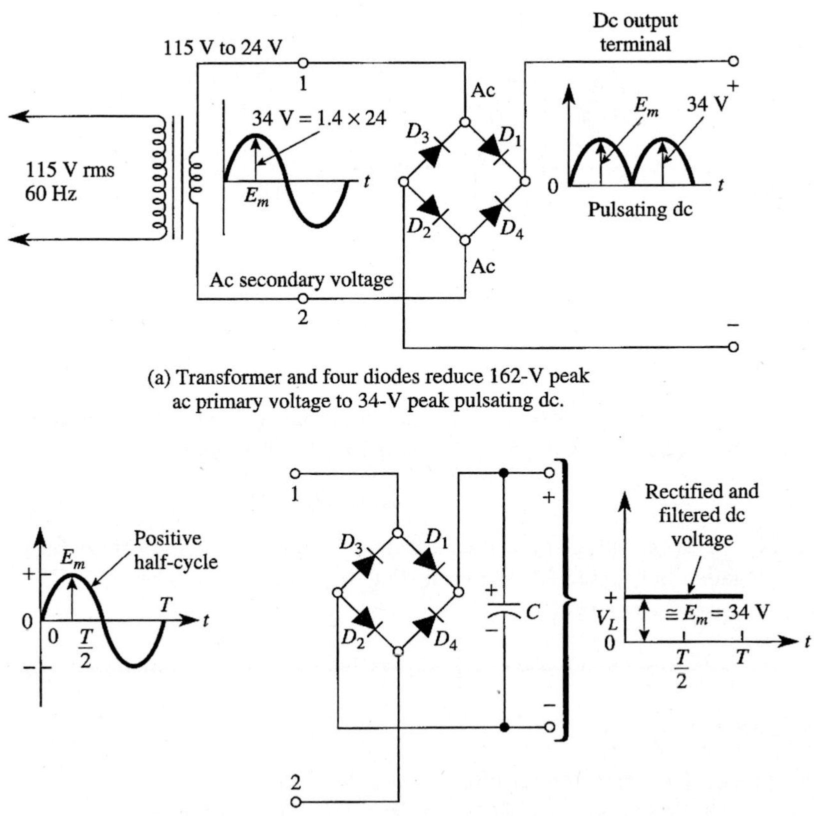

(a) Transformer and four diodes reduce 162-V peak ac primary voltage to 34-V peak pulsating dc.

(b) Capacitor C filters the pulsating dc in (a) to give a dc load voltage.

FIGURE 16-2 Transformer plus rectifier diodes plus filter capacitor equals unregulated power supply.

nated positive or negative. It is a *floating* supply. If you want a "positive" supply, you must "earth ground" the negative terminal.

The third wire (green) of the line cord extends earth ground from the "U"-shaped terminal of the wall outlet, usually to the metal chassis. This connection is to protect the user. Simply extend the green wire terminal to the negative terminal and call this terminal *power supply common.* All voltage measurements are with respect to power supply common, and it is designated in a schematic by a ground symbol. To make a negative supply, simply earth ground the positive terminal of the bridge.

16-1.4 Filter Capacitor

The pulsating dc voltage in Fig. 16-2(a) is not pure dc, so a filter capacitor is placed across the dc output terminals of the bridge rectifier [see Fig. 16-2(b)]. This capacitor smooths out the dc pulsations and gives an almost pure dc output load voltage, V_L, which is the unregulated voltage that supplies power to the load. The filter capacitor is typically a large electrolytic capacitor, 500 μF or more.

16-1.5 Load

In Fig. 16-2(b), nothing other than the filter capacitor is connected across the dc output terminals. The unregulated power supply is said to have no load. This means that the *no-load current,* or 0-load current, I_L, is drawn from the output terminals. Usually, the maximum expected load current, or full-load current, to be furnished by the supply is known. The load is modeled by resistor R_L as shown in Fig. 16-3(a). As stated in Section 16-0, the load voltage changes as the load current changes in an unregulated power supply. The manner in which this occurs is examined next. But the key idea to power supply analysis now becomes clear. The peak value of secondary ac voltage E_m determines the dc no-load voltage of V_L.

16-2 DC VOLTAGE REGULATION

16-2.1 Load Voltage Variations

A dc voltmeter connected across the output terminals in Fig. 16-2(b) measures the dc no-load voltage of V_L, or

$$V_{dc} \text{ no load} = E_m \tag{16-2}$$

From Example 16-1, V_{dc} no load is 34 V. An oscilloscope would also show the same value with no ac ripple voltage, as in Fig. 16-3(b). Now suppose that a load R_L was connected to draw a full-load dc current of $I_L = 1$ A, as in Fig. 16-3(a). An oscilloscope now shows that the load voltage V_L has a lower *average,* or dc value V_{dc}. Moreover, the load voltage has an ac ripple component, ΔV_o, superimposed on the dc value. The average value measured by a dc voltmeter is 24 V and is called V_{dc} full load. The peak-to-peak ripple voltage is called ΔV_o and measures 5 V in Fig. 16-3(b).

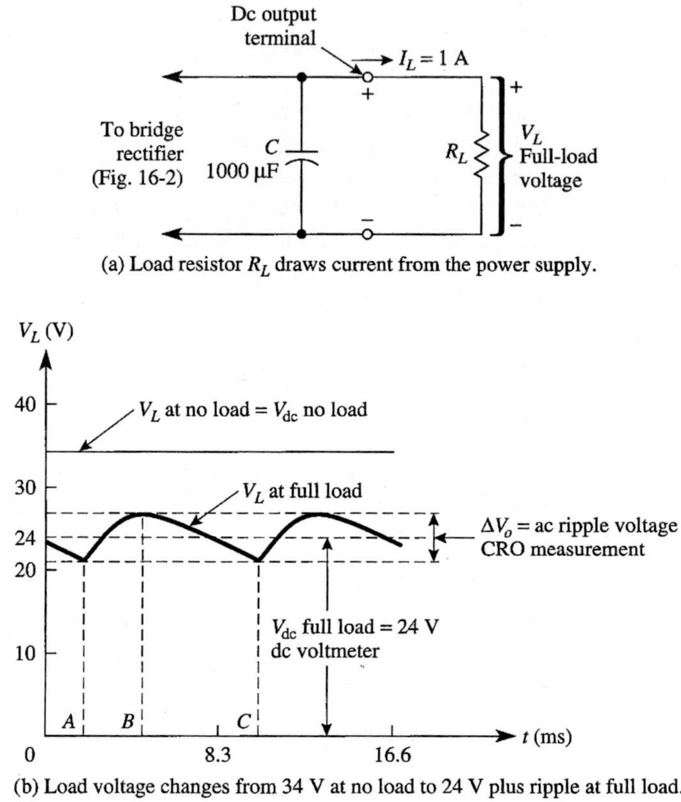

FIGURE 16-3 Variation of dc load voltage and ac ripple voltage from no-load current to full-load current.

There are two conclusions to be drawn from Fig. 16-3(b). First, the dc load voltage goes *down* as dc load current goes *up*; how much the load voltage drops can be estimated by a technique explained in Section 16-2.2. Second, the ac ripple voltage increases from 0 V at no-load current to a large value at full-load current. As a matter of fact, the ac ripple voltage increases directly with an increase in load current. The amount of ripple voltage can also be estimated, by a technique explained in Section 16-3.

16-2.2 DC Voltage Regulation Curve

In the unregulated power supply circuit Fig. 16-4(a), the load R_L is varied so that we can record corresponding values of dc load current and dc load voltage. The dc meters respond only to the *average* (dc) load current or voltage. If corresponding values of current and voltage are plotted, the result is the *dc voltage regulation curve* of Fig. 16-4(b). For example, point O represents the no-load condition, $I_L = 0$ and V_{dc} no load $= E_m = 34$ V. Point *A* represents the full-load condition, $I_L = 1$ A and V_{dc} full load $= 24$ V.

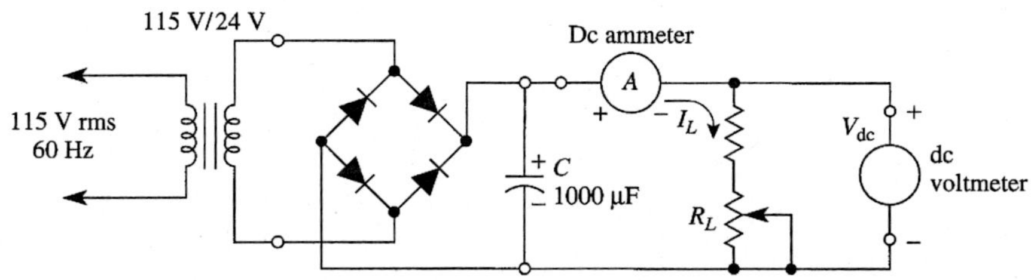

(a) Unregulated power supply performance measured with dc ammeter and voltmeter.

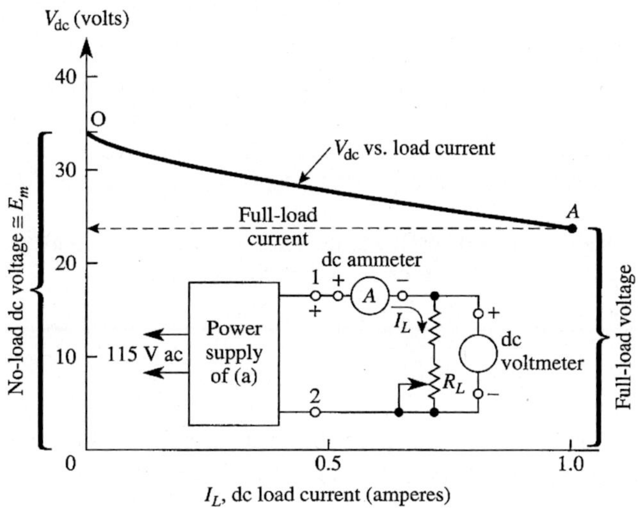

(b) Dc voltage regulation curve.

FIGURE 16-4 Dc load voltage varies with load current in (a) as shown by the voltage regulation curve in (b).

16-2.3 DC Model of a Power Supply

Figure 16-5(a) shows results obtained when making measurements of (a) no-load voltage with no-load current and (b) full-load voltage at full-load current. As you draw *more* dc current from the supply, something inside the supply causes an *increasing* internal voltage drop to leave less voltage available across the load. The simplest way to account for this behavior is to blame it on an *internal* or *output resistance* R_o. Accordingly, this dc behavior can be described by

$$V_{\text{dc NL}} = V_{\text{dc FL}} + I_{L\text{ FL}}\, R_o \qquad (16\text{-}3a)$$

or from Fig. 16-5(b)

$$V_{\text{dc FL}} = \frac{R_L}{R_o + R_L}(V_{\text{dc NL}}) \qquad (16\text{-}3b)$$

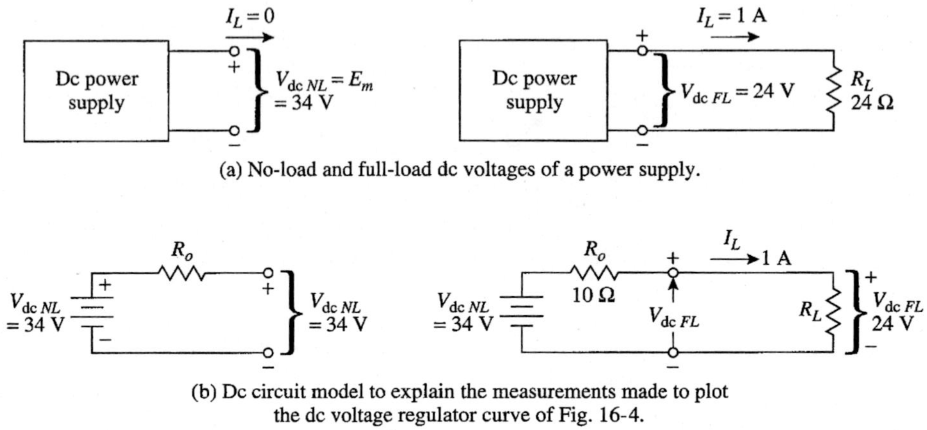

FIGURE 16-5 A dc model is developed to explain the dc voltage measurements for an unregulated power supply. $V_{\text{dc NL}} = V_{\text{dc FL}} + I_L R_o$.

We use the data of both the no-load and full-load measurements in Fig. 16-5 and the voltage regulation curve of Fig. 16-4 to measure R_o.

Example 16-2

(a) Calculate output resistance R_o from the no-load and full-load measurements of Figs. 16-4 and 16-5. (b) Predict dc output voltage at half-load where $I_L = 0.5$ A.

Solution (a) At no load, $I_L = 0$ and $V_{\text{dc NL}} = 34$ V; at full load, $I_L = 1$ A and $V_{\text{dc FL}} =$ 24 V. From Eq. (16-3a),

$$R_o = \frac{V_{\text{dc NL}} - V_{\text{dc FL}}}{I_{L\ \text{FL}}} = \frac{(34 - 24)\ \text{V}}{1\ \text{A}} = 10\ \Omega$$

(b) Use Eq. (16-3a) again, but modify it for $I_L = 0.5$ A.

$$V_{\text{dc}} = V_{\text{dc NL}} - I_L R_o = 34\ \text{V} - (0.5\ \text{A})(10\ \Omega)$$
$$= 34\ \text{V} - 5\ \text{V} = 29\ \text{V}$$

Note that the 5 V appears *internally* within the power supply to cause 5 V × 0.5 A = 2.5 W of heat.

R_o models the net effect of internal losses within the power supply. These losses occur because of the transformer, diodes, capacitor, and even wires going to the wall outlet. It is pointless to track down the contributions of each one. We are only interested in their net effect, R_o.

16-2.4 Percent Regulation

Another way to describe dc performance is by a specification called percent regulation. You measure the supply's no-load voltage and full-load voltage. Percent regulation is then calculated from

$$\%\ \text{regulation} = \frac{V_{\text{dc NL}} - V_{\text{dc FL}}}{V_{\text{dc FL}}} \times 100 \tag{16-4}$$

Example 16-3

Find percent regulation for the dc power supply data in Figs. 16-4 and 16-5.

Solution From the data, $V_{\text{dc NL}} = 34$ V and $V_{\text{dc FL}} = 24$ V. From Eq. (16-4),

$$\%\ \text{regulation} = \frac{(34 - 24)\ \text{V}}{24\ \text{V}} \times 100 = 41.7\%$$

Percent regulation tells you by *what percent the full-load voltage will rise* when you remove the load.

16-3 AC RIPPLE VOLTAGE

16-3.1 Predicting AC Ripple Voltage

Figure 16-6(b) shows how to measure both ac and dc performance of a power supply. Dc measurements (average values) are made with dc meters. Their measured dc values, I_L and V_{dc}, are summarized and plotted as the (dc) voltage regulation curve of Fig. 16-4(b).

The peak-to-peak ac ripple voltage ΔV_o is centered on V_{dc}. ΔV_o can be estimated from

$$\Delta V_o \simeq \frac{I_L}{200C} \tag{16-5a}$$

where ΔV_o is in volts, I_L in amperes, and C is the size of the filter capacitor in farads. If load voltage V_L is measured with an ordinary ac voltmeter, it will indicate the rms value of the ripple voltage V_{rms}. A coupling capacitor within the meter eliminates the dc component. V_{rms} is related to ΔV_o by the approximation

$$\Delta V_o \simeq 3.5\ V_{\text{rms}} \tag{16-5b}$$

We need one other characteristic of the power supply that will be used later in the chapter. It tells us how to design a power supply for a voltage regulator. It is called minimum instantaneous load voltage and it occurs at full load. As seen in Fig. 16-6(b),

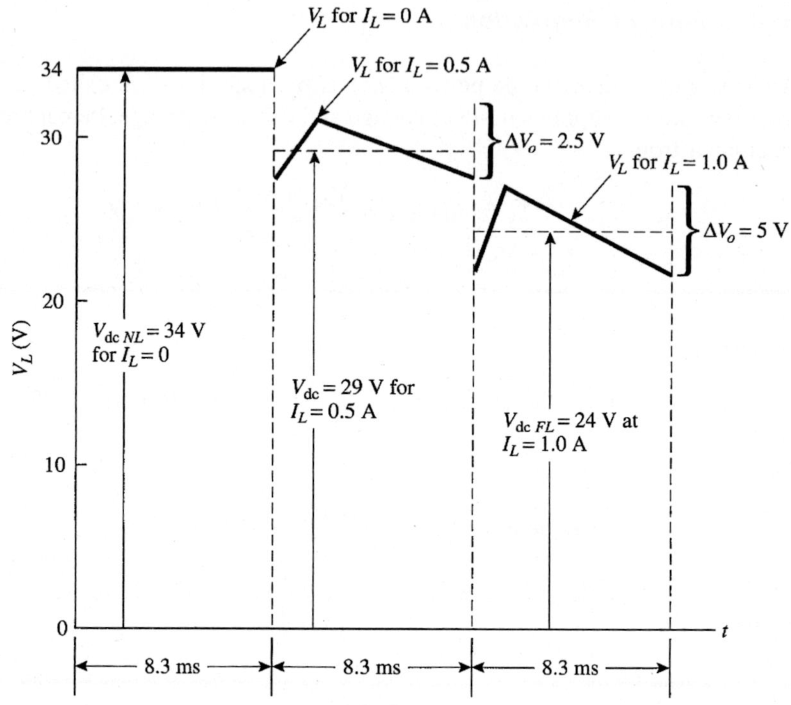

(a) A dc coupled oscilloscope shows both ac and dc components of the load voltage V_L.

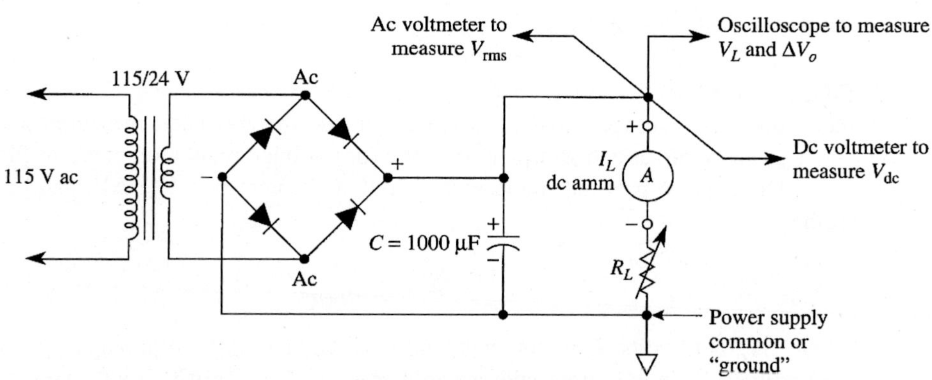

(b) Test circuit to measure ac and dc performance of an unregulated power supply.

FIGURE 16-6 This test circuit setup allows simultaneous measurement of both ac and dc performance of an unregulated power supply. An oscilloscope set on direct-coupled measures instantaneous load voltage V_L and peak-to-peak ripple voltage ΔL_o. Dc meters measure average (dc) load voltage V_{dc} and current I_L. An ac voltmeter measures the rms value of the ripple voltage V_{rms}.

$$\text{minimum } V_L = V_{\text{dc FL}} - \frac{\Delta V_o}{2} \tag{16-5c}$$

These principles will be illustrated by an example.

Example 16-4

A full-wave bridge rectifier has (1) a full-load current of 1 A, (2) a full-load voltage of 24 V, and (3) a filter capacitor of 1000 μF. Calculate (a) the peak-to-peak and rms value of ripple voltage at full load; (b) minimum instantaneous output voltage.

Solution (a) From Eq. (16-5a),

$$\Delta V_o = \frac{1 \text{ A}}{(200)(1000 \times 10^{-6} \text{ F})} = 5 \text{ V}$$

From Eq. (16-5b),

$$V_{\text{rms}} = \frac{\Delta V_o}{3.5} = \frac{5 \text{ V}}{3.5} = 1.43 \text{ V}$$

(b) From Eq. (16-5c),

$$\text{minimum } V_L = V_{\text{dc FL}} - \frac{\Delta V_o}{2} = 24 \text{ V} - \frac{5 \text{ V}}{2} = 21.5 \text{ V}$$

16-3.2 Ripple Voltage Frequency and Percent Ripple

The ripple voltage frequency of a full-wave rectifier with capacitor filter is 120 Hz, or twice the line voltage frequency. This is because the capacitor must charge and discharge twice for each cycle of line voltage. As shown in Fig. 16-6(a), the period for each cycle of ripple voltage is 8.3 ms.

The ac performance of a power supply can also be specified by a single percentage number. You measure the worst-case ripple voltage V_{rms}. This occurs at full-load current [see Fig. 16-6(a)]. Measure $V_{\text{dc FL}}$ and calculate *percent ripple* from

$$\% \text{ ripple} = \frac{V_{\text{rms}} \text{ at full load}}{V_{\text{dc FL}}} \times 100 \tag{16-6}$$

Example 16-5

Calculate the percent ripple for the power supply specified in Example 16-4.

Solution From Example 16-4, $V_{dc\ FL} = 24$ V and $V_{rms} = 1.43$ V. From Eq. (16-6),

$$\%\ \text{ripple} = \frac{1.43}{24} \times 100 = 6\%$$

16-3.3 Controlling Ripple Voltage

Equation (16-5a) tells us that ΔV_o depends directly on load current. Therefore, for the same-value filter capacitor, if the load current doubles, the ripple voltage doubles. Equation (16-5a) also shows that ΔV_o is inversely proportional to C. If C doubles, the ripple voltage is halved (for the same value of I_L). A nice rule of thumb is to substitute 1000 μF for C in Eq. (16-5a) to obtain

$$\Delta V_o = (5\text{ V})(I_L \text{ in amperes}) \tag{16-7}$$

when $C = 1000\ \mu$F.

If $I_L = 0.5$ A, then $\Delta V_o = 2.5$ V. Double C to 2000 μF if you want to reduce ΔV_o to 2.5 V/2 = 1.25 V.

16-4 DESIGN PROCEDURE FOR A FULL-WAVE BRIDGE UNREGULATED SUPPLY

16-4.1 Design Specification, General

The dc requirement for a power supply is specified typically as "I need 12 volts at 1 ampere." We now can define these values as (1) $V_{dc\ FL} = 12$ V and (2) $I_{L\ FL} = 1$ A. (Incidentally, R_L for full load is 12 V/1 A = 12 Ω.) The ac requirement can be specified in many ways, (1) % ripple less than 5%, (2) V_L minimum = 11 V, or (3) ΔV_o at full load must be less than 2.1 V.

Design Example 16-6

Design a full-wave bridge (FWB) rectifier to furnish 12 V at 1 A with less than 10% ripple. The design procedures use the principle of superposition. (a) Do the dc design. This gives you the transformer size and diode ratings. (b) Do the ac design. This gives you the capacitor size.

(a) **Dc design procedure showing a transformer and diodes.** Use Eqs. (16-3a) and (16-2) to find E_m. Then use Eq. (16-1) to rate the transformer's secondary. Since you have not purchased the parts yet, you cannot measure R_o. Therefore, as the designer, *you* must make an *informed estimate* of R_o. Guess that $R_o = 10$ Ω. Based on your *guess* (informed estimate), the rest is straightforward. Apply Eq. (16-3a),

$$V_{dc\ NL} = V_{dc\ FL} + I_{L\ FL}\ R_o = 12\ V + (1\ A)(10\ \Omega) = 22\ V$$

then Eq. (16-2),

$$E_m = V_{dc\ NL} = 22\ V$$

and finally, Eq. (16-1),

$$E_{rms} = \frac{E_m}{1.4} = 22\ V/1.4 = 15.5\ V$$

Your problem appears to worsen because you cannot buy a 115/15.5 V transformer from stock. So we learn the first basic lesson of power supply design. *You usually cannot get the dc voltage you want.* Even if you could get 12 V at 1 A, V_L would be 22 V at 0 A. The voltage regulator was invented to solve this problem. However, your company stocks 115 V/12.6 V CT transformers with two secondaries. Connect one 12.6-V winding in series aiding with a center tap and proper terminal of the other winding (6.3 V) to obtain $E_s = 6.3\ V + 12.6\ V = 18.9\ V$.

(You learned also that this is no way to power digital, linear, or automobile electronics, but it will give a fine opportunity to *analyze* a design later, in Example 16-7.) All that remains is to choose the transformer's ac current rating as follows: For a FWB rectifier, choose $I_{sec\ rms} = 1.8 I_{L\ FL}$. Therefore, choose $I_{secondary} = 2$ A.

Choosing the bridge diodes. Diode selection is easy. They are rated by average current, I_{av}, and peak inverse voltage, PIV. A prudent design guide is given as follows:

1. Choose a diode with $I_{av} \geq I_{L\ FL} = 1$ A.
2. Choose a diode with a PIV rating greater than $V_{dc\ NL} + 20\%\ V_{dc\ NL}$.

(b) **AC design procedure: Choosing filter capacitor.** The original specifications required less than 10% ripple at $V_{dc\ FL} = 12$ V. Even knowing that we *may not* realize this dc value, we proceed on the original specification. (You may get lucky and find an R_o of 15 Ω for the 115 V/18 V transformer that you purchase.)

First, calculate ΔV_o as follows. Apply Eq. (16-6),

$$V_{rms} = \frac{(\%\ \text{ripple})V_{dc\ NL}}{100} = \frac{10(12\ V)}{100} = 1.2\ V$$

then Eq. (16-5b),

$$\Delta V_o = 3.5(1.2\ V) = 4.2\ V$$

and finally, Eq. (16-5a),

$$C = \frac{I_L}{200\ \Delta V_o} = \frac{1\ A}{(200)(4.2\ V)} = 1190\ \mu F$$

Choose C greater than 1190 μF or $C = 1500$ μF. If necessary, you can construct C from three 500-μF caps in parallel.

Note: Electrolytic capacitors also have a voltage rating called "working volts dc" (WVDC). The filter capacitor should be rated the same as a diode:

$$\text{WVDC} = V_{dc\ NL} + 20\%\ V_{dc\ NL}$$

Design Summary We end up with an FWB rectifier that has the following parts:

1. Transformer: 115 V/18 V at 2.0 A
2. Four diodes: each 1 A at PIV = 25 V (or more)
3. Three capacitors: 500 μF connected in parallel, WVDC = 25 V (or more)

We do not have a satisfactory design, but that is often the nature of practical solutions for unregulated power supplies. However, let us use this design to analyze and predict performance of an FWB rectifier.

Analysis Example 16-7: FWB Unregulated Supply

Given the power supply design schematic of Fig. 16-7(a), (a) predict its dc performance by plotting a dc regulation curve and calculating percent regulation. *Assume:* $R_o = 7\ \Omega$, $I_{L\ FL} = 1.0$ A; (b) plot the no-load and full-load instantaneous voltage V_L that you would expect to see on an oscilloscope. Also calculate percent ripple.

Solution (a) Since $E_{rms} = 18\ V_{rms}$, calculate E_m and $V_{dc\ NL}$ from Eqs. (16-1) and (16-2):

$$V_{dc\ NL} = E_m = 1.4E_{rms} = 25.5\ \text{V}$$

Calculate $V_{dc\ FL}$ from Eq. (16-3a):

$$V_{dc\ FL} = V_{dc\ NL} - I_{L\ FL}\, R_o = 25.5\ \text{V} - (1\ \text{A})(7\ \Omega) = 18.5\ \text{V}$$

and percent regulation from Eq. (16-4):

$$\%\ \text{regulation} = \frac{V_{dc\ NL} - V_{dc\ FL}}{V_{dc\ FL}} \times 100 = \frac{(25.5 - 18.5)\ \text{V}}{18.5} \times 100 \simeq 38\%$$

(b) Calculate ΔV_o from Eq. (16-5a) and V_{rms} from Eq. (16-5b):

$$\Delta V_o \simeq \frac{I_L}{200C} = \frac{1\ \text{A}}{200(1000 \times 10^{-6}\ \text{F})} = 5\ \text{V}$$

$$V_{rms} = \frac{\Delta V_o}{3.5} = \frac{5\ \text{V}}{3.5} = 1.43\ \text{V}$$

Finally, calculate percent ripple from Eq. (16-6):

$$\%\ \text{ripple} = \frac{V_{rms}}{V_{dc\ FL}}\ 100 = \frac{1.43}{18.5} \times 100 = 7.7\%$$

Dc performance of the unregulated power supply is depicted in Fig. 16-7(b). Ac performance is summarized by the time plots at both no load and full load in Fig. 16-7(c). Note that $\Delta V_o = 0$ V at no-load current. Also, ΔV_o is (approximately) centered on $V_{dc\ FL}$.

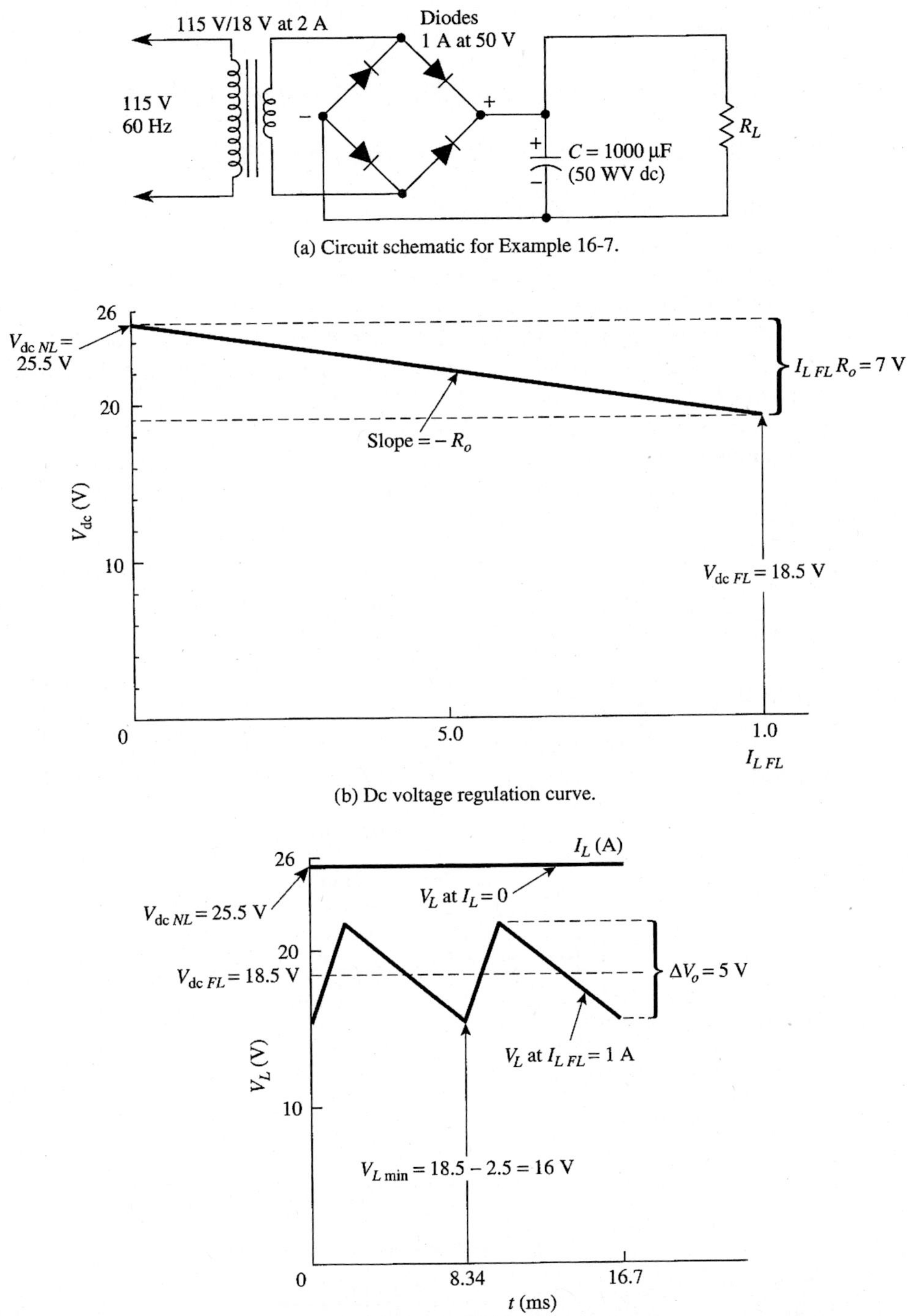

(a) Circuit schematic for Example 16-7.

(b) Dc voltage regulation curve.

(c) No-load and full-load instantaneous voltage.

FIGURE 16-7 The predicted dc performance for the filtered FWB rectifier in (a) is shown in (b) and its ac performance in (c).

16-5 BIPOLAR AND TWO-VALUE UNREGULATED POWER SUPPLIES

16-5.1 Bipolar or Positive and Negative Power Supplies

Many electronic devices need both positive (+) and negative (−) supply voltages. These voltages are measured with respect to a third common (or grounded) terminal. To obtain a positive and negative voltage, either two secondary transformer windings or one center-tapped secondary winding is needed.

A transformer rated at 115 V/24 V CT is shown in Fig. 16-8. Diodes D_1 and D_2 make terminal 1 positive with respect to center tap CT. Diodes D_3 and D_4 make terminal 2 negative with respect to the center tap. From Eq. (16-1) and Section 16-2.1, both no-load dc voltages are 1.41×12 V rms = 17 V. Capacitors $C+$ and $C-$, filter the positive and negative supply voltages, respectively. As shown in Sections 16-4 and 16-5, the ac ripple voltage and dc voltage regulation may be predicted for both no-load and full-load voltages.

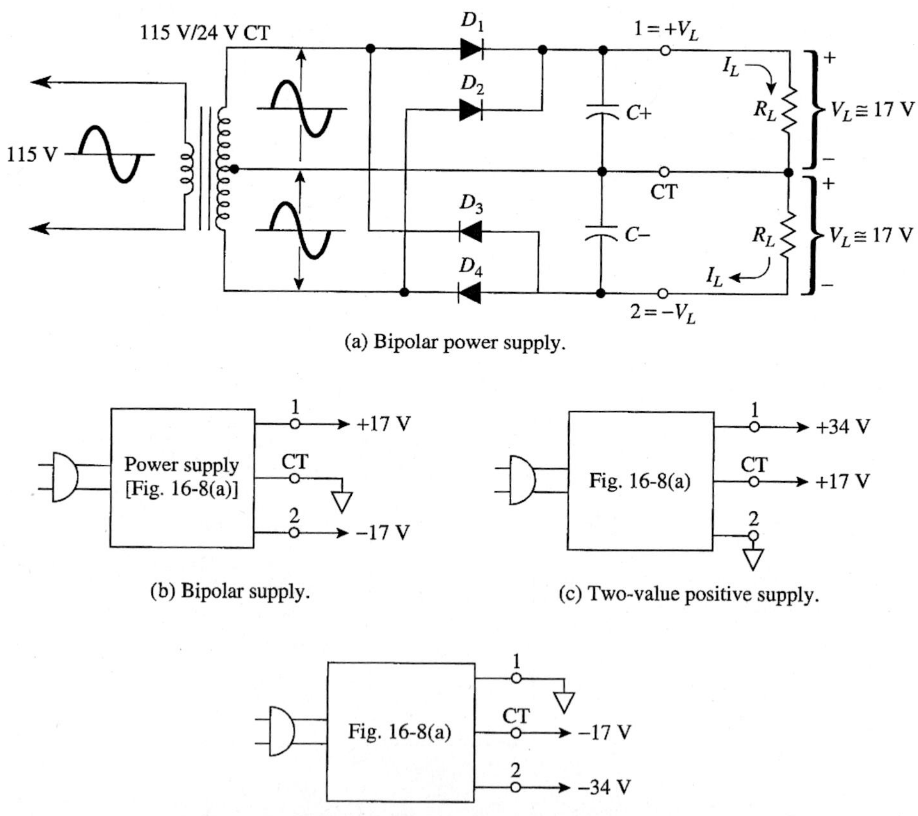

(a) Bipolar power supply.

(b) Bipolar supply.

(c) Two-value positive supply.

(d) Two-value negative supply.

FIGURE 16-8 Bipolar and two-value power supplies.

16-5.2 Two-Value Power Supplies

If the center tap of the power supply of Fig. 16-8 is grounded, we have a *bipolar* power supply. It is shown schematically in Fig. 16-8(b). If terminal 2 is grounded as in Fig. 16-8(c), we have a two-value positive supply. Finally, by grounding terminal 1 in Fig. 16-8(d), we get a two-value negative power supply. This indicates the versatility of the center-tapped transformer.

16-6 NEED FOR VOLTAGE REGULATION

Previous sections have shown that the unregulated power supply has two undesirable characteristics: the dc voltage decreases and the ac ripple voltage increases as load current increases. Both disadvantages can be minimized by adding a *voltage-regulator section* to the *unregulated supply* as in Fig. 16-9. The resulting power supply is classified as a *voltage-regulated supply.*

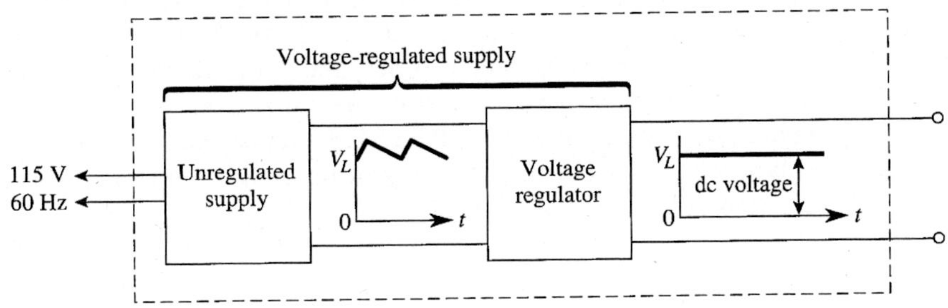

FIGURE 16-9 Unregulated supply plus a voltage regulator gives a voltage-regulated power supply.

16-7 THE HISTORY OF LINEAR VOLTAGE REGULATORS

16-7.1 The First Generation

An excellent dc voltage regulator can be built from an op amp, zener diode, two resistors or one potentiometer, and one or more transistors. In 1968, Fairchild Semiconductor Division integrated all of these components (plus others) into a single IC and called it the μA723 monolithic voltage regulator. Because of its flexibility, it has survived to the present day. It does, however, require a number of support components, has minimal internal protection circuitry, and requires the user to add boost transistors for more current capability and a resistor for limiting short-circuit current.

The race was on to make a three-terminal fixed-voltage regulator. National Semiconductor won with the LM309, in a close finish with Fairchild's 7800 series. The LM309 and μA7805 have three terminals. To use one, all you have to do is connect an unregulated supply between its input and common terminals. Then connect a load between the output and common and the design is complete. (Connect a decoupling capacitor across both input terminals and output terminals to improve performance.) These devices have internal protection circuitry that will be discussed later.

16-7.2 The Second Generation

The success of the +5-V regulators changed the philosophy of many system designers. There was no need to have a central regulator supplying current to each circuit board in the system and suffering the large I^2R loss. Now each printed circuit card could have its own on-board local regulator. The local regulator also protected its ICs against line voltage transients.

The +5-V regulator's success spawned an array of three-terminal regulators of 6, 8, 9, 12, 15, 18, and 24 V and their negative counterparts. Now, if you need a 15-V regulator to furnish 1 A, you simply buy one.

16-7.3 The Third Generation

Linear IC regulators were so popular that they created serious problems for original equipment manufacturers (OEMs). How do you stock all these sizes, and how do you make enough to suit the growing number of voltage requirements?

The LM117 was the first successful superior-performance adjustable positive IC voltage regulator. It was followed by the LM137 adjustable negative regulator. We will present only a few of the bewildering array of linear IC regulators. Space does not permit presentation of the switching regulators.

16-8 LINEAR IC VOLTAGE REGULATORS

16-8.1 Classification

Linear IC voltage regulators are classified by four characteristics:

1. *Polarity:* negative, positive, or dual tracking.
2. *Terminal count:* three-terminal or multiterminal.
3. *Fixed or adjustable output voltage:* standard fixed voltages are ±5, ±12, and ±15 V. Adjustable range is typically 1.2 to 37 V or −1.2 to −37 V.
4. *Output current:* Typical output current capabilities are 0.1, 0.2, 0.25, 0.5, 1.5, and 3 A and the new 5 and 10 A.

16-8.2 Common Characteristics

The instantaneous voltage at the input of an IC regulator must always exceed the dc output voltage by a value that is typically equal to 0.5 to 3 V. This requirement is called *minimum instantaneous input–output voltage, dropout voltage,* or simply *headroom.* As shown in Fig. 16-10(a), the LM340-15 voltage regulator has an output voltage of 15 V at a load of 1 A.

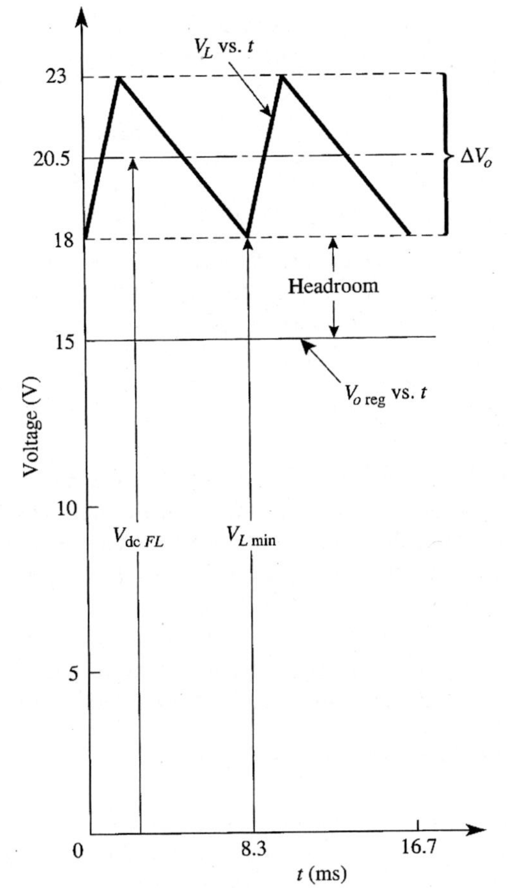

(a)

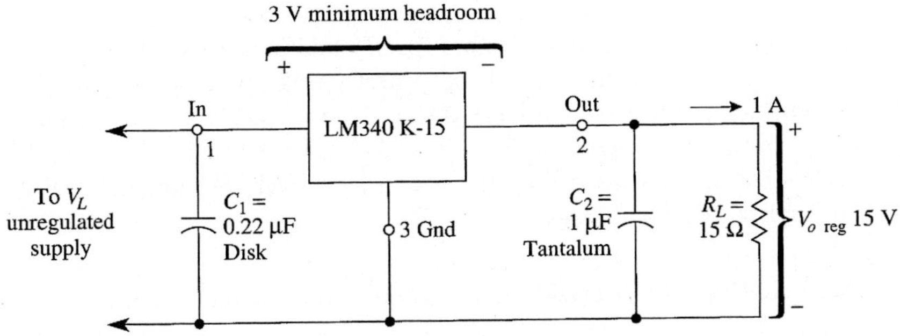

(b)

FIGURE 16-10 All voltage regulators need approximately 1 to 3 V between input and output terminals to ensure operation of the internal circuitry.

Suppose that the unregulated power supply that feeds the regulator has a 1000-μF capacitor and thus a ripple voltage of $\Delta V_o = 5$ V. As shown in Fig. 16-10(a), you need a minimum input voltage of

$$V_{L\,\text{min}} = V_{o\,\text{reg}} + \text{headroom} \tag{16-8}$$

or

$$V_{L\,\text{min}} = 15\text{ V} + 3\text{ V} = 18\text{ V}$$

This means that $V_{\text{dc FL}}$ must be 20.5 V at the very least [see Eq. (16-5c)]. Although you might be tempted to make $V_{\text{dc FL}}$ high to give plenty of headroom, you must remember that the worst regulator heat power is $I_{L\,\text{FL}}(V_{\text{dc FL}} - V_{o\,\text{reg}})$. So there is your trade-off. A higher $V_{\text{dc FL}}$ wastes more heat in the regulator.

16-8.3 Self-Protection Circuits

The internal circuitry of these devices senses the load current. If the load current exceeds a specified value, the output current is automatically limited until the overload is removed. They also measure both their input–output difference voltage and load current to be sure that no disallowed combination occurs. If it does, the regulator shuts down. This feature is called *safe area protection.*

Finally, these regulators even measure their own temperature to see if you heat-sinked them properly. If the internal die temperature exceeds 150° to 175°C, they shut down. If you remove the fault, the regulator goes back to work.

16-8.4 External Protection

Despite the well-designed internal protection circuitry, regulators can still be damaged by misuse, sabotage, or certain failures of external circuits. The measures you can take to safeguard against these eventualities are given in the data sheets of a particular regulator.

16-8.5 Ripple Reduction

Manufacturers of linear IC regulators specify their ac performance by a parameter called *ripple rejection.* It is the ratio of the peak-to-peak input ripple voltage $\Delta V_{o\,\text{unreg}}$ to the peak-to-peak output ripple voltage $\Delta V_{o\,\text{reg}}$. It is typically 60 dB or more. That is a reduction in ripple voltage of at least 1000:1. For example, if 5 V of ripple is at the regulator's input, less than 5 mV appears across the load. We now turn our attention to specific applications for IC regulators.

16-9 POWER SUPPLY FOR LOGIC CIRCUITS

16-9.1 The Regulator Circuit

A +5 V digital power supply for TTL logic or certain microprocessors is shown in Fig. 16-11. The K package of the LM340-05 is a steel TO-3 case and should be heat sinked for a case-to-ambient thermal resistance of a 6°C/W or less. This means that you should use a 0.002-in.-thick insulating thermal washer with thermal joint compound between the TO-3 case and its heat sink (or use the chassis as a heat sink).

The LM340K-05 can furnish up to 1.5 A. It has internal current limit at 2.1 A for pulse operation. It also has safe area protection that protects its output transistor. It has thermal shutdown protection at a junction temperature of 150°C to prevent burnout. The added diode protects the regulator against short circuits occurring at its input terminals.

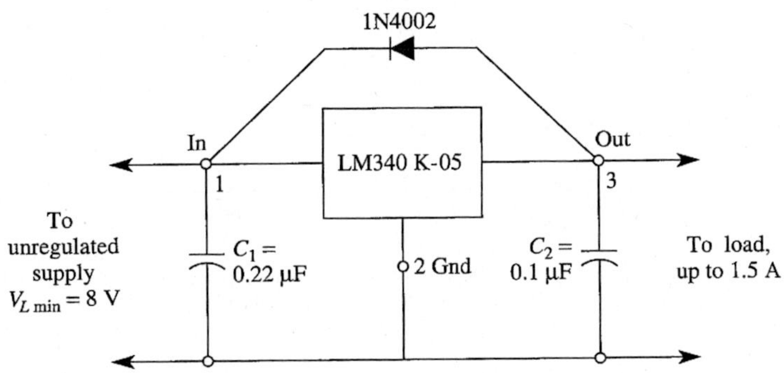

FIGURE 16-11 A TTL digital logic regulated power supply (5 V at 1 A.)

16-9.2 The Unregulated Supply

$V_{L\,min}$ for the unregulated FWB supply should allow 3 V of headroom and should be greater than (3 + 5) V = 8 V (see Fig. 16-11). Choose a 12.6-V transformer (1.8 × 1 A) at a 2-A current rating. From Sections 16-2 and 16-3: (1) $V_{dc\,NL} = 12.6\text{ V} \times 1.4 = 17.8$ V. (2) Assume that $R_o = 6\ \Omega$. If $I_{L\,FL} = 1$ A, then $V_{dc\,FL} = 17.8\text{ V} - 6\ \Omega(1\text{ A}) = 11.8$ V. (3) Pick a 1000-μF filter capacitor to give a $\Delta V_o = 5$ V. Thus, $V_{L\,min}$ should be at least 11.8 V − (5 V/2) = 9.3 V, leaving some margin. Pick the capacitor WVDC ≥ 25 V and diodes with a 25-V PIV rating. The diode current rating should exceed 1 A.

16-10 ±15-V POWER SUPPLIES FOR LINEAR APPLICATION

16-10.1 High-Current ±15-V Regulator

Figure 16-12(a) presents a bipolar ±15 V supply that can furnish 1 A from either (+) or (−) terminal. The LM340K-15 is a +15-V regulator with load current capability up to 1.5 A. To use it as a stand-alone +15-V supply, (1) remove the diodes, R_2, C_N, and the LM320-15; and (2) replace R_1 with a short circuit.

The LM320K-15 is a −15-V regulator with current capability up to 1.5 A. Both regulators have current limit, safe area, and thermal shutdown protection. They should be heat-sinked as directed in Section 16-10.2.

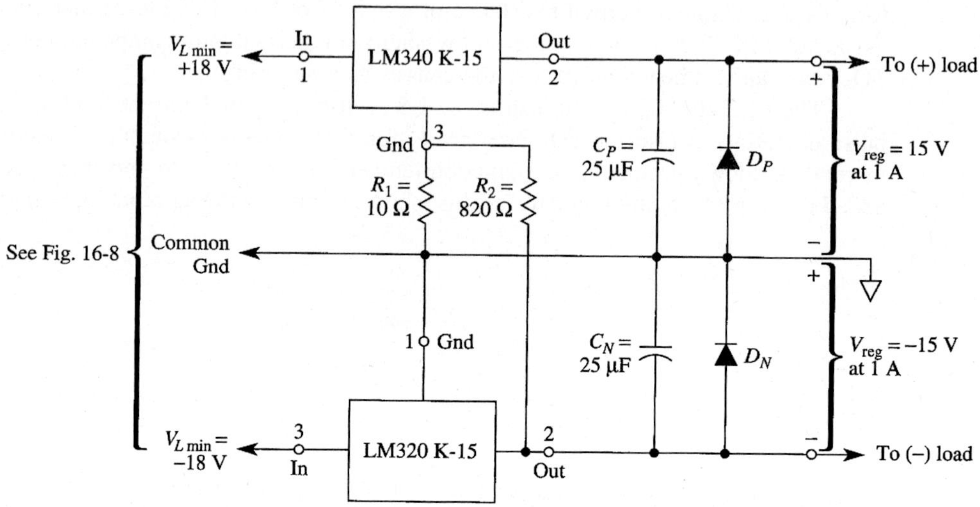

(a) Regulated ±15-V power supply for current up to ±1 A; the unregulated supply would be similar to that of Fig. 16-8.

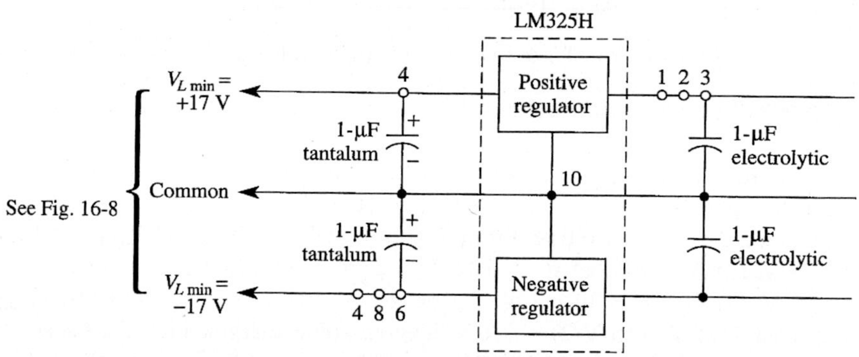

(b) Regulated ±15-V power supply for currents up to ±100 mA.

FIGURE 16-12 Two selections are presented for ±15-V supplies for op amps. The heavy-current 1-A version in (a) required two ICs. The modest, current version in (b) needs but a single IC.

Resistor R_1 is needed to ensure that the positive regulator starts up when the negative regulator has a heavy load. R_2 offsets the effect on ±15-V regulation caused by adding R_1.

16-10.2 Low-Current ± 15-V Regulator

Since one op amp rarely draws more than 5 mA, you need only a ±100-mA supply to power well over 20 op amps. For this reason an inexpensive low-power supply is shown in Fig. 16-12(b). The LM325H is a dual-tracking ±15-V supply in a 10-pin metal can package that can furnish ±100 mA. It has internal current limiting and thermal overload protection. (Buy a clip-on heat sink, or epoxy about a 2-in. × 2-in. piece of aluminum to the top surface.)

Note that the LM325 has two excellent voltage regulators packed into a single IC. The output capacitors provide energy storage to improve transient response. The input capacitors are needed if the unregulated supply is more than 4 in. from the LM325.

16-10.3 Unregulated Supply for the ±15-V Regulators

The unregulated supply required is shown in Fig. 16-8. Select:

1. $C+ = C- = 1000\ \mu$F minimum for both high- and low-current supplies, with WVDC = 30 V.
2. For the ±1-A supply, select a transformer of 115 V/36 V CT at 2 A. For the ±100-mA supply, select a 115 V/30 V CT at 0.2 A.
3. Diodes should be rated for $I_{av} \geq 1.0$ A for the high-current supply and ≥0.1 A for the low-current supply. PIV ratings for both should exceed 30 V (50 V is a standard size).

16-11 ADJUSTABLE THREE-TERMINAL POSITIVE VOLTAGE REGULATOR (THE LM317HV) AND NEGATIVE VOLTAGE REGULATOR (THE LM337HV)

There is a need for (1) regulated load voltages that are variable for laboratory supplies, (2) supply voltages that are *not* available as standard fixed-voltage regulators, (3) a very precisely adjustable supply voltage, or (4) providing a price-break lower cost for users who would like to stock a large quantity of one IC regulator type to furnish a variety of regulated output voltages.

The LM117 and LM137 families of adjustable three-terminal positive and negative voltage regulators, respectively, were developed. They are superb regulators with all the internal protection circuitry listed for the regulators in Sections 16-9 and 16-10. Since they are so versatile, they will encounter a variety of hostile applications, so it is prudent to add the external protection circuitry presented in Section 16-12.4.

16-12 LOAD VOLTAGE ADJUSTMENT

16-12.1 Adjusting the Positive Regulated Output Voltage

The LM317HV adjustable positive voltage regulator has only three terminals as shown in Fig. 16-13(a). Installation is simple, as shown in Fig. 16-13(b). The LM317 maintains a nominal 1.25 V between its output and adjust terminals. This voltage is called V_{ref} and can vary from chip to chip from 1.20 to 1.30 V. A 240-Ω resistor, R_1, is connected between these terminals to conduct a current of 1.2 V/240 Ω = 5 mA. This 5 mA flows through R_2. If R_2 is adjustable, the voltage drop across it, V_{R2}, will equal $R_2 \times 5$ mA. Output voltage of the regulator is set by V_{R2} plus the 1.2-V drop across R_1. In general terms, V_o is given by

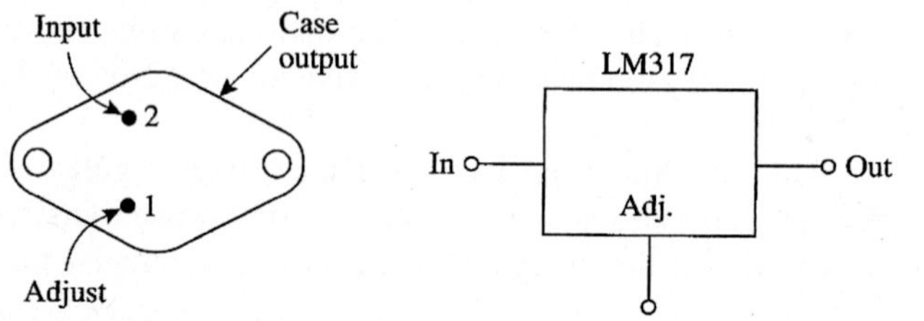

(a) LM317K connection diagram for TO-3 steel case and circuit schematic.

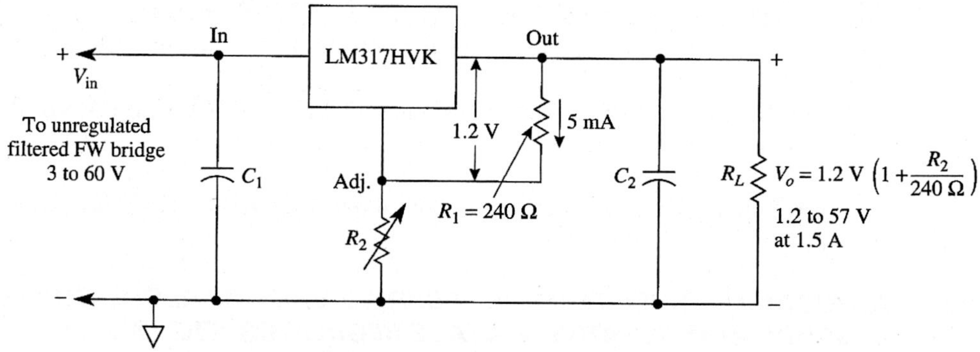

(b) Connecting the LM317HVK to act as an adjustable positive voltage regulator.

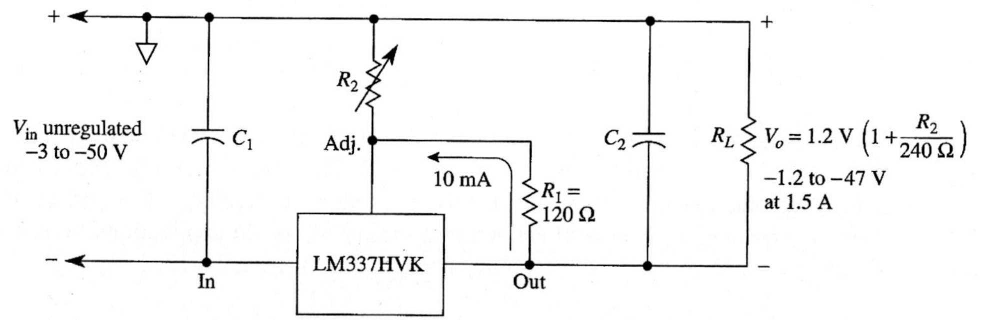

(c) Connecting the LM337HVK to act as an adjustable negative regulator.

FIGURE 16-13 Adjustable three-terminal positive (LM317) and negative (LM337) IC regulators are easy to use.

$$V_o = \frac{1.2\ \text{V}}{R_1}(R_1 + R_2) \tag{16-9a}$$

Normally, $R_1 = 240\ \Omega$. Thus any desired value of regulated output voltage is set by trimming R_2 to a value determined from

$$V_o = 1.2\ \text{V} + (5\ \text{mA})(R_2) \tag{16-9b}$$

For example, if you need a 5-V supply for TTL logic, make $R_2 = 760\ \Omega$. If you need a +15-V supply for an op amp or CMOS, make $R_2 = 2760\ \Omega$. $R_2 = 2160\ \Omega$ will give you 12-V car voltage. Make R_2 a 3-kΩ pot and you can adjust V_o to any voltage between 1.2 V (D-cell battery) and 16.2 V.

16-12.2 Characteristics of the LM317HVK

The LM317HVK will provide a regulated output current of up to 1.5 A, provided that it is not subjected to a power dissipation of more than about 15 W (TO-3 case). This means it should be electrically isolated from, and fastened to, a large heat sink such as the metal chassis of the power supply. A 5-in. × 5-in. piece of aluminum chassis stock also makes an adequate heat sink (see Section 16-10.1).

The LM317 requires a minimum "dropout" voltage of 3 V across its input and output terminals or it will drop out of regulation. Thus the upper limit of V_o is 3 V below the minimum input voltage from the unregulated supply.

It is good practice to connect bypass capacitors C_1 and C_2 (1-μF tantalum) as shown in Fig. 16-13(b). C_1 minimizes problems caused by long leads between the rectifier and LM317. C_2 improves transient response. Any ripple voltage from the rectifier will be reduced by a factor of over 1000 if R_2 is bypassed by a 1-μF tantalum capacitor or 10-μF aluminum electrolytic capacitor.

The LM317HVK protects itself against overheating, too much internal power dissipation, and too much current. When the chip temperature reaches 175°C, the 317 shuts down. If the product of output current and input-to-output voltage exceeds 15 to 20 W, or if currents greater than about 1.5 A are required, the LM317 also shuts down. When the overload condition is removed, the LM317 simply resumes operation. All of these protection features are made possible by the remarkable internal circuitry of the LM317.

16-12.3 Adjustable Negative-Voltage Regulator

An adjustable three-terminal *negative*-voltage regulator is also available [see the LM337HVK in Fig. 16-13(c)]. The negative regulator operates on the same principle as the positive regulator except that R_1 is a 120-Ω resistor and the maximum input voltage is reduced to 50 V.

V_o is given by Eq. (16-9b). If $R_1 = 120\ \Omega$, then V_o depends upon R_2 according to

$$V_o = 1.25\text{ V} + (10\text{ mA})R_2 \tag{16-10}$$

16-12.4 External Protection

It is standard practice to connect C_1 and C_2 to a regulator (see Fig. 16-14) for reasons stated in Section 16-12.2. Any regulator should have diode D_1 to protect it against input shorts; otherwise, load capacitance can pump current back into its output and destroy it.

Capacitor C_3 is added to greatly improve ac ripple voltage rejection. However, if a short circuit occurs across the regulator's output, C_3 will try to pump current back into the adjust terminal. Diode D_2 steers this current instead into the short circuit.

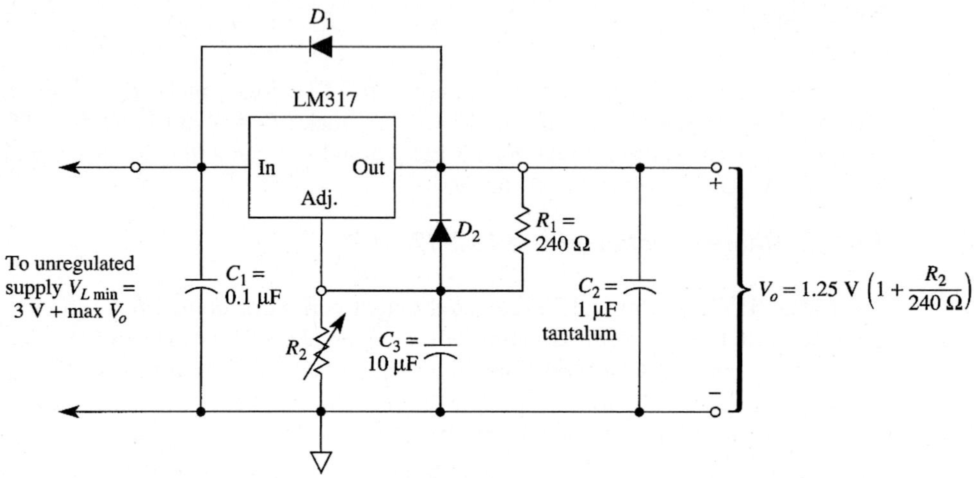

FIGURE 16-14 Variable-voltage positive regulated supply with external protection. D_1 protects the regulator from input short circuits. D_2 protects the regulator from output short circuits.

16-13 ADJUSTABLE LABORATORY-TYPE VOLTAGE REGULATOR

Standard LM317K and LM337K IC positive and negative regulators, respectively, are interconnected with support components in Fig. 16-15. They form an independently adjustable bipolar laboratory-type power supply. The steel K packages will easily furnish

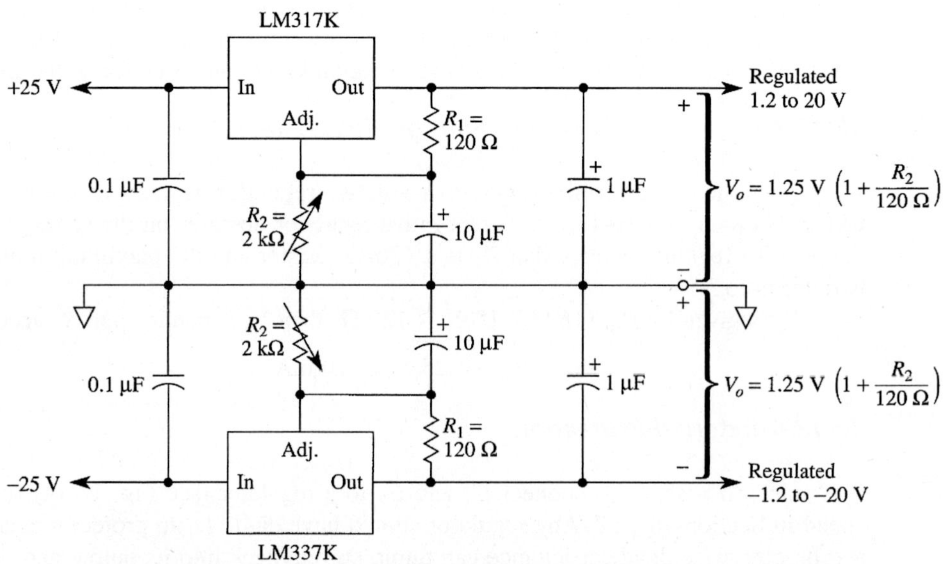

FIGURE 16-15 Adjustable bipolar laboratory-type voltage regulator. Positive and negative outputs can be adjusted independently for any voltage between 1.2 and 20 V.

1 A each if the regulators are heat-sinked properly (see Section 16-9.1). Variable resistor R_2 for each regulator may be adjusted for a regulated output voltage between approximately 1.2 and 20 V.

The unregulated supply has the circuitry of Fig. 16-8. A conservative design would select (1) a transformer, 115 V/50 VCT at 2 A, (2) diodes $I_{av} > 1$ A at PIV ≥ 50 V (IN 4002), and (3) capacitors of 1000 μF at WVDC ≥ 50 V.

16-14 OTHER LINEAR REGULATORS

With the large increase of battery-powered equipment and systems that operate from 3 V or 3.3 V instead of 5 V or other dc voltage levels, manufacturers such as Maxim and Linear Technology have responded with a variety of linear voltage regulators. Some systems use both the low voltage (3 V or 3.3 V) and a higher dc voltage. Regulators known as step-down dc–dc controllers are used in these applications. The output voltage of these chips is 3 V or 3.3 V, and a typical input voltage range from 3.6 V to 16 V. Other dc–dc controller chips have an adjustable output voltage. As with all voltage regulators these devices have short-circuit protection, thermal shutdown, and rail-to-rail and single-supply operation. The preceding sections have been written using some low-cost readily available regulators that you can use to power up general-purpose op amps, but look for these other regulators in low-power portable equipment.

PROBLEMS

16-1. A transformer is rated at 115 to 28 V rms at 1 A. What is the peak secondary voltage?

16-2. A 115/28 V transformer is used in Fig. 16-4(a). Find V_{dc} at no load.

16-3. As dc load current increases, what happens to **(a)** dc load voltage; **(b)** ac ripple voltage?

16-4. In Fig. 16-4, the transformer rating is 115 to 28 V at 1 A. What is V_{dc} at a full-load current of $I_L = 0.5$ A?

16-5. Dc measurements of a power supply give $V_{dc\ NL} = 17.8$ V and $V_{dc\ FL} = 13.8$ V at $I_{L\ FL} = 0.5$ A. Calculate: **(a)** R_o; **(b)** percent regulation.

16-6. What voltage readings would be obtained with an ac voltmeter for peak-to-peak ripple voltages of **(a)** 1 V; **(b)** 3 V?

16-7. The dc full-load voltage of a power supply is 28 V and the peak-to-peak ripple voltage is 6 V. Find the minimum instantaneous load voltage.

16-8. A 110 V/28 V CT transformer is installed in Fig. 16-8. What no-load dc voltage would be measured **(a)** between terminals 1 and 2; **(b)** from 1 to CT; **(c)** from 2 to CT?

16-9. Find the percent ripple in Example 16-7 if C is changed to 2000 μF.

16-10. Design an FWB unregulated power supply to output +15 V at 1 A, with less than 5% ripple. Assume that $R_o = 8\ \Omega$. Choose the transformer from available ratings of 115/12 V, 115/18 V, and 115/24 V—all at 2 A. Available capacitors are 500 μF and 1000 μF at WVDC = 50 V.

16-11. Given 115/25.2 V at a 3-A transformer and $C = 1000\ \mu$F in an FWB rectifier. Assume $R_o = 6\ \Omega$ and $I_{L\ FL} = 1$ A. Calculate **(a)** $V_{dc\ NL}$; **(b)** $V_{dc\ FL}$; **(c)** percent regulation; **(d)** ΔV_o; **(e)** ripple; **(f)** $V_{L\ min}$.

16-12. An unregulated power supply has $V_{dc\ NL} = 18$ V, $V_{dc\ FL} = 10$ V at 1 A, $\Delta V_o = 5$ V, and $C = 1000\ \mu$F. Calculate ΔV_o if you **(a)** double C to 2000 μF, or **(b)** reduce I to 0.5 A, or **(c)** both reduce I to 0.5 A and double C to 2000 μF.

16-13. An ac voltmeter indicates a value of 1.71 V rms across the FWB supply, and a dc voltmeter indicates 12 V dc. Draw the expected value of V_L that would be seen on a dc-coupled oscilloscope for a time interval of 16.7 ms (see Fig. 16-7).

16-14. If I_L measures 1 A in Problem 16-13, find **(a)** the value of C; **(b)** the value of R_L.

16-15. You need a regulated output voltage of 24.0 V. If $R_1 = 240\ \Omega$ in Fig. 16-13(a), find the required value of R_2.

16-16. Assume that the regulator of Problem 16-15 delivers a load current of 1.0 A. If its average dc input voltage is 30 V, show that the regulator must dissipate 6 W.

16-17. Find V_o if R_2 is short-circuited in **(a)** Fig. 16-13(a); **(b)** Fig. 16-13(b).

16-18. Adjustable resistor $R_2 = 0$ to 2500 Ω in Fig. 16-13(a). Find the upper and lower limits of V_o as R_2 is adjusted from 2500 Ω to 0 Ω.

16-19. Suppose that a 1200-Ω low-stop resistor, and a 2500-Ω pot are connected in series in place of the single resistor R_2 in Fig. 16-13(a). Find the upper and lower limits of V_o as the pot is adjusted from 2500 Ω to 0 Ω.

16-20. If the dropout voltage of an LM317 is 3 V, what is the minimum instantaneous input voltage for the regulator circuit of Problem 16-18?

16-21. Calculate the required values for R_2 in Fig. 16-15 to give outputs of ± 12 V.

APPENDIX 1

μA741 Frequency-Compensated Operational Amplifier*

*Courtesy of Fairchild Semiconductor, a Division of Fairchild Camera and Instrument Corporation.

Description

The μA741 is a high performance Monolithic Operational Amplifier constructed using the Fairchild Planar epitaxial process. It is intended for a wide range of analog applications. High common mode voltage range and absence of latch-up tendencies make the μA741 ideal for use as a voltage follower. The high gain and wide range of operating voltage provides superior performance in integrator, summing amplifier, and general feedback applications.

- **NO FREQUENCY COMPENSATION REQUIRED**
- **SHORT-CIRCUIT PROTECTION**
- **OFFSET VOLTAGE NULL CAPABILITY**
- **LARGE COMMON MODE AND DIFFERENTIAL VOLTAGE RANGES**
- **LOW POWER CONSUMPTION**
- **NO LATCH-UP**

Connection Diagram
10-Pin Flatpak

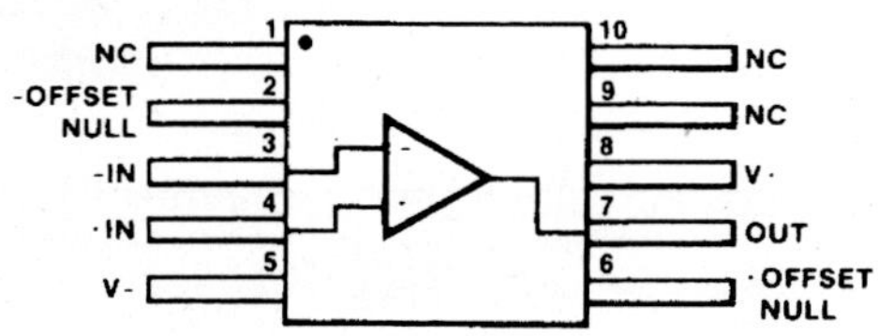

(Top View)

Order Information

Type	Package	Code	Part No.
μA741	Flatpak	3F	μA741FM
μA741A	Flatpak	3F	μA741AFM

Absolute Maximum Ratings

Supply Voltage	
μA741A, μA741, μA741E	±22 V
μA741C	±18 V
Internal Power Dissipation (Note 1)	
Metal Package	500 MW
DIP	310 mW
Flatpak	570 mW
Differential Input Voltage	±30 V
Input Voltage (Note 2)	±15 V
Storage Temperature Range	
Metal Package and Flatpak	−65°C to +150°C
DIP	−55°C to +125°C

Connection Diagram
8-Pin Metal Package

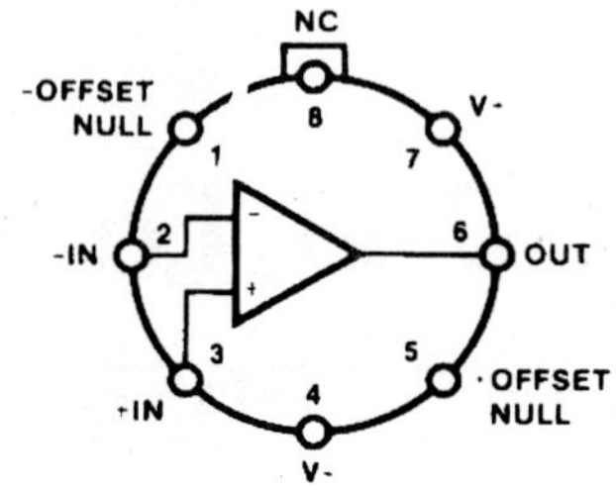

(Top View)

Pin 4 connected to case

Order Information

Type	Package	Code	Part No.
μA741	Metal	5W	μA741HM
μA741A	Metal	5W	μA741AHM
μA741C	Metal	5W	μA741HC
μA741E	Metal	5W	μA741EHC

Connection Diagram
8-Pin DIP

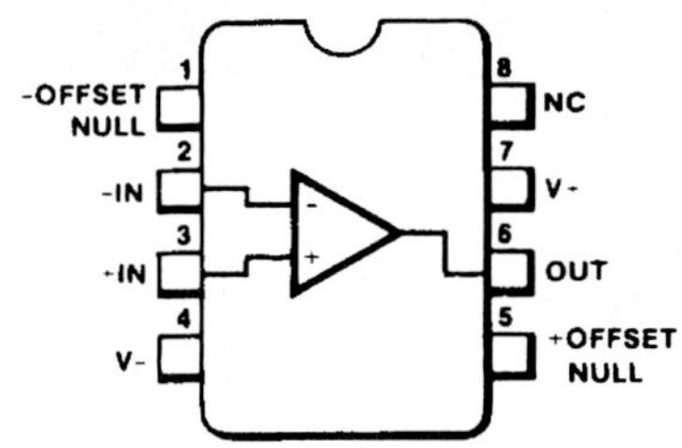

(Top View)

Order Information

Type	Package	Code	Part No.
μA741C	Molded DIP	9T	μA741TC
μA741C	Ceramic DIP	6T	μA741RC

Operating Temperature Range	
Military (μA741A, μA741)	−55°C to +125°C
Commercial (μA741E, μA741C)	0°C to +70°C
Pin Temperature (Soldering 60 s)	
Metal Package, Flatpak, and Ceramic DIP	300°C
Molded DIP (10 s)	260°C
Output Short Circuit Duration (Note 3)	Indefinite

Equivalent Circuit

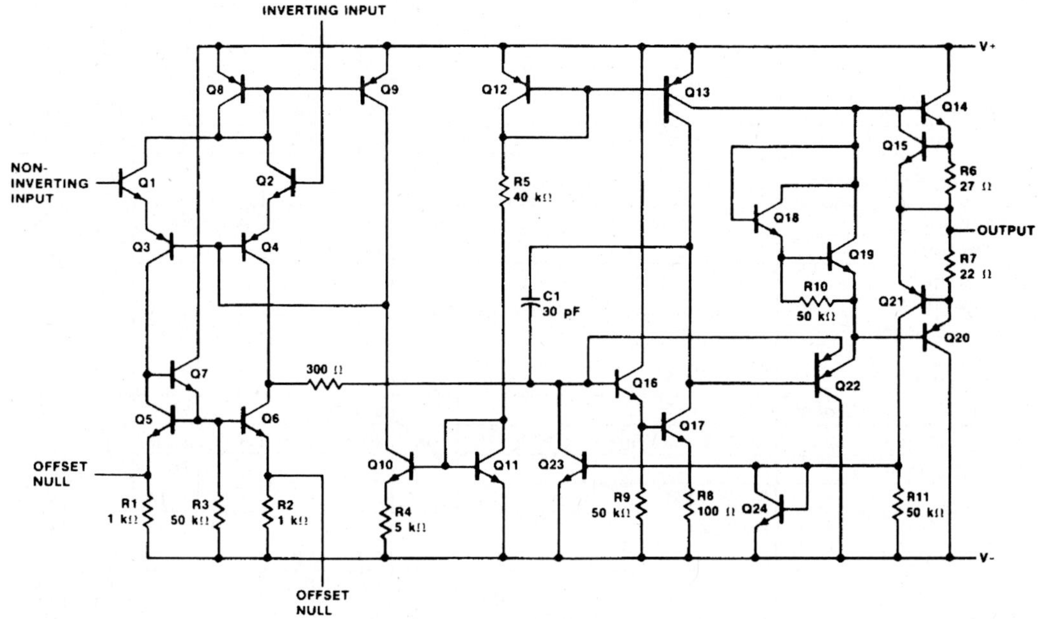

Notes

1. Rating applies to ambient temperatures up to 70°C. Above 70°C ambient derate linearly at 6.3 mW / °C for the metal package, 7.1 mW / °C for the flatpak, and 5.6 mW / °C for the DIP.
2. For supply voltages less than ± 15 V, the absolute maximum input voltage is equal to the supply voltage.
3. Short circuit may be to ground or either supply. Rating applies to +125°C case temperature or 75°C ambient temperature.

μA741 and μA741C
Electrical Characteristics $V_S = \pm 15$ V, $T_A = 25°C$ unless otherwise specified

Characteristic		Condition	μA741 Min	μA741 Typ	μA741 Max	μA741C Min	μA741C Typ	μA741C Max	Unit
Input Offset Voltage		$R_S \leq 10$ kΩ		1.0	5.0		2.0	6.0	mV
Input Offset Current				20	200		20	200	nA
Input Bias Current				80	500		80	500	nA
Power Supply Rejection Ratio		$V_S = +10, -20$ $V_S = +20, -10$ V, $R_S = 50\ \Omega$		30	150		30	150	μV/V
Input Resistance			.3	2.0		.3	2.0		MΩ
Input Capacitance				1.4			1.4		pF
Offset Voltage Adjustment Range				± 15			± 15		mV
Input Voltage Range						± 12	± 13		V
Common Mode Rejection Ratio		$R_S \leq 10$ kΩ				70	90		dB
Output Short Circuit Current				25			25		mA
Large Signal Voltage Gain		$R_L \geq 2$ kΩ, $V_{OUT} = \pm 10$ V	50k	200k		20k	200k		
Output Resistance				75			75		Ω
Output Voltage Swing		$R_L \geq 10$ kΩ				± 12	± 14		V
		$R_L \geq 2$ kΩ				± 10	± 13		V
Supply Current				1.7	2.8		1.7	2.8	mA
Power Consumption				50	85		50	85	mW
Transient Response (Unity Gain)	Rise Time	$V_{IN} = 20$ mV, $R_L = 2$ kΩ, $C_L \leq 100$ pF		.3			.3		μs
	Overshoot			5.0			5.0		%
Bandwidth (Note 4)				1.0			1.0		MHz
Slew Rate		$R_L \geq 2$ kΩ		.5			.5		V/μs

Notes

4. Calculated value from $BW(MHz) = \frac{0.35}{\text{Rise Time } (\mu s)}$
5. All $V_{CC} = 15$ V for μA741 and μA741C.
6. Maximum supply current for all devices
   25°C = 2.8 mA
   125°C = 2.5 mA
   −55°C = 3.3 mA

μA741 and μA741C
Electrical Characteristics (Cont.) The following specifications apply over the range of $-55°C \leq T_A \leq 125°C$ for μA741, $0°C \leq T_A \leq 70°C$ for μA741C

Characteristic	Condition	μA741			μA741C			Unit
		Min	Typ	Max	Min	Typ	Max	
Input Offset Voltage							7.5	mV
	$R_S \leq 10\ k\Omega$		1.0	6.0				mV
Input Offset Current							300	nA
	$T_A = +125°C$		7.0	200				nA
	$T_A = -55°C$		85	500				nA
Input Bias Current							800	nA
	$T_A = +125°C$		.03	.5				μA
	$T_A = -55°C$		.3	1.5				μA
Input Voltage Range		±12	±13					V
Common Mode Rejection Ratio	$R_S \leq 10\ k\Omega$	70	90					dB
Adjustment for Input Offset Voltage			±15			±15		mV
Supply Voltage Rejection Ratio	$V_S = +10, -20$; $V_S = +20, -10\ V$, $R_S = 50\ \Omega$		30	150				μV/V
Output Voltage Swing	$R_L \geq 10\ k\Omega$	±12	±14					V
	$R_L \geq 2\ k\Omega$	±10	±13		±10	±13		V
Large Signal Voltage Gain	$R_L = 2\ k\Omega$, $V_{OUT} = \pm 10\ V$	25k			15k			
Supply Current	$T_A = +125°C$		1.5	2.5				mA
	$T_A = -55°C$		2.0	3.3				mA
Power Consumption	$T_A = +125°C$		45	75				mW
	$T_A = -55°C$		60	100				mW

Notes

4. Calculated value from $BW(MHz) = \frac{0.35}{\text{Rise Time } (\mu s)}$
5. All V_{CC} = 15 V for μA741 and μA741C.
6. Maximum supply current for all devices
   25°C = 2.8 mA
   125°C = 2.5 mA
   −55°C = 3.3 mA

μA741A and μA741E
Electrical Characteristics $V_S = \pm 15$ V, $T_A = 25°C$ unless otherwise specified.

Characteristic		Condition	μA741A/E Min	Typ	Max	Unit
Input Offset Voltage		$R_S \leq 50\ \Omega$		0.8	3.0	mV
Average Input Offset Voltage Drift					15	μV/°C
Input Offset Current				3.0	30	nA
Average Input Offset Current Drift					0.5	nA/°C
Input Bias Current				30	80	nA
Power Supply Rejection Ratio		$V_S = +10, -20$; $V_S = +20$ V, -10 V, $R_S = 50\ \Omega$		15	50	μV/V
Output Short Circuit Current			10	25	40	mA
Power Consumption		$V_S = \pm 20$ V		80	150	mW
Input Impedance		$V_S = \pm 20$ V	1.0	6.0		MΩ
Large Signal Voltage Gain		$V_S = \pm 20$ V, $R_L = 2$ kΩ, $V_{OUT} = \pm 15$ V	50	200		V/mV
Transient Response (Unity Gain)	Rise Time			0.25	0.8	μs
	Overshoot			6.0	20	%
Bandwidth (Note 4)			.437	1.5		MHz
Slew Rate (Unity Gain)		$V_{IN} = \pm 10$ V	0.3	0.7		V/μs

The following specifications apply over the range of $-55°C \leq T_A \leq 125°C$ for the 741A, and $0°C \leq T_A \leq 70°C$ for the 741E.

Characteristic	Condition			Min	Typ	Max	Unit
Input Offset Voltage						4.0	mV
Input Offset Current						70	nA
Input Bias Current						210	nA
Common Mode Rejection Ratio	$V_S = \pm 20$ V, $V_{IN} = \pm 15$ V, $R_S = 50\ \Omega$			80	95		dB
Adjustment For Input Offset Voltage	$V_S = \pm 20$ V			10			mV
Output Short Circuit Current				10		40	mA
Power Consumption	$V_S = \pm 20$ V	μA741A	$-55°C$			165	mW
			$+125°C$			135	mW
		μA741E				150	mW
Input Impedance	$V_S = \pm 20$ V			0.5			MΩ
Output Voltage Swing	$V_S = \pm 20$ V	$R_L = 10$ kΩ		± 16			V
		$R_L = 2$ kΩ		± 15			V
Large Signal Voltage Gain	$V_S = \pm 20$ V, $R_L = 2$ kΩ, $V_{OUT} = \pm 15$ V			32			V/mV V/mV
	$V_S = \pm 5$ V, $R_L = 2$ kΩ, $V_{OUT} = \pm 2$ V			10			V/mV

Notes

4. Calculated value from: $BW(MHz) = \frac{0.35}{\text{Rise Time } (\mu s)}$
5. All $V_{CC} = 15$ V for μA741 and μA741C.
6. Maximum supply current for all devices
   25°C = 2.8 mA
   125°C = 2.5 mA
   −55°C = 3.3 mA

Typical Performance Curves for μA741A and μA741

Open Loop Voltage Gain as a Function of Supply Voltage

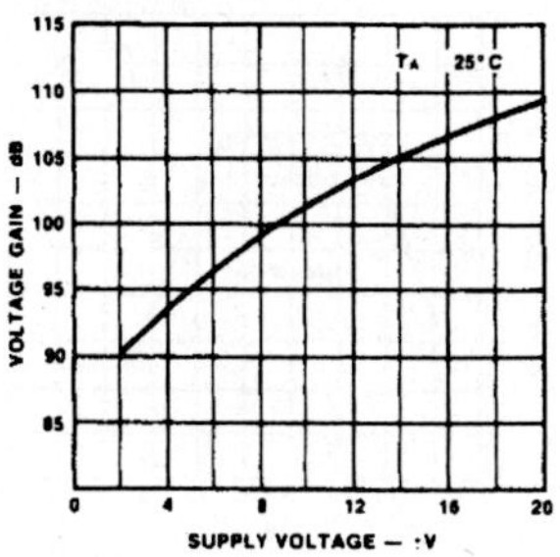

Output Voltage Swing as a Function of Supply Voltage

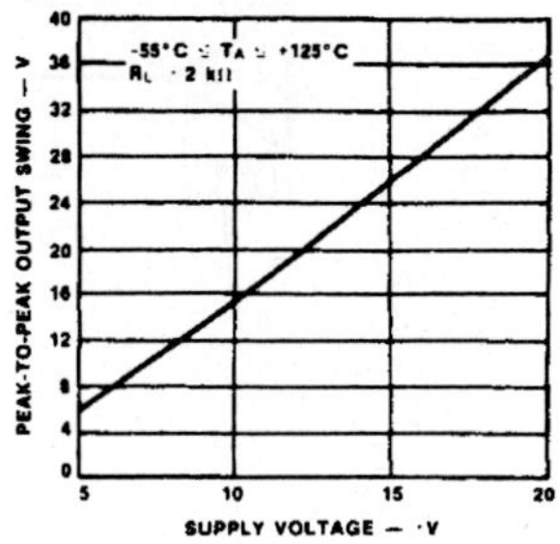

Input Common Mode Voltage as a Function of Supply Voltage

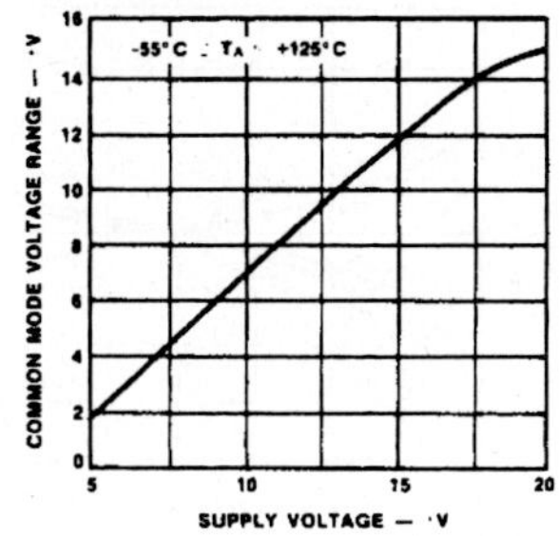

Typical Performance Curves for μA741E and μA741C

Open Loop Voltage Gain as a Function of Supply Voltage

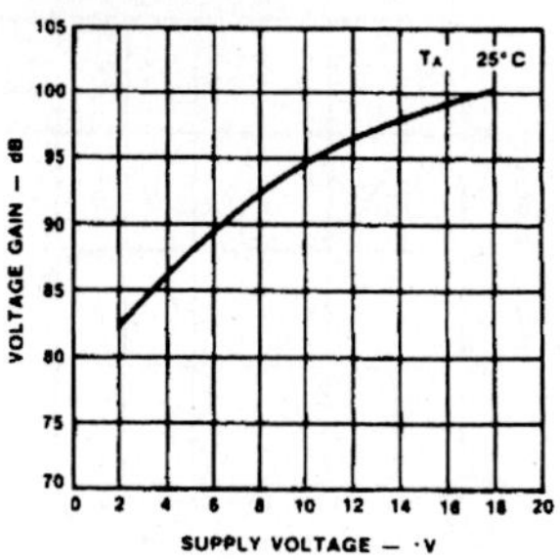

Output Voltage Swing as a Function of Supply Voltage

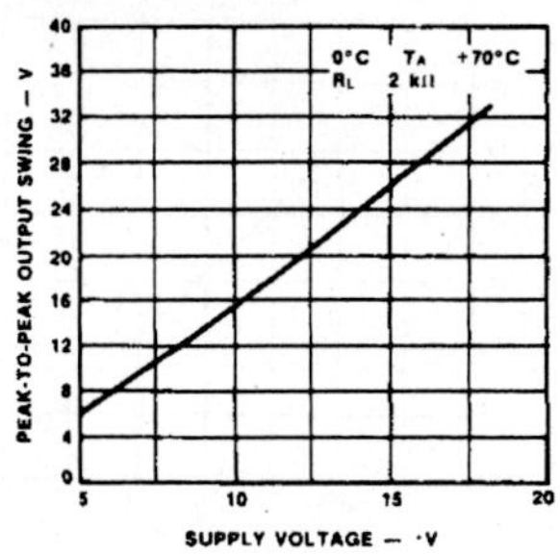

Input Common Mode Voltage Range as a Function of Supply Voltage

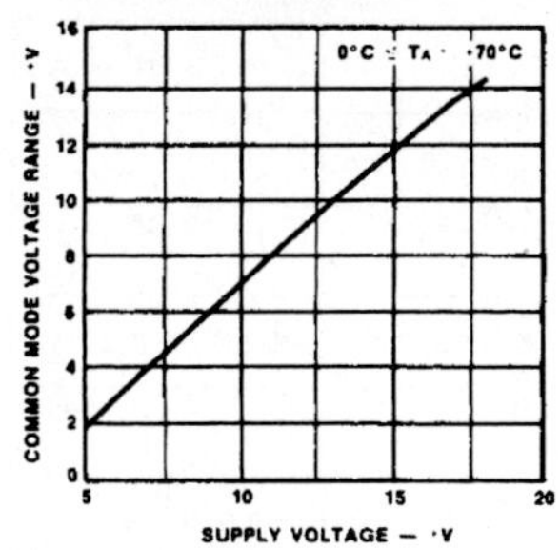

Transient Response

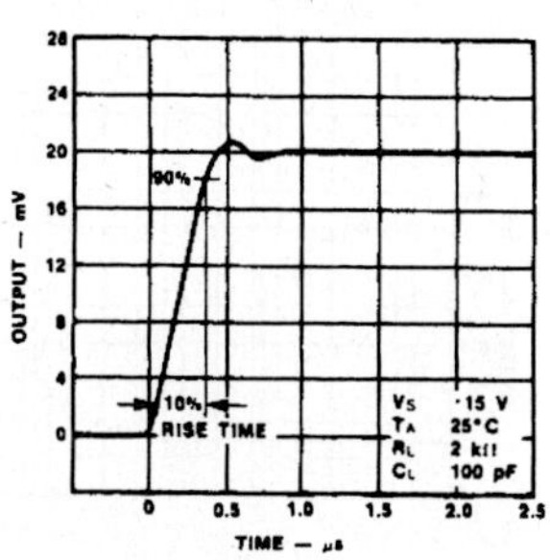

Transient Response Test Circuit

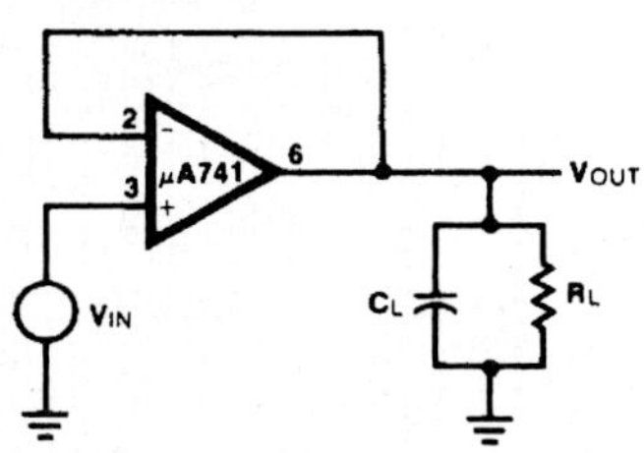

Common Mode Rejection Ratio as a Function of Frequency

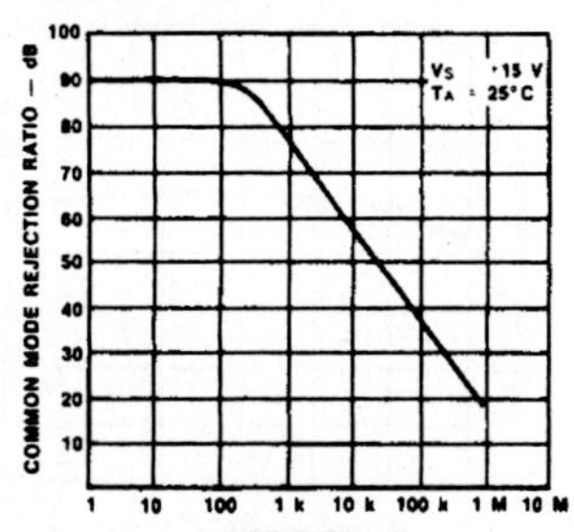

Typical Performance Curves for μA741E and μA741C (Cont.)

Frequency Characteristics as a Function of Supply Voltage

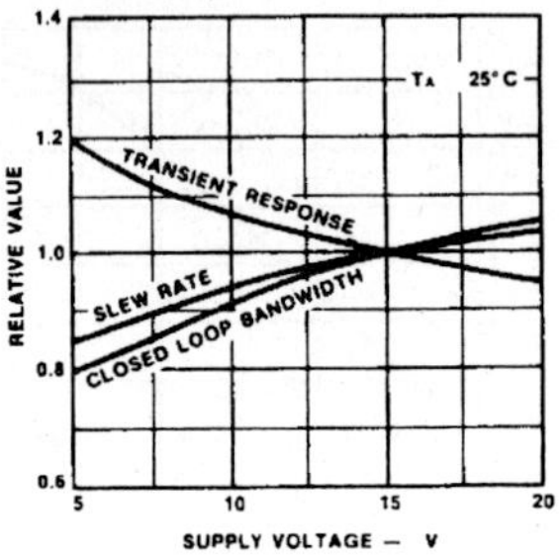

Voltage Offset Null Circuit

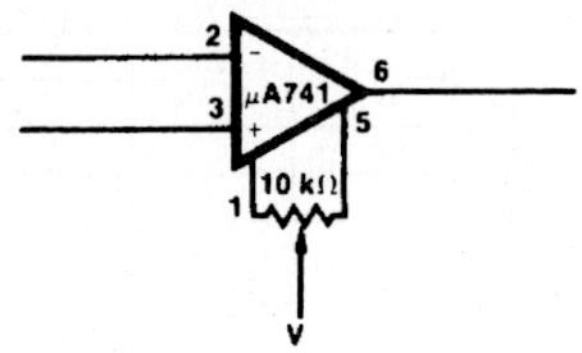

Voltage Follower Large Signal Pulse Response

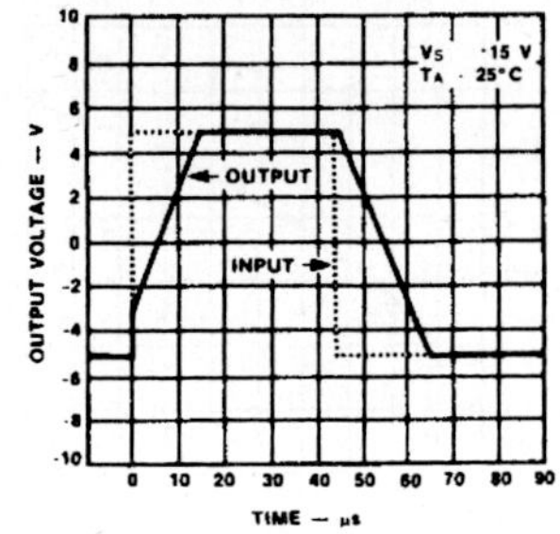

Typical Performance Curves for μA741A, μA741, μA741E and μA741C

Power Consumption as a Function of Supply Voltage

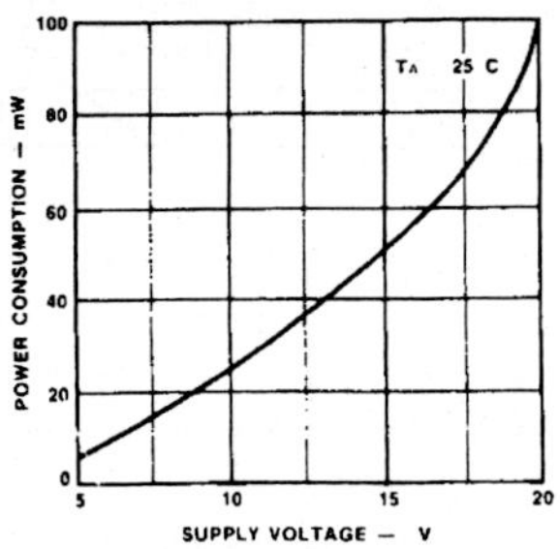

Open Loop Voltage Gain as a Function of Frequency

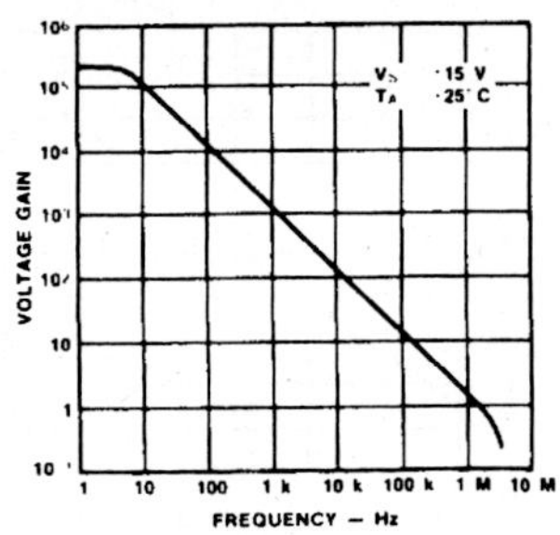

Open Loop Phase Response as a Function of Frequency

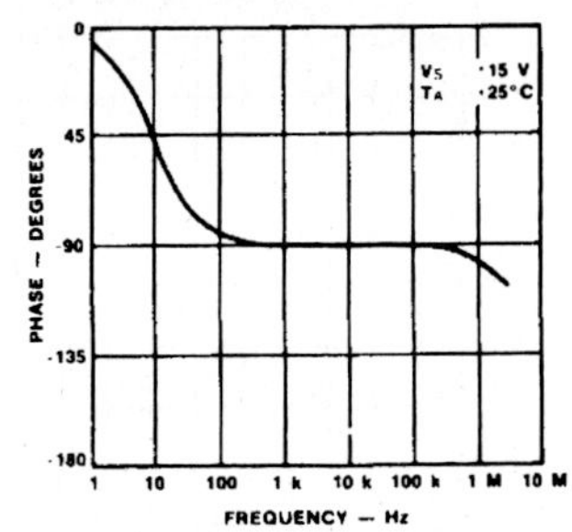

Input Offset Current as a Function of Supply Voltage

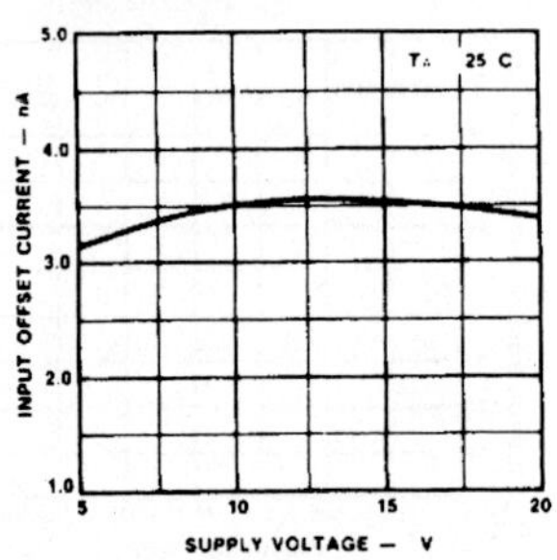

Input Resistance and Input Capacitance as a Function of Frequency

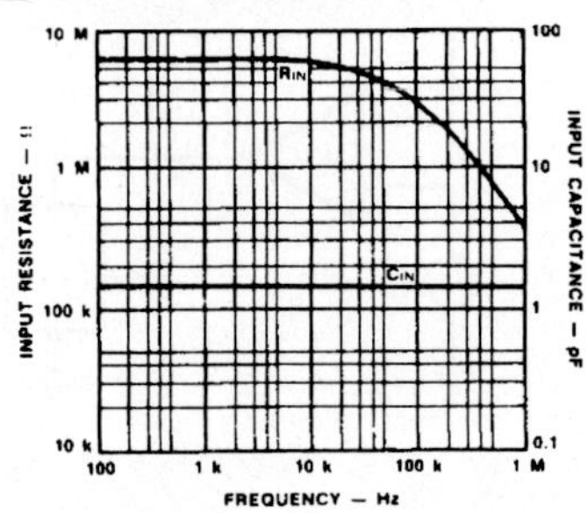

Output Resistance as a Function of Frequency

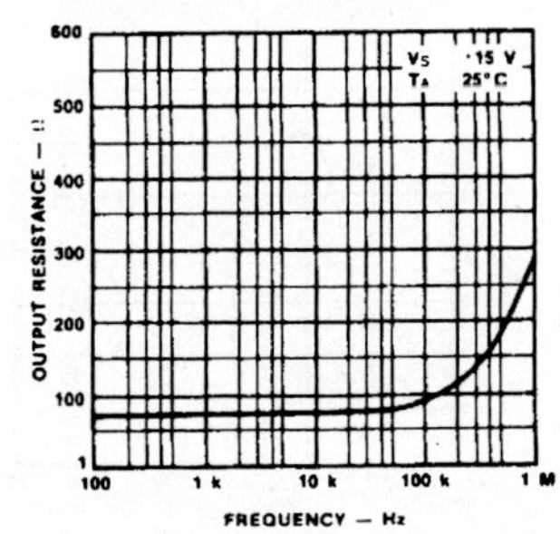

Typical Performance Curves for μA741A, μA741, μA741E and μA741C (Cont.)

Output Voltage Swing as a Function of Load Resistance

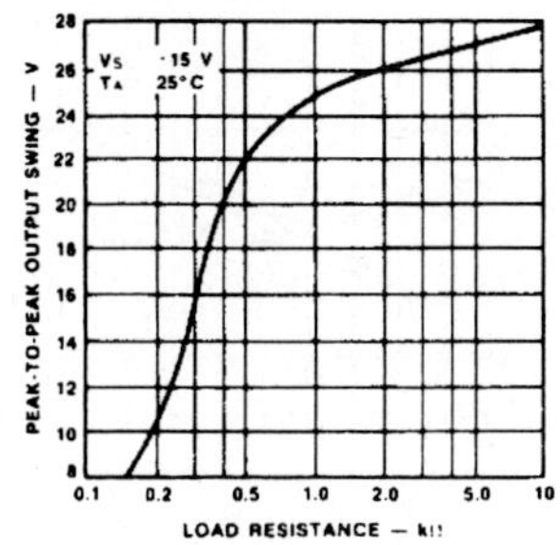

Output Voltage Swing as a Function of Frequency

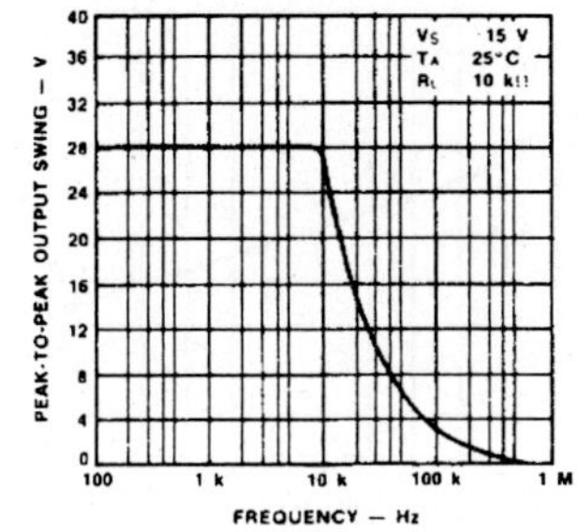

Absolute Maximum Power Dissipation as a Function of Ambient Temperature

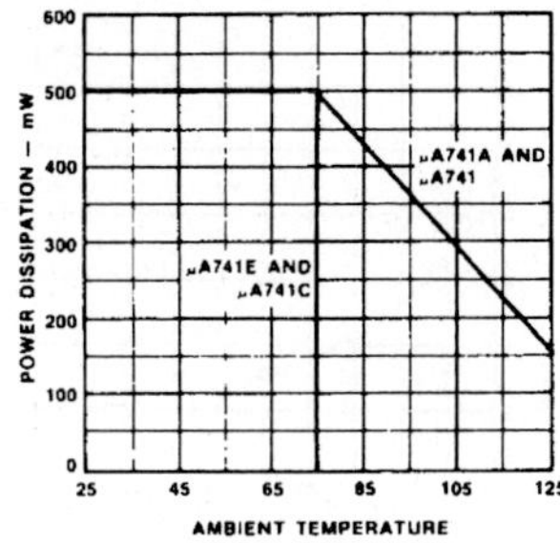

Input Noise Voltage as a Function of Frequency

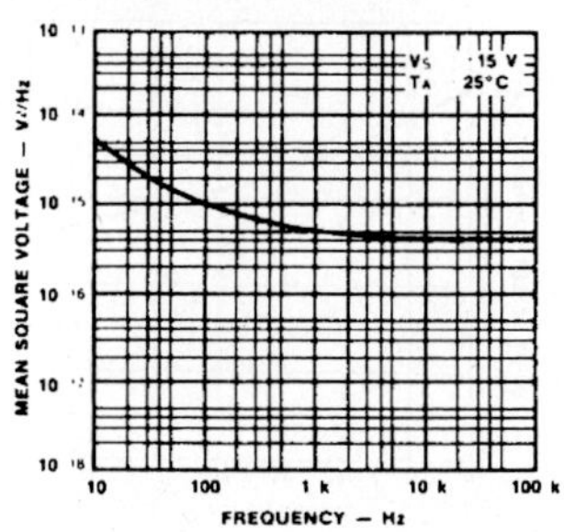

Input Noise Current as a Function of Frequency

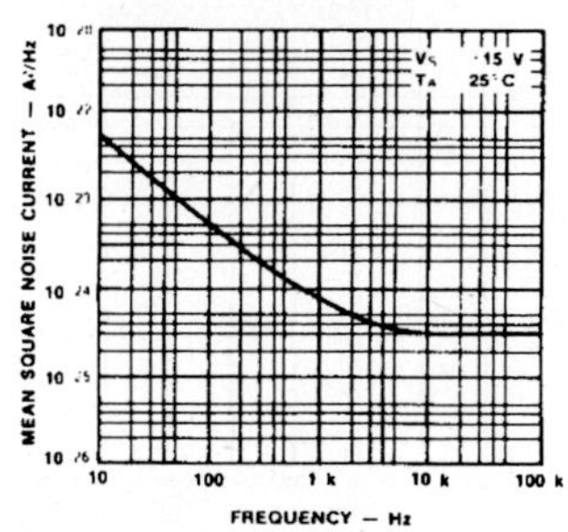

Broadband Noise for Various Bandwidths

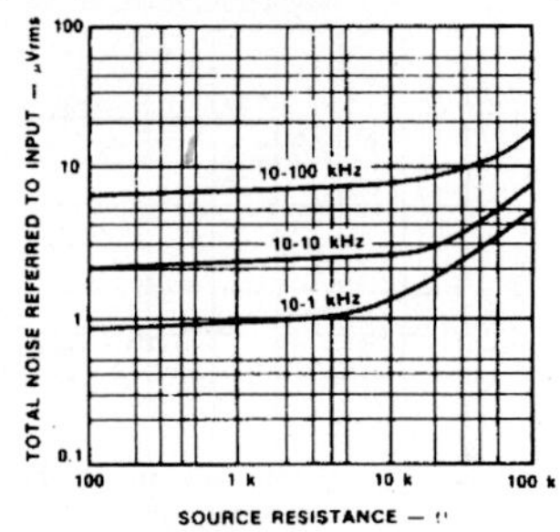

Typical Performance Curves for μA741A and μA741

Input Bias Current as a Function of Ambient Temperature

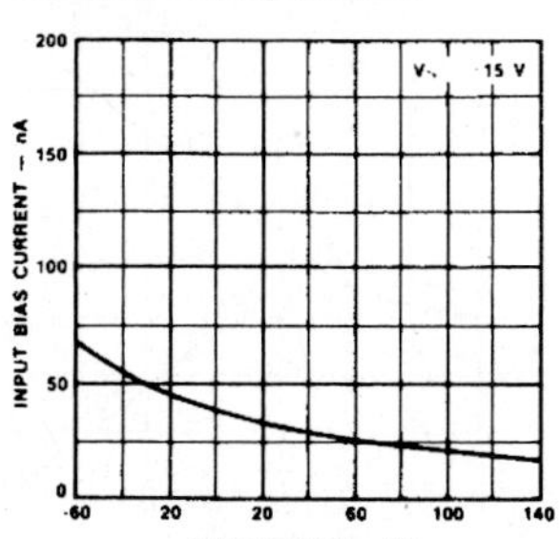

Input Resistance as a Function of Ambient Temperature

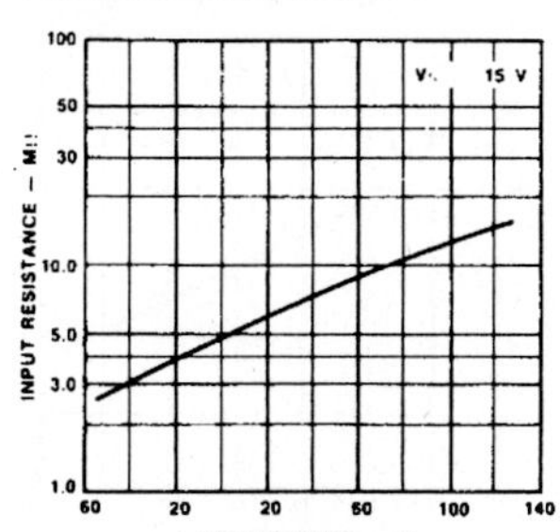

Output Short-Circuit Current as a Function of Ambient Temperature

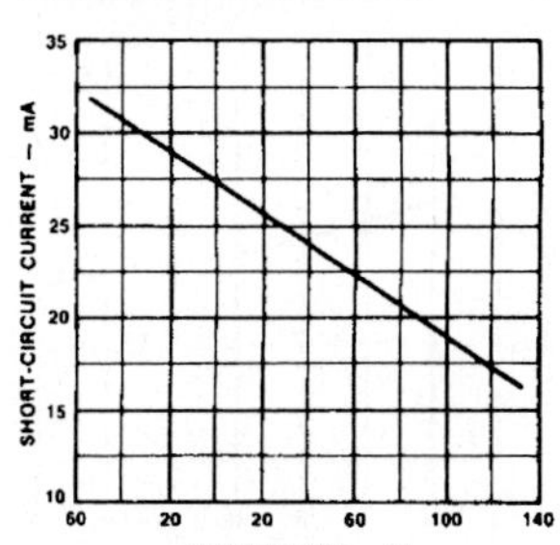

Typical Performance Curves for μA741A and μA741 (Cont.)

Input Offset Current as a Function of Ambient Temperature

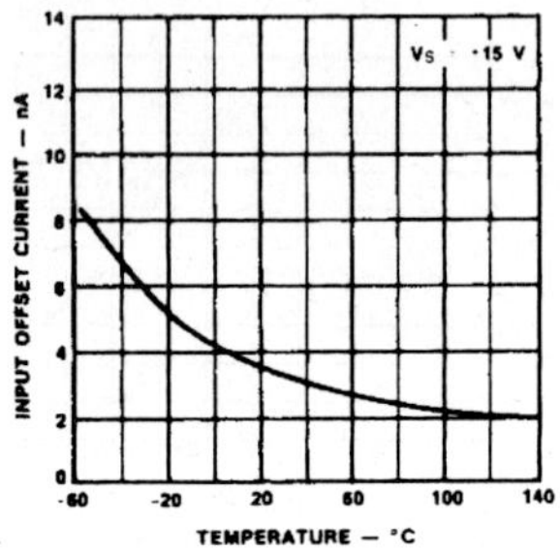

Power Consumption as a Function of Ambient Temperature

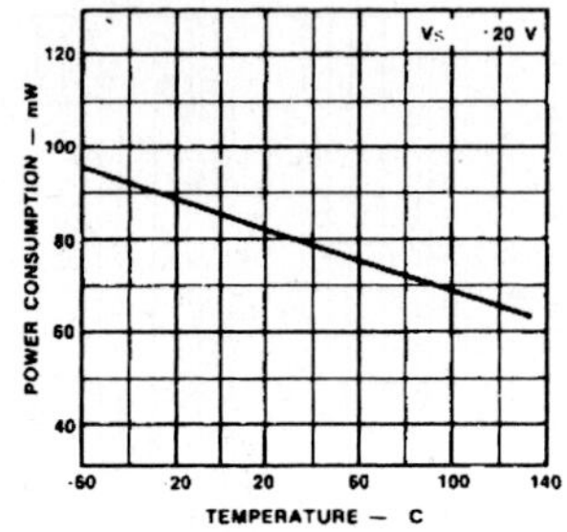

Frequency Characteristics as a Function of Ambient Temperature

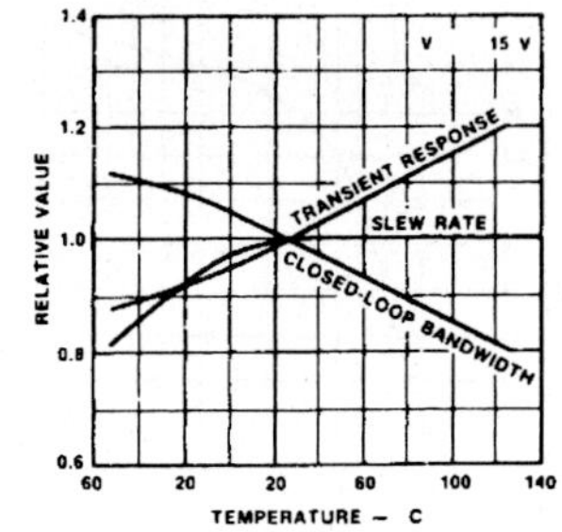

Typical Performance Curves for μA741E and μA741C

Input Bias Current as a Function of Ambient Temperature

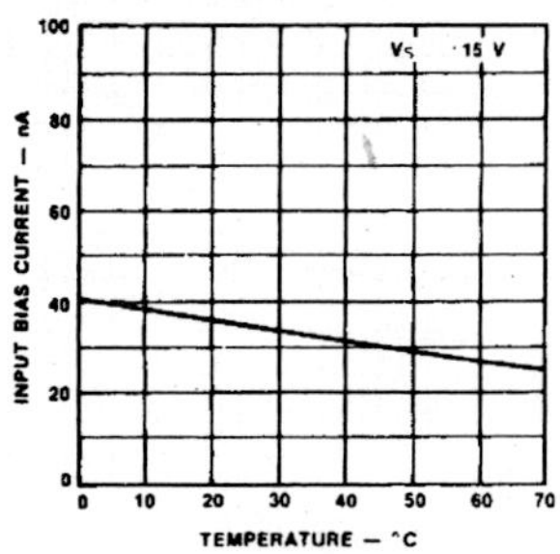

Input Resistance as a Function of Ambient Temperature

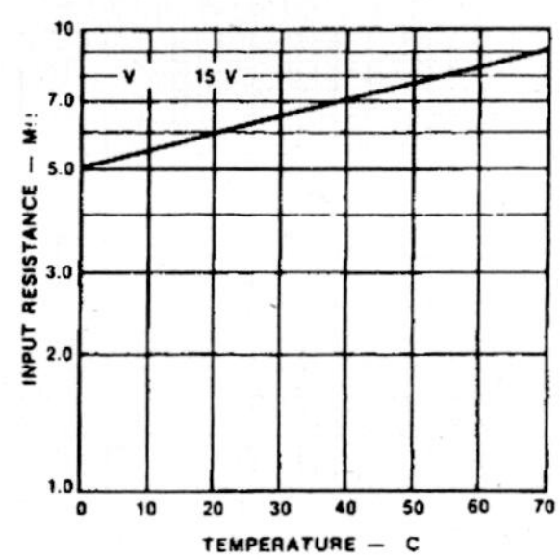

Input Offset Current as a Function of Ambient Temperature

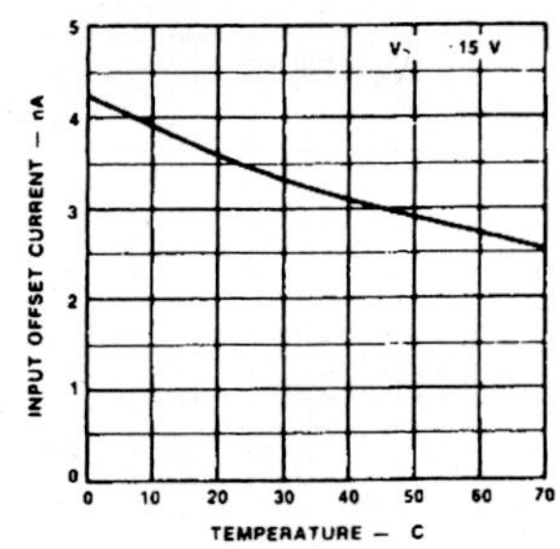

Power Consumption as a Function of Ambient Temperature

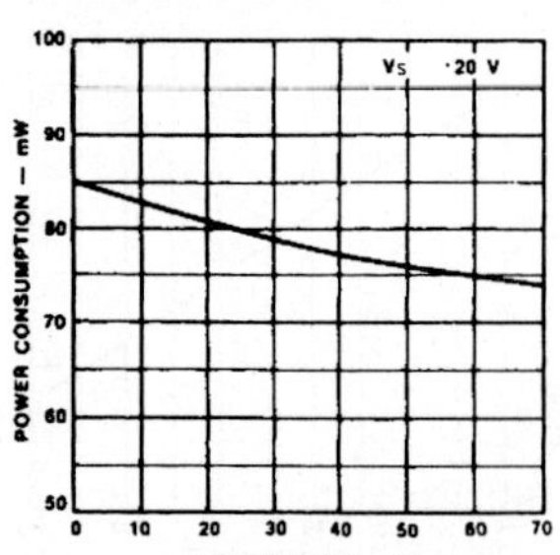

Output Short Circuit Current as a Function of Ambient Temperature

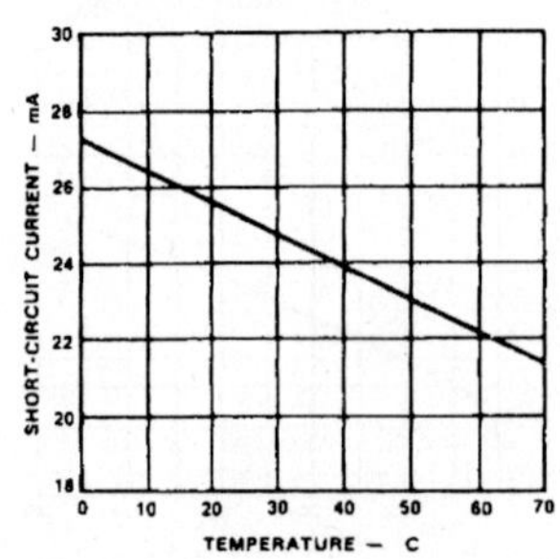

Frequency Characteristics as a Function of Ambient Temperature

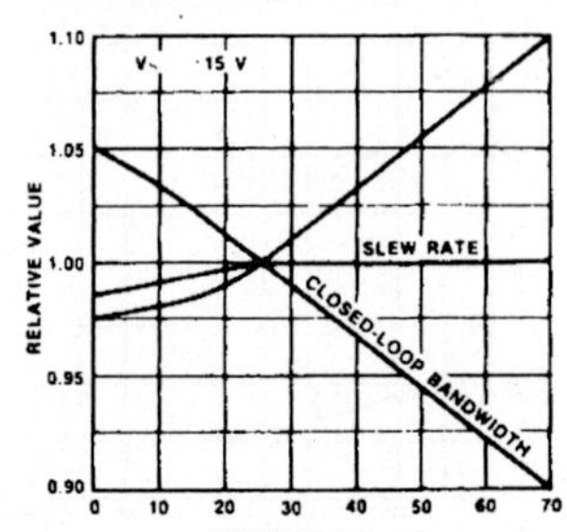

APPENDIX 2

LM301 Operational Amplifier*

*Courtesy of National Semiconductor Corporation.

General Description

The LM101A series are general purpose operational amplifiers which feature improved performance over industry standards like the LM709. Advanced processing techniques make possible an order of magnitude reduction in input currents, and a redesign of the biasing circuitry reduces the temperature drift of input current. Improved specifications include:

- Offset voltage 3 mV maximum over temperature (LM101A/LM201A)
- Input current 100 nA maximum over temperature (LM101A/LM201A)
- Offset current 20 nA maximum over temperature (LM101A/LM201A)
- Guaranteed drift characteristics
- Offsets guaranteed over entire common mode and supply voltage ranges
- Slew rate of 10V/μs as a summing amplifier

This amplifier offers many features which make its application nearly foolproof: overload protection on the input and output, no latch-up when the common mode range is exceeded, freedom from oscillations and compensation with a single 30 pF capacitor. It has advantages over internally compensated amplifiers in that the frequency compensation can be tailored to the particular application. For example, in low frequency circuits it can be overcompensated for increased stability margin. Or the compensation can be optimized to give more than a factor of ten improvement in high frequency performance for most applications.

In addition, the device provides better accuracy and lower noise in high impedance circuitry. The low input currents also make it particularly well suited for long interval integrators or timers, sample and hold circuits and low frequency waveform generators. Further, replacing circuits where matched transistor pairs buffer the inputs of conventional IC op amps, it can give lower offset voltage and drift at a lower cost.

The LM101A is guaranteed over a temperature range of −55°C to +125°C, the LM201A from −25°C to +85°C, and the LM301A from 0°C to 70°C.

Schematic ** and Connection Diagrams (Top Views)

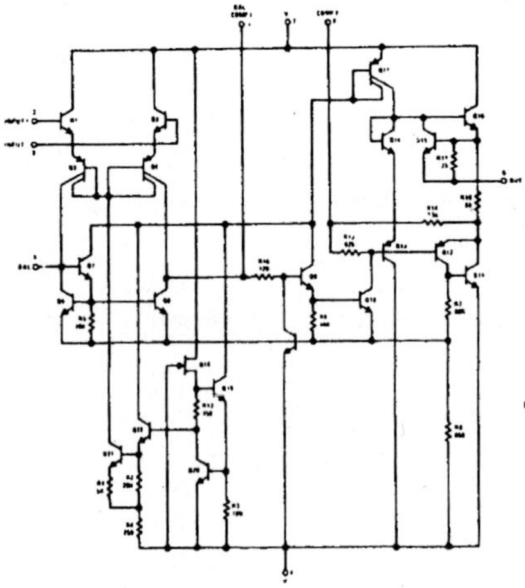

Metal Can Package

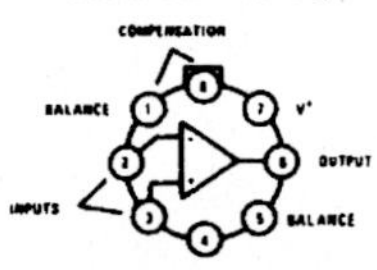

Order Number LM101AH, LM201AH or LM301AH
See NS Package H08C

Dual-In-Line Package

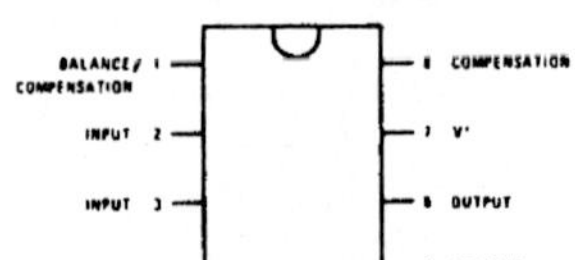

Order Number
LM101AJ, LM201AJ, LM301AJ
See NS Package J08A

Order Number LM301AN
See NS Package N08A

Dual-In-Line Package

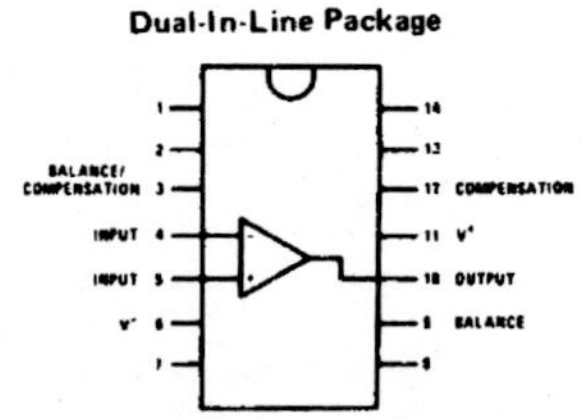

Note: Pin 6 connected to bottom of package.

Order Number LM101AJ-14
LM201AJ-14 or LM301AJ-14
See NS Package J14A

**Pin connections shown are for metal can.

Absolute Maximum Ratings

	LM101A/LM201A	LM301A
Supply Voltage	±22V	±18V
Power Dissipation (Note 1)	500 mW	500 mW
Differential Input Voltage	±30V	±30V
Input Voltage (Note 2)	±15V	±15V
Output Short Circuit Duration (Note 3)	Indefinite	Indefinite
Operating Temperature Range	−55°C to +125°C (LM101A) −25°C to +85°C (LM201A)	0°C to +70°C
Storage Temperature Range	−65°C to +150°C	−65°C to +150°C
Lead Temperature (Soldering, 10 seconds)	300°C	300°C

Electrical Characteristics (Note 4)

PARAMETER	CONDITIONS	LM101A/LM201A			LM301A			UNITS
		MIN	TYP	MAX	MIN	TYP	MAX	
Input Offset Voltage LM101A, LM201A, LM301A	$T_A = 25°C$ $R_S \leq 50\ k\Omega$		0.7	2.0		2.0	7.5	mV
Input Offset Current	$T_A = 25°C$		1.5	10		3.0	50	nA
Input Bias Current	$T_A = 25°C$		30	75		70	250	nA
Input Resistance	$T_A = 25°C$	1.5	4.0		0.5	2.0		MΩ
Supply Current	$T_A = 25°C$							
	$V_S = \pm 20V$		1.8	3.0				mA
	$V_S = \pm 15V$					1.8	3.0	mA
Large Signal Voltage Gain	$T_A = 25°C$, $V_S = \pm 15V$ $V_{OUT} = \pm 10V$, $R_L \geq 2\ k\Omega$	50	160		25	160		V/mV
Input Offset Voltage	$R_S \leq 50\ k\Omega$			3.0			10	mV
	$R_S \leq 10\ k\Omega$							mV
Average Temperature Coefficient of Input Offset Voltage	$R_S \leq 50\ k\Omega$		3.0	15		6.0	30	µV/°C
	$R_S \leq 10\ k\Omega$							µV/°C
Input Offset Current				20			70	nA
	$T_A = T_{MAX}$							nA
	$T_A = T_{MIN}$							nA
Average Temperature Coefficient of Input Offset Current	$25°C \leq T_A \leq T_{MAX}$		0.01	0.1		0.01	0.3	nA/°C
	$T_{MIN} \leq T_A \leq 25°C$		0.02	0.2		0.02	0.6	nA/°C
Input Bias Current				0.1			0.3	µA
Supply Current	$T_A = T_{MAX}$, $V_S = \pm 20V$		1.2	2.5				mA
Large Signal Voltage Gain	$V_S = \pm 15V$, $V_{OUT} = \pm 10V$, $R_L \geq 2k$	25			15			V/mV
Output Voltage Swing	$V_S = \pm 15V$							
	$R_L = 10\ k\Omega$	±12	±14		±12	±14		V
	$R_L = 2\ k\Omega$	±10	±13		±10	±13		V
Input Voltage Range	$V_S = \pm 20V$	±15						V
	$V_S = \pm 15V$		+15, −13		±12	+15, −13		V
Common-Mode Rejection Ratio	$R_S \leq 50\ k\Omega$	80	96		70	90		dB
	$R_S \leq 10\ k\Omega$							dB
Supply Voltage Rejection Ratio	$R_S \leq 50\ k\Omega$	80	96		70	96		dB
	$R_S \leq 10\ k\Omega$							dB

Note 1: The maximum junction temperature of the LM101A is 150°C, and that of the LM201A/LM301A is 100°C. For operating at elevated temperatures, devices in the TO-5 package must be derated based on a thermal resistance of 150°C/W, junction to ambient, or 45°C/W, junction to case. The thermal resistance of the dual-in-line package is 187°C/W, junction to ambient.

Note 2: For supply voltages less than ±15V, the absolute maximum input voltage is equal to the supply voltage.

Note 3: Continuous short circuit is allowed for case temperatures to 125°C and ambient temperatures to 75°C for LM101A/LM201A, and 70°C and 55°C respectively for LM301A.

Note 4: Unless otherwise specified, these specifications apply for C1 = 30 pF, $\pm 5V \leq V_S \leq \pm 20V$ and $-55°C \leq T_A \leq +125°C$ (LM101A), $\pm 5V \leq V_S \leq \pm 20V$ and $-25°C \leq T_A \leq +85°C$ (LM201A), $\pm 5V \leq V_S \leq \pm 15V$ and $0°C \leq T_A \leq +70°C$ (LM301A).

Guaranteed Performance Characteristics LM101A/LM201A

Input Voltage Range

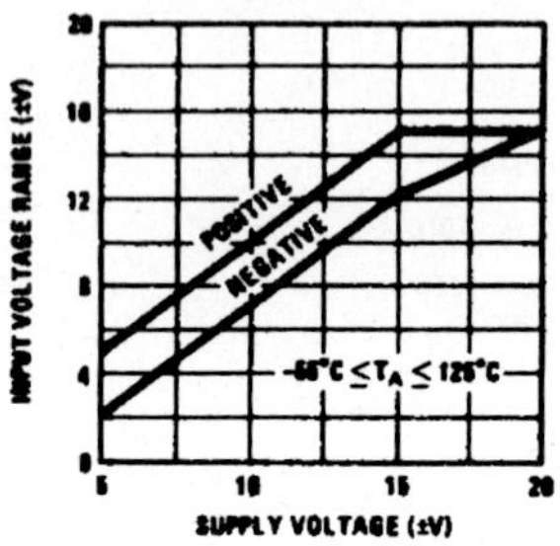

Output Swing

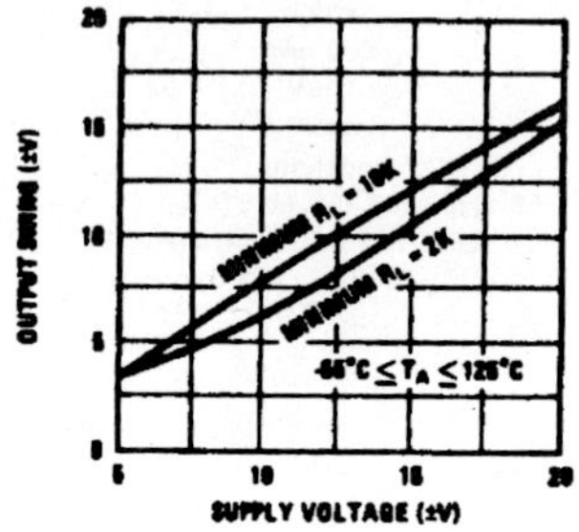

Voltage Gain

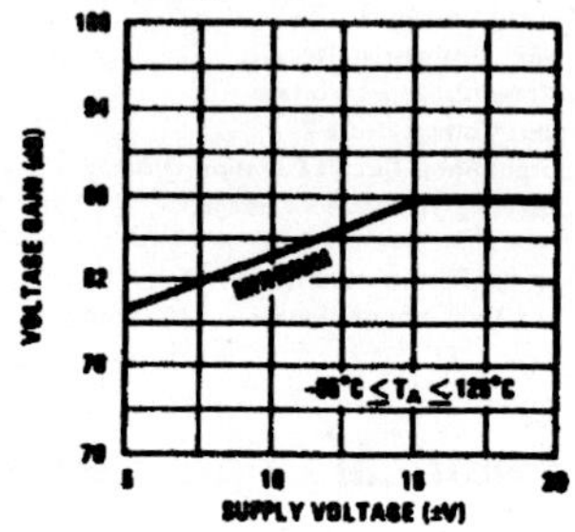

Guaranteed Performance Characteristics LM301A

Input Voltage Range

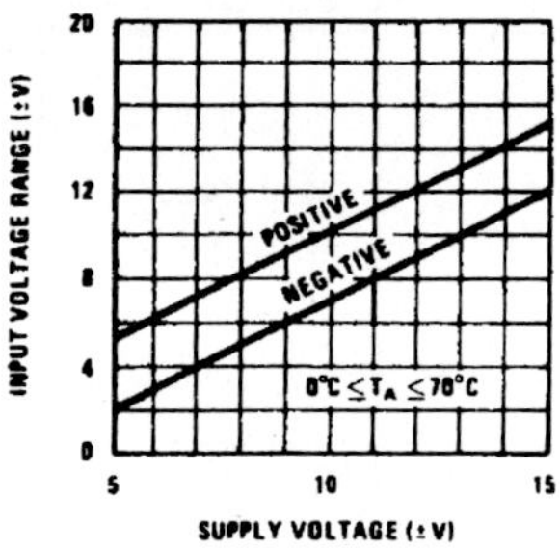

Output Swing

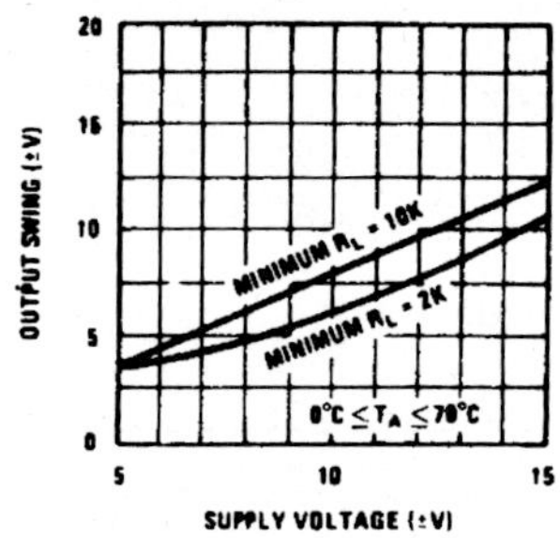

Voltage Gain

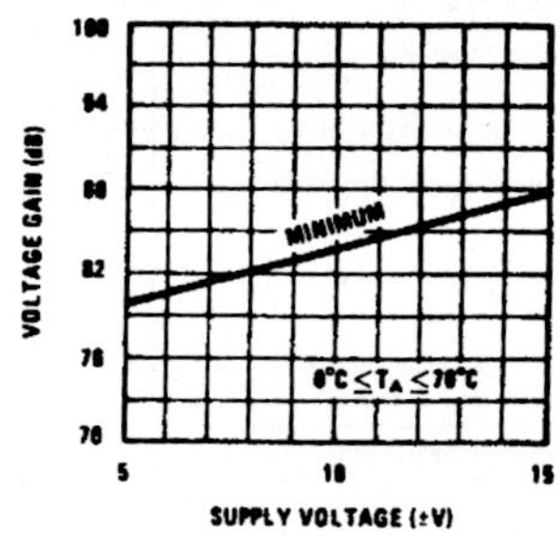

Typical Performance Characteristics

Supply Current

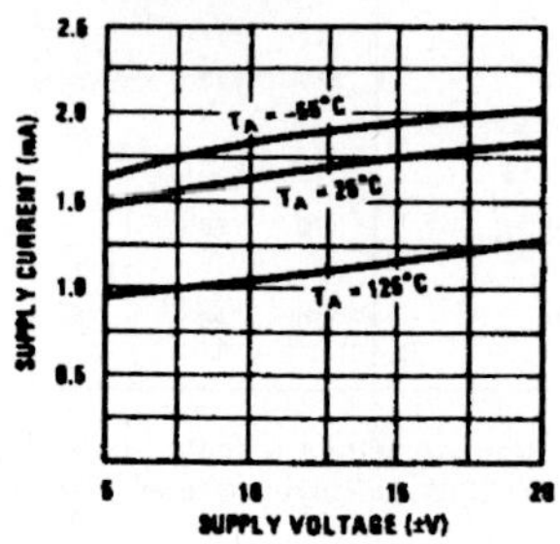

Voltage Gain

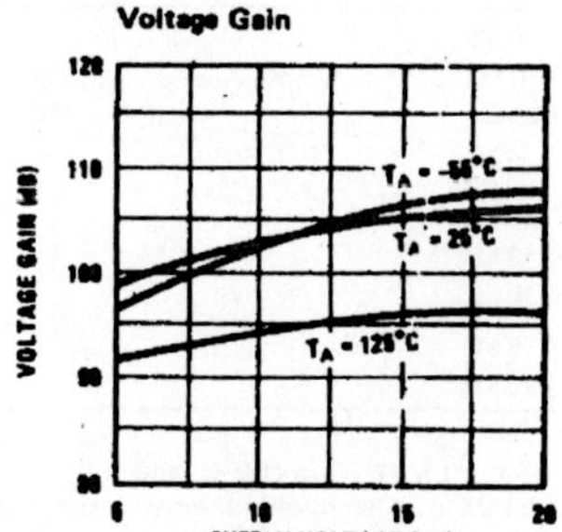

Maximum Power Dissipation

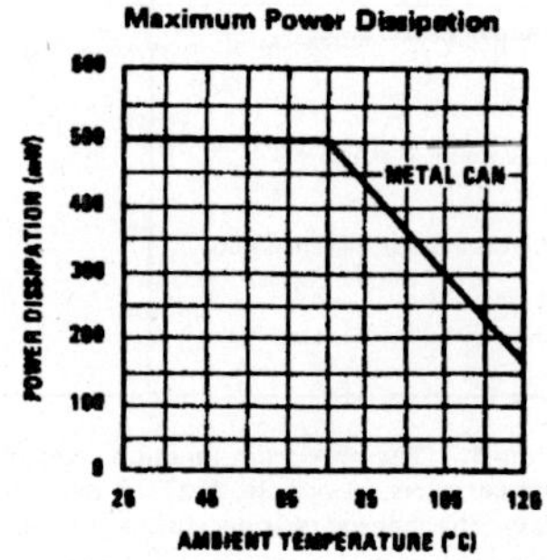

Typical Performance Characteristics (Continued)

Input Current, LM101A/LM201A/LM301A

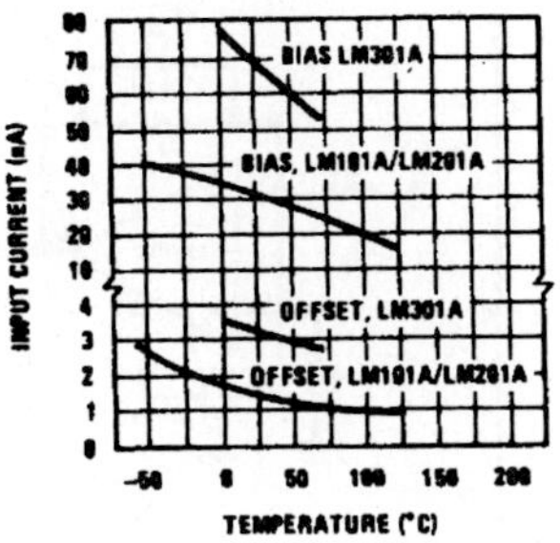

Current Limiting

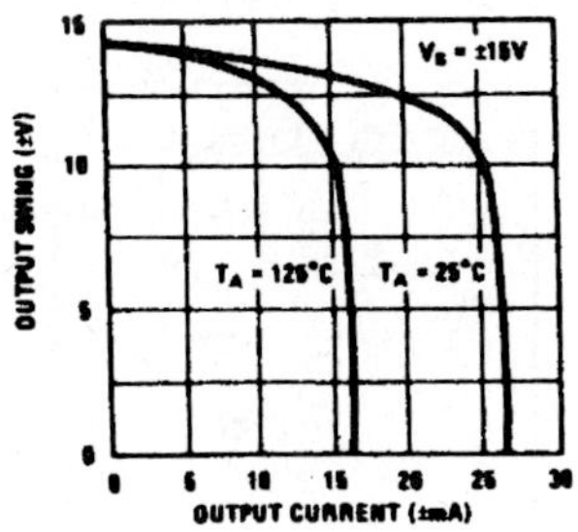

Input Noise Voltage

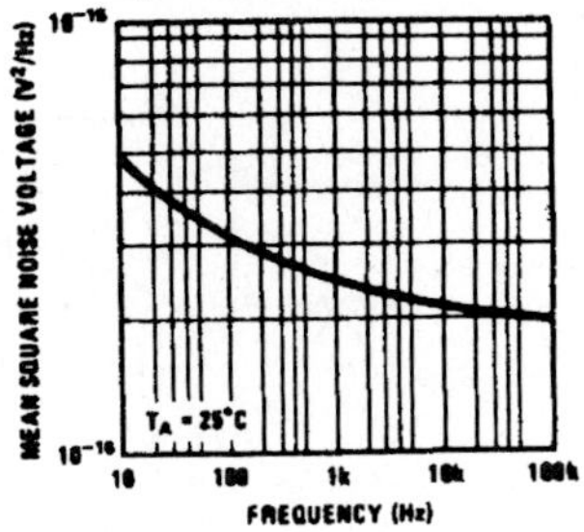

Input Noise Current

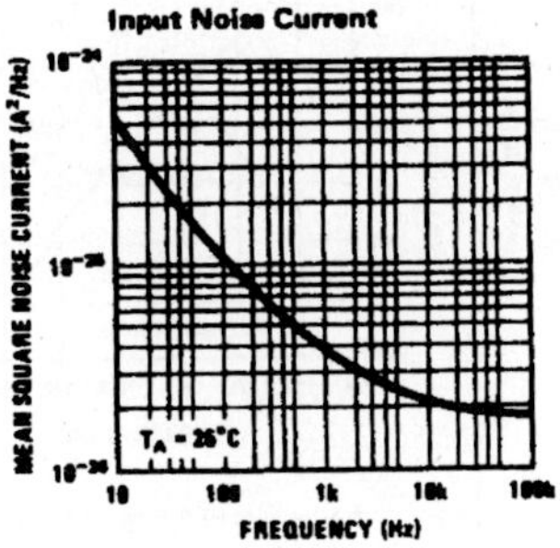

Common Mode Rejection

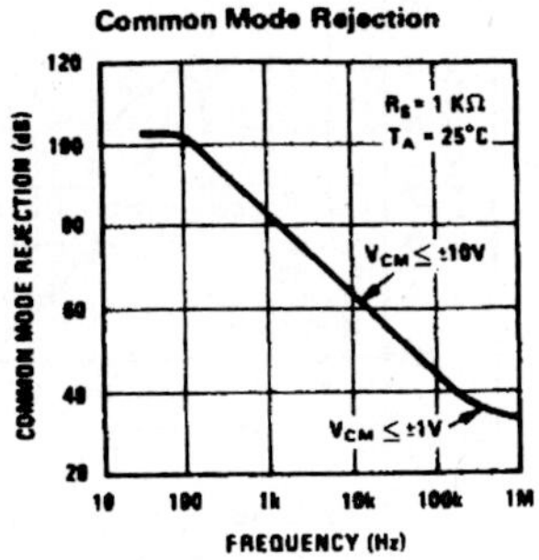

Power Supply Rejection

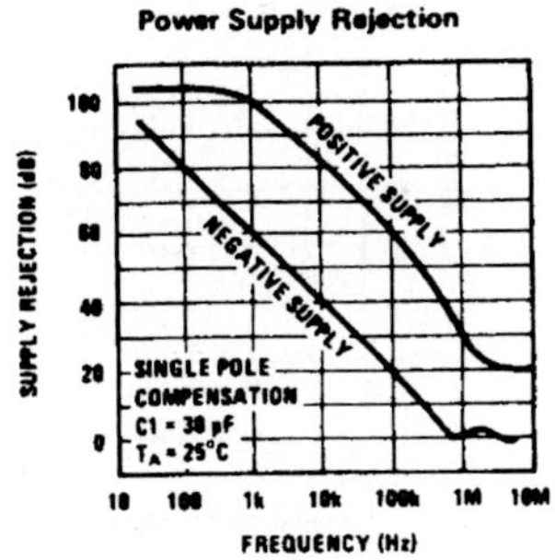

Closed Loop Output Impedance

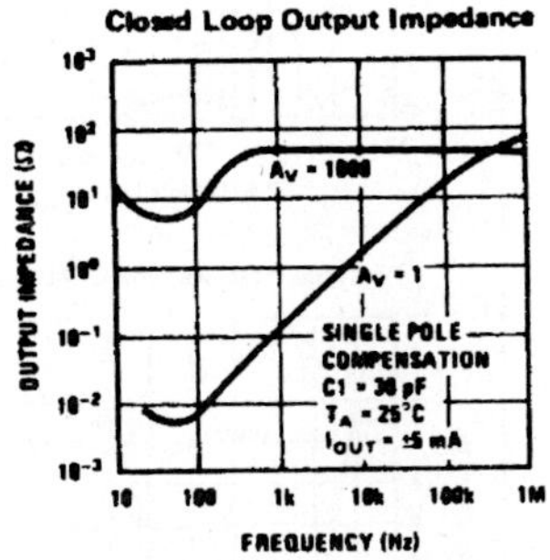

Typical Performance Characteristics for Various Compensation Circuits**

Single Pole Compensation

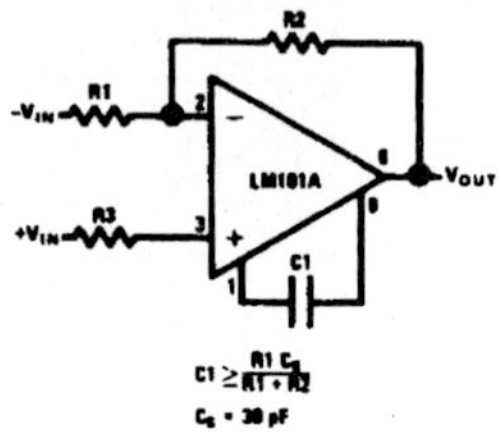

Two Pole Compensation

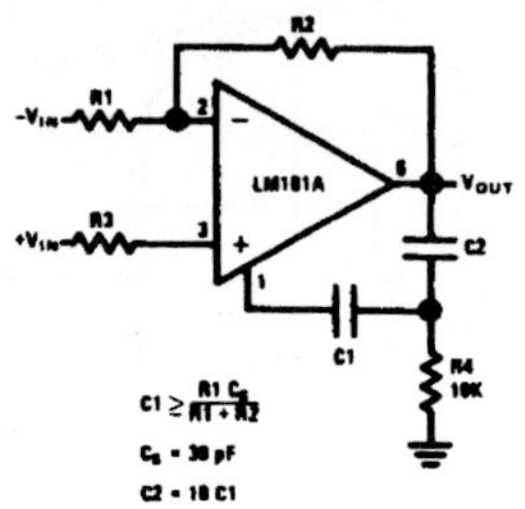

Feedforward Compensation

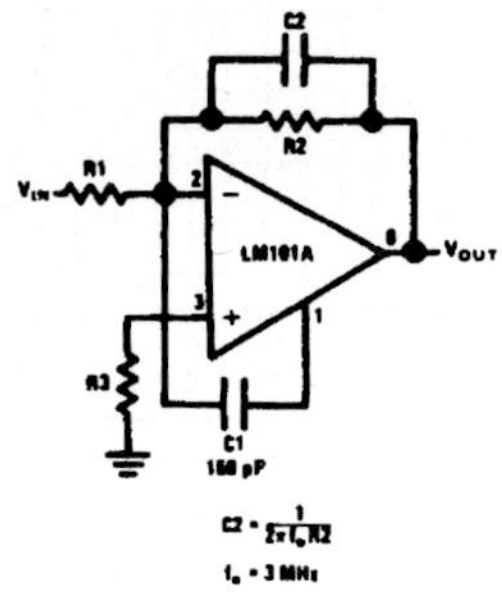

**Pin connections shown are for metal can.

Open Loop Frequency Response

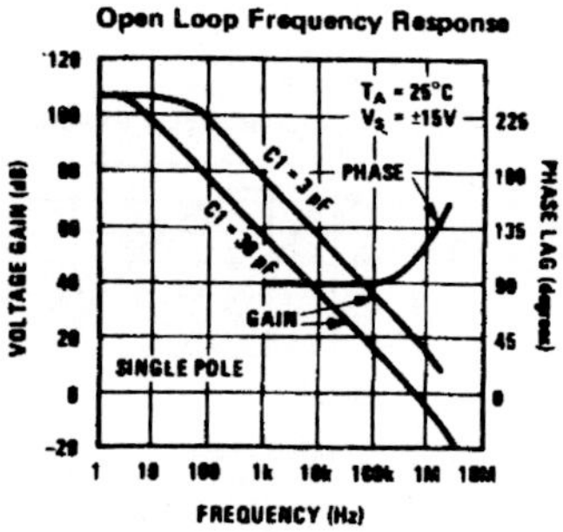

Open Loop Frequency Response

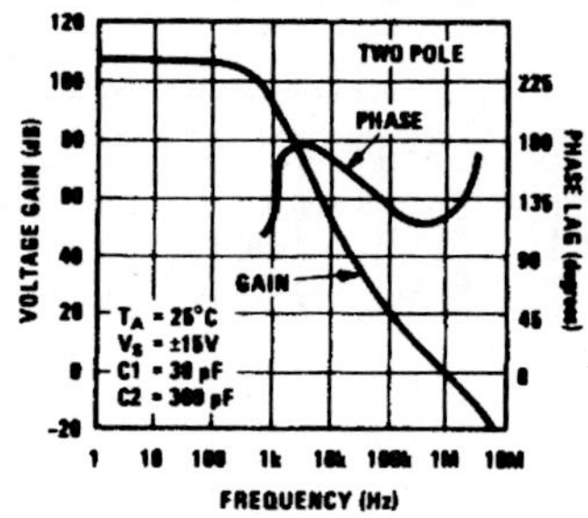

Open Loop Frequency Response

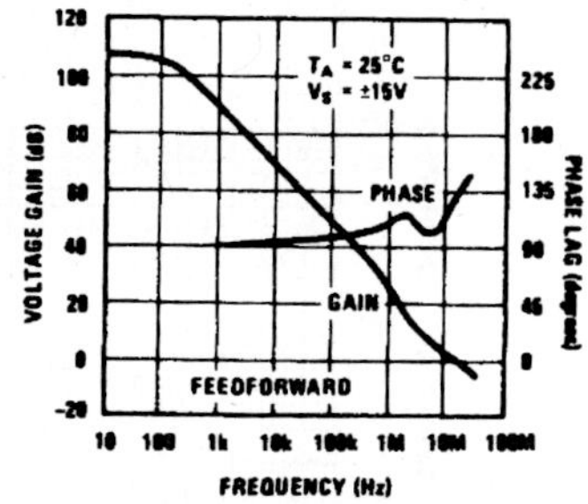

Large Signal Frequency Response

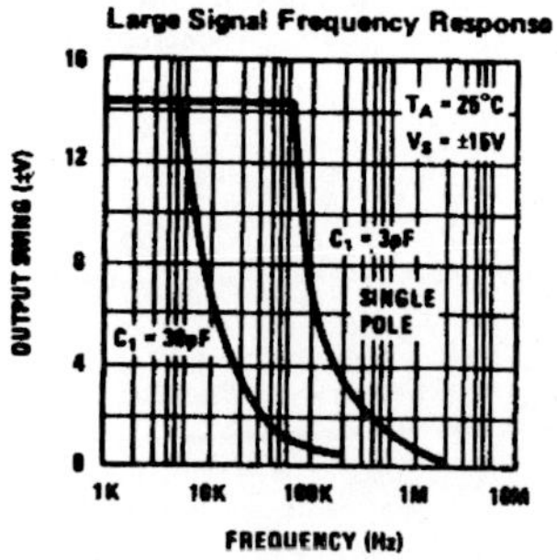

Large Signal Frequency Response

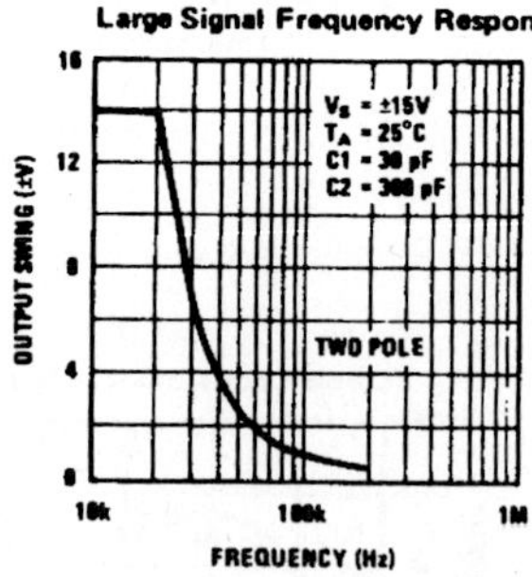

Large Signal Frequency Response

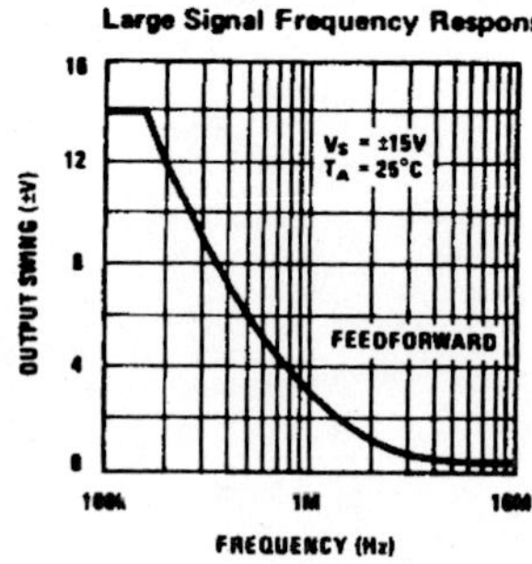

Voltage Follower Pulse Response

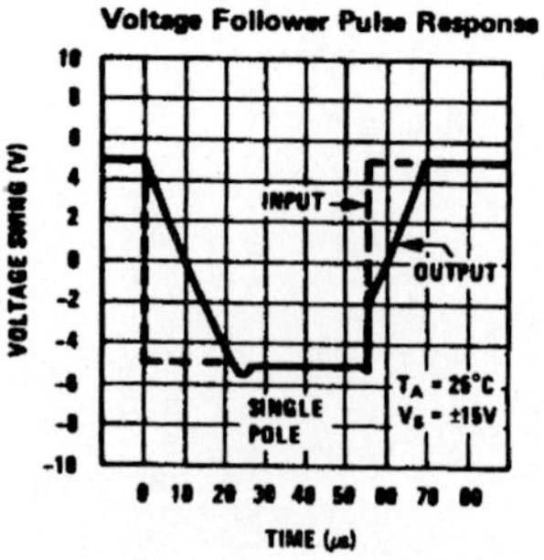

Voltage Follower Pulse Response

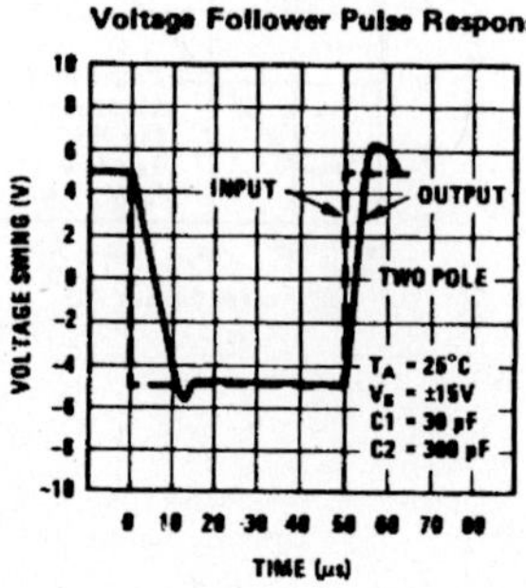

Inverter Pulse Response

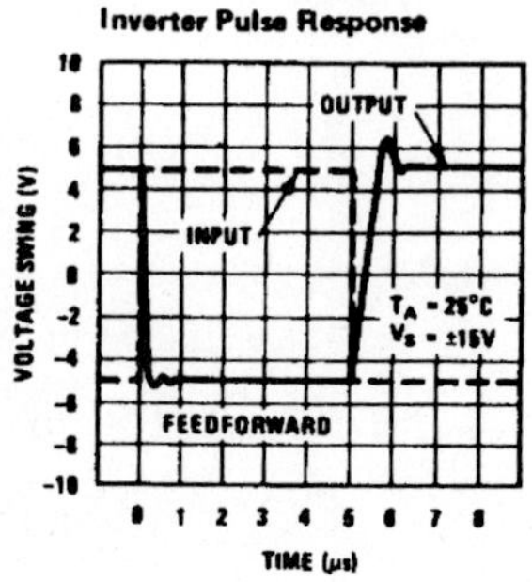

Typical Applications **

Variable Capacitance Multiplier

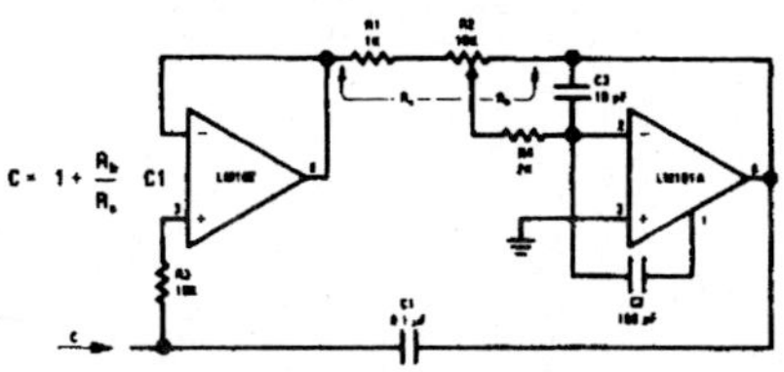

Simulated Inductor

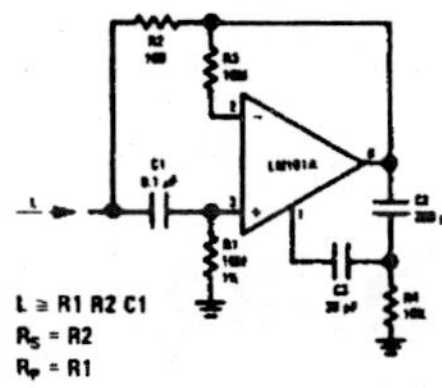

Fast Inverting Amplifier With High Input Impedance

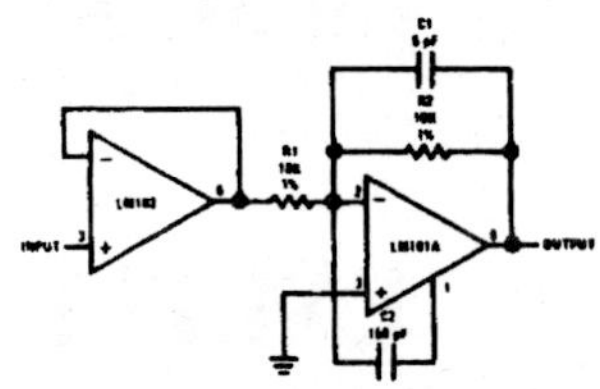

Inverting Amplifier with Balancing Circuit

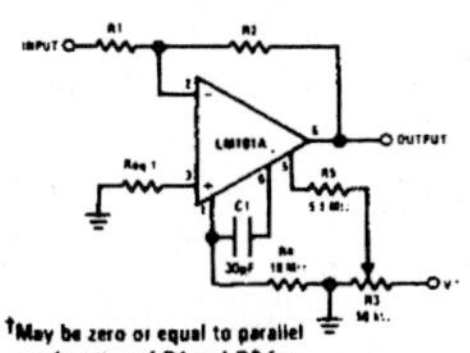

†May be zero or equal to parallel combination of R1 and R2 for minimum offset.

Sine Wave Oscillator

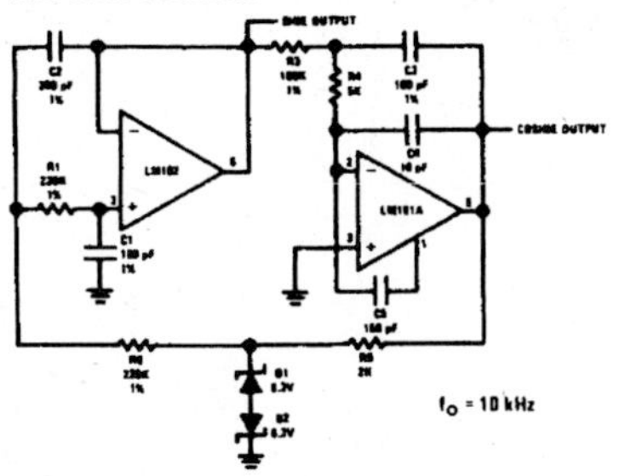

Integrator with Bias Current Compensation

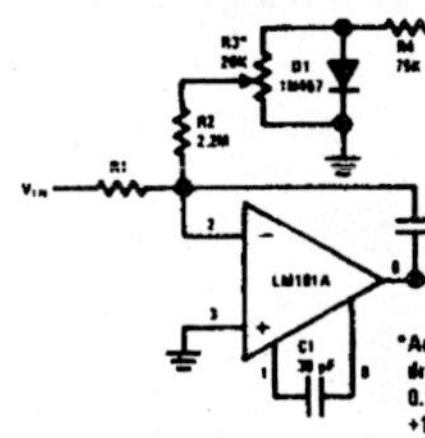

*Adjust for zero integrator drift. Current drift typically 0.1 nA/°C over −55°C to +125°C temperature range.

Application Hints **

Protecting Against Gross Fault Conditions

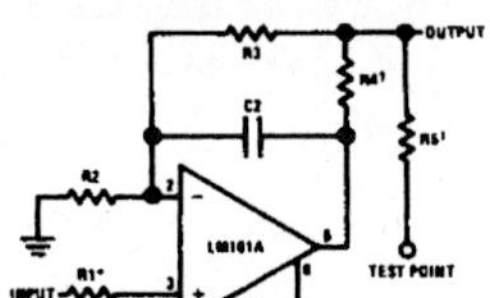

*Protects input

†Protects output

‡Protects output–not needed when R4 is used.

Compensating For Stray Input Capacitances Or Large Feedback Resistor

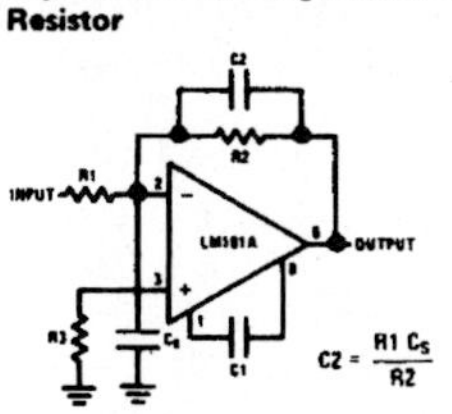

Isolating Large Capacitive Loads

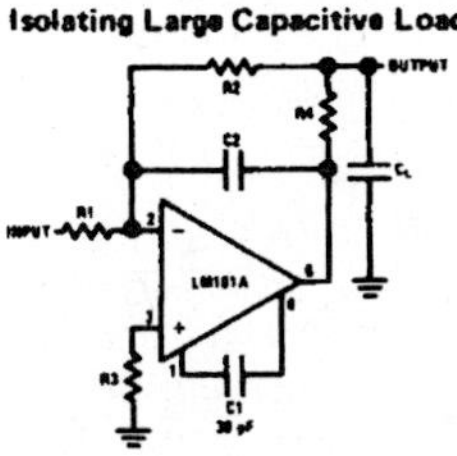

Although the LM101A is designed for trouble free operation, experience has indicated that it is wise to observe certain precautions given below to protect the devices from abnormal operating conditions. It might be pointed out that the advice given here is applicable to practically any IC op amp, although the exact reason why may differ with different devices.

When driving either input from a low-impedance source, a limiting resistor should be placed in series with the input lead to limit the peak instantaneous output current of the source to something less than 100 mA. This is especially important when the inputs go outside a piece of equipment where they could accidentally be connected to high voltage sources. Large capacitors on the input (greater than 0.1 μF) should be treated as a low source impedance and isolated with a resistor. Low impedance sources do not cause a problem unless their output voltage exceeds the supply voltage. However, the supplies go to zero when they are turned off, so the isolation is usually needed.

The output circuitry is protected against damage from shorts to ground. However, when the amplifier output is connected to a test point, it should be isolated by a limiting resistor, as test points frequently get shorted to bad places. Further, when the amplifier drives a load external to the equipment, it is also advisable to use some sort of limiting resistance to preclude mishaps.

Precautions should be taken to insure that the power supplies for the integrated circuit never become reversed–even under transient conditions. With reverse voltages greater than 1V, the IC will conduct excessive current, fuzing internal aluminum interconnects. If there is a possibility of this happening, clamp diodes with a high peak current rating should be installed on the supply lines. Reversal of the voltage between V^+ and V^- will always cause a problem, although reversals with respect to ground may also give difficulties in many circuits.

The minimum values given for the frequency compensation capacitor are stable only for source resistances less than 10 kΩ, stray capacitances on the summing junction less than 5 pF and capacitive loads smaller than 100 pF. If any of these conditions are not met, it becomes necessary to overcompensate the amplifier with a larger compensation capacitor. Alternately, lead capacitors can be used in the feedback network to negate the effect of stray capacitance and large feedback resistors or an RC network can be added to isolate capacitive loads.

Although the LM101A is relatively unaffected by supply bypassing, this cannot be ignored altogether. Generally it is necessary to bypass the supplies to ground at least once on every circuit card, and more bypass points may be required if more than five amplifiers are used. When feed-forward compensation is employed, however, it is advisable to bypass the supply leads of each amplifier with low inductance capacitors because of the higher frequencies involved.

**Pin connections shown are for metal can.

APPENDIX 3

LM311 Voltage Comparator*

*Courtesy of National Semiconductor Corporation.

General Description

The LM311 is a voltage comparator that has input currents more than a hundred times lower than devices like the LM306 or LM710C. It is also designed to operate over a wider range of supply voltages: from standard ±15V op amp supplies down to the single 5V supply used for IC logic. Its output is compatible with RTL, DTL and TTL as well as MOS circuits. Further, it can drive lamps or relays, switching voltages up to 40V at currents as high as 50 mA.

Features

- Operates from single 5V supply
- Maximum input current: 250 nA
- Maximum offset current: 50 nA
- Differential input voltage range: ±30V
- Power consumption: 135 mW at ±15V

Both the input and the output of the LM311 can be isolated from system ground, and the output can drive loads referred to ground, the positive supply or the negative supply. Offset balancing and strobe capability are provided and outputs can be wire OR'ed. Although slower than the LM306 and LM710C (200 ns response time vs 40 ns) the device is also much less prone to spurious oscillations. The LM311 has the same pin configuration as the LM306 and LM710C. See the "application hints" of the LM311 for application help.

Auxiliary Circuits**

**Note: Pin connections shown on schematic diagram and typical applications are for TO-5 package.

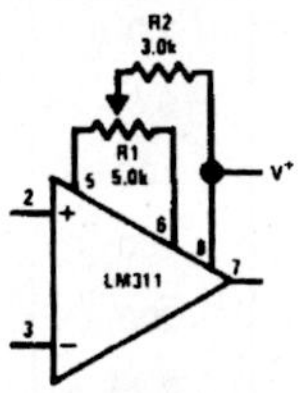

Offset Balancing

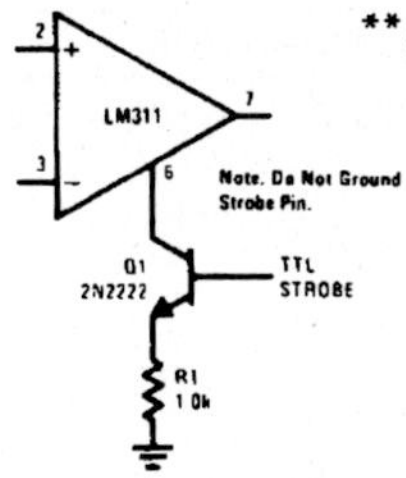

Strobing

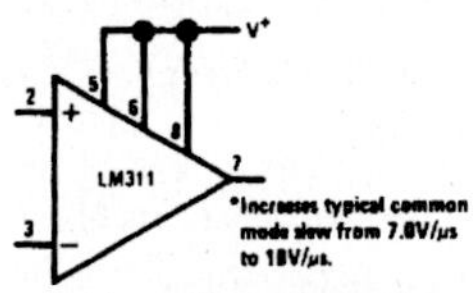

Increasing Input Stage Current*

Typical Applications**

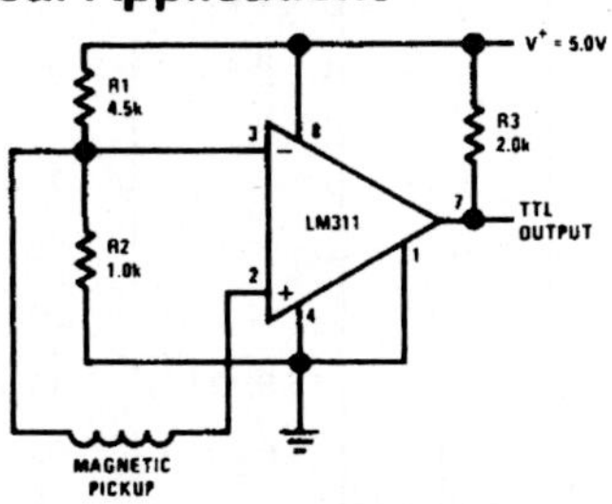

Detector for Magnetic Transducer

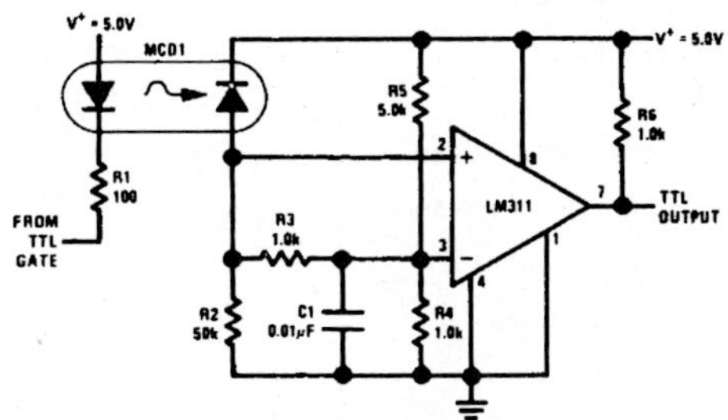

Digital Transmission Isolator

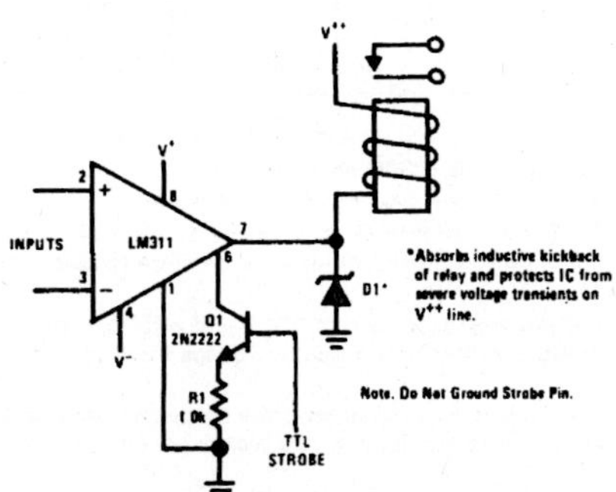

Relay Driver with Strobe

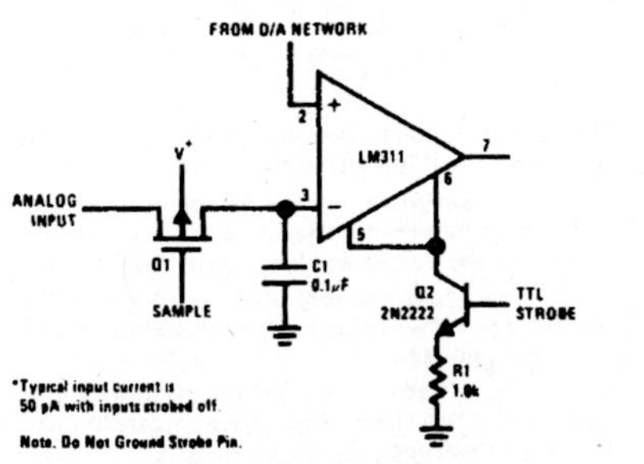

Strobing off Both Input* and Output Stages

Absolute Maximum Ratings

Total Supply Voltage (V_{84})	36V
Output to Negative Supply Voltage (V_{74})	40V
Ground to Negative Supply Voltage (V_{14})	30V
Differential Input Voltage	±30V
Input Voltage (Note 1)	±15V
Power Dissipation (Note 2)	500 mW
Output Short Circuit Duration	10 sec
Operating Temperature Range	0°C to 70°C
Storage Temperature Range	-65°C to 150°C
Lead Temperature (soldering, 10 sec)	300°C
Voltage at Strobe Pin	V^+–5V

Electrical Characteristics (Note 3)

PARAMETER	CONDITIONS	MIN	TYP	MAX	UNITS
Input Offset Voltage (Note 4)	T_A = 25°C, $R_S \leq$ 50k		2.0	7.5	mV
Input Offset Current (Note 4)	T_A = 25°C		6.0	50	nA
Input Bias Current	T_A = 25°C		100	250	nA
Voltage Gain	T_A = 25°C	40	200		V/mV
Response Time (Note 5)	T_A = 25°C		200		ns
Saturation Voltage	$V_{IN} \leq -10$ mV, I_{OUT} = 50 mA T_A = 25°C		0.75	1.5	V
Strobe ON Current	T_A = 25°C		3.0		mA
Output Leakage Current	$V_{IN} \geq$ 10 mV, V_{OUT} = 35V T_A = 25°C, I_{STROBE} = 3 mA		0.2	50	nA
Input Offset Voltage (Note 4)	$R_S \leq$ 50k			10	mV
Input Offset Current (Note 4)				70	nA
Input Bias Current				300	nA
Input Voltage Range		−14.5	13.8,−14.7	13.0	V
Saturation Voltage	$V^+ \geq$ 4.5V, V^- = 0 $V_{IN} \leq -10$ mV, $I_{SINK} \leq$ 8 mA		0.23	0.4	V
Positive Supply Current	T_A = 25°C		5.1	7.5	mA
Negative Supply Current	T_A = 25°C		4.1	5.0	mA

Note 1: This rating applies for ±15V supplies. The positive input voltage limit is 30V above the negative supply. The negative input voltage limit is equal to the negative supply voltage or 30V below the positive supply, whichever is less.

Note 2: The maximum junction temperature of the LM311 is 110°C. For operating at elevated temperatures, devices in the TO-5 package must be derated based on a thermal resistance of 150°C/W, junction to ambient, or 45°C/W, junction to case. The thermal resistance of the dual-in-line package is 100°C/W, junction to ambient.

Note 3: These specifications apply for V_S = ±15V and the Ground pin at ground, and 0°C < T_A < +70°C, unless otherwise specified. The offset voltage, offset current and bias current specifications apply for any supply voltage from a single 5V supply up to ±15V supplies.

Note 4: The offset voltages and offset currents given are the maximum values required to drive the output within a volt of either supply with 1 mA load. Thus, these parameters define an error band and take into account the worst-case effects of voltage gain and input impedance.

Note 5: The response time specified (see definitions) is for a 100 mV input step with 5 mV overdrive.

Note 6: Do not short the strobe pin to ground; it should be current driven at 3 to 5 mA.

Typical Performance Characteristics

Input Bias Current

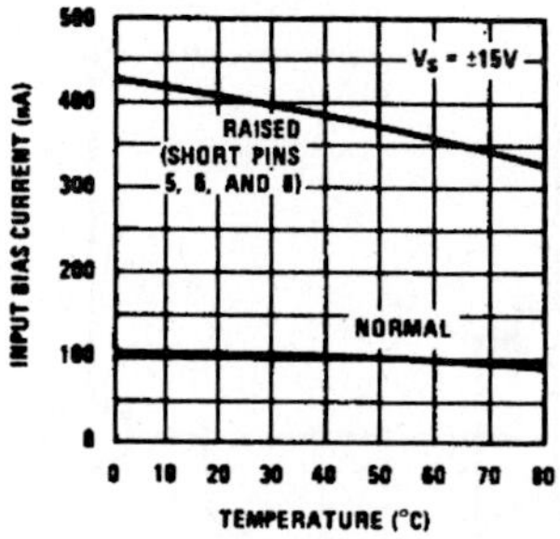

Input Offset Current

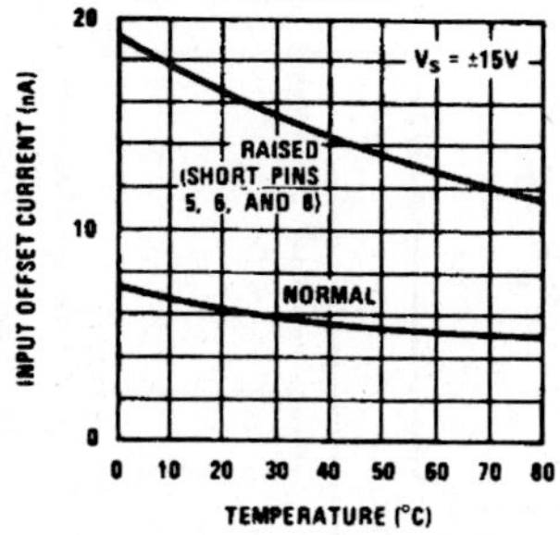

Offset Error

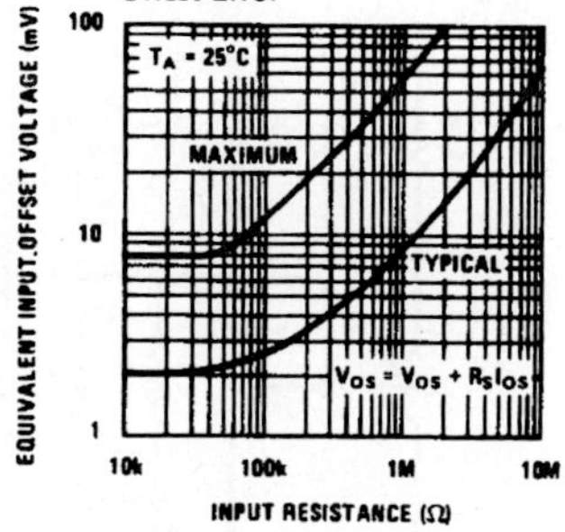

Input Characteristics

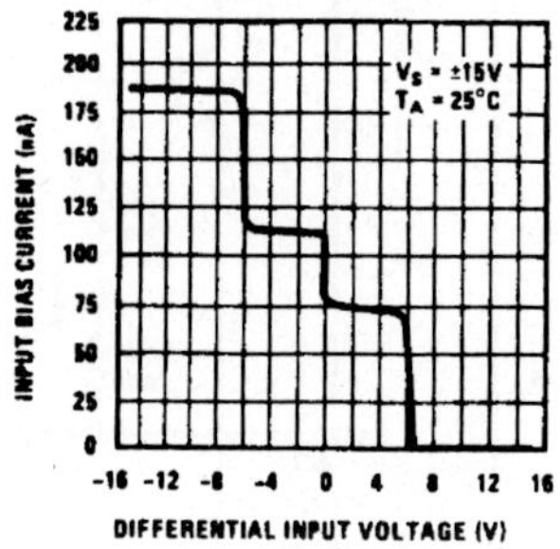

Common Mode Limits

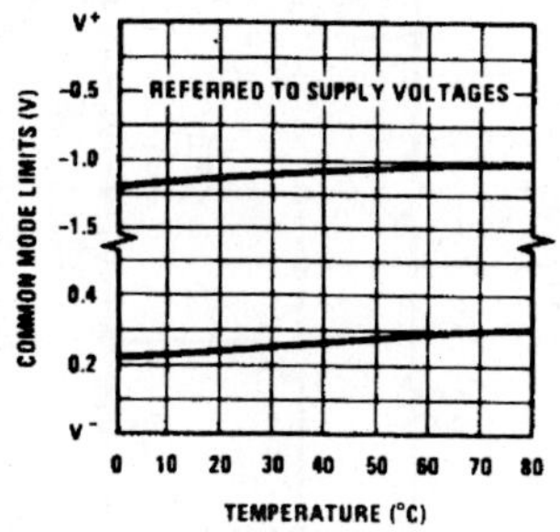

Transfer Function

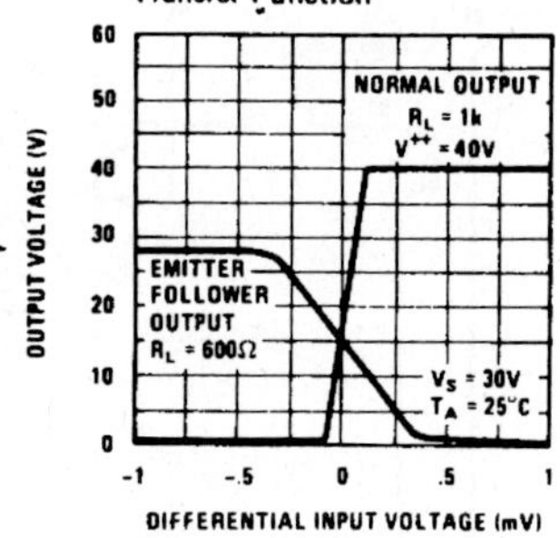

Response Time for Various Input Overdrives

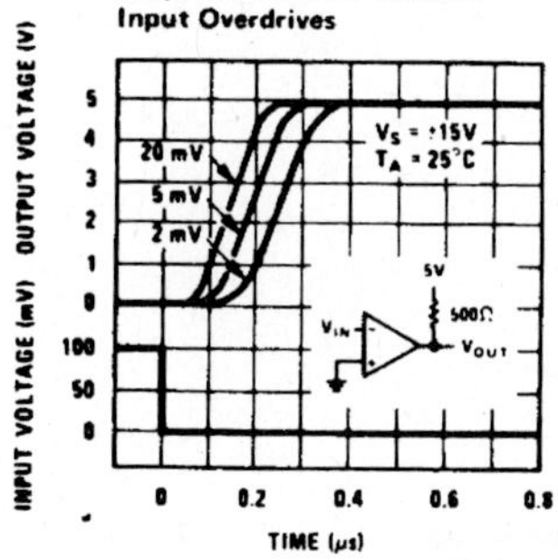

Response Time for Various Input Overdrives

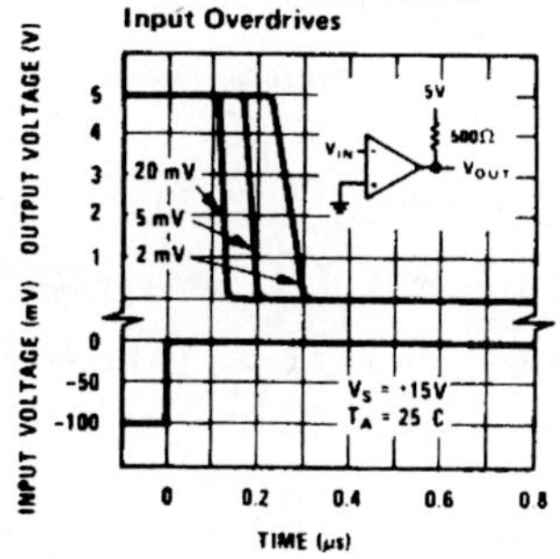

Output Saturation Voltage

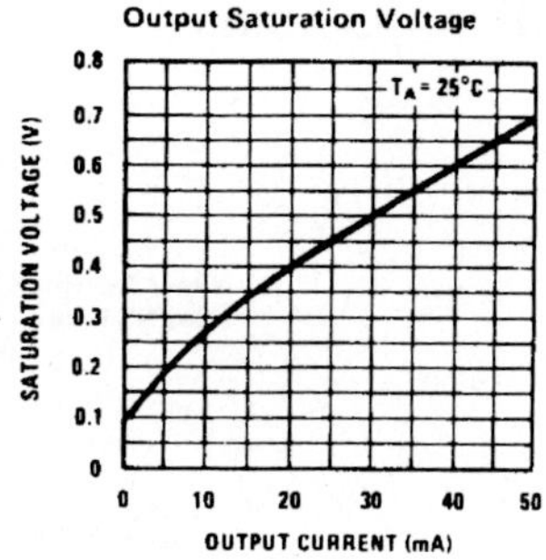

Response Time for Various Input Overdrives

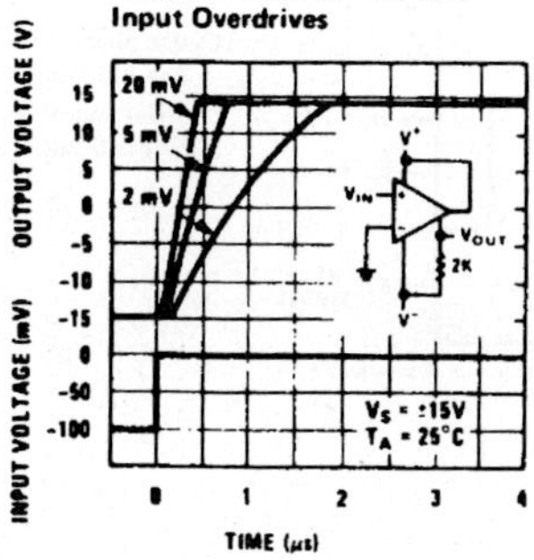

Response Time for Various Input Overdrives

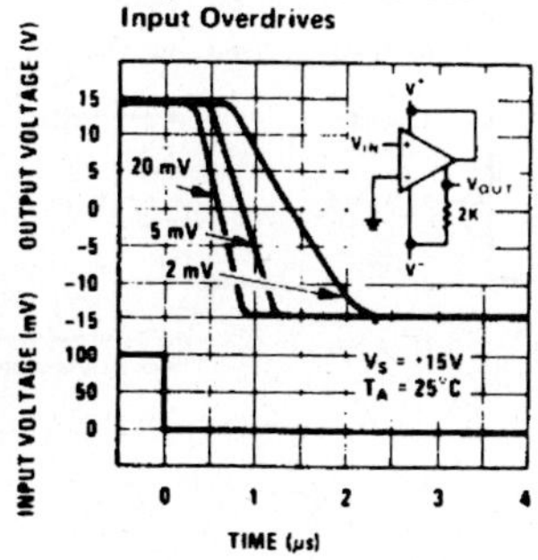

Output Limiting Characteristics

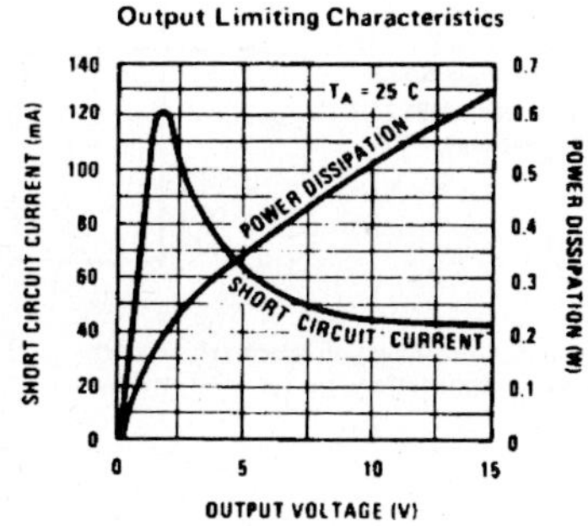

Schematic Diagram

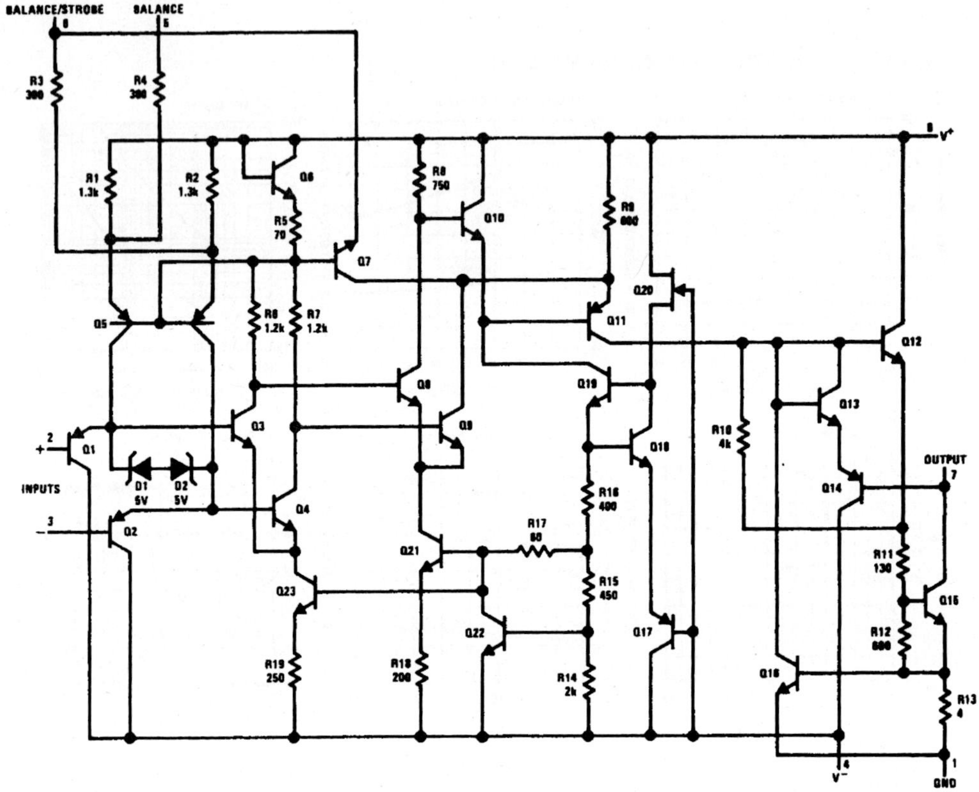

Connection Diagrams *

Metal Can Package

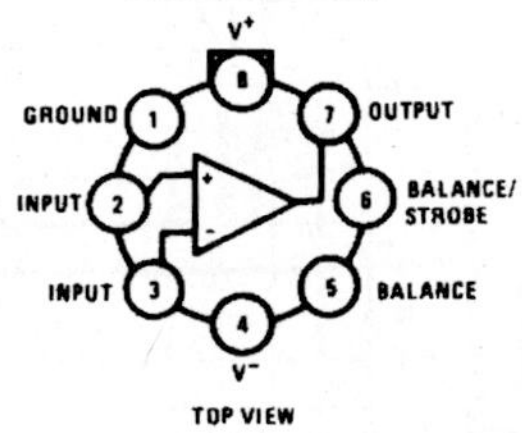

Order Number LM311H
See NS Package H08C

Dual-In-Line Package

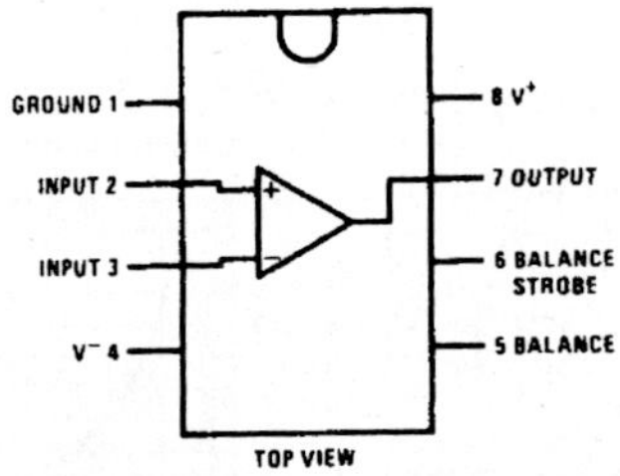

Order Number LM311N
See NS Package N08B
Order Number LM311J-8
See NS Package J08A

Dual-In-Line Package

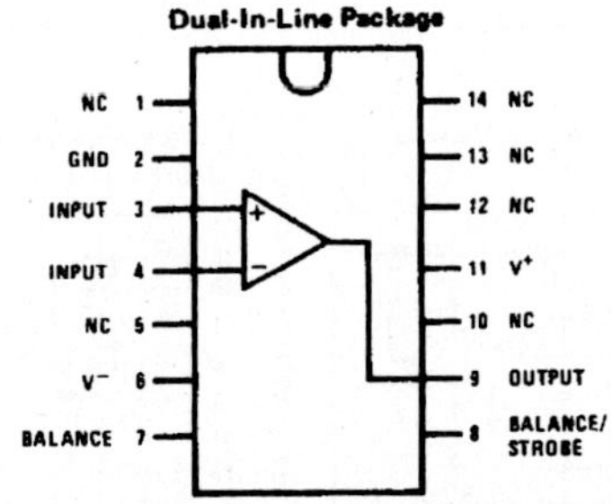

Note: Pin 6 connected to bottom of package.

Order Number LM311N-14
See NS Package N14A
Order Number LM311J
See NS Package J14A

*Pin connections shown on schematic diagram and typical applications are for TO-5 package.

Application Hints

CIRCUIT TECHNIQUES FOR AVOIDING OSCILLATIONS IN COMPARATOR APPLICATIONS

When a high-speed comparator such as the LM111 is used with fast input signals and low source impedances, the output response will normally be fast and stable, assuming that the power supplies have been bypassed (with 0.1 μF disc capacitors), and that the output signal is routed well away from the inputs (pins 2 and 3) and also away from pins 5 and 6.

However, when the input signal is a voltage ramp or a slow sine wave, or if the signal source impedance is high (1 kΩ to 100 kΩ), the comparator may burst into oscillation near the crossing-point. This is due to the high gain and wide bandwidth of comparators like the LM111. To avoid oscillation or instability in such a usage, several precautions are recommended, as shown in *Figure 1* below.

1. The trim pins (pins 5 and 6) act as unwanted auxiliary inputs. If these pins are not connected to a trim-pot, they should be shorted together. If they are connected to a trim-pot, a 0.01 μA capacitor C1 between pins 5 and 6 will minimize the susceptibility to AC coupling. A smaller capacitor is used if pin 5 is used for positive feedback as in *Figure 1*.

2. Certain sources will produce a cleaner comparator output waveform if a 100 pF to 1000 pF capacitor C2 is connected directly across the input pins.

3. When the signal source is applied through a resistive network, R_S, it is usually advantageous to choose an R_S' of substantially the same value, both for DC and for dynamic (AC) considerations. Carbon, tin-oxide, and metal-film resistors have all been used successfully in comparator input circuitry. Inductive wirewound resistors are not suitable.

4. When comparator circuits use input resistors (eg. summing resistors), their value and placement are particularly important. In all cases the body of the resistor should be close to the device or socket. In other words there should be very little lead length or printed-circuit foil run between comparator and resistor to radiate or pick up signals. The same applies to capacitors, pots, etc. For example, if R_s = 10 kΩ, as little as 5 inches of lead between the resistors and the input pins can result in oscillations that are very hard to damp. Twisting these input leads tightly is the only (second best) alternative to placing resistors close to the comparator.

5. Since feedback to almost any pin of a comparator can result in oscillation, the printed-circuit layout should be engineered thoughtfully. Preferably there should be a groundplane under the LM111 circuitry, for example, one side of a double-layer circuit card. Ground foil (or, positive supply or negative supply foil) should extend between the output and the inputs, to act as a guard. The foil connections for the inputs should be as small and compact as possible, and should be essentially surrounded by ground foil on all sides, to guard against capacitive coupling from any high-level signals (such as the output). If pins 5 and 6 are not used, they should be shorted together. If they are connected to a trim-pot, the trim-pot should be located, at most, a few inches away from the LM111, and the 0.01 μF capacitor should be installed. If this capacitor cannot be used, a shielding printed-circuit foil may be advisable between pins 6 and 7. The power supply bypass capacitors should be located within a couple inches of the LM111. (Some other comparators require the power-supply bypass to be located immediately adjacent to the comparator.)

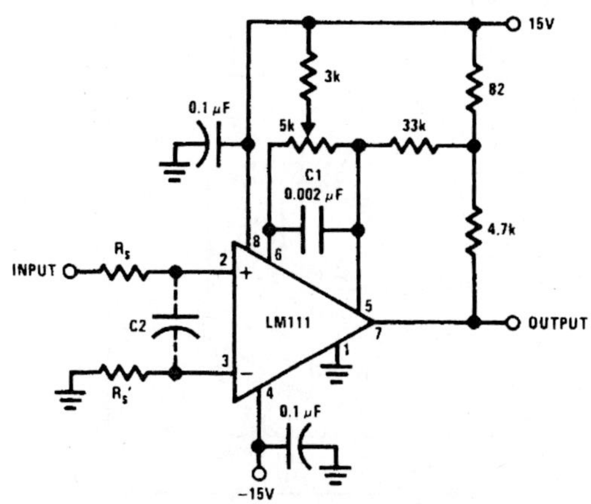

Pin connections shown are for LM111H in 8-lead TO-5 hermetic package

FIGURE 1. Improved Positive Feedback

Application Hints (Continued)

6. It is a standard procedure to use hysteresis (positive feedback) around a comparator, to prevent oscillation, and to avoid excessive noise on the output because the comparator is a good amplifier for its own noise. In the circuit of *Figure 2,* the feedback from the output to the positive input will cause about 3 mV of hysteresis. However, if R_S is larger than 100Ω, such as 50 kΩ, it would not be reasonable to simply increase the value of the positive feedback resistor above 510 kΩ. The circuit of *Figure 3* could be used, but it is rather awkward. See the notes in paragraph 7 below.

7. When both inputs of the LM111 are connected to active signals, or if a high-impedance signal is driving the positive input of the LM111 so that positive feedback would be disruptive, the circuit of *Figure 1* is ideal. The positive feedback is to pin 5 (one of the offset adjustment pins). It is sufficient to cause 1 to 2 mV hysteresis and sharp transitions with input triangle waves from a few Hz to hundreds of kHz. The positive-feedback signal across the 82Ω resistor swings 240 mV below the positive supply. This signal is centered around the nominal voltage at pin 5, so this feedback does not add to the V_{OS} of the comparator. As much as 8 mV of V_{OS} can be trimmed out, using the 5 kΩ pot and 3 kΩ resistor as shown.

8. These application notes apply specifically to the LM111, LM211, LM311, and LF111 families of comparators, and are applicable to all high-speed comparators in general, (with the exception that not all comparators have trim pins).

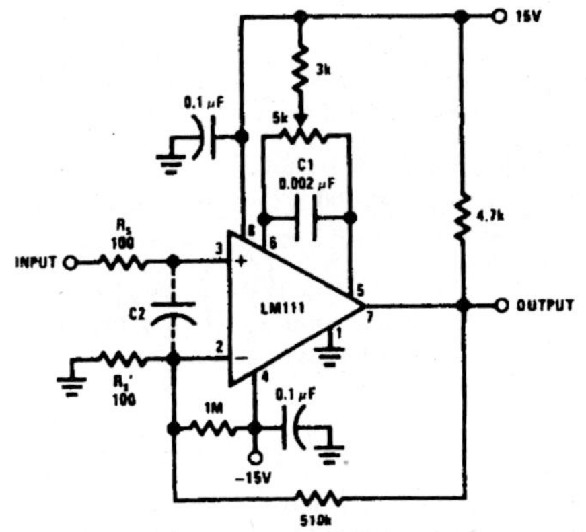

Pin connections shown are for LM111H in 8-lead TO-5 hermetic package

FIGURE 2. Conventional Positive Feedback

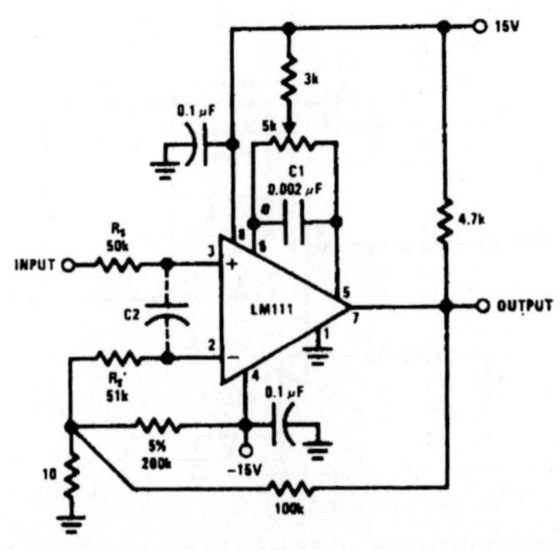

FIGURE 3. Positive Feedback With High Source Resistance

APPENDIX 4

LM117 3-Terminal Adjustable Regulator*

*Courtesy of National Semiconductor Corporation.

General Description

The LM117/LM217/LM317 are adjustable 3-terminal positive voltage regulators capable of supplying in excess of 1.5A over a 1.2V to 37V output range. They are exceptionally easy to use and require only two external resistors to set the output voltage. Further, both line and load regulation are better than standard fixed regulators. Also, the LM117 is packaged in standard transistor packages which are easily mounted and handled.

In addition to higher performance than fixed regulators, the LM117 series offers full overload protection available only in IC's. Included on the chip are current limit, thermal overload protection and safe area protection. All overload protection circuitry remains fully functional even if the adjustment terminal is disconnected.

Features

- Adjustable output down to 1.2V
- Guaranteed 1.5A output current
- Line regulation typically 0.01%/V
- Load regulation typically 0.1%
- Current limit constant with temperature
- **100% electrical burn-in**
- Eliminates the need to stock many voltages
- Standard 3-lead transistor package
- 80 dB ripple rejection

Normally, no capacitors are needed unless the device is situated far from the input filter capacitors in which case an input bypass is needed. An optional output capacitor can be added to improve transient response. The adjustment terminal can be bypassed to achieve very high ripple rejections ratios which are difficult to achieve with standard 3-terminal regulators.

Besides replacing fixed regulators, the LM117 is useful in a wide variety of other applications. Since the regulator is "floating" and sees only the input-to-output differential voltage, supplies of several hundred volts can be regulated as long as the maximum input to output differential is not exceeded.

Also, it makes an especially simple adjustable switching regulator, a programmable output regulator, or by connecting a fixed resistor between the adjustment and output, the LM117 can be used as a precision current regulator. Supplies with electronic shutdown can be achieved by clamping the adjustment terminal to ground which programs the output to 1.2V where most loads draw little current.

The LM117K, LM217K and LM317K are packaged in standard TO-3 transistor packages while the LM117H, LM217H and LM317H are packaged in a solid Kovar base TO-39 transistor package. The LM117 is rated for operation from −55°C to +150°C, the LM217 from −25°C to +150°C and the LM317 from 0°C to +125°C. The LM317T and LM317MP, rated for operation over a 0°C to +125°C range, are available in a TO-220 plastic package and a TO-202 package, respectively.

For applications requiring greater output current in excess of 3A and 5A, see LM150 series and LM138 series data sheets, respectively. For the negative complement, see LM137 series data sheet.

LM117 Series Packages and Power Capability

DEVICE	PACKAGE	RATED POWER DISSIPATION	DESIGN LOAD CURRENT
LM117	TO-3	20W	1.5A
LM217 LM317	TO-39	2W	0.5A
LM317T	TO-220	15W	1.5A
LM317M	TO-202	7.5W	0.5A
LM317LZ	TO-92	0.6W	0.1A

Typical Applications

1.2V–25V Adjustable Regulator

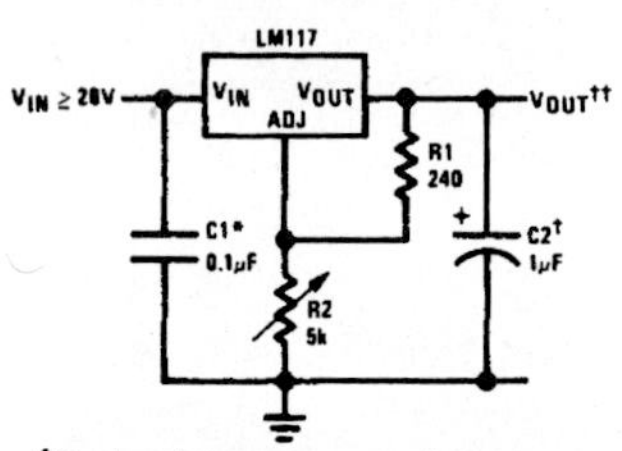

†Optional—improves transient response. Output capacitors in the range of 1 μF to 1000 μF of aluminum or tantalum electrolytic are commonly used to provide improved output impedance and rejection of transients.

*Needed if device is far from filter capacitors.

$$††V_{OUT} = 1.25V\left(1 + \frac{R2}{R1}\right)$$

Digitally Selected Outputs

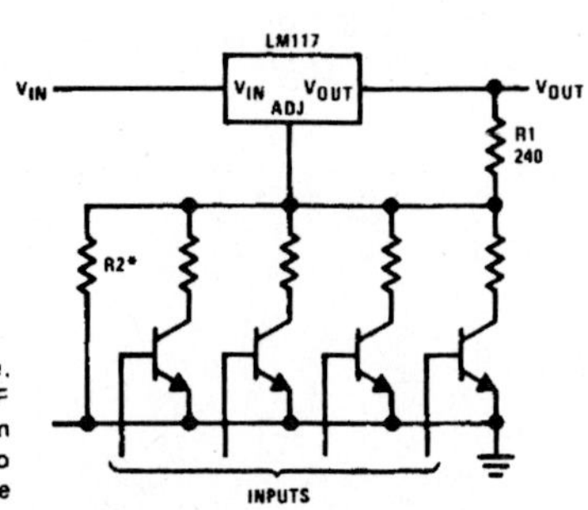

*Sets maximum V_{OUT}

5V Logic Regulator with Electronic Shutdown*

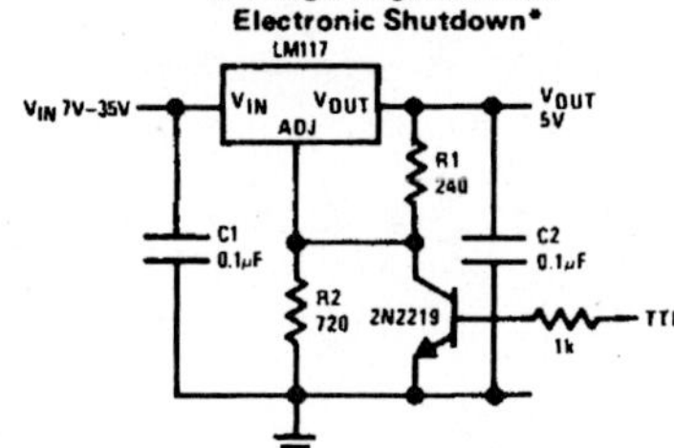

* Min output ≈ 1.2V

Absolute Maximum Ratings

Power Dissipation	Internally limited
Input–Output Voltage Differential	40V
Operating Junction Temperature Range	
LM117	−55°C to +150°C
LM217	−25°C to +150°C
LM317	0°C to +125°C
Storage Temperature	−65°C to +150°C
Lead Temperature (Soldering, 10 seconds)	300°C

Preconditioning

Burn-In in Thermal Limit — **100% All Devices**

Electrical Characteristics (Note 1)

PARAMETER	CONDITIONS	LM117/217			LM317			UNITS
		MIN	TYP	MAX	MIN	TYP	MAX	
Line Regulation	T_A = 25°C, 3V ≤ $V_{IN} - V_{OUT}$ ≤ 40V (Note 2)		0.01	0.02		0.01	0.04	%/V
Load Regulation	T_A = 25°C, 10 mA ≤ I_{OUT} ≤ I_{MAX}							
	V_{OUT} ≤ 5V, (Note 2)		5	15		5	25	mV
	V_{OUT} ≥ 5V, (Note 2)		0.1	0.3		0.1	0.5	%
Thermal Regulation	T_A = 25°C, 20 ms Pulse		0.03	0.07		0.04	0.07	%/W
Adjustment Pin Current			50	100		50	100	μA
Adjustment Pin Current Change	10 mA ≤ I_L ≤ I_{MAX} 3V ≤ (V_{IN}−V_{OUT}) ≤ 40V		0.2	5		0.2	5	μA
Reference Voltage	3V ≤ (V_{IN}−V_{OUT}) ≤ 40V, (Note 3) 10 mA ≤ I_{OUT} ≤ I_{MAX}, P ≤ P_{MAX}	1.20	1.25	1.30	1.20	1.25	1.30	V
Line Regulation	3V ≤ $V_{IN} - V_{OUT}$ ≤ 40V, (Note 2)		0.02	0.05		0.02	0.07	%/V
Load Regulation	10 mA ≤ I_{OUT} ≤ I_{MAX}, (Note 2)							
	V_{OUT} ≤ 5V		20	50		20	70	mV
	V_{OUT} ≥ 5V		0.3	1		0.3	1.5	%
Temperature Stability	T_{MIN} ≤ T_j ≤ T_{MAX}		1			1		%
Minimum Load Current	V_{IN}−V_{OUT} = 40V		3.5	5		3.5	10	mA
Current Limit	V_{IN}−V_{OUT} ≤ 15V							
	K and T Package	1.5	2.2		1.5	2.2		A
	H and P Package	0.5	0.8		0.5	0.8		A
	V_{IN}−V_{OUT} = 40V, T_j = +25°C							
	K and T Package	0.30	0.4		0.15	0.4		A
	H and P Package	0.15	0.07		0.075	0.07		A
RMS Output Noise, % of V_{OUT}	T_A = 25°C, 10 Hz ≤ f ≤ 10 kHz		0.003			0.003		%
Ripple Rejection Ratio	V_{OUT} = 10V, f = 120 Hz		65			65		dB
	C_{ADJ} = 10μF	66	80		66	80		dB
Long-Term Stability	T_A = 125°C		0.3	1		0.3	1	%
Thermal Resistance, Junction to Case	H Package		12	15		12	15	°C/W
	K Package		2.3	3		2.3	3	°C/W
	T Package					4		°C/W
	P Package					12		°C/W

Note 1: Unless otherwise specified, these specifications apply −55°C ≤ T_j ≤ +150°C for the LM117, −25°C ≤ T_j ≤ +150°C for the LM217, and 0°C ≤ T_j ≤ +125°C for the LM317; $V_{IN} - V_{OUT}$ = 5V; and I_{OUT} = 0.1A for the TO-39 and TO-202 packages and I_{OUT} = 0.5A for the TO-3 and TO-220 packages. Although power dissipation is internally limited, these specifications are applicable for power dissipations of 2W for the TO-39 and TO-202, and 20W for the TO-3 and TO-220. I_{MAX} is 1.5A for the TO-3 and TO-220 packages and 0.5A for the TO-39 and TO-202 packages.

Note 2: Regulation is measured at constant junction temperature, using pulse testing with a low duty cycle. Changes in output voltage due to heating effects are covered under the specification for thermal regulation.

Note 3: Selected devices with tightened tolerance reference voltage available.

Application Hints

In operation, the LM117 develops a nominal 1.25V reference voltage, V_{REF}, between the output and adjustment terminal. The reference voltage is impressed across program resistor R1 and, since the voltage is constant, a constant current I_1 then flows through the output set resistor R2, giving an output voltage of

$$V_{OUT} = V_{REF}\left(1 + \frac{R2}{R1}\right) + I_{ADJ}R2$$

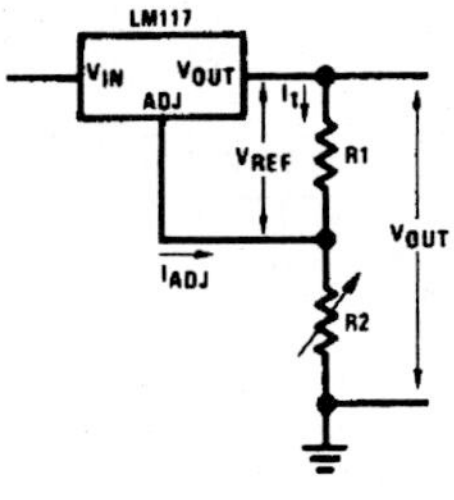

FIGURE 1.

Since the 100µA current from the adjustment terminal represents an error term, the LM117 was designed to minimize I_{ADJ} and make it very constant with line and load changes. To do this, all quiescent operating current is returned to the output establishing a minimum load current requirement. If there is insufficient load on the output, the output will rise.

External Capacitors

An input bypass capacitor is recommended. A 0.1µF disc or 1µF solid tantalum on the input is suitable input bypassing for almost all applications. The device is more sensitive to the absence of input bypassing when adjustment or output capacitors are used but the above values will eliminate the possibility of problems.

The adjustment terminal can be bypassed to ground on the LM117 to improve ripple rejection. This bypass capacitor prevents ripple from being amplified as the output voltage is increased. With a 10µF bypass capacitor 80 dB ripple rejection is obtainable at any output level. Increases over 10µF do not appreciably improve the ripple rejection at frequencies above 120 Hz. If the bypass capacitor is used, it is sometimes necessary to include protection diodes to prevent the capacitor from discharging through internal low current paths and damaging the device.

In general, the best type of capacitors to use are solid tantalum. Solid tantalum capacitors have low impedance even at high frequencies. Depending upon capacitor construction, it takes about 25µF in aluminum electrolytic to equal 1µF solid tantalum at high frequencies. Ceramic capacitors are also good at high frequencies; but some types have a large decrease in capacitance at frequencies around 0.5 MHz. For this reason, 0.01µF disc may seem to work better than a 0.1µF disc as a bypass.

Although the LM117 is stable with no output capacitors, like any feedback circuit, certain values of external capacitance can cause excessive ringing. This occurs with values between 500 pF and 5000 pF. A 1µF solid tantalum (or 25µF aluminum electrolytic) on the output swamps this effect and insures stability.

Load Regulation

The LM117 is capable of providing extremely good load regulation but a few precautions are needed to obtain maximum performance. The current set resistor connected between the adjustment terminal and the output terminal (usually 240Ω) should be tied directly to the output of the regulator rather than near the load. This eliminates line drops from appearing effectvely in series with the reference and degrading regulation. For example, a 15V regulator with 0.05Ω resistance between the regulator and load will have a load regulation due to line resistance of 0.05Ω x I_L. If the set resistor is connected near the load the effective line resistance will be 0.05Ω (1 + R2/R1) or in this case, 11.5 times worse.

Figure 2 shows the effect of resistance between the regulator and 240Ω set resistor.

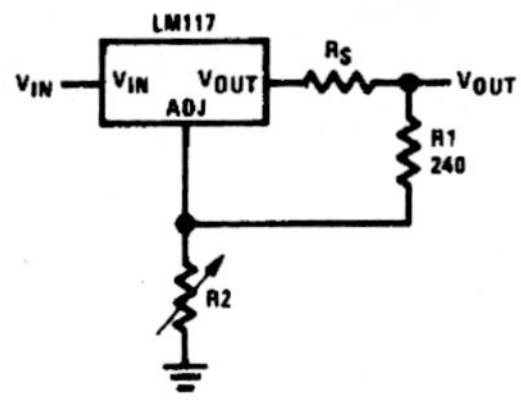

FIGURE 2. Regulator with Line Resistance in Output Lead

With the TO-3 package, it is easy to minimize the resistance from the case to the set resistor, by using two separate leads to the case. However, with the TO-5 package, care should be taken to minimize the wire length of the output lead. The ground of R2 can be returned near the ground of the load to provide remote ground sensing and improve load regulation.

Protection Diodes

When external capacitors are used with *any* IC regulator it is sometimes necessary to add protection diodes to prevent the capacitors from discharging through low current points into the regulator. Most 10µF capacitors have low enough internal series resistance to deliver 20A spikes when shorted. Although the surge is short, there is enough energy to damage parts of the IC.

When an output capacitor is connected to a regulator and the input is shorted, the output capacitor will discharge into the output of the regulator. The discharge

Application Hints (cont'd.)

current depends on the value of the capacitor, the output voltage of the regulator, and the rate of decrease of V_{IN}. In the LM117, this discharge path is through a large junction that is able to sustain 15A surge with no problem. This is not true of other types of positive regulators. For output capacitors of 25μF or less, there is no need to use diodes.

The bypass capacitor on the adjustment terminal can discharge through a low current junction. Discharge occurs when *either* the input or output is shorted. Internal to the LM117 is a 50Ω resistor which limits the peak discharge current. No protection is needed for output voltages of 25V or less and 10μF capacitance. *Figure 3* shows an LM117 with protection diodes included for use with outputs greater than 25V and high values of output capacitance.

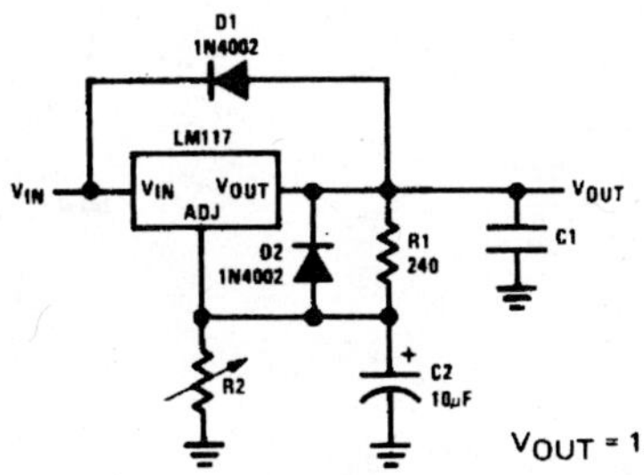

$$V_{OUT} = 1.25V\left(1 + \frac{R2}{R1}\right) + R2 I_{ADJ}$$

D1 protects against C1

D2 protects against C2

FIGURE 3. Regulator with Protection Diodes

Schematic Diagram

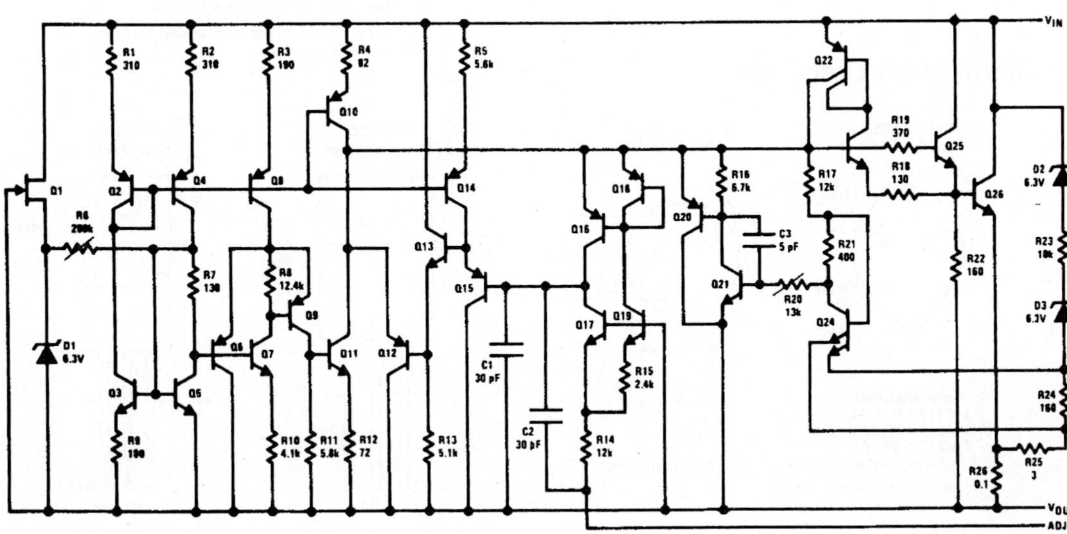

Typical Applications

AC Voltage Regulator

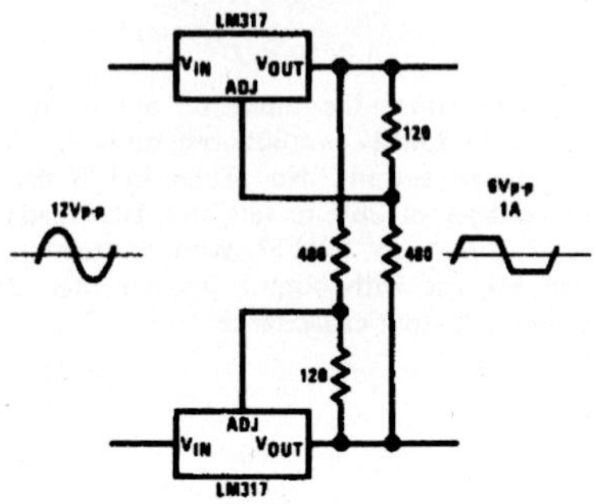

Adjustable 4A Regulator

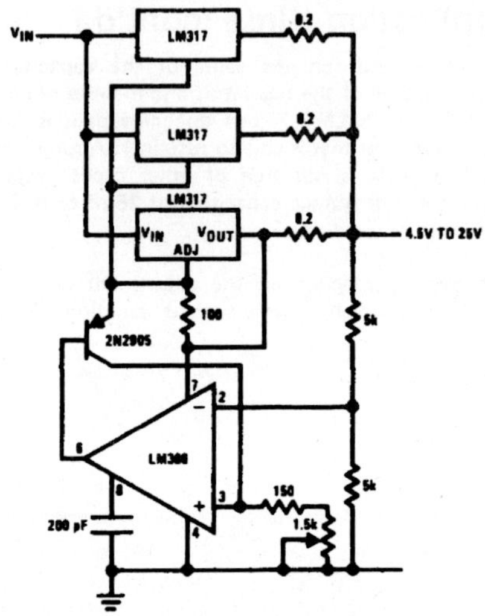

12V Battery Charger

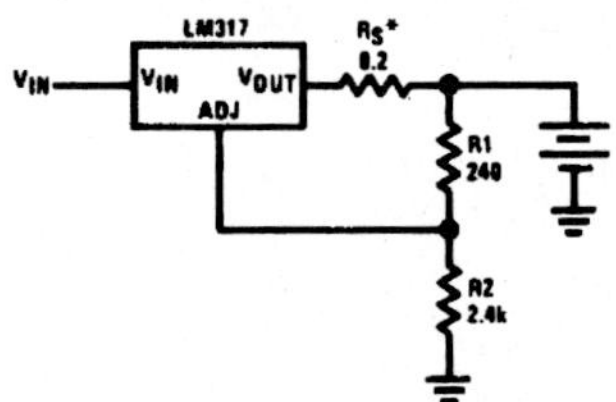

*R_S—sets output impedance of charger $Z_{OUT} = R_S \left(1 + \frac{R2}{R1}\right)$
Use of R_S allows low charging rates with fully charged battery.

Current Limited 6V Charger

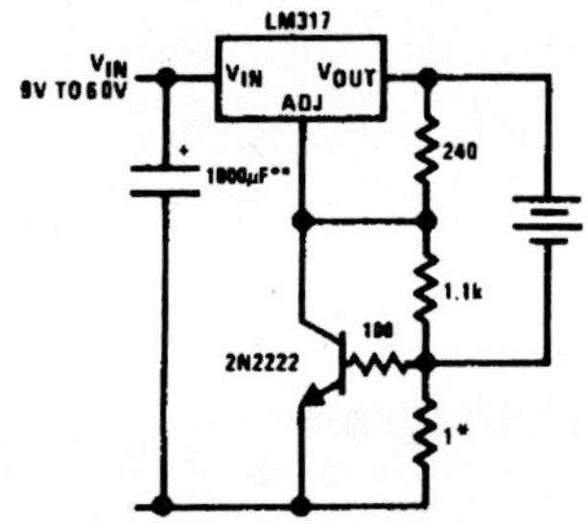

*Sets peak current (0.6A for 1Ω)

**The 1000µF is recommended to filter out input transients

50 mA Constant Current Battery Charger

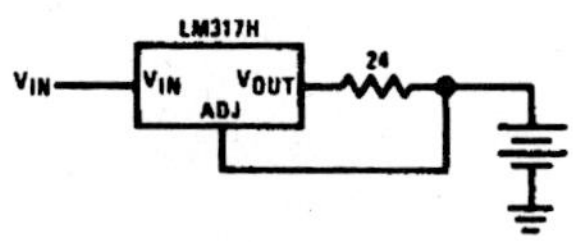

Connection Diagrams

(TO-3 STEEL)
Metal Can Package

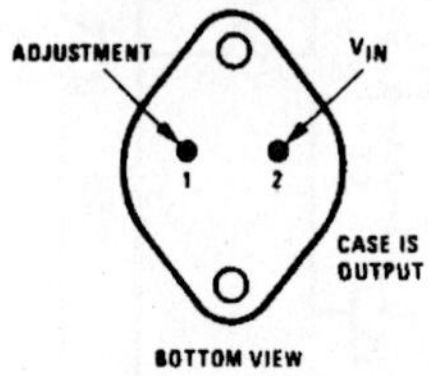

Order Number:
LM117K STEEL
LM217K STEEL
LM317K STEEL
See Package K02A

(TO-39)
Metal Can Package

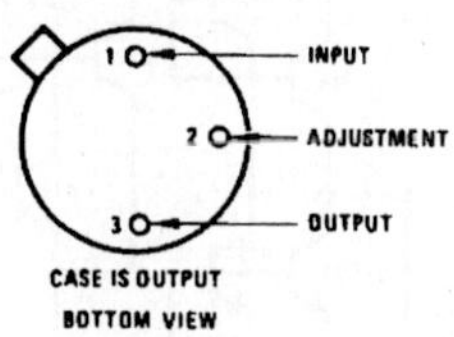

Order Number:
LM117H
LM217H
LM317H
See Package H03A

(TO-220)
Plastic Package

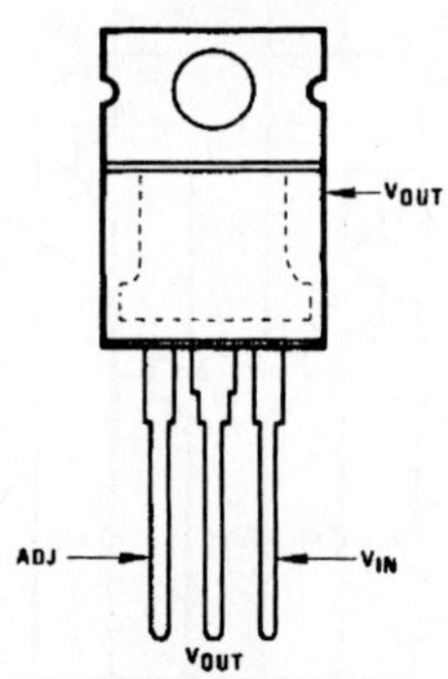

Order Number:
LM317T
See Package T03B

(TO-202)
Plastic Package

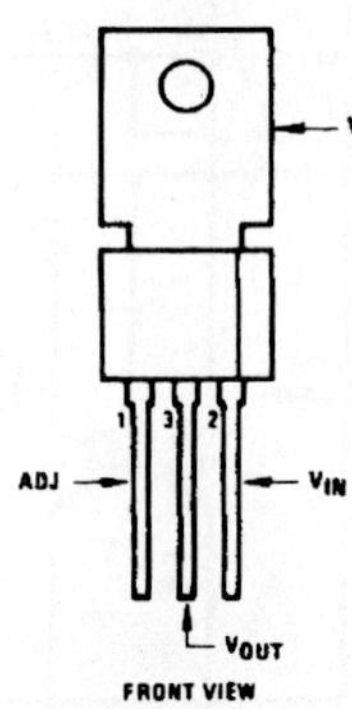

Order Number:
LM317MP
See Package P03A
Tab Formed Devices
LM317MP TB
See Package P03E

Answers to Selected Odd-Numbered Problems

CHAPTER 1

1-1. Mathematical operation
1-3. Part identification number
1-5. Package style
1-7. Pin 8
1-9. **(a)** The common connection to the positive and negative supplies or a ground symbol. **(b)** Make all voltage measurements with respect to power supply common.

CHAPTER 2

2-1. Two for supply power—+V and −V. Two for input signals—inverting and noninverting. One for output.
2-3. **(a)** pin 2 **(b)** pin 3 **(c)** pin 6
2-5. **(a)** $V_o = 0$ V **(b)** $I_{os} = 25$ mA (typical)

2-7.

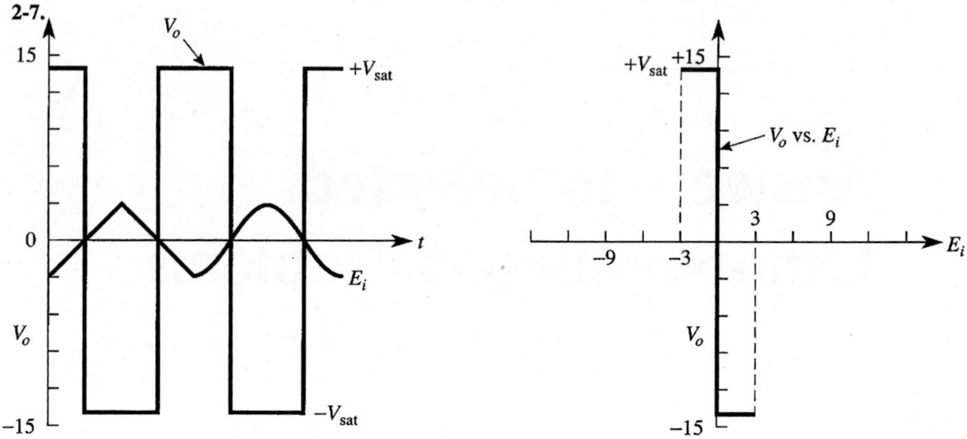

2-9.

Problem	E_i applied to	zero-crossing type:
2-7:	(−) input	inverting
2-8:	(+) input	noninverting

2-11. **(a)** If E_i is above V_{ref}, V_o will be above 0 V at $+V_{sat}$. **(b)** If signal is wired directly (or via a resistor) to the (+) input, the circuit is noninverting.

2-13. **(a)** Arbitrarily choose first a 0–5 V adjustment. **(b)** We now need a divider of 5 V/50 mV = 100 to 1. **(c)** Pick an isolating resistor equal to or greater than 10 × the 5-kΩ pot. Pick 100 kΩ.

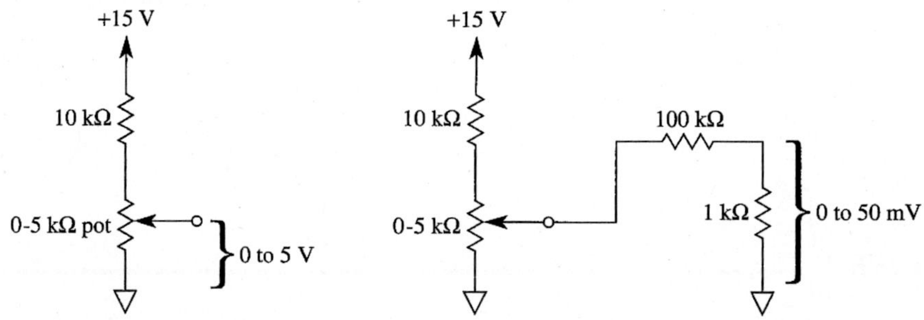

Therefore, the divider resistor is 100 kΩ ÷ 100, or 1 kΩ.

2-15.

V_{temp} (V)	T_{hi} (ms)
0	0
5	10
10	20

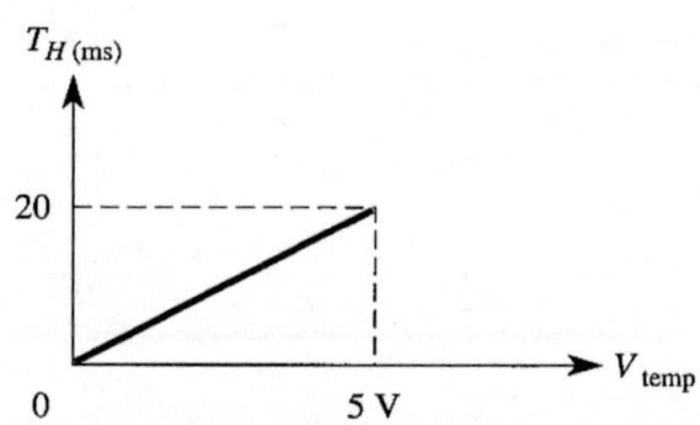

CHAPTER 3

3-1. Negative
3-3. If a circuit has negative feedback and V_o is not in saturation, **(a)** $E_d = 0$ V and **(b)** op amp inputs draw negligible signal current.
3-5. **(a)** $V_o = -10$ V, op amp I_o sinks 1.5 mA **(b)** $V_o = 4$ V, I_o sources 0.4 mA

3-7.

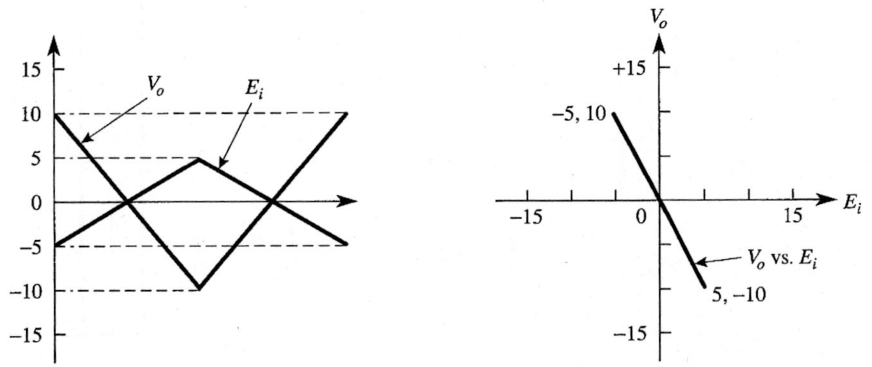

3-9. Noninverting amplifier: **(a)** $V_o = 10$ V; **(b)** $= V_o = -4$ V. In comparison, the gain magnitudes are equal.
3-11. **(a)** $R_i = 10$ kΩ **(b)** $R_f = 50$ kΩ

3-13.

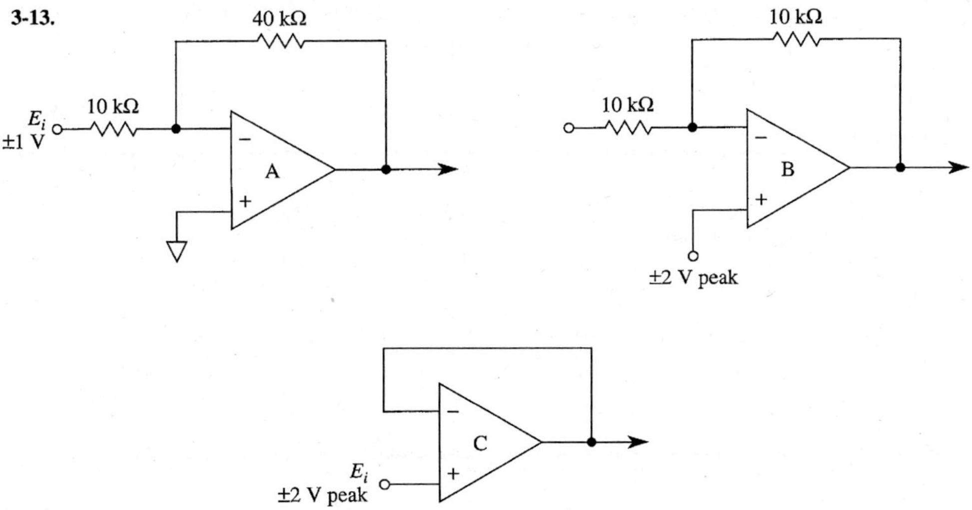

3-17. **(a)** $R_{i3} = 10$ kΩ, $R_{f3} = 50$ kΩ **(b)** $R_{i1} = 50$ kΩ **(c)** $R_{i2} = 16.67$ kΩ
3-19. Change Fig. 3-16(a) as follows: (1) Connect voltage followers (Fig. 3-10) between E_1 and its input and between E_2 and its input. (2) Change R_f to 50 kΩ. Then op amp B applies a gain of −5 to each channel. Thus, $V_o = (-E_1 \times -5) + E_2(-5) = 5E_1 - 5E_2 = 5(E_1 - E_2)$.
3-21. **(a)** −55°C to 150°C
(b) −40°C to 100°C
(c) TO-92 plastic package
TO-46 metal can package
(d) 400 μA to 5 mA
(e) ±1°C

CHAPTER 4

4-1. A fraction of the output voltage is applied to the (+) input.
4-3.

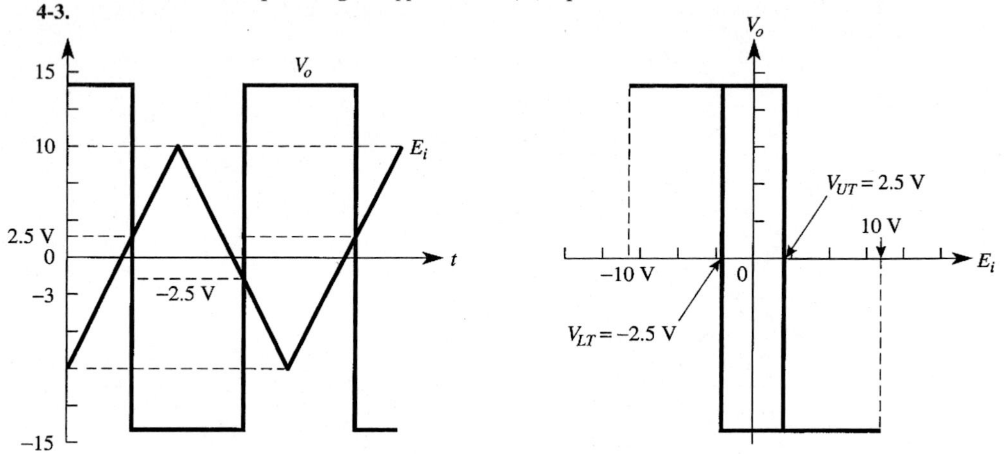

4-5. **(a)** 500 Hz **(b)** 10 V **(c)** 6 V **(d)** −6 V **(e)** 12 V
4-7. **(a)** $V_H = 1.5$ V, $V_{CTR} = -1.25$ V **(b)** $n = 20$ **(c)** −1.19 V **(d)** $R = 10$ kΩ, $nR = 200$ kΩ
4-9. $V_H = 1.5$ V, $V_{CTR} = 1.25$ V **(a)** $n = 17.3$ for $\pm V_{sat} = \pm 13$ V, $R = 10$ kΩ, $nR = 173$ kΩ **(b)** $V_{ref} = -15$ V, $mR = 100$ kΩ **4-11.** $V_{+in} \cong 9.9$ mV (open switch) or $V_{+in} \cong 0.4$ mV (closed switch) $V_o = 5$ V **4-13.** V_o = V++ **4-15.** 311

CHAPTER 5

5-1. **(a)** −1 mA **(b)** −2V **5-5.** **(a)** 4.5 kΩ **(b)** 3.18 kΩ **(c)** 1.59 kΩ
5-7. **(a)** *pnp* **(b)** 100 mA **(c)** 5 V **5-9.** 500 Ω
5-11. **(a)** −0.2 mA, −1.0 V, −2 V **(b)** 0.2 mA, 1 V, 2 V **5-13.** $I_L = 0$, $V_L = 0$; $V_o = +5$ V
5-15. See Figs. 5.7, 5.8, and 5.11. **5-17.** $I_L = 5$ mA **5-19.** $\theta = 64.2°$
5-21. $C = 1$ μF, $R_i = 159$ kΩ, $R_{resistors} = 300$ kΩ
5-23. 100°C = 373 K. The AD590 outputs 373 μA. In Fig. 5-20(a), current through R_f = 373 − 273 μA = 100 μA (right to left). VR_f = 100 μA × 10 kΩ = 1000 mV = V_o, $V_o = \frac{10 \text{ mV}}{°\text{C}} \times 100°\text{C} = 1000$ mV.

CHAPTER 6

6-1. **(a)** 6.9 V **(b)** −6.9 V **6-3.** yes; $R_f = 5$ kΩ **6-5.** The time for C to return to its initial state.
6-7. ±3 V; 1250 Hz **6-9.** 9.4 V; 250 Hz **6-11.** **(a)** Switched gain amplifier **(b)** Amplifier B; $V_o = V_{ref}$ **(c)** With pin 9 at 1 V, V_o will be a sine wave identical to V_{ref}. With pin 9 at −1 V, V_o will be inverted from V_{ref}. **6-13.** **(a)** 60° **(b)** 1.2 V **6-15.** Pick $C = 0.1$ μF. At 0.5 Hz, $R_i = 500$ kΩ; at 50 Hz, $R_i = 5$ kΩ

CHAPTER 7

7-1. 3 V

7-3.

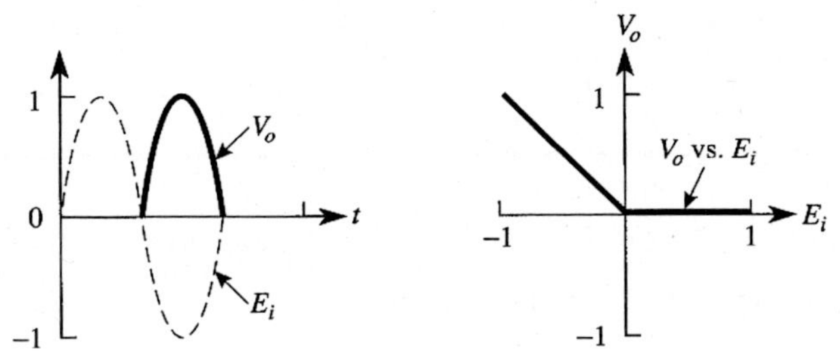

7-5. See Fig. 7-5. **7-7.** Peak detector or follow-and-hold circuit **7-9.** See Fig. 7-14.
7-11. See Figs. 7-18(b) and 7-17.

CHAPTER 8

8-1. $V_o = -1$ V **8-3.** $V_o = 0$ V **8-5.** $R_2 = R_4 = 100$ kΩ **8-7.** CMRR ≅ 5,623
8-9. **(a)** −200 mV **(b)** 0 V **8-11.** $V_o = -2\text{ V}\left(1 + \frac{2}{a}\right)$ **8-13.** **(a)** 0.143 V **(b)** $a = 0.1$
8-15. **(a)** $V_o = 5.0$ V **(b)** 2.5 V **8-17.** **(a)** Down **(b)** 10 mA **(c)** 1 V **(d)** 11 V
8-19. **(a)** V_o goes positive **(b)** V_o decreases **8-21.** **(a)** $V_o = 25$ mV **(b)** $V_o = 50$ mV **(c)** $V_o = 100$ mV

CHAPTER 9

9-3. $I_{B-} = 0.2\ \mu$A **9-5.** $V_o = 2.2$ V **9-7.** $V_o = -2.5$ mV **9-9.** $V_{io} \cong 2$ mV
9-11. $R_c = 5$ kΩ **9-13.** $\Delta V_o = \pm 50.5$ mV
9-15. **(a)** $V_{io} = 1$ mV **(b)** $I_{B-} = 0.2\ \mu$A **(c)** $I_B+ \simeq 0.1\ \mu$A **9-17.** $V_{io} \cong 38\ \mu$V

CHAPTER 10

10-1. 200,000 **10-3.** 5 MHz **10-9.** $A_{CL} = 990$
10-11. **(a)** $f_H = 100$ kHz **(b)** $A_{CL} = 70.7$ **10-13.** $f_{max} = 15.92$ kHz **10-15.** 6

CHAPTER 11

11-3. Band pass **11-5.** $R = 7.2$ kΩ **11-7.** $|V_o| = 0.707$ at f_c. There is a 45° phase angle at f_c for each capacitor. **11-9.** $f_c = 11.2$ kHz **11-11.** $\omega_c = 25$ krad/s **11-13.** $R = 8$ kΩ
11-15. $R_1 = 14$ kΩ, $R_2 = 7.07$ kΩ **11-17.** $R_3 = 6.35$ kΩ, $R_1 = 12.7$ kΩ, $R_2 = 3.17$ kΩ
11-19. **(a)** 10 Hz **(b)** 60 Hz **(c)** 6
11-21. Apply Eq. (11-3) **11-23.** $Q = 0.35$
11-25. **(a)** Connect the bandpass filter to an inverting adder as in Fig. 11-15. **(b)** $f_L = 92$ Hz, $f_H = 177$ Hz

CHAPTER 12

12-1. **(a)** $V_o = 2.5$ V **(b)** $V_o = -2.5$ V **(c)** $V_o = -2.5$ V **(d)** $V_o = 2.5$ V
12-5. **(a)** 3.2 V **(b)** 3.2 V peak at 800 Hz **12-7.** $V_{o\ dc} = 4.33$ V
12-9. Peak value = 2 V; sum frequency = 18 kHz; difference frequency = 12 kHz
12-11. Carrier = 100 kHz; lower = 90 kHz; upper = 110 kHz
12-13. 41 kHz, 1 kHz; 39 kHz, 1 kHz; 40 kHz, 0

CHAPTER 13

13-1. One-shot, free running **13-3.** 70 Hz **13-5.** 107 Hz **13-7.** 3.1 kΩ **13-9.** 6.95 ms
13-11. 1 ms $< t_{high} <$ 2 ms, $R_A = 15$ kΩ for $t_{high} = 1.65$ ms **13-13.** 170 ms
13-15. **(a)** 50 ms **(b)** 600 ms **(c)** 12.75 s **13-17.** 62.5 Hz

CHAPTER 14

14-1. **(a)** 1024 **(b)** 16,384 **14-3.** $V_{fs} = 9.99$ V **14-5.** D = 01011111
14-7. **(b)** $I_o = 0.125$ mA
(c) $\Delta V_o = 0.625$ V/bit
(d) $V_o = D \times 0.625$ V/bit
(e) $V_{fs} = 9.375$ V
14-9. Memory buffer register, selection logic **14-11.** **(a)** $\overline{CE}$ and $\overline{CS}$
14-13. 4000/4001; 8000/8001; C000/C001

CHAPTER 15

15-1. $D = 0111100000$ **15-3.** $V_{in} = 10.0204$ V. Caution: exceeds FSR.
15-5. **(a)** $T_1 = 83.33$ ms
(b) $V_{ref} = -50$ mV, $T_2 = 41.65$ ms
(c) $D = 500$
15-7. Conversion **15-9.** Ground pin 11
15-11. **(a)** ~ 10 μs
(b) 10^5 conversions
(c) See Example 15-7.
15-13. 255

CHAPTER 16

16-1. 38 V **16-3.** **(a)** Decreases **(b)** Increases **16-5.** **(a)** 8Ω **(b)** 29%
16-7. $V_{dcFL} = 25$ V **15-9.** 6.1%
16-11. **(a)** $E_m = 35.3$ V **(b)** $V_{dcFL} = 29.3$ V **(c)** 20% **(d)** 1.42 V **(e)** 4.9% **(f)** 26.8 V

16-13. **(a)** 6.0 V **(b)** ΔV_o centered on $V_{dc} = 12$ V **(c)** Waveshape

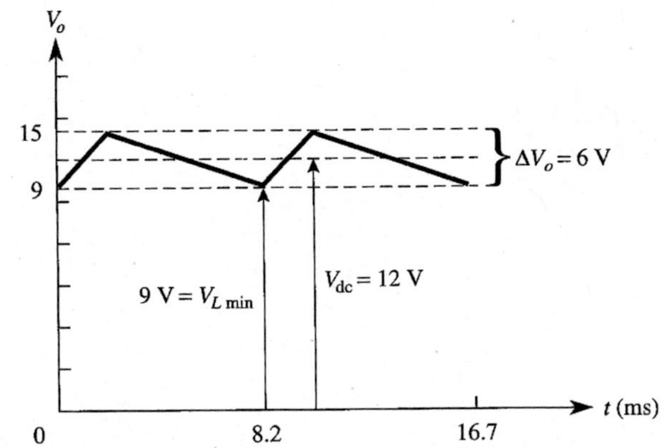

16-15. $R_2 = 4560\ \Omega$ **16-17.** **(a)** 1.2 V **(b)** −1.2 V **16-19.** 7.2 to 19.7 V
16-21. $R_2 = 1032\ \Omega$

Bibliography

Analog Devices, Inc., *Analog-Digital Conversion Handbook* (1972), *Analog-Digital Conversion Notes* (1977), *Non-linear Circuits Handbook* (1979), *Data Converter Reference Manual* Vol. II (1992), *Amplifier Reference Manual* (1992), *Applications Reference Manual* (1993), *Special Reference Manual* (1992).

BLH Electronics, *Strain Gages, SR4,* BLH, Waltham, Mass. (1979).

Burr-Brown International, *Product Data Book,* Tucson, Ariz. (1994); *Applications Handbook* (1994).

Coughlin, Robert F., and Villanucci, Robert F., *Introductory Operational Amplifiers and Linear ICs, Theory, and Experimentation,* Prentice-Hall, Inc., Englewood Cliffs, N.J. (1990).

Driscoll, Frederick F., *6800/68000 Microprocessors,* Breton Publishers, Boston (1987).

Driscoll, Frederick F., and Coughlin, Robert F., *Solid State Devices and Applications,* Prentice-Hall, Inc., Englewood Cliffs, N.J. (1975).

Driscoll, F., Coughlin, R., Villanucci, R., *Data Acquisition and Process Control with the M68HC11 Microcontroller,* 2nd Edition, Prentice Hall (2000).

Fairchild Instrument and Camera Corporation, *Linear Division Products* (1982), *Voltage Regulator Handbook* (1974), Fairchild, Mountain View, Calif.

Frederiksen, Thomas M., *Intuitive IC Op Amps,* National Semiconductor Technology Series, National Semiconductor Corporation, Santa Clara, Calif. (1984).

HARRIS CORPORATION, *Linear and Telecom IC,* Harris, Melbourne, Fla. (1994).

MORRISON, R., *Grounding and Shielding Techniques in Instrumentation,* John Wiley & Sons, Inc., New York (1967).

MOTOROLA SEMICONDUCTOR PRODUCTS, INC., *More Value Out of Integrated Operational Amplifier Data Sheets,* AN-273A, Motorola, Pheonix, Ariz. (1970).

NATIONAL SEMICONDUCTOR CORPORATION, *Linear Databook* (1989), *Linear Applications Handbook* (1991), *Special Functions Databook* (1989), *Voltage Regulator Handbook* (1986), *Audio Handbook* (no date), *Linear Supplement Databook* (1988), National Semiconductor Corporation, Santa Clara, Calif.

PEASE, ROBERT A., *Troubleshooting Analog Circuits,* Butterworth-Heinemann (1991).

PHILBRICK RESEARCHES, INC., *Applications Manual for Computing Amplifiers for Modeling, Measuring, Manipulating and Much Else,* Nimrod Press, Inc., Boston (1966).

SHEINGOLD, DANIEL H., *Transducer Interfacing Handbook,* Analog Devices, Inc., Norwood, Mass. (1980).

SIGNETICS CORPORATION, *Data Manual,* Signetics, Sunnyvale, Calif. (1977).

SMITH, J.I., *Modern Operational Circuit Design,* John Wiley & Sons, Inc., New York (1971).

VILLANUCCI, R., ET AL., *Electronics Techniques: Shop Practices and Construction,* 6th ed., Prentice-Hall, Inc., Englewood Cliffs, N.J. (1997).

Index